DMV Seminar
Band 20

Springer Basel AG

Hélène Esnault
Eckart Viehweg

Lectures on Vanishing Theorems

Springer Basel AG

Authors' addresses:

Hélène Esnault and Eckart Viehweg
Fachbereich 6, Mathematik
Universität GH Essen
Universitätsstr. 3
D–W–4300 Essen

A CIP catalogue record for this book is available from the Library of Congress, Washington D.C., USA

Deutsche Bibliothek Cataloging-in-Publication Data

Esnault, Hélène:
Lectures on vanishing theorems / Hélène Esnault ; Eckart Viehweg. – Basel ; Boston ; Berlin : Birkhäuser, 1992
(DMV-Seminar ; Bd. 20)
ISBN 978-3-7643-2822-1 ISBN 978-3-0348-8600-0 (eBook)
DOI 10.1007/978-3-0348-8600-0
NE: Viehweg, Eckart:; Deutsche Mathematiker-Vereinigung: DMV-Seminar

Originally published by Birkhäuser Verlag Basel in 1992

ISBN 978-3-7643-2822-1

Acknowledgement

These notes grew out of the DMV-seminar on algebraic geometry (Schloß Reisensburg, October 13–19, 1991). We thank the DMV (German Mathematical Society) for giving us the opportunity to organize this seminar and to present the theory of vanishing theorems to a group of younger mathematicians. We thank all the participants for their interest, for their useful comments and for the nice atmosphere during the seminar.

Table of Contents

Introduction

M. Kodaira's vanishing theorem, saying that the inverse of an ample invertible sheaf on a projective complex manifold X has no cohomology below the dimension of X and its generalization, due to Y. Akizuki and S. Nakano, have been proven originally by methods from differential geometry ([39] and [1]).

Even if, due to J.P. Serre's GAGA-theorems [56] and base change for field extensions the algebraic analogue was obtained for projective manifolds over a field k of characteristic $p = 0$, for a long time no algebraic proof was known and no generalization to $p > 0$, except for certain lower dimensional manifolds. Worse, counterexamples due to M. Raynaud [52] showed that in characteristic $p > 0$ some additional assumptions were needed.

This was the state of the art until P. Deligne and L. Illusie [12] proved the degeneration of the Hodge to de Rham spectral sequence for projective manifolds X defined over a field k of characteristic $p > 0$ and liftable to the second Witt vectors $W_2(k)$.

Standard degeneration arguments allow to deduce the degeneration of the Hodge to de Rham spectral sequence in characteristic zero, as well, a result which again could only be obtained by analytic and differential geometric methods beforehand. As a corollary of their methods M. Raynaud (loc. cit.) gave an easy proof of Kodaira vanishing in all characteristics, provided that X lifts to $W_2(k)$.

Short time before [12] was written the two authors studied in [20] the relations between logarithmic de Rham complexes and vanishing theorems on complex algebraic manifolds and showed that quite generally vanishing theorems follow from the degeneration of certain Hodge to de Rham type spectral sequences. The interplay between topological and algebraic vanishing theorems thereby obtained is also reflected in J. Kollár's work [41] and in the vanishing theorems M. Saito obtained as an application of his theory of mixed Hodge modules (see [54]).

It is obvious that the combination of [12] and [20] give another algebraic approach to vanishing theorems and it is one of the aims of these lecture notes to present it in all details. Of course, after the Deligne-Illusie-Raynaud proof of the original Kodaira and Akizuki-Nakano vanishing theorems, the main motivation to present the methods of [20] along with those of [12] is that they imply as well some of the known generalizations.

Generalizations have been found by D. Mumford [49], H. Grauert and O. Riemenschneider [25], C.P. Ramanujam [51] (in whose paper the method of coverings already appears), Y. Miyaoka [45] (the first who works with integral parts of $\mathbb{Q}$ divisors, in the surface case), by Y. Kawamata [36] and the second author [63]. All results mentioned replace the condition "ample" in Kodaira's result by weaker conditions. For Akizuki-Nakano type theorems A. Sommese (see for example [57]) got some improvement, as well as F. Bogomolov and A. Sommese (as explained in [6] and [57]) who showed the vanishing of the global sections in certain cases.

Many of the applications of vanishing theorems of Kodaira type rely on the surjectivity of the adjunction map

$$H^b(X, \mathcal{L} \otimes \omega_X(B)) \longrightarrow H^b(B, \mathcal{L} \otimes \omega_B)$$

where B is a divisor and $\mathcal{L}$ is ample or is belonging to the class of invertible sheaves considered in the generalizations.

J. Kollár [40], building up on partial results by Tankeev, studied the adjunction map directly and gave criteria for $\mathcal{L}$ and B which imply the surjectivity.

This list of generalizations is probably not complete and its composition is evidently influenced by the fact that all the results mentioned and some slight improvements have been obtained in [20] and [22] as corollaries of two vanishing theorems for sheaves of differential forms with values in "integral parts of $\mathbb{Q}$-divisors", one for the cohomology groups and one for restriction maps between cohomology groups.

In these notes we present the algebraic proof of Deligne and Illusie [12] for the degeneration of the Hodge to de Rham spectral sequence (Lecture 10). Beforehand, in Lectures 8 and 9, we worked out the properties of liftings of schemes and of the Frobenius morphism to the second Witt vectors [12] and the properties of the Cartier operator [34] needed in the proof. Even if some of the elegance of the original arguments is lost thereby, we avoid using the derived category. The necessary facts about hypercohomology and spectral sequences are shortly recalled in the appendix, at the end of these notes.

During the first seven lectures we take the degeneration of the Hodge to de Rham spectral sequence for granted and we develop the interplay between cyclic coverings, logarithmic de Rham complexes and vanishing theorems (Lectures 2 - 4).

We try to stay as much in the algebraic language as possible. Lectures 5 and 6 contain the geometric interpretation of the vanishing theorems obtained, i.e. the generalizations mentioned above. Due to the use of H. Hironaka's embedded resolution of singularities, most of those require the assumption that the manifolds considered are defined over a field of characteristic zero.

Raynaud's elegant proof of the Kodaira-Akizuki-Nakano vanishing theorem is reproduced in Lecture 11, together with some generalization. However, due to the non-availability of desingularizations in characterisitic p, those generalizations seem to be useless for applications in geometry over fields of characteristic $p > 0$.

In characteristic zero the generalized vanishing theorems for integral parts of $\mathbb{Q}$-divisors and J. Kollár's vanishing for restriction maps turned out to be powerful tools in higher dimensional algebraic geometry. Some examples, indicating "how to use vanishing theorems" are contained in the second half of Lecture 6, where we discuss higher direct images and the interpretation of vanishing theorems on non-compact manifolds, and in Lecture 7. Of course, this list is determined by our own taste and restricted by our lazyness. In particular, the applications of vanishing theorems in the birational classification theory and in the minimal model program is left out. The reader is invited to consult the survey's of S. Mori [46] and of Y. Kawamata, K. Matsuda and M. Matsuki [38].

There are, of course, more subjects belonging to the circle of ideas presented in these notes which we left aside:

- L. Illusie's generalizations of [12] to variations of Hodge structures [32].
- J.-P. Demailly's analytic approach to generalized vanishing theorems [13].
- M. Saito's results on "mixed Hodge modules and vanishing theorems" [54], related to J. Kollár's program [41].
- The work of I. Reider, who used unstability of rank two vector bundles (see [6]) to show that certain invertible sheaves on surfaces are generated by global sections [53] (see however (7.23)).
- Vanishing theorems for vector bundles.
- Generalizations of the vanishing theorems for integral parts of $\mathbb{Q}$-divisors ([2], [3], [42], [43] and [44]).

However, we had the feeling that we could not pass by the generic vanishing theorems of M. Green and R. Lazarsfeld [26]. The general picture of "vanishing theorems" would be incomplete without mentioning this recent development. We include in Lectures 12 and 13 just the very first results in this direction. In particular, the more explicit description and geometric interpretation of the "bad locus in $\mathrm{Pic}^0(X)$ ", contained in A. Beauville's paper [5] and Green and Lazarsfeld's second paper [27] on this subject is missing. During the preparation of these notes C. Simpson [58] found a quite complete description of such "degeneration loci".

The first Lecture takes possible proofs of Kodaira's vanishing theorem as a pretext to introduce some of the key words and methods, which will reappear throughout these lecture notes and to give a more technical introduction to its subject.

Methods and results due to P. Deligne and Deligne-Illusie have inspired and influenced our work. We cordially thank L. Illusie for his interest and several conversations helping us to understand [12].

§ 1 Kodaira's vanishing theorem, a general discussion

Let X be a projective manifold defined over an algebraically closed field k and let $\mathcal{L}$ be an invertible sheaf on X. By explicit calculations of the Čech-cohomology of the projective space one obtains:

1.1. Theorem (J. P. Serre [55]). *If $\mathcal{L}$ is ample and $\mathcal{F}$ a coherent sheaf, then there is some $\nu_0 \in \mathbb{N}$ such that*

$$H^b(X, \mathcal{F} \otimes \mathcal{L}^\nu) = 0 \quad \text{for } b > 0 \text{ and } \nu \geq \nu_0$$

In particular, for $\mathcal{F} = \mathcal{O}_X$, one obtains the vanishing of $H^b(X, \mathcal{L}^\nu)$ for $b > 0$ and ν sufficiently large.

If $\text{char}(k) = 0$, then "ν sufficiently large" can be made more precise. For example, it is enough to choose ν such that $\mathcal{A} = \mathcal{L}^\nu \otimes \omega_X^{-1}$ is ample, where $\omega_X = \Omega_X^n$ is the canonical sheaf of X, and to use:

1.2. Theorem (K. Kodaira [39]). *Let X be a complex projective manifold and $\mathcal{A}$ be an ample invertible sheaf. Then*

$$a) \quad H^b(X, \omega_X \otimes \mathcal{A}) = 0 \qquad \text{for } b > 0$$

$$b) \quad H^{b'}(X, \mathcal{A}^{-1}) = 0 \qquad \text{for } b' < n = \dim X.$$

Of course it follows from Serre-duality that a) and b) are equivalent. Moreover, since every algebraic variety in characteristic 0 is defined over a subfield of $\mathbb{C}$, one can use flat base change to extend (1.2) to manifolds X defined over any algebraically closed field of characteristic zero.

1.3. Theorem (Y. Akizuki, S. Nakano [1]). *Under the assumptions made in (1.2), let Ω_X^a denote the sheaf of a-differential forms. Then*

$$a) \quad H^b(X, \Omega_X^a \otimes \mathcal{A}) = 0 \qquad \text{for } a + b > n$$

$$b) \quad H^{b'}(X, \Omega_X^{a'} \otimes \mathcal{A}^{-1}) = 0 \qquad \text{for } a' + b' < n.$$

For a long time, the only proofs known for (1.2) and (1.3) used methods of complex analytic differential geometry, until in 1986 P. Deligne and L. Illusie found an elegant algebraic approach to prove (1.2) as well as (1.3), using characteristic p methods. About one year earlier, trying to understand several generalizations of (1.2), the two authors obtained (1.2) and (1.3) as a direct consequence of the decomposition of the de Rham-cohomology $H^k(Y, \mathbb{C})$ into a direct sum

$$\bigoplus_{b+a=k} H^b(Y, \Omega^a_Y)$$

or, equivalently, of the degeneration of the "Hodge to de Rham" spectral sequence, both applied to cyclic covers $\pi : Y \longrightarrow X$.

As a guide-line to the first part of our lectures, let us sketch two possible proofs of (1.2) along this line.

1. Proof: With Hodge decomposition for non-compact manifolds and topological vanishing: For sufficiently large N one can find a non-singular primedivisor H such that $\mathcal{A}^N = \mathcal{O}_X(H)$. Let $s \in H^0(X, \mathcal{A}^N)$ be the corresponding section. We can regard s as a rational function, if we fix some divisor A with $\mathcal{A} = \mathcal{O}_X(A)$ and take

$$s \in \mathbb{C}(X) \quad \text{with} \quad (s) + N \cdot A = H.$$

The field $L = \mathbb{C}(X)(\sqrt[N]{s})$ depends only on H. Let $\pi : Y \longrightarrow X$ be the covering obtained by taking the normalization of X in L (see (3.5) for another construction).

An easy calculation (3.13) shows that Y is non-singular as well as $D = (\pi^* H)_{red}$ and that $\pi : Y \longrightarrow X$ is unramified outside of D. One has

$$\pi^* \Omega^a_X(\log\ H) = \Omega^a_Y(\log\ D)$$

where $\Omega^a_X(\log\ H)$ denotes the sheaf of a-differential forms with logarithmic poles along H (see (2.1)). Moreover

$$\pi_* \mathcal{O}_Y = \bigoplus_{i=0}^{N-1} \mathcal{A}^{-i} \qquad \text{and}$$

$$\pi_* \Omega^n_Y(\log\ D) = \bigoplus_{i=0}^{N-1} \Omega^n_X(\log\ H) \otimes \mathcal{A}^{-i} = \bigoplus_{i=0}^{N-1} \Omega^n_X \otimes \mathcal{A}^{N-i}$$

Deligne [11] has shown that

$$H^k(Y - D, \mathbb{C}) \cong \bigoplus_{b+a=k} H^b(Y, \Omega^a_Y(\log\ D)).$$

Since $X - H$ is affine, the same holds true for $Y - D$ and hence $H^k(Y - D, \mathbb{C}) = 0$ for $k > n$. Altogether one obtains for $b > 0$

$$0 = H^b(Y, \Omega^n_Y(\log\ D)) = \bigoplus_{i=0}^{N-1} H^b(X, \Omega^n_X \otimes \mathcal{A}^{N-i}).$$

□

In fact, a similar argument shows as well that

$$H^b(X, \Omega^a_X(\log\ H) \otimes \mathcal{A}^{-1}) = 0$$

for $a+b > n$. We can deduce (1.3) from this statement by induction on dim X using the residue sequence (as will be explained in (6.4)).

The two ingredients of the first proof can be interpretated in a different way. First of all, since the de Rham complex on $Y - D$ is a resolution of the constant sheaf one can use GAGA [56] and Serre's vanishing to obtain the topological vanishing used above. Secondly, the decomposition of the de Rham cohomology of Y into the direct sum of (a,b)-forms, implies that the differential

$$d : \Omega^a_Y \longrightarrow \Omega^{a+1}_Y$$

induces the zero map

$$d : H^b(Y, \Omega^a_Y) \longrightarrow H^b(Y, \Omega^{a+1}_Y).$$

Using this one can give another proof of (1.2):

2. Proof: Closedness of global (p,q) forms and Serre's vanishing theorem: Let us return to the covering $\pi : Y \to X$ constructed in the first proof. The Galois-group G of $\mathbb{C}(Y)$ over $\mathbb{C}(X)$ is cyclic of order N. A generator σ of G acts on Y and D and hence on the sheaves $\pi_*\Omega^a_Y$ and $\pi_*\Omega^a_Y(\log\ D)$. Both sheaves decompose in a direct sum of sheaves of eigenvectors of σ and, if we choose the N-th root of unity carefully, the i-th summand of

$$\pi_*\Omega^a_Y(\log\ D) = \Omega^a_X(\log\ H) \otimes \pi_*\mathcal{O}_Y = \bigoplus_{i=0}^{N-1} \Omega^a_X(\log\ H) \otimes \mathcal{A}^{-i}$$

consists of eigenvectors with eigenvalue e^i. For $e^i \neq 1$ the eigenvectors of $\pi_*\Omega^a_Y$ and of $\pi_*\Omega^a_Y(\log\ D)$ coincide, the difference of both sheaves is just living in the invariant parts Ω^a_X and $\Omega^a_X(\log\ H)$. Moreover, the differential

$$d : \mathcal{O}_Y \longrightarrow \Omega^1_Y$$

is compatible with the G-action and we obtain a $\mathbb{C}$-linear map (in fact a connection)

$$\nabla_i : \mathcal{A}^{-i} \longrightarrow \Omega^1_X(\log\ H) \otimes \mathcal{A}^{-i}.$$

Both properties follow from local calculations. Let us show first, that

$$\pi_*\Omega^a_Y = \Omega^a_X \oplus \bigoplus_{i=1}^{N-1} \Omega^a_X(\log\ H) \otimes \mathcal{A}^{-i}\ .$$

Since H is non-singular one can choose local parameters $x_1, \ldots, x_n$ such that H is defined by $x_1 = 0$. Then

$$y_1 = \sqrt[N]{x_1} \text{ and } x_2, \ldots, x_n$$

are local parameters on Y. The local generators

$$N \cdot \frac{dx_1}{x_1}, dx_2, \ldots, dx_n \text{ of } \Omega^1_X(\log H)$$

lift to local generators

$$\frac{dy_1}{y_1}, dx_2, \ldots, dx_n \text{ of } \Omega^1_Y(\log D).$$

The a-form

$$\phi = s \cdot \frac{dy_1}{y_1} \wedge dx_2 \wedge \ldots \wedge dx_a$$

(for example) is an eigenvector with eigenvalue e^i if and only if the same holds true for s, i.e. if $s \in \mathcal{O}_X \cdot y_1^i$. If ϕ has no poles, s must be divisible by y_1. This condition is automatically satisfied as long as $i > 0$. For $i = 0$ it implies that s must be divisible by $y_1^N = x_1$.

The map ∇_i can be described locally as well. If

$$s = t \cdot y_1^i \in \mathcal{O}_X \cdot y_1^i$$

then on Y one has

$$ds = y_1^i \cdot dt + t \cdot dy_1^i$$

and therefore d respects the eigenspaces and ∇_i is given by

$$\nabla_i(s) = (dt + \frac{i}{N} \cdot t \frac{dx_1}{x_1}) \cdot y_1^i.$$

If $Res : \Omega^1_X(\log H) \longrightarrow \mathcal{O}_H$ denotes the residue map, one obtains in addition that

$$(Res \otimes id_{\mathcal{A}^{-1}}) \circ \nabla_1 : \mathcal{A}^{-1} \longrightarrow \mathcal{O}_H \otimes \mathcal{A}^{-1}$$

is the $\mathcal{O}_X$-linear map

$$s \longmapsto \frac{1}{N} s \mid_H .$$

Since $d : H^b(Y, \mathcal{O}_Y) \longrightarrow H^b(Y, \Omega^1_Y)$ is the zero map, the direct summand

$$\nabla_1 : H^b(X, \mathcal{A}^{-1}) \longrightarrow H^b(X, \Omega^1_X(\log D) \otimes \mathcal{A}^{-1})$$

is the zero map as well as the restriction map

$$N \cdot (Res \otimes id_{\mathcal{A}^{-1}}) \circ \nabla_1 : H^b(X, \mathcal{A}^{-1}) \longrightarrow H^b(H, \mathcal{O}_H \otimes \mathcal{A}^{-1}).$$

Hence, for all b we have a surjection

$$H^b(X, \mathcal{A}^{-N-1}) = H^b(X, \mathcal{O}_X(-H) \otimes \mathcal{A}^{-1}) \longrightarrow H^b(X, \mathcal{A}^{-1}).$$

Using Serre duality and (1.1) however, $H^b(X, \mathcal{A}^{-N-1}) = 0$ for $b < n$ and N sufficiently large.

□

Again, the proof of (1.2) gives a little bit more:

If $\mathcal{A}$ is an invertible sheaf such that $\mathcal{A}^N = \mathcal{O}_X(H)$ for a non-singular divisor H, then the restriction map

$$H^b(X, \mathcal{A}^{-1}) \longrightarrow H^b(H, \mathcal{O}_H \otimes \mathcal{A}^{-1})$$

is zero.

This statement is a special case of J. Kollár's vanishing theorem ([40], see (5.6,a)).

The main theme of the first part of these notes will be to extend the methods sketched above to a more general situation:
If one allows Y to be any cyclic cover of X whose ramification divisor is a normal crossing divisor, one obtains vanishing theorems for the cohomology (or for the restriction maps in cohomology) of a larger class of locally free sheaves.
Or, taking a more axiomatic point of view, one can consider locally free sheaves $\mathcal{E}$ with logarithmic connections

$$\nabla : \mathcal{E} \longrightarrow \Omega^1_X(\log\ H) \otimes \mathcal{E}$$

and ask which proporties of ∇ and H force cohomology groups of $\mathcal{E}$ to vanish. The resulting "vanishing theorems for integral parts of $\mathbb{Q}$-divisors" (5.1) and (6.2) will imply several generalizations of the Kodaira-Nakano vanishing theorem (see Lectures 5 and 6), especially those obtained by Mumford, Grauert and Riemenschneider, Sommese, Bogomolov, Kawamata, Kollár ......

However, the approach presented above is using (beside of algebraic methods) the Hodge theory of projective manifolds, more precisely the degeneration of the Hodge to de Rham spectral sequence

$$E_1^{ab} = H^b(Y, \Omega^a_Y(\log\ D)) \Longrightarrow \mathbb{H}^{a+b}(Y, \Omega^\bullet_Y(\log\ D))$$

again a result which for a long time could only be deduced from complex analytic differential geometry.

Both, the vanishing theorems and the degeneration of the Hodge to de Rham spectral sequence do not hold true for manifolds defined over a field of characteristic $p > 0$. However, if Y and D both lift to the ring of the second Witt-vectors (especially if they can be lifted to characteristic 0) and if $p \geq \dim X$, P. Deligne and L. Illusie were able to prove the degeneration (see [12]). In fact, contrary to characteristic zero, they show that the degeneration is induced by some local splitting:

If F_k and F_Y are the absolute Frobenius morphisms one obtains the geometric Frobenius by

$$\begin{array}{ccccc} Y & \xrightarrow{F} & Y' = Y \times_{\mathrm{Spec}\ k} \mathrm{Spec}\ k & \xrightarrow{\sigma} & Y \\ & \searrow & \downarrow & & \downarrow \\ & & \mathrm{Spec}\ k & \xrightarrow{F_k} & \mathrm{Spec}\ k \end{array}$$

with $F_Y = \sigma \circ F$. If we write $D' = (\sigma^* D)_{red}$ then, roughly speaking, they show that $F_*(\Omega^\bullet_Y(\log\ D))$ is quasi-isomorphic to the complex

$$\bigoplus_a \Omega^a_{Y'}(\log\ D')[-a]$$

with $\Omega^a_{Y'}(\log\ D')$ in degree a and with trivial differentials.

By base change for σ one obtains

$$\dim\ \mathbb{H}^k(Y, \Omega^\bullet_Y(\log\ D)) = \sum_{a+b=k} \dim\ H^b(Y', \Omega^a_{Y'}(\log\ D'))$$

$$= \sum_{a+b=k} \dim\ H^b(Y, \Omega^a_Y(\log\ D)).$$

Base change again allows to lift this result to characteristic 0.

Adding this algebraic proof, which can be found in Lectures 8 - 10, to the proof of (1.2) and its generalizations (Lectures 2 - 6) one obtains algebraic proofs of most of the vanishing theorems mentioned.

However, based on ideas of M. Raynaud, Deligne and Illusie give in [12] a short and elegant argument for (1.3) in characteristic p (and, by base change, in general):
By Serre's vanishing theorem one has for some $m \gg 0$

$$H^b(Y, \Omega^a_Y \otimes \mathcal{A}^{-p^\nu}) = 0 \quad \text{for} \quad \nu \geq (m+1)$$

and $a + b < n$, where $\mathcal{A}$ is ample on Y. One argues by descending induction on m:
As

$$\mathcal{A}^{p^{(m+1)}} = F^*(\mathcal{A}'^{p^m}) \ \text{ for } \ \mathcal{A}' = \sigma^* \mathcal{A}$$

and as $\Omega^\bullet_Y$ is a $\mathcal{O}_{Y'}$ complex, $\Omega^\bullet_Y \otimes \mathcal{A}^{-p^{(m+1)}}$ is a complex of $\mathcal{O}_{Y'}$ sheaves with

$$\mathbb{H}^k(Y, \Omega^\bullet_Y \otimes \mathcal{A}^{-p^{m+1}}) = 0 \quad \text{for} \quad k < n.$$

However one has

$$F_*(\Omega^\bullet_Y \otimes \mathcal{A}^{-p^{m+1}}) = \bigoplus_a \Omega^a_{Y'} \otimes \mathcal{A}'^{-p^m}[-a]$$

and

$$0 = H^b(Y', \Omega^a_{Y'} \otimes \mathcal{A}'^{-p^m}) = H^b(Y, \Omega^a_Y \otimes \mathcal{A}^{-p^m})$$

for $a + b < n$.

Unfortunately this type of argument does not allow to weaken the assumptions made in (1.2) or (1.3). In order to deduce the generalized vanishing theorems from the degeneration of the Hodge to de Rham spectral sequence in characteristic 0 we have to use H. Hironaka's theory of embedded resolution of singularities, at present a serious obstruction for carrying over arguments from characteristic 0 to characteristic p. Even the Grauert-Riemenschneider vanishing theorem (replace "ample" in (1.2) by "semi-ample of maximal Iitaka dimension") has no known analogue in characteristic p (see §11).

M. Green and R. Lazarsfeld observed, that "ample" in (1.2) can sometimes be replaced by "numerically trivial and sufficiently general". To be more precise, they showed that $H^b(X, \mathcal{N}^{-1}) = 0$ for a general element $\mathcal{N} \in Pic^0(X)$ if b is smaller than the dimension of the image of X under its Albanese map

$$\alpha : X \longrightarrow \text{Alb}(X).$$

By Hodge-duality (for Hodge theory with values in unitary rank one bundles) $H^b(X, \mathcal{N}^{-1})$ can be identified with $H^0(X, \Omega^b_X \otimes \mathcal{N})$. If $H^b(X, \mathcal{N}^{-1}) \neq 0$ for all $\mathcal{N} \in Pic^0(X)$ the deformation theory for cohomology groups, developed by Green and Lazarsfeld, implies that for all $\omega \in H^0(X, \Omega^1_X)$ the wedge product

$$H^0(X, \Omega^b_X \otimes \mathcal{N}) \longrightarrow H^0(X, \Omega^{b+1}_X \otimes \mathcal{N})$$

is non-trivial. This however implies that the image of X under the Albanese map, or equivalently the subsheaf of Ω^1_X generated by global sections is small. For example, if

$$S^b(X) = \{\mathcal{N} \in Pic^0(X);\ H^b(X, \mathcal{N}^{-1}) \neq 0\},$$

then the first result of Green and Lazarsfeld says that

$$\text{codim}_{Pic^0(X)}(S^b(X)) \geq \dim(\alpha(X)) - b.$$

It is only this part of their results we include in these notes, together with some straightforward generalizations due to H. Dunio [14] (see Lectures 12 and 13). The more detailed description of $S^b(X)$, due to Beauville [5], Green-Lazarsfeld [27] and C. Simpson [58] is just mentioned, without proof, at the end of Lecture 13.

§ 2 Logarithmic de Rham complexes

In this lecture we want to start with the definition and simple properties of the sheaf of (algebraic) logarithmic differential forms and of sheaves with logarithmic integrable connections, developed in [10]. The main examples of those will arise from cyclic covers (see Lecture 3). Even if we stay in the algebraic language, the reader is invited (see 2.11) to compare the statements and constructions with the analytic case.

Throughout this lecture X will be an algebraic manifold, defined over an algebraically closed field k, and $D = \sum_{j=1}^{r} D_j$ a reduced normal crossing divisor, i.e. a divisor with non-singular components D_j intersecting each other transversally.

We write $\tau : U = X - D \longrightarrow X$ and

$$\Omega_X^a(*D) = \lim_{\substack{\longrightarrow \\ \nu}} \Omega_X^a(\nu \cdot D) = \tau_* \Omega_U^a.$$

Of course $(\Omega_X^\bullet(*D), d)$ is a complex.

2.1. Definition. $\Omega_X^a(\log\ D)$ denotes the subsheaf of $\Omega_X^a(*D)$ of *differential forms with logarithmic poles* along D, i.e.: if $V \subseteq X$ is open, then

$$\Gamma(V, \Omega_X^a(\log\ D)) =$$

$$\{\ \alpha \in \Gamma(V, \Omega_X^a(*D));\ \alpha \text{ and } d\alpha \text{ have simple poles along } D\}.$$

2.2. Properties.

a)

$$(\Omega_X^\bullet(\log\ D), d) \hookrightarrow (\Omega_X^\bullet(*D), d).$$

is a subcomplex.

b)

$$\Omega_X^a(\log\ D) = \bigwedge^a \Omega_X^1(\log\ D)$$

c) $\Omega_X^a(\log\ D)$ is locally free. More precisely:
For $p \in X$, let us say with $p \in D_j$ for $j = 1, \ldots, s$ and $p \notin D_j$ for $j = s+1, \ldots, r$, choose local parameters $f_1, \ldots, f_n$ in p such that D_j is defined by $f_j = 0$ for $j = 1, \ldots, s$. Let us write

$$\delta_j = \begin{cases} \frac{df_j}{f_j} & \text{if} \quad j \leq s \\ df_j & \text{if} \quad j > s \end{cases}$$

and for $I = \{j_1, \dots, j_a\} \subset \{1, \dots, n\}$ with $j_1 < j_2 \dots < j_a$

$$\delta_I = \delta_{j_1} \wedge \dots \wedge \delta_{j_a}.$$

Then $\{\delta_I;\ \sharp I = a\}$ is a free system of generators for $\Omega^a_X(\log\ D)$.

PROOF: (see [10], II, 3.1 - 3.7). a) is obvious and b) follows from the explicite form of the generators given in c).
Since δ_j is closed, δ_I is a local section of $\Omega^a_X(\log\ D)$. By the Leibniz rule the $\mathcal{O}_X$-module Ω spanned by the δ_I is contained in $\Omega^a_X(\log\ D)$. Ω is locally free and, in order to show that $\Omega = \Omega^a_X(\log\ D)$ it is enough to consider the case $s = 1$. Each local section $\alpha \in \Omega^a_X(*D)$ can be written as

$$\alpha = \alpha_1 + \alpha_2 \wedge \frac{df_1}{f_1},$$

where α_1 and α_2 lie in $\Omega^a_X(*D)$ and $\Omega^{a-1}_X(*D)$ and where both are in the subsheaves generated over $\mathcal{O}(*D)$ by wedge products of $df_2, \dots, df_n$.
$\alpha \in \Omega^a_X(\log\ D)$ implies that

$$f_1 \cdot \alpha = f_1 \cdot \alpha_1 + \alpha_2 \wedge df_1 \ \in \Omega^a_X \ \text{ and } \ f_1 d\alpha = f_1 d\alpha_1 + d\alpha_2 \wedge df_1 \in \Omega^{a+1}_X.$$

Hence α_2 as well as $f_1\alpha_1$ are without poles. Since

$$d(f_1\alpha_1) = df_1 \wedge \alpha_1 + f_1 d\alpha_1 = df_1 \wedge \alpha_1 + f_1 d\alpha - d\alpha_2 \wedge df_1$$

the form $df_1 \wedge \alpha_1$ has no poles which implies $\alpha_1 \in \Omega^a_X$.

□

Using the notation from (2.2,c) we define

$$\alpha : \Omega^1_X(\log\ D) \longrightarrow \bigoplus_{j=1}^{s} \mathcal{O}_{D_j}$$

by

$$\alpha(\sum_{j=1}^{n} a_j \delta_j) = \bigoplus_{j=1}^{s} a_j|_{D_j}.$$

For $a \geq 1$ we have correspondingly a map

$$\beta_1 : \Omega^a_X(\log\ D) \longrightarrow \Omega^{a-1}_{D_1}(\log\ (D - D_1)|_{D_1})$$

given by:
If φ is a local section of $\Omega^a_X(\log\ D)$, we can write

$$\varphi = \varphi_1 + \varphi_2 \wedge \frac{df_1}{f_1}$$

where φ_1 lies in the span of the δ_I with $1 \notin I$ and

$$\varphi_2 = \sum_{1 \in I} a_I \delta_{I-\{1\}}.$$

Then

$$\beta_1(\varphi) = \beta_1(\varphi_2 \wedge \frac{df_1}{f_1}) = \sum a_I \delta_{I-\{1\}}|_{D_1}.$$

Of course, β_i will denote the corresponding map for the i-th component. Finally, the natural restriction of differential forms gives

$$\gamma_1 : \Omega^a_X(\log\,(D - D_1)) \longrightarrow \Omega^a_{D_1}(\log\,(D - D_1)|_{D_1}).$$

Since the sheaf on the left hand side is generated by

$$\{f_1 \cdot \delta_I;\ 1 \in I\} \cup \{\delta_I;\ 1 \notin I\}$$

we can describe γ_1 by

$$\gamma_1(\sum_{1 \in I} f_1 a_I \delta_I + \sum_{1 \notin I} a_I \delta_I) = \sum_{1 \notin I} a_I \delta_I|_{D_1}.$$

Obviously one has

2.3. Properties. One has three exact sequences:
a)

$$0 \to \Omega^1_X \longrightarrow \Omega^1_X(\log\ D) \xrightarrow{\alpha} \bigoplus_{j=1}^{r} \mathcal{O}_{D_j} \to 0.$$

b)

$$0 \to \Omega^a_X(\log\,(D - D_1)) \longrightarrow \Omega^a_X(\log\ D) \xrightarrow{\beta_1} \Omega^{a-1}_{D_1}(\log\,(D - D_1)|_{D_1}) \to 0.$$

c)

$$0 \to \Omega^a_X(\log\ D)(-D_1) \longrightarrow \Omega^a_X(\log\,(D - D_1)) \xrightarrow{\gamma_1} \Omega^a_{D_1}(\log\,(D - D_1)|_{D_1}) \to 0.$$

By (2.2,b) $(\Omega^\bullet_X(\log\ D), d)$ is a complex. It is the most simple example of a logarithmic de Rham complex.

2.4. Definition. Let $\mathcal{E}$ be a locally free coherent sheaf on X and let

$$\nabla : \mathcal{E} \longrightarrow \Omega^1_X(\log\ D) \otimes \mathcal{E}$$

be a k-linear map satisfying

$$\nabla(f \cdot e) = f \cdot \nabla(e) + df \otimes e.$$

One defines
$$\nabla_a : \Omega^a_X(\log\ D) \otimes \mathcal{E} \longrightarrow \Omega^{a+1}_X(\log\ D) \otimes \mathcal{E}$$
by the rule
$$\nabla_a(\omega \otimes e) = d\omega \otimes e + (-1)^a \cdot \omega \wedge \nabla(e).$$
We assume that $\nabla_{a+1} \circ \nabla_a = 0$ for all a. Such ∇ will be called an *integrable logarithmic connection along* D, or just a connection. The complex
$$(\Omega^\bullet_X(\log\ D) \otimes \mathcal{E}, \nabla_\bullet)$$
is called the *logarithmic de Rham complex* of $(\mathcal{E}, \nabla)$.

2.5. Definition. For an integrable logarithmic connection
$$\nabla : \mathcal{E} \longrightarrow \Omega^1_X(\log\ D) \otimes \mathcal{E}$$
we define the residue map along D_1 to be the composed map
$$Res_{D_1}(\nabla) : \mathcal{E} \xrightarrow{\nabla} \Omega^1_X(\log\ D) \otimes \mathcal{E} \xrightarrow{\beta_1' = \beta_1 \otimes id_{\mathcal{E}}} \mathcal{O}_{D_1} \otimes \mathcal{E}.$$

2.6. Lemma.
a) $Res_{D_1}(\nabla)$ *is* $\mathcal{O}_X$*-linear and it factors through*
$$\mathcal{E} \xrightarrow{restr.} \mathcal{O}_{D_1} \otimes \mathcal{E} \longrightarrow \mathcal{O}_{D_1} \otimes \mathcal{E}$$
where restr. the restriction of $\mathcal{E}$ *to* D_1*. By abuse of notations we will call the second map* $Res_{D_1}(\nabla)$ *again.*
b) One has a commutative diagram
$$\begin{array}{ccc} \Omega^a_X(\log\ (D - D_1)) \otimes \mathcal{E} & \xrightarrow{(\nabla_a)\circ(incl.)} & \Omega^{a+1}_X(\log\ D) \otimes \mathcal{E} \\ \downarrow{\scriptstyle \gamma_1 \otimes id_{\mathcal{E}}} & & \downarrow{\scriptstyle \beta_1 \otimes id_{\mathcal{E}} = \beta_1'} \\ \Omega^a_{D_1}(\log\ (D - D_1)\ |_{D_1}) \otimes \mathcal{E} & \xrightarrow{((-1)^a \cdot id) \otimes Res_{D_1}(\nabla)} & \Omega^a_{D_1}(\log\ (D - D_1)\ |_{D_1}) \otimes \mathcal{E} \end{array}$$

PROOF: a) We have
$$\nabla(g \cdot e) = g \cdot \nabla(e) + dg \otimes e \quad \text{and} \quad \beta_1'(\nabla(g \cdot e)) = g \cdot \beta_1'(\nabla(e)).$$
If f_1 divides g then $g \cdot \beta_1'(\nabla(e)) = 0$.
b) For $\omega \in \Omega^a_X(\log\ (D - D_1))$ and $e \in \mathcal{E}$ we have
$$\begin{aligned} \beta_1'(\nabla_a(\omega \otimes e)) &= \beta_1'(d\omega \otimes e + (-1)^a \cdot \omega \wedge \nabla(e)) \\ &= \beta_1'((-1)^a \cdot \omega \wedge \nabla(e)). \end{aligned}$$

If $\omega = f_1 \cdot a_I \cdot \delta_I$ for $1 \in I$, then

$$(-1)^a \omega \wedge \nabla(e) \in \Omega_X^{a+1}(\log\ D)(-D_1)$$

and $\beta_1'(\nabla_a(\omega \otimes\ e)) = 0$. On the other hand, $\gamma_1(\omega) \otimes e = 0$ by definition.
If $\omega = a_I \delta_I$ for $1 \notin I$, then

$$\gamma_1(\omega) \otimes e = a_I \cdot \delta_I \mid_{D_1} \otimes e$$

and

$$\beta_1'((-1)^a \omega \wedge \nabla(e)) = (-1)^a \omega|_{D_1} \otimes\ Res_{D_1}(\nabla)(e).$$

□

2.7. Lemma. *Let*

$$B = \sum_{j=1}^{r} \mu_j D_i$$

be any divisor and $(\nabla, \mathcal{E})$ as in (2.4). Then ∇ induces a connection ∇^B with logarithmic poles on

$$\mathcal{E} \otimes \mathcal{O}_X(B) = \mathcal{E}(B)$$

and the residues satisfy

$$Res_{D_j}(\nabla^B) = Res_{D_j}(\nabla) - \mu_j \cdot id_{D_j}.$$

PROOF: A local section of $\mathcal{E}(B)$ is of the form

$$\sigma = \prod_{j=1}^{s} f_j^{-\mu_j} \cdot e$$

and

$$\nabla^B(\sigma) = \prod_{j=1}^{s} f_j^{-\mu_j} \nabla(e) + d(\prod_{j=1}^{s} f_j^{-\mu_j}) \otimes e =$$

$$= \prod_{j=1}^{s} f_j^{-\mu_j} \nabla(e) + \sum_{k=1}^{s} (\prod_{j=1}^{s} f_j^{-\mu_j}) \cdot (-\mu_k) \frac{df_k}{f_k} \otimes e.$$

Hence $\nabla^B : \mathcal{E}(B) \longrightarrow \Omega_X^1(\log\ D) \otimes \mathcal{E}(B)$ is well defined. One obtains

$$Res_{D_1}(\nabla^B(\sigma)) = \prod_{j=1}^{s} f_j^{-\mu_j}\ Res_{D_1}(\nabla(e)) + \prod_{j=1}^{s} f_j^{-\mu_j}(-\mu_1) \otimes e \mid_{D_1}.$$

□

2.8. Definition. a) We say that $(\nabla, \mathcal{E})$ satisfies the condition $(*)$ if for all divisors

$$B = \sum_{j=1}^{r} \mu_j D_j \geq D$$

and all $j = 1 \ldots r$ one has an isomorphism of sheaves

$$Res_{D_j}(\nabla^B) = Res_{D_j}(\nabla) - \mu_j \cdot id_{D_j} : \mathcal{E}\mid_{D_j} \longrightarrow \mathcal{E}\mid_{D_j} .$$

b) We say that $(\nabla, \mathcal{E})$ satisfies the condition (!) if for all divisors

$$B = \sum_{j=1}^{r} -\nu_j D_j \leq 0$$

and all $j = 1, \ldots, r$

$$Res_{D_j}(\nabla^B) = Res_{D_j}(\nabla) + \nu_j \cdot \, id_{D_j} : \mathcal{E}\mid_{D_j} \longrightarrow \mathcal{E}\mid_{D_j}$$

is an isomorphism of sheaves.

In other words, $(*)$ means that no $\mu_j \in \mathbb{Z}$, $\mu_j \geq 1$, is an eigenvalue of $Res_{D_j}(\nabla)$ and (!) means the same for $\mu_j \in \mathbb{Z}$, $\mu_j \leq 0$. We will see later, that $(*)$ and (!) are only of interest if char $(k) = 0$.

2.9. Properties.
a) Assume that $(\mathcal{E}, \nabla)$ satisfies $(*)$ and that $B = \sum \mu_j D_j \geq 0$. Then the natural map

$$(\Omega^{\bullet}_X(\log\ D) \otimes \mathcal{E}, \nabla_{\bullet}) \longrightarrow (\Omega^{\bullet}_X(\log\ D) \otimes \mathcal{E}(B), \nabla^B_{\bullet})$$

between the logarithmic de Rham complexes is a quasi-isomorphism.
b) Assume that $(\mathcal{E}, \nabla)$ satisfies (!) and that $B = \sum -\mu_j D_j \leq 0$. Then the natural map

$$(\Omega^{\bullet}_X(\log\ D) \otimes \mathcal{E}(B), \nabla^B_{\bullet}) \longrightarrow (\Omega^{\bullet}_X(\log\ D) \otimes \mathcal{E}, \nabla_{\bullet})$$

between the logarithmic de Rham complexes is a quasi-isomorphism.

(2.9) follows from the definition of $(*)$ and (!) and from:

2.10. Lemma. *For $(\mathcal{E}, \nabla)$ as in (2.4) assume that*

$$Res_{D_1}(\nabla) : \mathcal{E}\mid_{D_1} \longrightarrow \mathcal{E}\mid_{D_1}$$

is an isomorphism. Then the inclusion of complexes

$$(\Omega^{\bullet}_X(\log\ D) \otimes \mathcal{E}(-D_1), \nabla^{-D_1}_{\bullet}) \longrightarrow (\Omega^{\bullet}_X(\log\ D) \otimes \mathcal{E}, \nabla_{\bullet})$$

is an quasi-isomorphism.

PROOF: Consider the complexes $\mathcal{E}^{(\nu)}$:

$$\mathcal{E}(-D_1) \longrightarrow \Omega^1_X(\log\ D) \otimes \mathcal{E}(-D_1) \longrightarrow \dots \longrightarrow \Omega^{\nu-1}_X(\log\ D) \otimes \mathcal{E}(-D_1) \longrightarrow$$
$$\longrightarrow \Omega^{\nu}_X(\log\,(D-D_1)) \otimes \mathcal{E} \longrightarrow \Omega^{\nu+1}_X(\log\ D) \otimes \mathcal{E} \longrightarrow \dots \longrightarrow \Omega^n_X(\log\ D) \otimes \mathcal{E}$$

We have an inclusion

$$\mathcal{E}^{(\nu+1)} \longrightarrow \mathcal{E}^{(\nu)}$$

and, by (2.6,b) the quotient is the complex

$$0 \longrightarrow \Omega^{\nu}_{D_1}(\log\,(D-D_1)|_{D_1}) \otimes \mathcal{E} \xrightarrow{(-1)^{\nu}\otimes Res_{D_1}(\nabla)} \Omega^{\nu}_{D_1}(\log\,(D-D_1)|_{D_1}) \otimes \mathcal{E} \longrightarrow 0$$

Since the quotient has no cohomology all the $\mathcal{E}^{(\nu)}$ are quasi-isomorphic, especially $\mathcal{E}^{(0)}$ and $\mathcal{E}^{(n)}$, as claimed.

□

2.11. The analytic case

At this point it might be helpful to consider the analytic case for a moment: $\mathcal{E}$ is a locally free sheaf over the sheaf of analytic functions $\mathcal{O}_X$,

$$\nabla : \mathcal{E} \longrightarrow \Omega^1_X(\log\ D) \otimes \mathcal{E}$$

is a holomorphic and integrable connection. Then $\ker(\nabla\ |_U) = V$ is a local constant system. If $(*)$ holds true, i.e. if the residues of ∇ along the D_j do not have strictly positive integers as eigenvalues, then (see [10], II, 3.13 and 3.14)

$$(\Omega^{\bullet}_X(\log\ D) \otimes \mathcal{E}, \nabla_{\bullet})$$

is quasi-isomorphic to $R\tau_* V$. By Poincaré-Verdier duality (see [20], Appendix A) the natural map

$$\tau_! V^{\vee} \longrightarrow (\Omega^{\bullet}_X(\log\ D) \otimes \mathcal{E}^{\vee}(-D), \nabla^{\vee}_{\bullet})$$

is a quasi-isomorphism. Hence (!) implies that the natural map

$$\tau_! V \longrightarrow (\Omega^{\bullet}_X(\log\ D) \otimes \mathcal{E},\ \nabla_{\bullet})$$

is a quasi-isomorphism as well. In particular, topological properties of U give vanishing theorems for

$$\mathbb{H}^l(X, \Omega^{\bullet}_X(\log\ D) \otimes \mathcal{E})$$

and for some l. More precisely, if we choose $r(U)$ to be the smallest number that satisfies:

> For all local constant systems V on U one has $H^l(U,V) = 0$ for $l > n + r(U)$,

then one gets:

2.12. Corollary.
a) If $(\mathcal{E}, \nabla)$ satisfies $(*)$, then for $l > n + r(U)$

$$\mathbb{H}^l(X, \Omega^{\bullet}_X(\log\ D) \otimes \mathcal{E}) = H^l(U, V) = 0.$$

b) If $(\mathcal{E}, \nabla)$ satisfies (!), then for $l < n - r(U)$

$$\mathbb{H}^l(X, \Omega^{\bullet}_X(\log\ D) \otimes \mathcal{E}) = H^l_c(U, V) = 0.$$

By GAGA (see [56]), (2.12) remains true if we consider the complex of algebraic differential forms over the complex projective manifold X, even if the number $r(U)$ is defined in the analytic topology.

(2.12) is of special interest if both, $(*)$ and (!), are satisfied, i.e. if none of the eigenvalues of $Res_{D_j}(\nabla)$ is an integer. Examples of such connections can be obtained, analytically or algebraically, by cyclic covers.

If U is affine (or a Stein manifold) one has $r(U) = 0$. For U affine there is no need to use GAGA and analytic arguments. Considering blowing ups and the Leray spectral sequence one can obtain (2.12) for algebraic sheaves from:

2.13. Corollary. *Let X be a projective manifold defined over the algebraically closed field k. Let B be an effective ample divisor, $D = B_{red}$ a normal crossing divisor and $(\mathcal{E}, \nabla)$ a logarithmic connection with poles along D (as in (2.4)).*
a) *If $(\mathcal{E}, \nabla)$ satisfies $(*)$, then for $l > n$*

$$\mathbb{H}^l(X, \Omega^\bullet_X(\log\ D) \otimes \mathcal{E}) = 0.$$

b) *If $(\mathcal{E}, \nabla)$ satisfies (!), then for $l < n$*

$$\mathbb{H}^l(X, \Omega^\bullet_X(\log\ D) \otimes \mathcal{E}) = 0.$$

PROOF: (2.9) allows to replace $\mathcal{E}$ by $\mathcal{E}(N \cdot B)$ in case a) or by $\mathcal{E}(-N \cdot B)$ in case b) for $N > 0$. By Serre's vanishing theorem (1.1) we can assume that

$$H^b(X, \Omega^a_X(\log\ D) \otimes \mathcal{E}) = 0$$

for $a + b = l$. The Hodge to de Rham spectral sequence (see (A.25)) implies (2.13).

□

§ 3 Integral parts of $\mathbb{Q}$-divisors and coverings

Over complex manifolds the Riemann Hilbert correspondence obtained by Deligne [10] is an equivalence between logarithmic connections $(\mathcal{E}, \nabla)$ and representations of the fundamental group $\pi_1(X - D)$. For applications in algebraic geometry the most simple representations, i.e. those who factor through cyclic quotient groups of $\pi_1(X - D)$, turn out to be useful. The induced invertible sheaves and connections can be constructed directly as summands of the structure sheaves of cyclic coverings. Those constructions remain valid for all algebraically closed fields.

Let X be an algebraic manifold defined over the algebraically closed field k.

3.1. Notation. a) Let us write Div(X) for the group of divisors on X and

$$\mathrm{Div}_{\mathbb{Q}}(X) = \mathrm{Div}(X) \otimes_{\mathbb{Z}} \mathbb{Q}.$$

Hence a $\mathbb{Q}$-*divisor* $\Delta \in \mathrm{Div}_{\mathbb{Q}}(X)$ is a sum

$$\Delta = \sum_{j=1}^{r} \alpha_j D_j$$

of irreducible prime divisors D_j with coefficients $\alpha_j \in \mathbb{Q}$.
b) For $\Delta \in \mathrm{Div}_{\mathbb{Q}}(X)$ we write

$$[\Delta] = \sum_{j=1}^{r} [\alpha_j] \cdot D_j$$

where for $\alpha \in \mathbb{Q}$, $[\alpha]$ denotes the integral part of α, defined as the only integer such that

$$[\alpha] \leq \alpha < [\alpha] + 1.$$

$[\Delta]$ will be called the integral part of Δ.
c) For an invertible sheaf $\mathcal{L}$, an effective divisor

$$D = \sum_{j=1}^{r} \alpha_j Dj$$

and a positive natural number N, assume that $\mathcal{L}^N = \mathcal{O}_X(D)$. Then we will write for $i \in \mathbb{N}$

$$\mathcal{L}^{(i,D)} = \mathcal{L}^i(-[\frac{i}{N}D]) = \mathcal{L}^i \otimes \mathcal{O}_X(-[\frac{i}{N} \cdot D]).$$

Usually N and D will be fixed and we just write $\mathcal{L}^{(i)}$ instead of $\mathcal{L}^{(i,D)}$.
d) If

$$D = \sum_{j=1}^{r} \alpha_j D_j$$

is a normal crossing divisor, we will write, for simplictiy,

$$\Omega^a_X(\log\ D) \ \text{ instead of } \ \Omega^a_X(\log\ D_{red}).$$

In spite of their strange definition the sheaves $\mathcal{L}^{(i)}$ will turn out to be related to cyclic covers in a quite natural way. We will need this to prove:

3.2. Theorem. *Let X be a projective manifold,*

$$D = \sum_{j=1}^{r} \alpha_j D_j$$

be an effective normal crossing divisor, $\mathcal{L}$ an invertible sheaf and $N \in \mathbb{N} - \{0\}$ prime to char(k), *such that $\mathcal{L}^N = \mathcal{O}_X(D)$. Then for $i = 0, \ldots, N-1$ the sheaf $\mathcal{L}^{(i)^{-1}}$ has an integrable logarithmic connection*

$$\nabla^{(i)} : \mathcal{L}^{(i)^{-1}} \longrightarrow \Omega^1_X(\log\ D^{(i)}) \otimes \mathcal{L}^{(i)^{-1}}$$

$$\textit{with poles along} \quad D^{(i)} = \sum_{\substack{j=1 \\ \frac{i\cdot\alpha_j}{N} \notin \mathbb{Z}}}^{r} D_j,$$

satisfying:
a) *The residue of $\nabla^{(i)}$ along D_j is given by multiplication with*

$$(i \cdot \alpha_j - N \cdot [\frac{i \cdot \alpha_j}{N}]) \cdot N^{-1} \quad \in k.$$

b) *Assume that either* char$(k) = 0$, *or, if* char$(k) = p \neq 0$, *that X and D admit a lifting to $W_2(k)$ (see (8.11)) and that $p \geq$* dim X. *Then the spectral sequence*

$$E_1^{ab} = H^b(X, \Omega^a_X(\log\ D^{(i)}) \otimes \mathcal{L}^{(i)^{-1}}) \Longrightarrow \mathbb{H}^{a+b}(X, \Omega^\bullet_X(\log\ D^{(i)}) \otimes \mathcal{L}^{(i)^{-1}})$$

associated to the logarithmic de Rham complex

$$(\Omega^\bullet_X(\log\ D^{(i)}) \otimes \mathcal{L}^{(i)^{-1}}, \nabla^{(i)}_\bullet)$$

degenerates in E_1.
c) *Let A and B be reduced divisors (both having the lifting property (8.11) if* char$(k) = p \neq 0$) *such that B, A and $D^{(i)}$ have pairwise no commom components and such that $A + B + D^{(i)}$ is a normal crossing divisor. Then $\nabla^{(i)}$ induces a logarithmic connection*

$$\mathcal{O}_X(-B) \otimes \mathcal{L}^{(i)^{-1}} \longrightarrow \Omega^1_X(\log\ (A + B + D^{(i)}))(-B) \otimes \mathcal{L}^{(i)^{-1}}$$

and under the assumptions of b) the spectral sequence

$$E_1^{ab} = H^b(X, \Omega^a_X(\log\ (A + B + D^{(i)}))(-B) \otimes \mathcal{L}^{(i)^{-1}}) \Longrightarrow$$
$$\mathbb{H}^{a+b}(X, \Omega^\bullet_X(\log\ (A + B + D^{(i)}))(-B) \otimes \mathcal{L}^{(i)^{-1}})$$

degenerates in E_1 as well.

3.3. Remarks. a) In (3.2), whenever one likes, one can assume that $i = 1$. In fact, one just has to replace $\mathcal{L}$ by $\mathcal{L}' = \mathcal{L}^i$ and D by $D' = i \cdot D$ .Then

$$\mathcal{L}'^N = \mathcal{O}_X(i \cdot D) = \mathcal{O}_X(D')$$

and

$$\mathcal{L}'^{(1,D')} = \mathcal{L}'(-[\frac{D'}{N}]) = \mathcal{L}^i(-[\frac{i}{N}D]).$$

b) Next, one can always assume that $0 < \alpha_j < N$. In fact, if $\alpha_1 \geq N$, then

$$\mathcal{L}' = \mathcal{L}(-D_1) \quad \text{and} \quad D' = D - N \cdot D_1$$

give the same sheaves as $\mathcal{L}$ and D:

$$\mathcal{L}'^{(i,D')} = \mathcal{L}^i(-i \cdot D_1 - [\frac{i}{N} \cdot D']) = \mathcal{L}^i(-[\frac{i}{N} \cdot D]).$$

c) In particular, for $i = 1$ and $0 < a_j < N$ we have

$$\mathcal{L}^{(1)} = \mathcal{L} \quad \text{and} \quad D^{(1)} = D.$$

Nevertheless, in the proof of (3.2) we stay with the notation, as started.
d) Finally, for $i \geq N$ one has

$$\mathcal{L}^{(i,D)} = \mathcal{L}^i(-[\frac{i}{N} \cdot D]) = \mathcal{L}^{i-N}(-[\frac{i-N}{N} \cdot D]) = \mathcal{L}^{(i-N,D)}.$$

The "$\mathcal{L}^{(i)}$" are the most natural notation for "integral parts of $\mathbb{Q}$- divisors" if one wants to underline their relations with coverings. In the literature one finds other equivalent notations, more adapted to the applications one has in mind:

3.4. Remarks.
a) Sometimes the integral part $[\Delta]$ is denoted by $\lfloor \Delta \rfloor$.
b) One can also consider the round up $\{\Delta\} = \lceil \Delta \rceil$ given by

$$\{\Delta\} = -[-\Delta]$$

or the fractional part of Δ given by

$$< \Delta >= \Delta - [\Delta].$$

c) For $\mathcal{L}$, N and D as in (3.1,c) one can write

$$\mathcal{L} = \mathcal{O}_X(C)$$

for some divisor C. Then

$$\Delta = C - \frac{1}{N} \cdot D \ \in \mathrm{Div}_{\mathbb{Q}}(X)$$

has the property that $N \cdot \Delta$ is a divisor linear equivalent to zero. One has

$$\mathcal{L}^{(i,D)} = \mathcal{O}_X(i \cdot C - [\frac{i}{N} \cdot D]) = \mathcal{O}_X(-[-i \cdot \Delta]) = \mathcal{O}_X(\{i \cdot \Delta\}).$$

d) On the other hand, for $\Delta \in \mathrm{Div}_Q(X)$ and $N > 0$ assume that $N \cdot \Delta$ is a divisor linear equivalent to zero. Then one can choose a divisor C such that $C - \Delta$ is effective. For $\mathcal{L} = \mathcal{O}_X(C)$ and $D = N \cdot C - N \cdot \Delta \ \in \mathrm{Div}(X)$ one has

$$\mathcal{L}^N = \mathcal{O}_X(N \cdot C) = \mathcal{O}_X(D)$$

and

$$\mathcal{L}^{(i,D)} = \mathcal{O}_X(i \cdot C - [\frac{i}{N} D]) =$$

$$\mathcal{O}_X(-[-i \cdot C + \frac{i}{N} \cdot D]) = \mathcal{O}_X(\{i \cdot \Delta\}).$$

e) Altogether, (3.2) is equivalent to:

For $\Delta \in \mathrm{Div}_{\mathbb{Q}}(X)$ such that $N \cdot \Delta$ is a divisor linear equivalent to zero, assume that $< \Delta >$ is supported in D and that D is a normal crossing divisor. Then $\mathcal{O}_X(\{\Delta\})$ has a logarithmic integrable connection with poles along D which satisfies a residue condition similar to (3.2,a) and the E_1-degeneration.

We leave the exact formulation and the translation as an exercise.

3.5. Cyclic covers. Let $\mathcal{L}, N$ and

$$D = \sum_{j=1}^{r} \alpha_j D_j$$

be as in (3.1,c) and let $s \in H^0(X, \mathcal{L}^N)$ be a section whose zero divisor is D. The dual of

$$s : \mathcal{O}_X \longrightarrow \mathcal{L}^N \text{ , i.e. } s^\vee : \mathcal{L}^{-N} \longrightarrow \mathcal{O}_X,$$

defines a $\mathcal{O}_X$-algebra structure on

$$\mathcal{A}' = \bigoplus_{i=0}^{N-1} \mathcal{L}^{-i}.$$

In fact,

$$\mathcal{A}' = \bigoplus_{i=0}^{\infty} \mathcal{L}^{-i}/I$$

where I is the ideal-sheaf generated locally by

$$\{s^\vee(l) - l, \ l \text{ local section of } \mathcal{L}^{-N}\}.$$

Let

$$Y' = \mathbf{Spec}_X(\mathcal{A}') \xrightarrow{\pi'} X$$

be the spectrum of the $\mathcal{O}_X$-algebra $\mathcal{A}'$, as defined in [30], page 128, for example.

Let $\pi : Y \to X$ be the finite morphism obtained by normalizing $Y' \to X$. To be more precise, if Y' is reducible, Y will be the disjoint union of the normalizations of the components of Y' in their function fields. We will call Y *the cyclic cover obtained by taking the n-th root out of s* (or *out of D*, if $\mathcal{L}$ is fixed).

Obviously one has:

3.6. Claim. Y is uniquely determined by:
a) $\pi : Y \to X$ is finite.
b) Y is normal.
c) There is a morphism $\phi : \mathcal{A}' \to \pi_* \mathcal{O}_Y$ of $\mathcal{O}_X$-algebras, isomorphic over some dense open subscheme of X.

3.7. Notations. For D, N and $\mathcal{L}$ as in (3.1,c) let us write

$$\mathcal{A} = \bigoplus_{i=0}^{N-1} \mathcal{L}^{(i)^{-1}}.$$

The inclusion

$$\mathcal{L}^{-i} \longrightarrow \mathcal{L}^{(i)^{-1}} = \mathcal{L}^{-i}([\frac{i}{N} \cdot D])$$

gives a morphism of $\mathcal{O}_X$-modules

$$\phi : \mathcal{A}' \longrightarrow \mathcal{A}.$$

3.8. Claim. $\mathcal{A}$ has a structure of an $\mathcal{O}_X$-algebra, such that ϕ is a homomorphism of algebras.

Proof: The multiplication in $\mathcal{A}'$ is nothing but the multiplication

$$\mathcal{L}^{-i} \times \mathcal{L}^{-j} \longrightarrow \mathcal{L}^{-i-j}$$

$$\text{composed with } s^\vee : \mathcal{L}^{-i-j} \longrightarrow \mathcal{L}^{-i-j+N} ,$$

in case that $i + j \geq N$. For $i, j \geq 0$ one has

$$[\frac{i}{N} \cdot D] + [\frac{j}{N} \cdot D] \leq [\frac{i+j}{N} \cdot D]$$

and, for $i + j \geq N$, one has

$$\mathcal{L}^{(i+j)} = \mathcal{L}^{i+j}(-[\frac{i+j}{N} \cdot D]) = \mathcal{L}^{i+j-N}(-[\frac{i+j-N}{N} \cdot D]) = \mathcal{L}^{(i+j-N)}.$$

This implies that the multiplication of sections

$$\mathcal{L}^{(i)^{-1}} \times \mathcal{L}^{(j)^{-1}} \longrightarrow \mathcal{L}^{-i-j}([\frac{i}{N}D] + [\frac{j}{N}D]) \longrightarrow \mathcal{L}^{(i+j)^{-1}}$$

is well defined, and that for $i + j \geq N$ the right hand side is nothing but $\mathcal{L}^{(i+j-N)^{-1}}$.

□

3.9.

Assume that N is prime to char(k), e a fixed primitive N-th root of unit and $G =< \sigma >$ the cyclic group of order N. Then G acts on $\mathcal{A}$ by $\mathcal{O}_X$-algebra homomorphisms defined by:

$$\sigma(l) = e^i \cdot l \quad \text{for a local section} \quad l \quad \text{of} \quad \mathcal{L}^{(i)^{-1}} \subset \mathcal{A}.$$

Obviously the invariants under this G-action are

$$\mathcal{A}^G = \mathcal{O}_X.$$

3.10. Claim. Assume that N is prime to char(k). Then

$$\mathcal{A} = \pi_* \mathcal{O}_Y \quad \text{or (equivalently)} \quad Y = \mathbf{Spec}(\mathcal{A}) .$$

3.11. Corollary (see [16]). *The cyclic group G acts on Y and on $\pi_*\mathcal{O}_Y$. One has $Y/G = X$ and the decomposition*

$$\pi_* \mathcal{O}_Y = \bigoplus_{i=0}^{N-1} \mathcal{L}^{(i)^{-1}}$$

is the decomposition in eigenspaces.

Proof of 3.10.: For any open subvariety X_0 in X with $\mathrm{codim}_X(X - X_0) \geq 2$ and for $Y_0 = \pi^{-1}(X_0)$ consider the induced morphisms

$$\begin{array}{ccc} Y_0 & \xrightarrow{\iota'} & Y \\ {\scriptstyle \pi_0}\downarrow & & \downarrow{\scriptstyle \pi} \\ X_0 & \xrightarrow{\iota} & X \end{array}$$

Since Y is normal one has $\iota'_* \mathcal{O}_{Y_0} = \mathcal{O}_Y$ and $\pi_* \mathcal{O}_Y = \iota_* \pi_{0*} \mathcal{O}_{Y_0}$. Since $\mathcal{A}$ is locally free, (3.10) follows from

$$\pi_{0*} \mathcal{O}_{Y_0} = \mathcal{A}|_{X_0}.$$

Especially we may choose $X_0 = X - \mathrm{Sing}(D_{red})$ and, by abuse of notations, assume from now on that D_{red} is non-singular.

As remarked in (3.6) the equality of $\mathcal{A}$ and $\pi_* \mathcal{O}_Y$ follows from:

3.12. Claim. $\mathbf{Spec}\ (\mathcal{A}) \longrightarrow X$ is finite and $\mathbf{Spec}(\mathcal{A})$ is normal.

Proof: (3.12) is a local statement and to prove it we may assume that $X = \mathrm{Spec}\ B$ and that D consists of just one component, say $D = \alpha_1 \cdot D_1$. Let us fix isomorphisms $\mathcal{L}^i \simeq \mathcal{O}_X$ for all i and assume that D_1 is the zero set of $f_1 \in B$. For some unit $u \in B^*$ the section $s \in H^0(X, \mathcal{L}^N) \simeq B$ is identified with $f = u \cdot f_1^{\alpha_1}$. For completeness, we allow D (or α_1) to be zero.
The $\mathcal{O}_X$-algebra $\mathcal{A}'$ is given by the B-algebra

$$H^0(X, \mathcal{A}') = \bigoplus_{i=0}^{N-1} H^0(X, \mathcal{L}^{-i})$$

which can be identified with the quotient of the ring of polynomials

$$A' = B[t]/_{t^N - f} = \bigoplus_{i=0}^{N-1} B \cdot t^i.$$

In this language

$$A = \bigoplus_{i=0}^{N-1} B \cdot t^i \cdot f_1^{-[\frac{i}{N}\alpha_1]} = \bigoplus_{i=0}^{N-1} H^0(X, \mathcal{L}^{(i)^{-1}}) = H^0(X, \mathcal{A})$$

and $\phi : \mathcal{A}' \to \mathcal{A}$ induces the natural inclusion $A' \hookrightarrow A$.

Hence (3.12) follows from the first part of the following claim.

□

3.13. Claim. Using the notations introduced above, assume that N is prime to char(k). Then one has
a) Spec A is non-singular and π : Spec $A \longrightarrow$ Spec B is finite.
b) If $\alpha_1 = 0$, then Spec $A \longrightarrow$ Spec B is non-ramified (hence étale).
c) if α_1 is prime to N, we have a defining equation $g \in A$ for $\Delta_1 = (\pi^* D_1)_{red}$ with

$$g^N = u^a \cdot f_1 \quad \text{for some} \quad a \in \mathbb{N}.$$

d) If Γ is a divisor in Spec B such that $D + \Gamma$ has normal crossings, then $\pi^*(D + \Gamma)$ has normal crossings as well.

PROOF: Let us first consider the case $\alpha_1 = 0$. Then

$$A' = A = B[t]/_{t^N - u}$$

for $u \in B^*$. A is non-singular, as follows, for example, from the Jacobi-criterion, and A is unramified over B. Hence

$$\text{Spec } A \longrightarrow \text{Spec } B$$

is étale in this case and a), b) and d) are obvious.

If $\alpha_1 = 1$, then again

$$A' = A = B[t]/_{t^N - u \cdot f_1}.$$

For $p \in$ Spec B, choose $f_2, \ldots, f_n$ such that $f_1 \cdot u, f_2, \ldots, f_n$ is a local parameter-system in p. Then $t, f_2, \ldots, f_n$ will be a local parameter system, for $q = \pi^{-1}(p)$.

Similar, if α_1 is prime to N, and if c) holds true, g and $f_2, \cdots, f_n$ will be a local parameter system in q and, composing both steps, Spec A will always be non-singular and d) holds true.

Let us consider the ring

$$R = B[t_0, t_1]/_{t_0^N - u, t_1^N - f_1}.$$

Identifying t with $t_0 \cdot t_1^{\alpha_1}$ we obtain A' as a subring of R. Spec R is non-singular over p and Spec $R \longrightarrow$ Spec B is finite.

The group $H =< \sigma_0 > \oplus < \sigma_1 >$ with $ord\ (\sigma_0) = ord\ (\sigma_1) = N$ operates on R by

$$\sigma_\nu(t_\mu) = \begin{cases} t_\mu & \text{if } \nu \neq \mu \\ e \cdot t_\mu & \text{if } \nu = \mu \end{cases}$$

Let H' be the kernel of the map $\gamma : H \longrightarrow G =< \sigma >$ given by $\gamma(\sigma_0) = \sigma$ and $\gamma(\sigma_1) = \sigma^{\alpha_1}$. The quotient

$$\text{Spec } (R)/_{H'} = \text{Spec } R^{H'}$$

is normal and finite over Spec B.
One has $(\sigma_0^\mu, \sigma_1^\nu) \in H'$, if and only if $\mu + \nu\alpha_1 \equiv 0 \bmod N$. Hence $R^{H'}$ is generated by monomials $t_0^a \cdot t_1^b$ where $a, b \in \{0, \ldots, N-1\}$ satisfy:

(*) $\quad a\mu + b\nu \equiv 0 \bmod N$ for all (μ, ν) with $\mu + \alpha_1\nu \equiv 0 \bmod N$.

Obviously, (*) holds true for (a, b) if $b \equiv a \cdot \alpha_1 \mod N$. On the other hand, choosing ν to be a unit in $\mathbb{Z}/_N$, (*) implies that $b \equiv a \cdot \alpha_1 \mod N$.

Hence, for all (a, b) satisfying (*) we find some k with $b = a \cdot \alpha_1 + k \cdot N$. Since $a, b \in \{0, \ldots, N-1\}$ we have

$$\frac{a \cdot \alpha_1}{N} \geq -k = \frac{a \cdot \alpha_1}{N} - \frac{b}{N} > \frac{a \cdot \alpha_1}{N} - 1$$

or $k = -[\frac{a \cdot \alpha_1}{N}]$.

Therefore one obtains

$$R^{H'} = \bigoplus_{a=0}^{N-1} t_0^a \cdot t_1^{a \cdot \alpha_1 - N \cdot [\frac{a \cdot \alpha_1}{N}]} \cdot B = \bigoplus_{a=0}^{N-1} (t_0 \cdot t_1^{\alpha_1})^a \cdot f_1^{-[\frac{a \cdot \alpha_1}{N}]} \cdot B$$

$$\text{and hence} \qquad R^{H'} = \bigoplus_{a=0}^{N-1} t^a \cdot f_1^{-[\frac{a \cdot \alpha_1}{N}]} \cdot B = A.$$

If α_1 is prime to N, we can find $a \in \{0, \ldots, N-1\}$ with $a \cdot \alpha_1 = 1 + l \cdot N$ for $l \in \mathbb{Z}$. Then

$$a \cdot \alpha_1 - N \cdot [\frac{a \cdot \alpha_1}{N}] = 1$$

and $g = t^a \cdot f_1^{-[\frac{a\cdot\alpha_1}{N}]}$ satisfies

$$g^N = u^a \cdot f_1^{a\cdot\alpha_1 - N[\frac{a\cdot\alpha_1}{N}]} = u^a \cdot f_1.$$

□

3.14. Remarks.
a) If Y is irreducible, for example if D is reduced, the local calculation shows Y is nothing but the normalisation of X in $k(X)(\sqrt[N]{f})$, where f is a rational function giving the section s.
b) $\pi' : Y' \longrightarrow X$ can be as well described in the following way (see [30], p. 128-129):
Let $\mathbf{V}(\mathcal{L}^{-\alpha}) = \text{Spec }(\bigoplus_{i=0}^{\infty} \mathcal{L}^{-\alpha})$ be the geometric rank one vector bundle associated to $\mathcal{L}^{-\alpha}$. The geometric sections of $\mathbf{V}(\mathcal{L}^{-\alpha}) \longrightarrow X$ correspond to $H^0(X, \mathcal{L}^{\alpha})$. Hence s gives a section σ of $\mathbf{V}(\mathcal{L}^{-N})$ over X. We have a natural map

$$\tau : \mathbf{V}(\mathcal{L}^{-1}) \longrightarrow \mathbf{V}(\mathcal{L}^{-N})$$

and $Y' = \tau^{-1}(\sigma(X))$.

The local computation in (3.13) gives a little bit more information than asked for in (3.12):

3.15. Lemma. *Keeping the notations and assumptions from (3.5) assume that N is prime to* char(k). *Then one has*
a) Y is reducible, if and only if for some $\mu > 1$, dividing N, there is a section s' in $H^0(X, \mathcal{L}^{\frac{N}{\mu}})$ with $s = s'^{\mu}$.
b) $\pi : Y \to X$ is étale over $X - D_{red}$ and Y is non-singular over $X - \text{Sing}(D_{red})$.
c) For $\Delta_j = (\pi^* D_j)_{red}$ we have

$$\pi^* D = \sum_{j=1}^{r} \frac{N \cdot \alpha_j}{gcd(N, \alpha_j)} \cdot \Delta_j.$$

d) If Y is irreducible then the components of Δ_j have over D_j the ramification index

$$e_j = \frac{N}{gcd(N, \alpha_j)}.$$

PROOF: For a) we can consider the open set Spec $B \subset X - D_{red}$. Hence Spec $B[t]/_{t^N - u}$ is in Y dense and open. Y is reducible if and only if $t^N - u$ is reducible in $B[t]$, which is equivalent to the existence of some $u' \in B$ with $u = u'^{\mu}$.
b) has been obtained in (3.13) part a) and b).
For c) and d) we may assume that $D = \alpha_1 \cdot D_1$ and, splitting the covering in

two steps, that either N divides α_1 or that N is prime to α_1.
In the first case, we can as well choose α_1 to be zero (by 3.3,b) and c) as well as d) follow from part b).
If α_1 is prime to N,then $\pi^*D = e_1 \cdot \alpha_1 \cdot \Delta_1$. Since π^*D is the zero locus of $f = t^N$, N divides $e_1 \cdot \alpha_1$. On the other hand, since e_1 divides deg $(Y/X) = N$, one has $e_1 = N$ in this case.

□

3.16. Lemma. *Keeping the notations from (3.5) assume that N is prime to* char(k) *and that D_{red} is non-singular. Then one has:*
a) (Hurwitz's formula) $\pi^*\Omega^b_X(\log\ D) = \Omega^b_Y(\log\ (\pi^*D))$.
b) The differential d on Y induces a logarithmic integrable connection

$$\pi_*(d) : \bigoplus_{i=0}^{N-1} \mathcal{L}^{(i)^{-1}} \longrightarrow \pi_*\Omega^1_Y(\log\ (\pi^*D)) = \bigoplus_{i=0}^{N-1} \Omega^1_X(\log\ D) \otimes \mathcal{L}^{(i)^{-1}},$$

compatible with the direct sum decomposition.
c) If $\nabla^{(i)} : \mathcal{L}^{(i)^{-1}} \longrightarrow \Omega^1_X(\log\ D) \otimes \mathcal{L}^{(i)^{-1}}$ denotes the i-th component of $\pi_*(d)$ then $\nabla^{(i)}$ is a logarithmic integrable connection with residue

$$Res_{D_j}(\nabla^{(i)}) = (\frac{i \cdot \alpha_j}{N} - [\frac{i \cdot \alpha_j}{N}]) \cdot id_{\mathcal{O}_{D_j}}.$$

d) One has

$$\pi_*(\Omega^b_Y) = \bigoplus_{i=0}^{N-1} \Omega^b_X(\log\ D^{(i)}) \otimes \mathcal{L}^{(i)^{-1}} \quad for\ \ D^{(i)} = \sum_{\substack{j=1 \\ \frac{i \cdot \alpha_j}{N} \in \mathbb{Q}-\mathbb{Z}}}^{r} D_j.$$

e) The differential

$$\pi_*(d) : \pi_*\mathcal{O}_Y = \bigoplus_{i=0}^{N-1} \mathcal{L}^{(i)^{-1}} \longrightarrow \pi_*(\Omega^1_Y) = \bigoplus_{i=0}^{N-1} \Omega^1_X(\log\ D^{(i)}) \otimes \mathcal{L}^{(i)^{-1}}$$

decomposes into a direct sum of

$$\nabla^{(i)} : \mathcal{L}^{(i)^{-1}} \longrightarrow \Omega^1_X(\log\ D^{(i)}) \otimes \mathcal{L}^{(i)^{-1}}$$

PROOF: Again we can argue locally and assume that $X =$ Spec B and $D = \alpha_1 D_1$ as in (3.12).
If $\alpha_1 = 0$, or if N divides α_1, then f_1 is a defining equation for $\Delta_1 = (\pi^*D_1)_{red}$ and the generators for $\Omega^b_X(\log\ D)$ are generators for $\Omega^b_Y(\log\ \pi^*D)$ as well.
For α_1 prime to N, we have by (3.13,c) a defining equation g for $\Delta_1 = (\pi^*D_1)_{red}$ with $g^N = u^a \cdot f_1$. Hence

$$N \cdot \frac{dg}{g} = \frac{df_1}{f_1} + a \cdot \frac{du}{u}$$

and, since $N \in k^*$ and $a \cdot \frac{du}{u} \in \Omega^1_X$, one finds that $\frac{df_1}{f_1}$ and $\pi^*\Omega^1_X$ generate $\Omega^1_Y(\log\ \pi^*D)$.

We can split π in two coverings of degree $N \cdot \gcd(N, \alpha_1)^{-1}$ and $\gcd(N, \alpha_1)$. Hence we obtain a) for $b = 1$. The general case follows.

The group G acts on $\pi_*\Omega^b_Y$ and $\pi_*\Omega^b_Y(\log\ \pi^*D))$ compatibly with the inclusion, and the action on the second sheaf is given by $id \otimes \sigma$ if one writes

$$\pi_*\Omega^b_Y(\log\ (\pi^*D)) = \Omega^b_X(\log\ D) \otimes \pi_*\mathcal{O}_Y.$$

Let l be a local section of $\Omega^b_X(\log\ D) \otimes \mathcal{L}^{(i)^{-1}}$ written as

$$l = \phi \cdot g_i \quad \text{for} \quad \phi \in \Omega^b_X(\log\ D) \quad \text{and} \quad g_i = t^i \cdot f_1^{-[\frac{i \cdot \alpha_1}{N}]}.$$

Since

$$g_i^N = u^i \cdot f_1^{i \cdot \alpha_1 - N \cdot [\frac{i\alpha_1}{N}]}$$

has a zero along Δ_1 if and only if

$$\frac{i \cdot \alpha_1}{N} \notin \mathbb{Z},$$

we find that l lies in Ω^b_Y in this case.
On the other hand, if g_i is a unit, l lies in Ω^b_Y if and only if ϕ has no pole along D and we obtain d).

We have

$$N\frac{dg_i}{g_i} = i \cdot \frac{du}{u} + (i \cdot \alpha_1 - N[\frac{i \cdot \alpha_1}{N}])\frac{df_1}{f_1}$$

or

$$dg_i = (\frac{i}{N}\frac{du}{u} + (\frac{i}{N}\alpha_1 - [\frac{i \cdot \alpha_1}{N}])\frac{df_1}{f_1}) \cdot g_i.$$

Hence,

$$d(g_i \cdot \phi) \in \Omega^{b+1}_X(\log\ D) \otimes \mathcal{L}^{(i)^{-1}},$$

and (π_*d) respects the direct sum decomposition. Obviously, the Leibniz rule for d implies that (π_*d) as well as the components $\nabla^{(i)}$ are connections and b) and e) hold true.

Finally, for c), let $\phi \in \mathcal{O}_X$. Then by the calculations given above, we find

$$Res_{D_1}(\nabla^{(i)})(g_i \cdot \phi) = (\frac{i}{N}\alpha_1 - [\frac{i \cdot \alpha_1}{N}])g_i \cdot \phi\,|_{D_1}\,.$$

□

PROOF OF 3.2,A:
If X is projective and

$$D = \sum_{j=1}^{r} \alpha_j D_j$$

a normal crossing divisor we found the connection

$$\nabla^{(i)} : \mathcal{L}^{(i)^{-1}} \longrightarrow \Omega^1_X(\log\ D^{(i)}) \otimes \mathcal{L}^{(i)^{-1}}$$

with the residues as given in (3.2,a) over the open submanifold $X - \mathrm{Sing}(D_{red})$. Of course, $\nabla^{(i)}$ extends to X since

$$\mathrm{codim}_X(\mathrm{Sing}(D_{red})) \geq 2.$$

□

Over a field k of characteristic zero, to prove the E_1-degeneration, as stated in (3.2,b) or (3.2,c) one can apply the degeneration of the logarithmic Hodge to de Rham spectral sequence (see (10.23) for example) to some desingularization of Y. We will sketch this approach in (3.22). One can as well reduce (3.2,b) to the more familiar degeneration of the Hodge spectral sequence

$$E_1^{ab} = H^b(T, \Omega^a_T) \Longrightarrow \mathbb{H}^{a+b}(T, \Omega^\bullet_T)$$

for projective manifolds T by using the following covering Lemma, due to Y. Kawamata [35]:

3.17. Lemma. *Keeping the notations from (3.5) assume that N is prime to* char(k) *and that D is a normal crossing divisor. Then there exists a manifold T and a finite morphism*

$$\delta : T \longrightarrow Y$$

such that:
a) The degree of δ divides a power of N.
b) If A and B are reduced divisors such that $D + A + B$ has at most normal crossings and if $A + B$ has no common component with D, then we can choose T such that $(\pi \circ \delta)^*(D + A + B)$ is a normal crossing divisor and $(\pi \circ \delta)^*A$ as well as $(\pi \circ \delta)^*B$ are reduced.

PROOF OF (3.2) IN CHARACTERISTIC ZERO, ASSUMING THE E_1 DEGENERATION OF THE HODGE TO DE RHAM SPECTRAL SEQUENCE:
Let $X_0 = X - \mathrm{Sing}(D_{red})$, $Y_0 = \pi^{-1}(X_0)$ and $T_0 = \delta^{-1}(Y_0)$.
$\delta_*\Omega^\bullet_{T_0}$ contains $\Omega^\bullet_{Y_0}$ as direct summand. Since $(\pi \circ \delta)$ is flat $(\pi \circ \delta)_*\Omega^\bullet_T$ will contain

$$\bigoplus_{i=0}^{N-1} \Omega^\bullet_X(\log\ D^{(i)}) \otimes \mathcal{L}^{(i)^{-1}}$$

as a direct summand. The E_1-degeneration of the spectral sequence

$$E_1^{ab} = H^b(T, \Omega_T^a) \Longrightarrow \mathbb{H}^{a+b}(T, \Omega_T^\bullet)$$

implies (3.2,b) for each $i \in \{0, \ldots, N-1\}$. Finally, if A and B are the divisors considered in (3.2,c), $A' = (\pi \circ \delta)^* A$ and $B' = (\pi \circ \delta)^* B$,

$$\Omega_X^\bullet(\log\,(A + B + D^{(i)}))(-B) \otimes \mathcal{L}^{(i)^{-1}}$$

is a direct summand of

$$(\pi \circ \delta)_* \Omega_T^\bullet(\log\,(A' + B'))(-B')$$

and we can use the E_1-degeneration of

$$E_1^{ab} = H^b(T, \Omega_T^a(\log\,(A' + B'))(-B')) \Longrightarrow \mathbb{H}^{a+b}(T, \Omega_T^\bullet(\log\,(A' + B'))(-B')).$$

□

3.18. Remarks.
a) If $A = B = 0$ the degeneration of the spectral sequence, used to get (3.2,b), follows from classical Hodge theory. In general, i.e. for (3.2,c), one has to use the Hodge theory for open manifolds developed by Deligne [11].
In these lectures (see (10.23)) we will reproduce the algebraic proof of Deligne and Illusie for the degeneration.
b) If char $(k) \neq 0$ and if $X, \mathcal{L}$ and D admit a lifting to $W_2(k)$ (see (8.11)), then the manifold T constructed in (3.17) will again admit a lifting to $W_2(k)$. Hence the proof of (3.2,b and c) given above shows as well:

Assuming the degeneration of the Hodge to de Rham spectral sequence (proved in (10.21)) theorem (3.2) holds true under the additional assumption that $\mathcal{L}$ lifts to $W_2(k)$ as well.

c) Using (3.2,a) we will give a direct proof of (3.2,b and c) at the end of §10, without using (3.17), for a field k of characteristic $p \neq 0$. By reduction to characteristic p one obtains a second proof of (3.2) in characteristic zero.
d) In Lectures 4 - 7, we will assume (3.2) to hold true.

To prove (3.17) we need:

3.19. Lemma (Kawamata [35]). *Let X be a quasi-projective manifold, let*

$$D = \sum_{j=1}^{r} D_j$$

be a reduced normal crossing divisor, and let

$$N_1, \ldots, N_r \in \mathbb{N} - \{0\}$$

be prime to char(k). *Then there exists a projective manifold* Z *and a finite morphism* $\tau : Z \to X$ *such that:*
a) For $j = 1, \ldots r$ *one has* $\tau^* D_j = N_j \cdot (\tau^* D_j)_{red}$.
b) $\tau^*(D)$ *is a normal crossing divisor.*
c) The degree of τ *divides some power of* $\prod_{j=1}^r N_j$.
d) If X *and* D *satisfy the lifting property (8.11) the same holds true for* Z.

PROOF: If we replace the condition that $D = \sum_{j=1}^r D_j$ is the decomposition of D into irreducible (non-singular) components by the condition that $D = \sum_{j=1}^r D_j$ for non-singular divisors $D_1 \ldots D_r$ we can construct Z by induction and hence assume that $N_1 = N$ and $N_2 = \ldots = N_r = 1$.

Let $\mathcal{A}$ be an ample invertible sheaf such that $\mathcal{A}^N(-D_1)$ is generated by its global sections. Choose $n = \dim\ X$ general divisors $H_1, \ldots, H_n$ with

$$\mathcal{O}_X(H_i) = \mathcal{A}^N(-D_1).$$

The divisor $D + \sum_{i=1}^n H_i$ will be a reduced normal crossing divisor. Let

$$\tau_i : Z_i \longrightarrow X$$

be the cyclic cover obtained by taking the N-th root out of $H_i + D_1$. Then Z_i satisfies the properties a), c) and d) asked for in (3.19) but, Z_i might have singularities over $H_i \cap D_1$ and $\tau_i^*(D)$ might have non-normal crossings over $H_i \cap D_1$. Let Z be the normalization of

$$Z_1 \times_X Z_2 \times_X \ldots \times_X Z_n.$$

Z can inductively be constructed as well in the following way:
Let $Z^{(\nu)}$ be the normalization of $Z_1 \times_X \ldots \times_X Z_\nu$ and $\tau^{(\nu)} : Z^{(\nu)} \to X$ the induced morphism. Then, outside of the singular locus of $Z^{(\nu)}$, the cover $Z^{(\nu+1)}$ is obtained from $Z^{(\nu)}$ by taking the N-th root out of

$$\tau^{(\nu)^*}(H_{\nu+1} + D_1) = \tau^{(\nu)^*}(H_{\nu+1}) + N \cdot (\tau^{(\nu)^*} D_1)_{red}.$$

This is the same as taking the N-th root out of $\tau^{(\nu)^*}(H_{\nu+1})$ by (3.2,b) and (3.10). Since this divisor has no singularities, we find by (3.15,b) that the singularities of $Z^{(\nu+1)}$ lie over the singularities of $Z^{(\nu)}$, hence inductively over $H_1 \cap D_1$. However, as Z is independent of the numbering of the H_i, the singularities of Z are lying over

$$\bigcap_{i=1}^n (H_i \cap D_1) = (\bigcap_{i=1}^n H_i) \cap D_1 = \emptyset.$$

□

PROOF OF (3.17): Let $\tau : Z \to X$ be the covering constructed in (3.19) for

$$D_{red} = \sum_{j=1}^{r} D_j$$

and $N = N_1 = \ldots = N_r$. Let T be the normalization of $Z \times_X Y$. Then T is obtained again by taking the N-th root out of $\tau^* D$. Since $\tau^* D = N \cdot D'$ for some divisor D' on Z, we can use (3.3,b), (3.10) and (3.15,b) to show that T is étale over Z.

For part c), we apply the same construction to the manifold Z, given for the divisor $D + A + B$, where the prescribed multiplicities for the components of A and B are one.

□

Generalizations and variants in the analytic case

(3.17) is a special case of the more general covering lemma of Kawamata:

3.20. Lemma. *Let X be a projective manifold,* $\mathrm{char}(k) = 0$ *and let $\pi : Y \to X$ be a finite cover such that the ramification locus $D = \Delta(Y/X)$ in X has normal crossings. Then there exists a manifold T and a finite morphism $\delta : T \to Y$. Moreover, one can assume that $\pi \circ \delta : T \to X$ is a Galois cover.*

For the proof see [35]. As shown in [63] (3.16) can be generalized as well:

3.21. Lemma. *(Generalized Hurwitz's formula) For $\pi : Y \to X$ as in* (3.20) *let $\delta : Z \to Y$ be a desingularization such that $(\pi \circ \delta)^* D = D'$ is a normal crossing divisor. Then one has an inclusion*

$$\delta^* \pi^* \Omega^a_X(\log\ D) \longrightarrow \Omega^a_Z(\log\ D')$$

giving an isomorphism over the open subscheme U in Z where $(\pi \circ \delta)\,|_Z$ is finite.

If Y in (3.20) is normal, it has at most quotient singularities (see (3.24) for a slightly different argument). In particular, Y has rational singularities (see [62] or (5.13)), i.e.:

$$R^b \delta_* \mathcal{O}_Z = 0 \quad \text{for} \quad b > 0.$$

One can even show (see [17]):

3.22. Lemma. *For Y normal and $\pi : Y \to X$, $\delta : Z \to Y$ as in (3.21) and $\tau = \pi \circ \delta$ one has:*

$$R^b \tau_* \Omega^a_Z(\log\ D') = \begin{cases} \Omega^a_X(\log\ D) \otimes \bigoplus_{i=0}^{N-1} \mathcal{L}^{(i)^{-1}} & \text{for } b = 0 \\ 0 & \text{for } b > 0 \end{cases}$$

For $b = 0$ this statement follows directly from (3.21). For $b > 0$, however, the only way we know to get (3.22) is to use GAGA and the independence from the choosen compactification of the mixed Hodge structure of the open manifold $Z - D'$ (see Deligne [10]).

Using (3.21) and (3.22) one finds again (see [20]):
The degeneration of the spectral sequence

$$E_1^{ab} = H^b(Z, \Omega_Z^a(\log\ D')) \Longrightarrow H^{a+b}(Z, \Omega_Z^\bullet(\log\ D'))$$

implies (3.2,b).

Let us end this section with the following

3.23. Corollary. *Under the assumptions of 3.2 assume that $k = \mathbb{C}$. Then*

$$\dim\ (H^b(X, \Omega_X^a(\log\ D^{(i)}) \otimes \mathcal{L}^{(i)^{-1}})) = \dim\ (H^a(X, \Omega_X^b(\log\ D^{(N-i)}) \otimes \mathcal{L}^{(N-i)^{-1}})).$$

PROOF: By GAGA we can assume that we consider the analytic sheaf of differentials. The Hodge duality on the covering T constructed in (3.17) is given by conjugation. Since under conjugation e^i goes to e^{N-i} for a primitive N-th root of unity, we obtain (3.23).

□

Let us end this section by showing that the cyclic cover Y constructed in (3.5) has at most quotient singularities. Slightly more generally one has the following lemma which, as mentioned above, also follows from (3.20).

3.24. Lemma. *Let X be a quasi-projective manifold, Y a normal variety and let $\pi : Y \longrightarrow X$ be a separable finite cover. Assume that the ramification divisor*

$$D = \sum_{j=1}^{m} D_j = \Delta(Y/X)$$

of π in X is a normal crossing divisor and that for all j and all components B_j^i of $\pi^{-1}(D_j)$ the ramification index $e(B_j^i)$ is prime to char k.
Then Y has at most quotient singularities, i.e. each point $y \in Y$ has a neighbourhood of the form T/G where T is nonsingular and G a finite group acting on T.

PROOF: One can assume that X is affine. For $j = 1, \cdots, m$ define

$$n_j = lcm\{e(B_j^i);\ B_j^i \text{ component of } \pi^{-1}(D_j)\}.$$

Let $\tau : Z \longrightarrow X$ be the cyclic cover obtained by taking sucessively the n_j-th root out of D_j. In other terms, Z is the normalization of the fibered product of the different coverings of X obtained by taking the n_j root out of D_j or, equivalently, τ is the composition of

$$Z = Z_m \xrightarrow{\tau_m} Z_{m-1} \xrightarrow{\tau_{m-1}} \cdots \longrightarrow Z_1 \xrightarrow{\tau_1} Z_0 = X$$

where $\tau_j : Z_j \longrightarrow Z_{j-1}$ is the cover obtained by taking the n_j-th root out of $(\tau_1 \circ \tau_2 \circ \cdots \circ \tau_{j-1})^*(D_j)$. By (3.15,b) Z_j is non singular. Z is Galois over X with Galois group

$$G = \prod_{j=1}^{m} \mathbb{Z}/n_j \cdot \mathbb{Z}.$$

Let T be the normalization of $Z \times_X Y$ and $\delta : T \longrightarrow Y$ the induced morphism. Each component T_0 of T is Galois over Y with a subgroup of G as Galois group.

The morphism $\delta_0 = \delta|_{T_0}$ is obtained by taking sucessively the $\frac{n_j}{\alpha_j}$-th root out of

$$\pi^{-1}(D_j) = \sum_{i=1}^{r_j} \frac{e(B_j^i)}{\alpha_j} \cdot B_j^i$$

for

$$\alpha_j = gcd\{e(B_j^i);\ B_j^i \text{ component of } \pi^{-1}(D_j)\}.$$

By (3.15) all components of $\delta^{-1}(B_j^i)$ have ramification index

$$\frac{\frac{n_j}{\alpha_j}}{gcd\{\frac{n_j}{\alpha_j}, \frac{e(B_j^i)}{\alpha_j}\}} = \frac{n_j}{e(B_j^i)}$$

over Y. Hence they are ramified over X with order n_j. In other terms, the induced morphism $T_0 \longrightarrow Z$ is unramified and T_0 is a non-singular Galois cover of Y.

□

§ 4 Vanishing theorems, the formal set-up.

Theorem 3.2 , whose proof has been reduced to the E_1-degeneration of a Hodge to de Rham spectral-sequences, implies immediately several vanishing theorems for the cohomology of the sheaves $\mathcal{L}^{(i)}$.
To underline that in fact the whole information needed is hidden in (3.2) and (2.9) we consider in this lecture a more general situation and we state the assumptions explicitly, which are needed to obtain the vanishing of certain cohomology groups.

(4.2) and (4.8) are of special interest for applications whereas the other variants can been skipped at the first reading.

4.1. Assumptions. Let X be a projective manifold defined over an algebraically closed field k and let

$$D = \sum_{j=1}^{r} D_j$$

be a reduced normal crossing divisor. Let $\mathcal{E}$ be a locally free sheaf on X of finite rank and let

$$\nabla : \mathcal{E} \longrightarrow \Omega^1_X(\log\ D) \otimes \mathcal{E}$$

be an integrable connection with logarithmic poles along D.
We will assume in the sequel that ∇ *satisfies the* E_1*-degeneration* i.e. that *the Hodge to de Rham spectral sequence (A.25)*

$$E_1^{ab} = H^b(X, \Omega^a_X(\log\ D) \otimes \mathcal{E}) \Longrightarrow \mathbb{H}^{a+b}(X, \Omega^\bullet_X(\log\ D) \otimes \mathcal{E})$$

degenerates in E_1.

4.2. Lemma (Vanishing for restriction maps I). *Assume that* ∇ *satisfies the condition (!) of (2.8), i.e. that for all* $\mu \in \mathbb{N}$ *and for* $j = 1, \ldots, r$ *the map*

$$Res_{D_j}(\nabla) + \mu \cdot id_{\mathcal{O}_{D_j}} : \mathcal{E}\mid_{D_j} \longrightarrow \mathcal{E}\mid_{D_j}$$

is an isomorphism. Assume that ∇ *satisfies the* E_1*-degeneration (4.1). Then for all effective divisors*

$$D' = \sum_{j=1}^{r} \mu_j D_j$$

and all b *the natural map*

$$H^b(X, \mathcal{O}_X(-D') \otimes \mathcal{E}) \longrightarrow H^b(X, \mathcal{E})$$

is surjective.

Proof: By (2.9,b) the map

$$\Omega^\bullet_X(\log\ D) \otimes \mathcal{E}(-D') \longrightarrow \Omega^\bullet_X(\log\ D) \otimes \mathcal{E}$$

is a quasi-isomorphism and hence induces an isomorphism of the hypercohomology groups. Let us consider the exact sequences of complexes

$$\begin{array}{ccccccccc}
0 \longrightarrow & \Omega^{\bullet \geq 1}_X(\log\ D) \otimes \mathcal{E} & \longrightarrow & \Omega^\bullet_X(\log\ D) \otimes \mathcal{E} & \longrightarrow & \mathcal{E} & \longrightarrow 0 \\
& \uparrow & & \uparrow & & \uparrow & \\
0 \longrightarrow & \Omega^{\bullet \geq 1}_X(\log\ D) \otimes \mathcal{E}(-D') & \longrightarrow & \Omega^\bullet_X(\log\ D) \otimes \mathcal{E}(-D') & \longrightarrow & \mathcal{E}(-D') & \longrightarrow 0.
\end{array}$$

By assumption, the spectral sequence for $\Omega^\bullet_X(\log\ D) \otimes \mathcal{E}$ degenerates in E_1, which implies that the morphism α in the following diagram is surjective (see (A.25)).

$$\begin{array}{ccc}
\mathbb{H}^b(X, \Omega^\bullet_X(\log\ D) \otimes \mathcal{E}) & \xrightarrow{\alpha} & H^b(X, \mathcal{E}) \\
\uparrow = & & \uparrow \beta \\
\mathbb{H}^b(X, \Omega^\bullet_X(\log\ D) \otimes \mathcal{E}(-D')) & \longrightarrow & H^b(X, \mathcal{E}(-D'))
\end{array}$$

Hence β is surjective as well.

□

4.3. Variant. *If in (4.2)*

$$D' = \sum_{j=1}^{s} \mu_j D_j \geq 0 \quad \textit{for} \quad s \leq r,$$

then it is enough to assume that for $j = 1, \ldots, s$ *and for* $0 \leq \mu \leq \mu_j - 1$

$$Res_{D_j}(\nabla) + \mu \cdot id_{\mathcal{O}_{D_j}}$$

is an isomorphism.

PROOF: By (2.10) this is enough to give the quasi-isomorphism

$$\Omega^\bullet_X(\log\ D) \otimes \mathcal{E}(-D') \longrightarrow \Omega^\bullet_X(\log\ D) \otimes \mathcal{E}$$

needed in the proof of (4.2).

□

4.4. Lemma (Dual version of (4.2) and (4.3)). *Assume that*

$$\nabla : \mathcal{E} \longrightarrow \Omega^1_X(\log\ D) \otimes \mathcal{E}$$

satisfies the E_1*-degeneration and that for* $j = 1, \ldots, s$ *and* $1 \leq \mu \leq \mu_j$

$$Res_{D_j}(\nabla) - \mu \cdot id_{\mathcal{O}_{D_j}}$$

is an isomorphism (for example, if ∇ *satisfies the condition* $(*)$ *from (2.8,a)). Then for*

$$D' = \sum_{j=1}^{s} \mu_j D_j$$

and all b *the map*

$$H^b(X, \omega_X(D) \otimes \mathcal{E}) \longrightarrow H^b(X, \omega_X(D + D') \otimes \mathcal{E})$$

is injective.

PROOF: Consider the diagram

$$\begin{array}{ccc} H^b(X, \omega_X(D + D') \otimes \mathcal{E}) & \longrightarrow & \mathbb{H}^{n+b}(X, \Omega^\bullet_X(\log\ D) \otimes \mathcal{E}(D')) \\ \uparrow \beta & & \uparrow \gamma \\ H^b(X, \omega_X(D) \otimes \mathcal{E}) & \xrightarrow{\alpha} & \mathbb{H}^{n+b}(X, \Omega^\bullet_X(\log\ D) \otimes \mathcal{E}). \end{array}$$

α is injective by the E_1-degeneration (see (A.25)) , γ is an isomorphism by (2.10) and hence β is injective.

□

The lemma (4.2) or its variant (4.3) implies that for all b the natural restriction maps

$$H^b(X,\mathcal{E}) \longrightarrow H^b(D',\mathcal{O}_{D'}\otimes\mathcal{E})$$

are the zero maps. For higher differential forms this remains true, if D' is a non-singular divisor:

4.5. Lemma (Vanishing for restriction maps II). *Assume that*

$$\nabla : \mathcal{E} \longrightarrow \Omega^1_X(\log\ D)\otimes\mathcal{E}$$

satisfies E_1-degeneration. Let D' be a non-singular subdivisor of D and assume that for all components D_j of D' the map $Res_{D_j}(\nabla)$ is an isomorphism. (For example this follows from condition (!) in (2.8,b)).
Then the restriction (see (2.3))

$$H^b(X,\Omega^a_X(\log\ (D-D'))\otimes\mathcal{E}) \longrightarrow H^b(D',\Omega^a_{D'}(\log\ (D-D')\mid_{D'})\otimes\mathcal{E})$$

is zero for all a and b.

PROOF: As we have seen in (2.6,b) the restriction map factors through

$$H^b(\nabla_a) : H^b(X,\Omega^a_X(\log\ D)\otimes\mathcal{E}) \longrightarrow H^b(X,\Omega^{a+1}_X(\log\ D)\otimes\mathcal{E})$$

provided $Res_{D_j}(\nabla)$ is an isomorphism on the different components D_j of D'. By E_1-degeneration, $H^b(\nabla_a)$ is the zero map (see (A.25)).

□

Before we are able to state the global vanishing for $\mathcal{E}$ or $\Omega^a_X(\log\ D)\otimes\mathcal{E}$ we need some more notations.

4.6. Definition. Let $U \subset X$ be an open subscheme and let B be an effective divisor with $B_{red} = X - U$. Then we define the *(coherent) cohomological dimension* of (X,B) to be the least integer α such that for all coherent sheaves $\mathcal{F}$ and all $k > \alpha$ one finds some $\nu_0 > 0$ with $H^k(X,\mathcal{F}(\nu\cdot B)) = 0$ for all multiples ν of ν_0. Finally, for the reduced divisor $D = X - U$, we write

$$cd(X,D) = \text{Min}\{\ \alpha\ ;\ \text{there exists some effective divisor } B \text{ with } B_{red} = D,\ \text{such that } \alpha \text{ is the cohomological dimension of } (X,B)\}.$$

4.7. Examples.
a) For $D = X - U$ the embedding $\iota : U \to X$ is affine and for a coherent sheaf $\mathcal{G}$ on X we have

$$H^b(U,\mathcal{G}\mid_U) = H^b(X,\iota_*(\mathcal{G}\mid_U)) = \lim_{\substack{\longrightarrow \\ \alpha\in\mathbb{N}}} H^b(X,\mathcal{G}\otimes\mathcal{O}_X(\alpha\cdot B)),$$

where B is any effective divisor with $B_{red} = D$. In particular, if $b > cd(X,D)$ we find

$$H^b(U,\mathcal{G}\mid_U) = 0$$

b) By Serre duality one obtains as well that for $b < n - cd(X, D)$ we can find $B > 0$ such that for a locally free sheaf $\mathcal{G}$ and all multiples ν of some $\nu_0 > 0$ one has

$$\dim H^b(X, \mathcal{G} \otimes \mathcal{O}_X(-\nu \cdot B)) = 0.$$

c) If D is the support of an effective ample divisor, then Serre's vanishing theorem (see (1.1)) implies $cd(X, D) = 0$. We are mostly interested in this case, hopefully an excuse for the clumsy definition given in (4.6).

4.8. Lemma (Vanishing for cohomology groups).
Assume that X is projective and that

$$\nabla : \mathcal{E} \longrightarrow \Omega^1_X(\log\ D) \otimes \mathcal{E}$$

satisfies the E_1-degeneration (see (4.1)).
a) If ∇ satisfies the condition $(*)$ of (2.8) and if $a + b > n + cd(X, D)$, then

$$H^b(X, \Omega^a_X(\log\ D) \otimes \mathcal{E}) = 0.$$

b) If ∇ satisfies the condition (!) of (2.8) and if $a + b < n - cd(X, D)$, then

$$H^b(X, \Omega^a_X(\log\ D) \otimes \mathcal{E}) = 0.$$

Proof: Let us choose $\alpha \in \mathbb{Z}$ with $\alpha \geq 0$ in case a) and with $\alpha \leq 0$ in case b). For $B \geq D$, (2.9) tells us that

$$\Omega^\bullet_X(\log\ D) \otimes \mathcal{E} \quad \text{and} \quad \Omega^\bullet_X(\log\ D) \otimes \mathcal{E}(\alpha \cdot B)$$

are quasi-isomorphic. In both cases we have a spectral sequence

$$E_1^{ab} = H^b(X, \Omega^a_X(\log\ D) \otimes \mathcal{E}(\alpha \cdot B)) \Longrightarrow$$

$$\Longrightarrow \mathbb{H}^{a+b}(X, \Omega^\bullet_X(\log\ D) \otimes \mathcal{E}(\alpha \cdot B)) = \mathbb{H}^{a+b}(X, \Omega^\bullet_X(\log\ D) \otimes \mathcal{E}).$$

By assumption this spectral sequence degenerates for $\alpha = 0$ and, for arbitrary α we have (see (A.16))

$$\sum_{a+b=l} \dim\ H^b(X, \Omega^a_X(\log\ D) \otimes \mathcal{E}) = \dim\ \mathbb{H}^l(X, \Omega^\bullet_X(\log\ D) \otimes \mathcal{E})$$

$$\leq \sum_{a+b=l} \dim\ H^b(X, \Omega^a_X(\log\ D) \otimes \mathcal{E}(\alpha \cdot B)).$$

By definition of $cd(X, D)$ we can choose B such that the right hand side is zero for $l > n + cd(X, D)$ and all $\alpha > 0$ in case a), or $l < n - cd(X, D)$ and all $\alpha < 0$ in case b).

□

The same argument shows:

4.9. Variant. *In (4.8) we can replace a) and b) by:*
c) Let D_* and $D_!$ be effective divisors, both smaller than D, and assume that

i) For all components D_j of D_* and all $\mu \in \mathbb{N} - \{0\}$

$$Res_{D_j}(\nabla) - \mu \cdot id_{\mathcal{O}_{D_j}}$$

is an isomorphism.

ii) For all components D_j of $D_!$ and all $\mu \in \mathbb{N}$

$$Res_{D_j}(\nabla) + \mu \cdot id_{\mathcal{O}_{D_j}}$$

is an isomorphism.

Then

$$H^b(X, \Omega^a_X(\log\ D) \otimes \mathcal{E}) = 0$$

for $a + b > n + cd(X, D_*)$ and for $a + b < n - cd(X, D_!)$.

The analytic case

As we have seen in the proof of (4.8) the condition (*) implies that

$$\mathbb{H}^l(X, \Omega^\bullet_X(\log\ D) \otimes \mathcal{E}) = 0 \quad \text{for} \quad l > n + cd(X, D).$$

For $k = \mathbb{C}$, this is not the best possible result. In fact, as mentioned in (2.12), (*) implies that over $\mathbb{C}$

$$\mathbb{H}^l(X, \Omega^\bullet_X(\log\ D) \otimes \mathcal{E}) = 0 \ \text{for} \quad l > n + r(X - D)$$

where $r(X - D)$ is the least number α such that $H^l(X - D, V) = 0$ for all locally constant systems V on $X - D$ and $l > n + \alpha$.

(2.12) and the E_1-degeneration asked for in (4.8) and (4.9) imply immediately that "$cd(\ )$" in 4.8 and 4.9 can be replaced by "$r(\ )$".
As we will see, $r(X - D)$ might be smaller than $cd(X, D)$. For the results which follow we only know, at present, proofs by analytic methods.

4.10. Definition. Let U be an algebraic irreducible variety and $g : U \to W$ a morphism. Then define

$$r(g) = \text{Max}\{ \ \dim\ \Gamma - \dim\ g(\Gamma) - \text{codim}\ \Gamma; \ \Gamma \text{ irreducible closed subvariety of } U \ \}$$

4.11. Properties.

a)

$$r(g) = \text{Max}\{ \ \dim\ (\text{generic fibre of } g\,|_\Gamma) - \text{codim}\ \Gamma; \ \Gamma \text{ irreducible closed subvariety of } U \ \}$$

b) If b denotes the maximal fibre dimension for g, then

$$r(g) \leq \text{Max}\{\dim\ U - \dim\ W; \ \ b - 1\}.$$

c) If $U' \subseteq U$ is open and dense, then $r(g\,|_{U'}) \leq r(g)$.
d) If $\Delta \subseteq U$ is closed then $r(g\,|_\Delta) \leq r(g) + \text{codim}_U\ (\Delta)$.

PROOF: a) and c) are obvious and b) follows from a). For d) one remarks that for $\Gamma \subset \Delta$ one has

$$\text{codim}_\Delta\ (\Gamma) = \text{codim}_U\ (\Gamma) - \text{codim}_U\ (\Delta).$$

□

4.12. Lemma. *(Improvement of 4.8 using analytic methods)*
Let X be a projective manifold defined over an algebraically closed field k of characteristic zero. Assume that

$$\nabla : \mathcal{E} \longrightarrow \Omega^1_X(\log\ D) \otimes \mathcal{E}$$

is an integrable connection satisfying the E_1-degeneration and let

$$g : X - D \longrightarrow W$$

be a proper surjective morphism to an affine variety W.
a) If ∇ satisfies the condition (*) of (2.8) then

$$H^b(X, \Omega^a_X(\log\ D) \otimes \mathcal{E}) = 0$$

for $a + b > n + r(g)$.
b) If ∇ satisfies the condition (!) of (2.8) then

$$H^b(X, \Omega^a_X(\log\ D) \otimes \mathcal{E}) = 0$$

for $a + b < n - r(g)$.

PROOF: By flat base chance we can replace k by any other field k', such that $X, D, \mathcal{E}$ are defined over k'. Hence, we may assume that $k = \mathbb{C}$.
By GAGA (see [56]) we may assume in (4.12) that all the sheaves and ∇ are analytic. Then, by (2.12) and by the E_1-degeneration it is enough to show:

4.13. Lemma. *Let U be an analytic manifold, W be an affine manifold and $g : U \to W$ be a proper morphism. Then, for all local constant systems V on U and $l > \dim\ (U) + r(g)$ one has $H^l(U, V) = 0$.*

PROOF (SEE [22]): The sheaves $R^a g_* V$ are analytically constructible sheaves ([61]) and their support

$$S_a = \text{Supp}(R^a g_* V)$$

must be a Stein space, hence

$$H^b(W, R^a g_* V) = 0 \ \text{ for } \ b > \dim\ S_a.$$

However, the general fibre of $g\ |_{g^{-1}(S_a)}$ must have a dimension larger than or equal to $\frac{a}{2}$. Hence

$$2 \cdot (\dim\ g^{-1}(S_a) - \dim\ S_a) \geq a$$

and $H^b(W, R^a g_* V) = 0$ for

$$a + b > n + r(g) \geq 2 \cdot \dim\ g^{-1}(S_a) - \dim\ S_a \geq a + \dim\ S_a.$$

By the Leray spectral sequence (A.27)

$$E_2^{ba} = H^b(W, R^a g_* V) \Longrightarrow H^{a+b}(U, V)$$

one obtains (4.13).

□

4.14. Remark. If W is affine and $g : U \to W$ obtained by blowing up a point, then for X and D as in (4.6) one has $cd(X, D) = \dim\ U - 1$, whereas $r(g) = \dim\ U - 2$.

§ 5 Vanishing theorems for invertible sheaves

In this lecture we will deduce several known generalizations of the Kodaira-Nakano vanishing theorem by applying the vanishing (5.1) obtained for "integral parts of $\mathbb{Q}$-divisors" from (3.2), combined with (4.2). Needless to say that in all those corollaries of (5.1) one loses some information and that it might be more reasonable to try to work with (5.1) or correspondingly with (6.2) directly, whenever it is possible.

Let us remind you, that the proof of (3.2) is not yet complete. The necessary arguments needed to show the E_1-degeneration will only be presented in Lecture 10.

Very quickly we will have to restrict ourselves to characteristic zero. One reason is that the condition (*) and (!) are too much to ask for in characteristic $p \neq 0$. But more substantially, most of our proofs will start with "blow up B to get a normal crossing divisor", hence with an application of H. Hironaka's theorem on the existence of desingularizations.

Let us start with (4.2). For simplicity, we restrict ourselves to $i = 1$ and $\mathcal{L} = \mathcal{L}^{(1)}$. By (3.3) we are not losing any information.

5.1. Vanishing for restriction maps related to $\mathbb{Q}$-divisors:

Let X be a projective manifold defined over an algebraically closed field k, let $\mathcal{L}$ be an invertible sheaf, $N \in \mathbb{N} - \{0\}$ and let

$$D = \sum_{j=1}^{r} \alpha_j D_j$$

be a normal crossing divisor with $0 < \alpha_j < N$ and $\mathcal{L}^N = \mathcal{O}_X(D)$. Let

$$D' = \sum_{j=1}^{r} \mu_j D_j$$

be an effective divisor. Then one has:
a) *If* char $(k) = 0$ *then for all b the natural morphism*

$$H^b(X, \mathcal{L}^{-1}(-D')) \longrightarrow H^b(X, \mathcal{L}^{-1})$$

is surjective and hence the morphism

$$H^b(X, \omega_X \otimes \mathcal{L}) \longrightarrow H^b(X, \omega_X(D') \otimes \mathcal{L})$$

injective.
b) *If* char $(k) = 0$ *and if* C *is a reduced divisor without common component with* D *such that* $D + C$ *is a normal crossing divisor, then for all* b *the natural morphism*

$$H^b(X, \mathcal{L}^{-1}(-C - D')) \longrightarrow H^b(X, \mathcal{L}^{-1}(-C))$$

is surjective and hence the morphism

$$H^b(X, \omega_X(C) \otimes \mathcal{L}) \longrightarrow H^b(X, \omega_X(D' + C) \otimes \mathcal{L})$$

injective.
c) *If* char $(k) = p \neq 0$, *then a) and b) hold true under the additional assumptions:*

> i) X *and* D *(as well as* C*) satisfy the lifting property (8.11) and* dim $(X) \leq p$.
>
> ii) N *is prime to* p.
>
> iii) *For all* j *and* $0 \leq \mu \leq \mu_j - 1$ *one has* $\alpha_j + \mu \cdot N \not\equiv 0$ *mod* p.

PROOF: By (3.2,c) $\mathcal{O}_X(-C) \otimes \mathcal{L}^{-1}$ has a logarithmic connection ∇ with poles along $C + D_{red}$ satisfying E_1-degeneration, and $Res_{D_j}(\nabla) = \frac{\alpha_j}{N}$. Hence (5.1) follows from (4.2) and (4.3) or (4.4).

□

5.2. Corollary (Kodaira [39], Deligne, Illusie [12]). *Let* X *be a projective manifold and* $\mathcal{L}$ *an invertible sheaf. If* char $(k) = p > 0$, *then assume in addition that* X *and* $\mathcal{L}$ *admit a lifting to* $W_2(k)$ *(8.11) and that* dim $X \leq p$. *Then, if* $\mathcal{L}$ *is ample,*

$$H^b(X, \mathcal{L}^{-1}) = 0 \quad \textit{for} \quad b < n = \dim (X)$$

PROOF: Choose N, prime to $p = \text{char}(k)$, such that

$$H^b(X, \mathcal{L}^{-N-1}) = H^{n-b}(X, \omega_X \otimes \mathcal{L}^{N+1}) = 0$$

for $b < n$, and such that $\mathcal{L}^N$ is generated by global sections. If D is a general section of $\mathcal{L}^N$, then we can apply (5.1) and find

$$H^b(X, \mathcal{L}^{-1}(-D)) \longrightarrow H^b(X, \mathcal{L}^{-1})$$

to be surjective. Since the group on the left hand side is zero, we are done.

□

If char $(k) = p$, then we will see in (11.3) that it is sufficient to assume that X lifts to $W_2(k)$. The condition that $\mathcal{L}$ lifts as well is not necessary.

5.3. Definition. Let X be a projective variety and $\mathcal{L}$ be an invertible sheaf on X. If $H^0(X, \mathcal{L}^\nu) \neq 0$, the sections of $\mathcal{L}$ define a rational map

$$\phi_\nu = \phi_{\mathcal{L}^\nu} : X \longrightarrow \mathbb{P}(H^0(X, \mathcal{L}^\nu)).$$

The *Iitaka-dimension* $\kappa(\mathcal{L})$ of $\mathcal{L}$ is given by

$$\kappa(\mathcal{L}) = \begin{cases} -\infty & \text{if } H^0(X, \mathcal{L}^\nu) = 0 \text{ for all } \nu \\ \\ \operatorname{Max}\{\dim \phi_\nu(X);\ H^0(X, \mathcal{L}^\nu) \neq 0\} & \text{otherwise} \end{cases}$$

5.4. Properties. For X and $\mathcal{L}$ as above one has:

a) $\kappa(\mathcal{L}) \in \{-\infty, 0, 1, \ldots, \dim\ X\}$.

b) If $H^0(X, \mathcal{L}^\nu) \neq 0$ for some $\nu > 0$ then one can find $a, b \in \mathbb{R}$, $a, b > 0$, such that

$$a \cdot \mu^{\kappa(\mathcal{L})} \leq \dim\ H^0(X, \mathcal{L}^{\nu \cdot \mu}) \leq b \cdot \mu^{\kappa(\mathcal{L})} \quad \text{for all} \quad \mu \in \mathbb{N} - \{0\}.$$

c) If $\kappa(\mathcal{L}) \neq -\infty$, then

$$\kappa(\mathcal{L}) = \text{ tr.deg } \Big(\bigoplus_{\mu \geq 0} H^0(X, \mathcal{L}^\mu)\Big) - 1.$$

d) One has $\kappa(\mathcal{L}) = \dim\ X$, if and only if for some $\nu > 0$ and some effective divisor C the sheaf $\mathcal{L}^\nu(-C)$ is ample.

e) If $\mathcal{A}$ is very ample and A the zero divisor of a general section of $\mathcal{A}$, then

$$\kappa(\mathcal{L}\,|_A) \geq Min\ \{\kappa(\mathcal{L}), \dim\ A\}.$$

PROOF: a), b) and c) are wellknown and their proof can be found, for example, in [46], §1.
For d) let A be an ample effective divisor. For $n = \dim\ X$ and some $\nu \in \mathbb{N} - \{0\}$ one finds $a, b \in \mathbb{R}$, $a, b > 0$, with $a \cdot \mu^n < \dim\ H^0(X, \mathcal{L}^{\nu \cdot \mu})$ and $\dim\ H^0(X, \mathcal{L}^{\nu \cdot \mu}|_A) < b \cdot \mu^{n-1}$. Hence, the exact sequence

$$0 \longrightarrow H^0(X, \mathcal{L}^{\nu \cdot \mu}(-A)) \longrightarrow H^0(X, \mathcal{L}^{\nu \cdot \mu}) \longrightarrow H^0(A, \mathcal{L}^{\nu \cdot \mu}\,|_A)$$

shows that for some μ we have $\mathcal{O}_X(A)$ as a subsheaf of $\mathcal{L}^{\nu \cdot \mu}$. On the other hand, if $\mathcal{A} \subset \mathcal{L}^\nu$ is ample then

$$n = \kappa(\mathcal{A}) \leq \kappa(\mathcal{L}^\nu) = \kappa(\mathcal{L}).$$

If $\kappa(\mathcal{L}) = n$ in e), then, using d) for example, $\kappa(\mathcal{L}\,|_A) = n-1$.
If $\kappa(\mathcal{L}) < n$, then $H^0(X, \mathcal{O}_X(-A) \otimes \mathcal{L}^\nu) = 0$ for all ν and b) implies

$$\kappa(\mathcal{L}\,|_A) \geq \kappa(\mathcal{L}).$$

□

For our purposes we can take (5.4,b) as definition of $\kappa(\mathcal{L})$, and we only need to know (5.4,d) and (5.4,e).

5.5. Definition. An invertible sheaf $\mathcal{L}$ on X is called

a) *semi-ample*, if for some $\mu > 0$ the sheaf $\mathcal{L}^\mu$ is generated by global sections.

b) *numerically effective (nef)* if for all curves C in X one has

$$\deg\,(\mathcal{L}\,|_C) \geq 0.$$

The proof of (5.2) can be modified to give in characteristic zero a stronger statement:

5.6. Corollary. *Let X be a projective manifold defined over a field k of characteristic zero and let $\mathcal{L}$ be an invertible sheaf.*

a) (Kollár [40])
If $\mathcal{L}$ is semi-ample and B an effective divisor with $H^0(X, \mathcal{L}^\nu(-B)) \neq 0$ for some $\nu > 0$, then the natural maps

$$H^b(X, \mathcal{L}^{-1}(-B)) \longrightarrow H^b(X, \mathcal{L}^{-1})$$

are surjective for all b, or, equivalently, the adjunction map

$$H^b(X, \mathcal{L} \otimes \omega_X(B)) \longrightarrow H^b(B, \mathcal{L} \otimes \omega_B)$$

is surjective for all b.

b) (Grauert-Riemenschneider [25])
If $\mathcal{L}$ is semi-ample and $\kappa(\mathcal{L}) = n = \dim\ X$, then

$$H^b(X, \mathcal{L}^{-1}) = 0 \quad \textit{for} \quad b < n.$$

Proof: Obviously a) and b) are compatible with blowing ups $\tau : X' \to X$. In fact, using the Leray spectral sequence (A.27) we just have to remember that

$$R^b\tau_*\tau^*\mathcal{L} = \mathcal{L} \otimes R^b\tau_*\mathcal{O}_{X'} = \begin{cases} \mathcal{L} & \text{for} \quad b = 0 \\ 0 & \text{for} \quad b \neq 0 \end{cases}$$

(See (5.13) for a generalization).
Hence we may assume in a) that $\mathcal{L}^\nu = \mathcal{O}_X(B+C)$ for an effective normal crossing divisor $B+C$. We can choose some μ, with

$$[\frac{B+C}{\mu}] = 0$$

and such that $\mathcal{L}^\mu$ is generated by its global sections. If D_1 is a general divisor of $\mathcal{L}^\mu$, i.e. the zero set of a general $s \in H^0(X, \mathcal{L}^\mu)$, then $D = D_1 + B + C$ has normal crossings and $[\frac{D}{\mu}] = 0$. Hence, for $N = \nu + \mu$ and $D' = B$ the assumptions of (5.1,a) hold true and we obtain a).

For b), let us choose some divisor C and some ν such that $\mathcal{L}^\nu(-C) = \mathcal{A}$ is ample. Replacing $\mathcal{A}$ by some multiple, we may assume by Serre's vanishing theorem (1.1) that

$$H^b(X, \mathcal{L}^{-1} \otimes \mathcal{A}^{-1}) = H^{n-b}(X, \omega_X \otimes \mathcal{L} \otimes \mathcal{A}) = 0$$

for $b < n$, and that $\mathcal{A} = \mathcal{O}_X(B)$ for some divisor B. By a)

$$H^b(X, \mathcal{O}_X(-B) \otimes \mathcal{L}^{-1}) \longrightarrow H^b(X, \mathcal{L}^{-1})$$

is surjective. One obtains b), since the left hand side is zero.

□

It is not difficult to modify both parts of this proof to include in b) the case that $\mathcal{L}$ is nef and $\kappa(\mathcal{L}) = \dim X$. Moreover, considering very ample divisors on X and using induction on $\dim(X)$, one can as well remove the assumption "$\kappa(\mathcal{L}) = \dim(X)$" and obtain the vanishing for $b < \kappa(\mathcal{L})$.
We leave the details to the reader. Those techniques will appear in (5.12) anyway, when we prove a more general statement.

5.7. Lemma. *For an invertible sheaf $\mathcal{L}$ on a projective manifold X the following two conditions are equivalent:*

a) *$\mathcal{L}$ is numerically effective.*

b) *For an ample sheaf $\mathcal{A}$ and all $\nu > 0$ the sheaf $\mathcal{L}^\nu \otimes \mathcal{A}$ is ample.*

PROOF: By Seshadri's criterion $\mathcal{A}'$ is ample if and only if for some $\epsilon > 0$ and all curves C in X

$$\deg\, (\mathcal{A}' \mid_C) \geq \epsilon \cdot m(C)$$

where $m(C)$ is the maximal multiplicity of points on C.

□

5.8. Lemma. *For $X, \mathcal{L}$ as in (5.7), assume that $\mathcal{L}$ is numerically effective (and, if* char $(k) = p \neq 0$*, that X and $\mathcal{L}$ satisfy the lifting property (8.11) and that* $\dim X \leq p$*). Then one has:*

a) $\kappa(\mathcal{L}) = n = \dim X$, *if and only if* $c_1(\mathcal{L})^n > 0$ *(where* $c_1(\mathcal{L})$ *is the Chern class of* $\mathcal{L}$*).*

b) *For* $b \geq 0$ *and for all invertible sheaves* $\mathcal{F}$ *one has a constant* $c_b > 0$ *with*

$$\dim H^b(X, \mathcal{F} \otimes \mathcal{L}^\nu) \leq c_b \cdot \nu^{n-b} \quad \textit{for all} \quad \nu \in \mathbb{N}.$$

PROOF: a) follows from b) and from the Hirzebruch-Riemann-Roch theorem which tells us that $\chi(X, \mathcal{L}^\nu)$ is a polynomial of deg n with highest coefficient

$$\frac{1}{n!} \cdot c_1(\mathcal{L})^n.$$

For b) we assume by induction on $\dim X$, that it holds true for all hypersurfaces H in X. We can choose an H, which satisfies

$$H^b(X, \mathcal{O}_X(H) \otimes \mathcal{L}^\nu \otimes \mathcal{F}) = 0.$$

In fact, we just have to choose H such that $\mathcal{F} \otimes \omega_X^{-1} \otimes \mathcal{O}_X(H)$ is ample. Then by (5.7)

$$\mathcal{F} \otimes \omega_X^{-1} \otimes \mathcal{O}_X(H) \otimes \mathcal{L}^\nu$$

will be ample for all $\nu \geq 0$ and the vanishing required holds true by (5.2). From the exact sequence

$$0 \longrightarrow \mathcal{F} \otimes \mathcal{L}^\nu \longrightarrow \mathcal{F} \otimes \mathcal{L}^\nu \otimes \mathcal{O}_X(H) \longrightarrow \mathcal{F} \otimes \mathcal{O}_H(H) \otimes \mathcal{L}^\nu \longrightarrow 0$$

we obtain an isomorphism

$$H^{b-1}(H, \mathcal{F} \otimes \mathcal{O}_H(H) \otimes \mathcal{L}^\nu) \simeq H^b(X, \mathcal{F} \otimes \mathcal{L}^\nu)$$

for $b > 1$ and a surjection

$$H^0(H, \mathcal{F} \otimes \mathcal{O}_H(H) \otimes \mathcal{L}^\nu) \longrightarrow H^1(X, \mathcal{F} \otimes \mathcal{L}^\nu).$$

By induction we find c_b for $b \geq 1$ and, since $H^0(X, \mathcal{F} \otimes \mathcal{L}^\nu)$ is bounded above by a polynomial of deg ν, for $b = 0$ as well.

□

Even if $\mathcal{L}$ is nef, there is in general no numerical characterisation of $\kappa(\mathcal{L})$. For example, there are numerically effective invertible sheaves $\mathcal{L}$ with $\kappa(\mathcal{L}) = -\infty$. Following Kawamata [37], one defines:

5.9. Definition. Let $\mathcal{L}$ be a numerically effective invertible sheaf. Then the *numerical Iitaka-dimension* is defined as

$$\nu(\mathcal{L}) = \text{Min}\ \{\nu \in \mathbb{N} - \{0\};\ \ c_1(\mathcal{L})^\nu \text{ numerically trivial }\} - 1.$$

5.10. Properties. Let $X, \mathcal{L}$ be as in (5.8). Then one has:

a) $\nu(\mathcal{L}) \geq \kappa(\mathcal{L})$.

b) If $\mathcal{L}$ is semi-ample then

$$\nu(\mathcal{L}) = \kappa(\mathcal{L}).$$

PROOF: If $\nu(\mathcal{L}) = \dim\ X$ or $\kappa(\mathcal{L}) = \dim\ X$, then (5.8,a) gives

$$\nu(\mathcal{L}) = \kappa(\mathcal{L}) = n = \dim\ X.$$

Hence we can assume both to be smaller than n. By (5.4,e) we have for a general hyperplane section H of X

$$\kappa(\mathcal{L}\mid_H) \geq \kappa(\mathcal{L})$$

and obviously

$$\nu(\mathcal{L}\mid_H) = \nu(\mathcal{L}).$$

By induction on dim (X) one obtains a).
For b) we may assume that $\mathcal{L} = \tau^*\mathcal{M}$ for a morphism $\tau : X \to Z$ with $\mathcal{M}$ ample and with dim $(Z) = \kappa(\mathcal{L})$. Then

$$c_1(\mathcal{L})^\nu = \tau^* c_1(\mathcal{M})^\nu = 0$$

if and only if $\nu > \dim\ Z$.

□

The following lemma, due to Y. Kawamata [37], is more difficult to prove, and we postpone its proof to the end of this lecture.

5.11. Lemma. *For an invertible sheaf $\mathcal{N}$ on a projective manifold X, defined over a field k of characteristic zero, the following two conditions are equivalent:*
a) $\mathcal{N}$ is numerically effective and $\nu(\mathcal{N}) = \kappa(\mathcal{N})$
b) There exist a blowing up $\tau : Z \to X$, some $\mu_0 \in \mathbb{N} - \{0\}$ and an effective divisor C on Z such that

$$\tau^*\mathcal{N}^\mu \otimes \mathcal{O}_Z(-C)$$

is semi-ample for all $\mu \in \mathbb{N} - \{0\}$ divisible by μ_0.

5.12. Corollary. *Let X be a projective manifold defined over a field k of characteristic zero, let $\mathcal{L}$ be an invertible sheaf on X, let*

$$D = \sum_{j=1}^{r} \alpha_j D_j$$

be a normal crossing divisor and $N \in \mathbb{N}$. Assume that

$$0 < \alpha_j < N \ \ for \ \ j = 1, \ldots, r.$$

Then one has:
a) *If $\mathcal{L}^N(-D)$ is semi-ample and B an effective divisor such that*

$$H^0(X, (\mathcal{L}^N(-D))^\nu \otimes \mathcal{O}_X(-B)) \neq 0$$

for some $\nu > 0$, then for all b the maps

$$H^b(X, \mathcal{L}^{-1}(-B)) \longrightarrow H^b(X, \mathcal{L}^{-1})$$

are surjective.
b) *In a) one can replace "semi-ample" by the assumption that $\mathcal{L}^N(-D)$ is numerically effective and*

$$\kappa(\mathcal{L}^N(-D)) = \nu(\mathcal{L}^N(-D)).$$

c) (Kawamata [36] - Viehweg [63])
If $\mathcal{L}^N(-D)$ is numerically effective and

$$c_1(\mathcal{L}^N(-D))^n > 0,$$

then

$$H^b(X, \mathcal{L}^{-1}) = 0 \quad \textit{for} \quad b < n.$$

d) (Kawamata [36] - Viehweg [63])
If $\mathcal{L}^N(-D)$ is numerically effective, then

$$H^b(X, \mathcal{L}^{-1}) = 0 \quad \textit{for} \quad b < \kappa(\mathcal{L}).$$

e) *Part d) remains true if one replaces $\kappa(\mathcal{L})$ by $\kappa(\mathcal{L} \otimes \mathcal{N}^{-1})$ for a numerically effective invertible sheaf $\mathcal{N}$.*

Again, the assumptions are compatible with blowing ups, except for "$0 < \alpha_j < N$". We need:

5.13. Claim. Let $\tau : X' \to X$ be a proper birational morphism and $\mathcal{M} = \tau^*\mathcal{L}$. Assume that $\Delta = \tau^*D$ has normal crossings. Then for

$$\mathcal{M}^{(i)} = \mathcal{M}(-[\frac{i \cdot \Delta}{N}])$$

one has

$$R^b\tau_*\mathcal{M}^{(i)^{-1}} = \begin{cases} \mathcal{L}^{(i)^{-1}} & \text{for} \quad b = 0 \\ 0 & \text{for} \quad b \neq 0. \end{cases}$$

PROOF: We may assume that X is affine and that $\mathcal{L} = \mathcal{O}_X$. For $b = 0$ claim (5.13) follows from the inequality

$$[\frac{i \cdot \Delta}{N}] = [\frac{i \cdot \tau^*D}{N}] \geq \tau^*[\frac{i \cdot D}{N}].$$

In general, for $b \geq 0$, (5.13) follows from the rationality of the singularities of the cyclic covers Y and Y' obtained by taking the N-th rooth out of D and Δ. In fact, let Y'' be a desingularization of Y' and let

$$\begin{array}{ccccc} Y'' & \xrightarrow{\sigma} & Y' & \xrightarrow{\delta} & Y \\ & & \pi' \downarrow & & \pi \downarrow \\ & & X' & \xrightarrow{\tau} & X \end{array}$$

be the induced morphisms. If Y' has rational singularities, one has by definition $R^a\sigma_*\mathcal{O}_{Y''} = 0$ for $a > 0$. Hence, if Y has rational singularities as well,

$$R^a(\delta \circ \sigma)_*\mathcal{O}_{Y''} = R^a\delta_*\mathcal{O}_{Y'} = 0$$

and

$$R^a\tau_*(\pi'_*\mathcal{O}_{Y'}) = 0,$$

which implies (5.13).

By (3.24) we know that Y and Y' have quotient singularities. This implies that Y and Y' have rational singularities (see for example [62]). Let us recall the proof:

Let Y be any normal variety with quotient singularities and $\varphi : Z \longrightarrow Y$ the corresponding Galois cover with Z non singular. Let $\delta : Y' \longrightarrow Y$ be a desingularization such that $D' = \delta^*(\Delta(Z/Y))$ is a normal crossing divisor, where $\Delta(Z/Y)$ denotes the set of ramified points in Y. If Z' is the normalization of Y' in the function field of Z, (3.24) tells us that Z' has at most quotient singularities. Let finally $\gamma : Z'' \longrightarrow Z'$ be a desingularization. Altogether we obtain

$$\begin{array}{ccccc} Z'' & \xrightarrow{\gamma} & Z' & \xrightarrow{\delta'} & Z \\ & & \varphi' \downarrow & & \varphi \downarrow \\ & & Y' & \xrightarrow{\delta} & Y \end{array}$$

where Z'', Z and Y' are nonsingular.

Let us assume that for all quotient singularities and for all a with $a_0 > a > 0$ we know that the a-th higher direct image of the structure sheaf of the desingularization is zero. Then the Leray spectral sequence gives an injection

$$R^{a_0}\delta'_*\mathcal{O}_{Z'} = R^{a_0}\delta'_*(\gamma_*\mathcal{O}_{Z''}) \hookrightarrow R^{a_0}(\delta' \circ \gamma)_*\mathcal{O}_{Z''}.$$

Since $\delta' \circ \gamma$ is a birational proper morphism of non singular varieties

$$R^{a_0}(\delta' \circ \gamma)_*\mathcal{O}_{Z''} = 0.$$

Since the finite morphisms φ and φ' have no higher direct images one obtains

$$R^{a_0}\delta_*(\varphi'_*\mathcal{O}_{Z'}) = \varphi_*(R^{a_0}\delta'_*\mathcal{O}_{Z'}) = 0.$$

However, $\mathcal{O}_{Y'}$ is a direct summand of $\varphi'_*\mathcal{O}_{Z'}$ and hence $R^{a_0}\delta_*\mathcal{O}_{Y'} = 0$.

□

Proof of 5.12: let us first reduce b) to a):
Applying (5.13) and replacing $\mathcal{M}$ by $\mathcal{M}^{(1)}$, we can assume that the morphism $\tau : Z \to X$ in (5.11,b), applied to $\mathcal{N} = \mathcal{L}^N(-D)$, is an isomorphism and that for the divisor C in (5.11,b) $D+C$ is a normal crossing divisor. $\mathcal{L}^{N\cdot\mu}(-\mu\cdot D-C)$ is semi-ample for all μ divisible by μ_0. Choosing μ large enough, the multiplicities of $\mu \cdot D + C$ will be bounded above by $N \cdot \mu$. Moreover, we can assume that

$$H^0(X, \mathcal{L}^{N\cdot\mu}(-\mu \cdot D - C)) \neq 0$$

and hence, replacing μ again by some multiple, that

$$H^0(X, (\mathcal{L}^{N\cdot\mu}(-\mu{\cdot}D - C))^\nu \otimes \mathcal{O}_X(-B)) \neq 0$$

for some $\nu > 0$. Hence a) implies b).

To prove a), let us write

$$\mathcal{L}^{\nu\cdot N}(-\nu \cdot D) = \mathcal{O}_X(B + B')$$

or

$$\mathcal{L}^{\nu\cdot N} = \mathcal{O}_X(\nu \cdot D + B + B')$$

Blowing up, again, we can assume $D + B + B'$ to be a normal crossing divisor. For μ sufficiently large, we can assume that $(\mathcal{L}^N(-D))^\mu$ is generated by global sections. If H is zero set of a general section, then

$$\mathcal{L}^{(\nu+\mu)\cdot N} = \mathcal{O}_X(H + (\nu + \mu) \cdot D + B + B').$$

If μ is large enough, the multiplicities of the components of

$$D' = H + (\nu + \mu) \cdot D + B + B'$$

are smaller than $(\nu + \mu) \cdot N$ and, applying (5.1,a) for $N' = (\nu + \mu) \cdot N$ and $\mathcal{L}^{N'} = \mathcal{O}_X(D')$, we obtain (5.12,a).

Let us remark next, that d), under the additional condition that $\kappa(\mathcal{L}) = n$, implies c):
In fact, $c_1(\mathcal{L}^N(-D))^n > 0$ implies by (5.8,a) that

$$n = \kappa(\mathcal{L}^N(-D)) \leq \kappa(\mathcal{L}^N) = \kappa(\mathcal{L}).$$

To prove d), for $\kappa(\mathcal{L}) = n$, we can apply (5.4,d). Hence we find a divisor $C > 0$ and $\mu > 0$ with $\mathcal{L}^{\mu}(-C)$ ample. Then by (5.7)

$$\mathcal{L}^{N\cdot\nu+\mu}(-\nu\cdot D - C)$$

is ample for all ν and, by Serre's vanishing theorem (1.1)

$$H^b(X, \mathcal{L}^{-1} \otimes (\mathcal{L}^{-N\cdot\nu-\mu}(\nu{\cdot}D + C))^{\eta}) = 0 \quad \text{for} \quad b < n$$

and for η sufficiently large. As in the proof of (5.6) or by (5.13) this condition is compatible with blowing ups. Hence we may assume $D + C$ to be a normal crossing divisor and, choosing ν large enough, we may again assume that the multiplicities of $D' = \nu{\cdot}D + C$ are smaller than $N' = N{\cdot}\nu + \mu$. Replacing D' and N' by some high multiple we can assume in addition that $\mathcal{L}^{N'}(-D')$ is generated by global section and that

$$H^b(X, \mathcal{L}^{-N'-1}(D')) = 0 \quad \text{for} \quad b < n.$$

For an effective divisor B with $\mathcal{O}_X(B) = \mathcal{L}'^{N}(-D')$ we can apply a) and find

$$0 = H^0(X, \mathcal{L}^{-1}(-B)) \longrightarrow H^0(X, \mathcal{L}^{-1})$$

to be surjective.
For $\kappa(\mathcal{L}) <$ dim X part d) is finally reduced to the case $\kappa(\mathcal{L}) =$ dim X by induction:
Let H be a general hyperplane such that

$$H^b(X, \mathcal{O}_X(-H) \otimes \mathcal{L}^{-1}) = 0 \quad \text{for} \quad b < n.$$

The exact sequence

$$0 \longrightarrow \mathcal{O}_X(-H) \otimes \mathcal{L}^{-1} \longrightarrow \mathcal{L}^{-1} \longrightarrow \mathcal{L}^{-1}\,|_H \longrightarrow 0$$

give isomorphisms

$$H^b(X, \mathcal{L}^{-1}) \simeq H^b(H, \mathcal{L}^{-1}\,|_H)$$

for $b < n - 1$. Since $\kappa(\mathcal{L}) \leq n - 1$ we have $\kappa(\mathcal{L}\,|_H) \geq \kappa(\mathcal{L})$ and both groups vanish for $b < \kappa(\mathcal{L})$ by induction on $\dim X$.
e) follows by the same argument: If $\kappa(\mathcal{L} \otimes \mathcal{N}^{-1}) = \dim X$, then (5.4,c) and (5.7) imply that $\kappa(\mathcal{L}) = \dim X$ as well. For $\kappa(\mathcal{L} \otimes \mathcal{N}^{-1}) < \dim X$ again

$$\kappa(\mathcal{L}|_H \otimes \mathcal{N}^{-1}|_H) \geq \kappa(\mathcal{L} \otimes \mathcal{N}^{-1})$$

and by induction one obtains e).

□

Let us end this section by proving Kawamata's lemma (5.11) which was needed to reduce (5.12,b) to (5.12,a):

PROOF OF (5.11): Let us assume b). Since $\tau^*\mathcal{N}^\mu$ is nef, if and only if $\mathcal{N}$ is nef, and since $\nu(\mathcal{N}) = \nu(\tau^*\mathcal{N})$, we can assume that τ is an isomorphism. Moreover, we can assume $\mu_0 = 1$.
"$\mathcal{N}^\mu(-C)$ semi-ample for all $\mu > 0$" implies that $\mathcal{N}$ is nef. One obtains from (5.10)

$$\nu(\mathcal{N}) \geq \kappa(\mathcal{N}) \geq \kappa(\mathcal{N}^\mu(-C)) = \nu(\mathcal{N}^\mu(-C)).$$

For $\nu = \nu(\mathcal{N})$ the leading term in μ of

$$c_1(\mathcal{N}^\mu(-C))^\nu = (\mu \cdot c_1(\mathcal{N}) - C)^\nu$$

is $\mu^\nu \cdot c_1(\mathcal{N})^\nu$. Since this term intersects $H_1 \cdot \ldots \cdot H_{n-\nu}$ strictly positively, for general hyperplanes $H_1, \ldots, H_{n-\nu}$, we find $\nu \leq \nu(\mathcal{N}^\mu(-C))$.

To show the other direction, let $\phi_{\mu_0} : X \to Y$ be the rational map

$$X \longrightarrow \phi_{\mu_0}(X) = Y \subset \mathbb{P}(H^0(X, \mathcal{N}^{\mu_0})).$$

We can and we will assume that ϕ_{μ_0} has a connected general fibre, that dim $(Y) = \kappa(\mathcal{N})$ and, blowing X up if necessary, that ϕ_{μ_0} is a morphism. For some effective divisor D we have

$$\mathcal{N}^{\mu_0}(-D) = \phi_{\mu_0}^*\mathcal{L}, \quad \text{for } \mathcal{L} \text{ ample on } Y.$$

If F is a general fibre of ϕ_{μ_0}, then $D\,|_F$ is nef.

5.14. Claim. $D\,|_F$ is zero.

Assuming (5.14) we can blow up Y and X and assume that $D = \phi_{\mu_0}^*\Delta$ for some divisor Δ on Y. For example, blowing up Y one can assume that ϕ_{μ_0} factors over a flat morphism $\phi' : X' \to Y$ and that $\mathcal{N}$ is the pullback of a sheaf $\mathcal{N}'^{\mu_0}(-D') = \phi'^*\mathcal{L}$ for some semi-ample sheaf $\mathcal{L}$ and by (5.14) $D' \subseteq \phi'^*\Delta$ for some divisor Δ on Y. Since $\mathcal{N}'$ is numerically effective $D' \cdot C \geq 0$ for all curves C in X' contained in a fibre of ϕ'. This is only possible if $D' = \phi'^*\Delta$. Let us denote by $\tau : X \to Y$ the morphism obtained. We have $\mathcal{N}^{\mu_0} = \tau^*\mathcal{M}$ for some sheaf $\mathcal{M}$ on Y. Of course $\kappa(\mathcal{M}) = \dim Y$ and (5.12,b) holds true for $\mathcal{M}$ on Y, i.e. $\mathcal{M}^\mu(-\Gamma)$ is ample for some divisor $\Gamma > 0$ and all $\mu >> 0$. Then

$$\mathcal{N}^{\mu \cdot \mu_0}(-\tau^*\Gamma)$$

is semi-ample for all $\mu >> 0$.

□

PROOF OF (5.14): We may assume that $\mu_0 = 1$. For $\phi = \phi_1$, one has

$$c_1(\mathcal{N}) = c_1(\phi^*\mathcal{L}) + D = \phi^* c_1(\mathcal{L}) + D.$$

D is effective, hence $c_1(\mathcal{N})^{\nu_1} \cdot c_1(\phi^*\mathcal{L})^{\nu_2} \cdot D$ are semi-positive cycles, i.e. for $n = \nu_1 + \nu_2 + 1 + r$ one has

$$H_1 \cdot \ldots \cdot H_r \cdot c_1(\mathcal{N})^{\nu_1} \cdot c_1(\phi^*\mathcal{L})^{\nu_2} \cdot D \geq 0.$$

By definition

$$0 \equiv c_1(\mathcal{N})^{\nu+1} = c_1(\mathcal{N})^{\nu} \cdot (c_1(\phi^*\mathcal{L}) + D) \quad \text{for} \quad \nu = \nu(\mathcal{N}).$$

Since $c_1(\phi^*\mathcal{L})$ is also represented by an effective divisor, this is only possible if

$$0 \equiv c_1(\mathcal{N})^{\nu} \cdot c_1(\phi^*\mathcal{L}) = c_1(\mathcal{N})^{\nu-1} \cdot c_1(\phi^*\mathcal{L}) \cdot (c_1(\phi^*\mathcal{L}) + D).$$

The same argument shows that $c_1(\phi^*\mathcal{L})^2 \cdot c_1(\mathcal{N})^{\nu-1} \equiv 0$ and after ν steps we get

$$c_1(\phi^*\mathcal{L})^{\nu} \cdot c_1(\phi^*\mathcal{L}) + c_1(\phi^*\mathcal{L})^{\nu} \cdot D \equiv 0$$

and hence

$$c_1(\phi^*\mathcal{L})^{\nu} \cdot D = F \cdot D = 0.$$

□

§ 6 Differential forms and higher direct images

The title of this lecture is a little bit misleading. We want to apply the vanishing theorems for differential forms with values in invertible sheaves of integral parts of $\mathbb{Q}$-divisors (which follow directly from (3.2), (4.8) and (4.13)) to some more concrete situations.

For invertible sheaves themselves one obtains thereby different proofs of (5.2), (5.6,b), (5.12,c) and (5.12,d) but, as far as we can see, nothing more. For $\Omega^a_X \otimes \mathcal{L}^{-1}$ we obtain the Kodaira-Nakano vanishing theorem and some generalizations. Finally we consider the vanishing for higher direct images, which can be reduced, as usually, to the global vanishing theorems. As a straightforward application one obtains vanishing theorems for certain non compact manifolds.

In Lecture 5 we could at least point out some of the intermediate steps which remain true in characteristic $p \neq 0$. However, since $(*)$ and (!) only make sense in characteristic 0, we can as well assume throughout this chapter:

6.1. Assumptions. X is a projective manifold defined over an algebraically closed field k of characteristic zero and $\mathcal{L}$ is an invertible sheaf on X.

Global vanishing theorems in characteristic $p > 0$ will appear, as far as it is possible, in Lecture 11.

6.2. Global vanishing theorem for integral parts of $\mathbb{Q}$-divisors.

For $X, \mathcal{L}$ as in (6.1) let

$$D = \sum_{j=1}^{r} \alpha_j D_i$$

be a normal crossing divisor, $N \in \mathbb{N}$ with $0 < a_j < N$ for $j = 1 \ldots r$ and $\mathcal{L}^N = \mathcal{O}_X(D)$. Then one has:
a)

$$H^b(X, \Omega_X^a(\log\ D) \otimes \mathcal{L}^{-1}) = 0,$$

for $a + b < n - cd(X, D)$ and for $a + b > n + cd(X, D)$.
b) *Let A and B be reduced divisors such that $D + A + B$ has normal crossings and such that A, B and D have pairwise no common component. Then*

$$H^b(X, \Omega_X^a(\log\ (A + B + D))(-B) \otimes \mathcal{L}^{-1}) = 0$$

for $a + b < n - cd(X, D + B)$ and for $a + b > n + cd(X, D + A)$.
c) *If there exists a proper morphism*

$$g : X - D \longrightarrow W$$

for an affine variety W, then one can replace $cd(X, D)$ by $r(g)$ in a) (see (4.10)).

PROOF: By 3.2 $\mathcal{L}^{-1}(-B)$ has a logarithmic integrable connection ∇ with poles along $A + B + D$, such that the E_1-degeneration holds true. Moreover, $Res_{D_j}(\nabla) \notin \mathbb{Z}$ for $j = 1, \ldots, r$. For a component A_j of A we have $Res_{A_j}(\nabla) = 0$, and for a component B_j of B we have $Res_{B_j}(\nabla) = 1$. Hence a) follows from (4.8), b) from (4.9) and finally c) from (4.13).

□

6.3. Corollary. *For $X, \mathcal{L}$ as in (6.1), assume that $\mathcal{L}^N = \mathcal{O}_X(D)$ for a normal crossing divisor $D = \sum_{j=1}^r \alpha_j D_i$ with $0 < \alpha_j < N$ and assume that there exists an ample effective divisor B with $B_{red} = D_{red}$. Then $H^b(X, \mathcal{L}^{-1}) = 0$ for $b < n$.*

PROOF: Apply (6.2,a), and (4.7,c).

□

2ND PROOF OF (5.12,C), (5.12,D) AND (5.12.E).: As we have seen in Lecture 5, it is enough to proof (5.12,d) for $\kappa(\mathcal{L}^N(-D)) = \dim\ X$. Moreover, we may blow up, whenever we like.

We can write (replacing N and D by some multiple)

$$\mathcal{L}^N(-D) = \mathcal{A}(\Gamma)$$

for some effective divisor Γ and some ample sheaf $\mathcal{A}$. Blowing up, we can replace everything by some high multiple and subtract some effective divisor E from the pullback of $\mathcal{A}$ such that the sheaf obtained remains ample. Hence one can assume $D + \Gamma$ to be a normal crossing divisor. Since $\mathcal{L}^N(-D)$ is numerically effective, we can replace $\mathcal{A}$ by $\mathcal{A} \otimes \mathcal{L}^N(-D)$ and repeating this we can assume that the multiplicities of $D + \Gamma$ are bounded by N. Finally, replacing again everything by some multiple we are reduced to the case that $\mathcal{A}$ is very ample. Writing

$$\mathcal{L}^N = \mathcal{O}_X(D') \quad \text{for} \quad D' = D + \Gamma + H,$$

H a general divisor for $\mathcal{A}$, we can apply (6.3).

□

6.4. Corollary (Akizuki-Kodaira-Nakano [1]). *For $X, \mathcal{L}$ as in (6.1) assume that $\mathcal{L}$ is ample. Then*

$$H^b(X, \Omega_X^a \otimes \mathcal{L}^{-1}) = 0 \quad \textit{for} \quad a + b < n.$$

Moreover, if $A + B$ is a reduced normal crossing divisor, the same holds true for

$$H^b(X, \Omega_X^a(\log (A + B))(-B) \otimes \mathcal{L}^{-1}).$$

PROOF: We can write $\mathcal{L}^N = \mathcal{O}_X(D)$ for a non-singular divisor D and we may even assume that $D + A + B$ is a reduced normal crossing divisor. Moreover, for N large enough, $D + B$ and $D + A$ will both be ample and

$$cd(X, D + B) = cd(D + A) = 0.$$

By (2.3,b) one has an exact sequence

$$\ldots \longrightarrow H^{b-1}(D, \Omega_D^{a-1}(\log (A + B) \mid_D)(-B \mid_D) \otimes \mathcal{L}^{-1}) \longrightarrow$$
$$\longrightarrow H^b(X, \Omega_X^a(\log (A + B))(-B) \otimes \mathcal{L}^{-1}) \longrightarrow$$
$$\longrightarrow H^b(X, \Omega_X^a(\log (A + B + D))(-B) \otimes \mathcal{L}^{-1}) \longrightarrow \ldots$$

By induction on dim X we can assume that the first group is zero for $a + b < n + 1$ and by (6.2,b) the last group is zero for $a + b \neq n$.

□

If one tries to weaken "$\mathcal{L}$ ample" in this proof, one has to replace it by some condition compatible with the induction step.

6.5. Definition. An invertible sheaf $\mathcal{L}$ is called *l-ample* if the following two conditions hold true:

a) $\mathcal{L}^N$ is generated by global sections, for some $N \in \mathbb{N} - \{0\}$, and hence $\phi_N : X \longrightarrow \mathbb{P}(H^0(X, \mathcal{L}^N))$ a morphism.

b) For N as in a) $l \geq \text{Max}\{\dim\ \phi_N^{-1}(z); z \in \phi_N(X)\}$.

6.6. Corollary (A. Sommese [57], generalized in [22]). *For $X, \mathcal{L}$ as in (6.1) assume $\mathcal{L}$ to be l-ample. Then*

$$H^b(X, \Omega_X^a \otimes \mathcal{L}^{-1}) = 0$$

for $a + b <$ Min $\{\kappa(\mathcal{L}),\ \dim\ X - l + 1\}$.

PROOF: Using the notation from (6.5), we have seen in (4.11,b) that

$$r(\phi_N) \leq \text{Max } \{\dim \ X - \kappa(\mathcal{L}), \ l-1 \ \}.$$

Hence (6.6) is a special case of the following more technical statement .

□

6.7. Corollary. *For $X, \mathcal{L}$ as in (6.1) let $\tau : X \to Y$ be a morphism and let E be an effective normal crossing divisor with $\tau^{-1}(\tau(X-E)) = X - E$. If for some ample sheaf $\mathcal{A}$ on Y and for some $\nu > 0$ one has $\mathcal{L}^\nu = \tau^*\mathcal{A}$, then*

$$H^b(X, \Omega_X^a(\log \ E) \otimes \mathcal{L}^{-1}) = 0$$

for $a + b < \dim \ X - r(\tau \mid_{X-E})$.

PROOF: If I is the ideal sheaf of $\tau(E)$, then $\mathcal{A}^\mu \otimes I$ will be generated by global sections for some $\mu \gg 0$. Hence, we may assume that $\mathcal{L}^N(-E)$ is generated by global section for $N = \nu \cdot \mu$. Moreover, we can assume that N is larger than the multiplicities of the components of E. If D is the divisor of a general section of $\mathcal{L}^N(-E)$, then $D + E$ is a normal crossing divisor. We have for

$$\tau : D \longrightarrow \tau(D), \ \mathcal{L}\mid_D \quad \text{and} \quad E\mid_D$$

the same assumptions as those asked for in 6.7. Moreover, by (4.11,d) we have

$$r(\tau \mid_{X-E}) + 1 \geq r(\tau \mid_{D-E}).$$

By induction on dim X we may assume that

$$H^{b-1}(D, \Omega_D^{a-1}(\log \ E\mid_D) \otimes \mathcal{L}^{-1}) = 0$$

for

$$a + b < n - r(\tau \mid_{X-E}) \ \leq \ n + 1 - r(\tau \mid_{D-E}).$$

The exact sequence (see (2.3,b))

$$0 \to \Omega_X^a(\log \ E) \otimes \mathcal{L}^{-1} \to \Omega_X^a(\log \ (D+E)) \otimes \mathcal{L}^{-1} \to \Omega_D^{a-1}(\log \ E\mid_D) \otimes \mathcal{L}^{-1} \to 0$$

implies that for those a, b the map

$$H^b(X, \Omega_X^a(\log \ E) \otimes \mathcal{L}^{-1}) \longrightarrow H^b(X, \Omega_X^a(\log \ (D+E)) \otimes \mathcal{L}^{-1})$$

is injective. However, since

$$r(\tau \mid_{X-E}) \geq r(\tau \mid_{X-(D+E)})$$

(6.2,c) tells us that

$$H^b(X, \Omega_X^a(\log \ (D+E)) \otimes \mathcal{L}^{-1}) = 0$$

for $a + b < \dim \ X - r(\tau \mid_{X-E})$. □

6.8. Remarks.

a) The reader will have noticed that (6.7) does not use the full strength of (6.2). Several other applications and extensions of (6.2) can be found in the literature, (see for example [2] [3] [43] [44]).

b) Sometimes it is nicer to use the dual version of (6.7). Since

$$\bigwedge^n \Omega^1_X(\log\ E) = \omega_X \otimes \mathcal{O}(E_{red})$$

we find the dual of $\Omega^a_X(\log\ E)$ to be

$$\omega_X^{-1} \otimes \Omega_X^{n-a}(\log\ E)(-E_{red})$$

and by Serre duality (6.7) is equivalent to the vanishing of

$$H^b(X, \Omega^a_X(\log\ E)(-E_{red}) \otimes \mathcal{L})$$

for

$$a + b > \dim\ X + r(\tau\ |_{X-E}).$$

c) For (6.7) we used the invariant $r(g)$ and lemma (4.12), the latter being proved by analytic methods. However, playing around with de Rham complexes and their hypercohomology, one should be able to find an algebraic analogue of those arguments.

In [63] the second author used the Hodge duality (as in (3.23)) to reduce vanishing for $H^b(X, \mathcal{L}^{-1})$ to the Bogomolov-Sommese vanishing theorem. The latter fits nicely into the scheme explained in this lecture, (see the proof of (13.10,a)).

6.9. Corollary (F. Bogomolov, A. Sommese). *For $X, \mathcal{L}$ as in (6.1) and for a normal crossing divisor B one has*

$$H^0(X, \Omega^a_X(\log\ B) \otimes \mathcal{L}^{-1}) = 0$$

for $a < \kappa(\mathcal{L})$.

PROOF: (6.9) is compatible with blowing ups and we can assume that

$$\phi_N : X \longrightarrow \mathbb{P}(H^0(X, \mathcal{L}^N))$$

is a morphism. For N large enough, we can choose D such that $\phi_N\ |_{X-D}$ has equidimensional fibres of dimension $n - \kappa(\mathcal{L})$ and such that $\mathcal{L}^N = \mathcal{O}_X(D)$. Moreover we may assume $B + D$ to be a normal crossing divisor. By (6.2,c)

$$H^0(X, \Omega^a_X(\log\ (B + D)) \otimes \mathcal{L}^{(1)^{-1}}) = 0$$

for $a < \kappa(\mathcal{L})$. As $\Omega^a_X(\log\ B) \otimes \mathcal{L}^{-1}$ is a subsheaf of $\Omega^a_X(\log\ (B + D)) \otimes \mathcal{L}^{(1)^{-1}}$, one obtains (6.8).

□

Global vanishing theorems always give rise to the vanishing of certain direct image sheaves.

6.10. Notations. Let X be a manifold, defined over an algebraically closed field of characteristic zero and let $f : X \to Z$ be a proper surjective morphism. Let $\mathcal{L}$ be an invertible sheaf on X. We will call $\mathcal{L}$
a) *f-numerically effective* if for all curves C in X with dim $f(C) = 0$ one has deg $(\mathcal{L}\mid_C) \geq 0$
b) *f-semi-ample* if for some $N > 0$ the natural map $f^* f_* \mathcal{L}^N \longrightarrow \mathcal{L}^N$ is surjective.

The relative Grauert-Riemenschneider vanishing theorem says, that for a birational morphism $f : X \to Z$ one has $R^b f_* \omega_X = 0$ for $b > 0$. As a generalization one obtains:

6.11. Corollary.
a) *For $f : X \to Z$ as in (6.10) let $\mathcal{L}$ be an invertible sheaf such that $\mathcal{L}^N(-D)$ is f-numerically effective for a normal crossing divisor*

$$D = \sum_{j=1}^{r} \alpha_j D_j.$$

Then

$$R^b f_*(\mathcal{L}^{(1)} \otimes \omega_X) = 0 \quad \textit{for} \quad b > \text{dim } X - \text{dim } Z - \kappa(\mathcal{L}\mid_F)$$

where F is a general fibre of f.
b) *In particular, if $f : X \to Z$ is birational and if $D = f^*\Delta$ is a normal crossing divisor for some effective Cartier divisor Δ on Z, then*

$$R^b f_*(\omega_X \otimes \mathcal{O}_X(-[\frac{D}{N}])) = 0 \quad \textit{for} \quad b > 0.$$

Proof: Obviously, since $\mathcal{O}_X(-D) = f^*\mathcal{O}_Z(-\Delta)$ is f numerically effective, b) is a special case of a).
As usual, in a), we can add the assumption $0 < \alpha_j < N$ for $j = 1, \ldots r$, and we will have $\mathcal{L}^{(1)} = \mathcal{L}$.

The statement being local in Z, we can assume Z to be affine or, compactifying X and Z, we can assume Z to be projective. By (5.13) we are allowed to blow X up and we can assume that X is projective as well.
The assumptions made imply that for $\mathcal{A}$ ample invertible on Z

$$f^*\mathcal{A}^\nu \otimes \mathcal{L}^N(-D)$$

will be numerically effective for $\nu >> 0$ and that

$$\kappa(f^*\mathcal{A}^\nu \otimes \mathcal{L}) \geq \kappa(\mathcal{L}\mid_F) + \text{dim } (Z).$$

Using Serre's vanishing (1.1) we can assume that for all $c > 0$ and $b \geq 0$

$$H^c(Z, \mathcal{A}^\nu \otimes R^b f_*(\mathcal{L} \otimes \omega_X)) = 0$$

and that $H^0(Z, \mathcal{A}^\nu \otimes R^b f_*(\mathcal{L} \otimes \omega_X))$ generates the sheaf $\mathcal{A}^\nu \otimes R^b f_*(\mathcal{L} \otimes \omega_X)$. By the Leray spectral sequence (A.27) we obtain

$$H^b(X, f^*\mathcal{A}^\nu \otimes \mathcal{L} \otimes \omega_X) = H^0(Z, \mathcal{A}^\nu \otimes R^b f_*(\mathcal{L} \otimes \omega_X))$$

and by (5.12,d) this group is zero for

$$b > \dim\ X - \dim\ (Z) - \kappa(\mathcal{L}\,|_F) \geq \dim\ X - \kappa(f^*\mathcal{A}^\nu \otimes \mathcal{L}).$$

□

In the special case for which $\mathcal{L}^N(-D)$ is f-semi-ample the vanishing of

$$R^b f_*(\mathcal{L} \otimes \omega_X) \quad \text{for} \quad b > \dim\ X - \dim Z$$

follows as well from the next statement, due to J. Kollár, [40].

6.12. Corollary of 5.12,a) (J. Kollár). *In addition to the assumption of (6.11,a) we even assume that $\mathcal{L}^N(-D)$ is f-semi-ample. Then*

$$R^b f_*(\mathcal{L}^{(1)} \otimes \omega_X)$$

has no torsion for $b \geq 0$.

PROOF: As above we can assume X and Z to be projective and $\mathcal{L}^N(-D)$ to be semi-ample. Moreover, we may assume $\mathcal{L}^N(-D)$ to contain $f^*\mathcal{A}$ for a very ample sheaf $\mathcal{A}$ on Z, that $\mathcal{L} = \mathcal{L}^{(1)}$, that

$$H^c(X, R^b f_*(\mathcal{L} \otimes \omega_X)) = 0 \quad \text{for} \quad c > 0$$

and that $R^b f_*(\mathcal{L} \otimes \omega_X)$ is generated by its global sections. If $R^b f_*(\mathcal{L} \otimes \omega_X)$ has torsion for some b, then the map

$$R^b f_*(\mathcal{L} \otimes \omega_X) \longrightarrow R^b f_*(\mathcal{L} \otimes \omega_X) \otimes \mathcal{O}_Z(A)$$

has a non-trivial kernel $\mathcal{K}$ for some effective ample divisor A on Z. We may assume that $\mathcal{O}_Z(A) = \mathcal{A}$. Replacing $\mathcal{L}$ by $\mathcal{L} \otimes f^*\mathcal{A}^\nu$ again, we can assume that $H^0(Z, \mathcal{K}) \neq 0$. For $B = f^*A$,this implies that $H^0(Z, \mathcal{K})$ lies in the kernel of

$$H^b(X, \mathcal{L} \otimes \omega_X) \longrightarrow H^b(X, \mathcal{L} \otimes \omega_X(B))$$

and hence that

$$H^{n-b}(X, \mathcal{L}^{-1}(-B)) \longrightarrow H^{n-b}(X, \mathcal{L}^{-1})$$

is not surjective, contradicting (5.12,a).

□

Some of the vanishing theorems mentioned and a partial degeneration of the Hodge to de Rham spectral sequence remain true for certain non-compact manifolds. One explanation for those results, obtained by I. Bauer and S. Kosarew in [4] and [42] by different methods, is the following lemma.

6.13. Lemma. *Let Z be a projective variety in characteristic zero, $U \subseteq Z$ be an open non-singular subvariety, $\delta : X \to Z$ be a desingularization such that $\iota : U \simeq \delta^{-1}(U) \to X$. Assume that $X - \iota(U) = E$ for a reduced normal crossing divisor E. Then, for $a + b < \dim X - \dim \delta(E) - 1$ and all invertible sheaves $\mathcal{M}$ on Z one has*

$$H^b(X, \Omega^a_X(\log E) \otimes \delta^*\mathcal{M}) = H^b(U, \Omega^a_X \otimes \mathcal{M} |_U).$$

Proof: Let $\mathcal{A}$ be an ample invertible sheaf on Z and let E' be an effective exceptional divisor, such that $\mathcal{O}_X(-E')$ is relatively ample for δ.
For fixed $\nu \geq 0$ we can choose $\mathcal{A}$ large enough, such that for all a, b

$$R^b\delta_*(\Omega^a_X(\log E) \otimes \mathcal{O}_X(-E - \nu \cdot E')) \otimes \mathcal{A}$$

is generated by global sections and

$$H^c(Z, R^b\delta_*(\Omega^a_X(\log E) \otimes \mathcal{O}_X(-E - \nu \cdot E')) \otimes \mathcal{A}) = 0$$

for $c > 0$. Moreover, for $\nu > 0$, we may assume that $\tau^*\mathcal{A}(-\nu \cdot E')$ is ample. Using Serre duality (as explained in (6.8,b)) and the Leray spectral sequence (A.27) one finds

$$H^{n-b}(X, \Omega^{n-a}_X(\log E) \otimes \delta^*\mathcal{A}^{-1}(\nu \cdot E'))^* =$$
$$H^b(X, \Omega^a_X(\log E)(-E) \otimes \delta^*\mathcal{A}(-\nu \cdot E')) =$$
$$H^0(Z, R^b\delta_*(\Omega^a_X(\log E) \otimes \mathcal{O}_X(-E - \nu \cdot E')) \otimes \mathcal{A}).$$

By (6.4), for $\nu > 0$, or by (6.7), for $\nu = 0$, we find

$$R^b\delta_*(\Omega^a_X(\log E) \otimes \mathcal{O}_X(-E - \nu \cdot E')) = 0$$

for $a + b > \dim X$. For those a and b and for

$$\mathcal{K}_\nu = \mathcal{O}_X / \mathcal{O}_X(-\nu \cdot E')$$

we have

$$R^b\delta_*(\Omega^a_X(\log E)(-E) \otimes \mathcal{K}_\nu) = 0.$$

One obtains for $a + b > \dim X + \dim \tau(E)$ from the Leray spectral sequence (A.27) that

$$H^b(X, \delta^*\mathcal{M}^{-1} \otimes \Omega^a_X(\log E)(-E) \otimes \mathcal{K}_\nu) = 0.$$

Hence for all $\nu \geq 0$ and $a + b > \dim X + \dim \tau(E) + 1$ the map

$$H^b(X, \delta^*\mathcal{M}^{-1} \otimes \Omega^a_X(\log E) \otimes \mathcal{O}_X(-E - \nu \cdot E')) \longrightarrow$$

$$\longrightarrow H^b(X, \delta^*\mathcal{M}^{-1} \otimes \Omega_X^a(\log\ E)(-E))$$

is bijective. By Serre duality again,

$$H^b(X, \delta^*\mathcal{M} \otimes \Omega_X^a(\log\ E)) \longrightarrow H^b(X, \delta^*\mathcal{M} \otimes \Omega_X^a(\log\ E) \otimes \mathcal{O}_X(\nu \cdot E'))$$

is an isomorphism for $a + b < \dim\ X - \dim\ \tau(E) - 1$ and, taking the limit for $\nu \in \mathbb{N}$, we obtain (6.13).

□

6.14. Corollary (I. Bauer, S. Kosarew [4]). *Let Z be a projective variety in characteristic zero, $U \subseteq Z$ be an open non-singular subvariety. Then, for $k < n - \dim\ (Z - U) - 1$ one has*

$$\dim \mathbb{H}^k(U, \Omega_U^\bullet) = \sum_{a+b=k} \dim H^b(U, \Omega_U^a).$$

Proof: We can choose a desingularisation $\delta : X \to Z$ and E as in (6.13). Then we have a natural map of spectral sequences

$$\begin{array}{ccc} E_1^{ab} = H^b(U, \Omega_U^a) & \Longrightarrow & \mathbb{H}^{a+b}(U, \Omega_U^\bullet) \\ \uparrow \varphi_{a,b} & & \uparrow \varphi \\ E_1'^{ab} = H^b(X, \Omega_X^a(\log\ E)) & \Longrightarrow & \mathbb{H}^{a+b}(X, \Omega_X^\bullet(\log\ E)). \end{array}$$

Since $E_1'^{ab}$ degenerates in E_1 and since $\varphi_{a,b}$ are isomorphisms for

$$a + b < n - \dim\ (Z - U) - 1$$

the second spectral sequence has to degenerate for those a, b.

□

6.15. Corollary (see also I. Kosarew, S. Kosarew [42]).
For Z and U as in (6.14) let $\mathcal{L}$ be an l-ample invertible sheaf on Z. Then

$$H^b(U, \Omega_U^a \otimes \mathcal{L}^{-1}|_U) = 0$$

for

$$a + b < Min\ \{\kappa(\mathcal{L}), \dim\ X - l + 1,\ \dim\ X - \dim\ (Z - U) - 1\ \}$$

Proof: For X, E as in (6.13) and $\mathcal{M} = \delta^*\mathcal{L}$ we have by (6.7)

$$H^b(X, \mathcal{M}^{-1} \otimes \Omega_X^a(\log\ E)) = 0$$

for $a + b < \dim\ X - r(\tau|_U)$ where

$$\tau : X \xrightarrow{\ \delta\ } Z \xrightarrow{\ \phi_N\ } \phi_N(Z)$$

is the composition of δ with the map given by global sections of $\mathcal{L}^N$.
However,

$$r(\tau\mid_U) \leq r(\phi_N) \leq \text{Max}\{\dim X - \kappa(\mathcal{L}), l-1\}.$$

□

6.16. Remark. For $k = \mathbb{C}$, the reason for which certain coherent sheaves $\mathcal{F}$ on Z satisfy $H^b(Z, \mathcal{F} \otimes \mathcal{H}) = 0$ for $\mathcal{H}$ ample, seems to be related to the existence of connections. This point of view, which is exploited in J. Kollár's work on vanishing theorems [40], [41] and extended by M. Saito (see [54] and the references given there), should imply that sheaves arising as natural subquotients of $\mathcal{O}_Z \otimes R^k f_* V$ for a morphism $f : X \to Z$ of manifolds and a locally constant system V, sometimes have vanishing properties as the one stated above.
J. Kollár proved, for example, that for a morphism $f : X \to Z$, where X and Z are projective varieties and X non-singular, one has

$$H^c(Z, R^b f_* \omega_X \otimes \mathcal{H}) = 0$$

for $c > 0$ and $\mathcal{H}$ ample on Z.

Slightly more generally one has

6.17. Corollary (of (5.12,b)). *Let $f : X \to Z$ be a surjective morphism of projective varieties defined over an algebraically closed field of characteristic zero, with X non-singular. Let $\mathcal{L}$ be an invertible sheaf on X,*

$$D = \sum_{j=1}^{r} \alpha_j D_j$$

a normal crossing divisor and $N \in \mathbb{N}$ with

$$0 < \alpha_j < N \quad \text{for} \quad j = 1, \ldots, r.$$

a) *If $\mathcal{L}^N(-D)$ is semi-ample and $\mathcal{K}$ a numerically effective invertible sheaf on Z with $\kappa(\mathcal{K}) = \dim Z$, then for $c > 0$ and $b \geq 0$*

$$H^c(Z, \mathcal{K} \otimes R^b f_*(\omega_X \otimes \mathcal{L})) = 0.$$

b) If $\mathcal{L}^N(-D)$ is numerically effective, $\kappa(\mathcal{L}^N(-D)) = \nu(\mathcal{L}^N(-D))$, and if $(\mathcal{L}^N(-D))^\mu$ contains $f^*\mathcal{H}$ for some ample sheaf $\mathcal{H}$ on Z and some $\mu > 0$, then for $c > 0$ and all b

$$H^c(Z, R^b f_*(\omega_X \otimes \mathcal{L})) = 0.$$

Proof: By (5.10,b), replacing $\mathcal{L}$ by $\mathcal{L} \otimes f^*\mathcal{K}$, a) follows from b).
If H is the zero divisor of a general section of $\mathcal{H}^\mu$ for $\mu \gg 0$ and if $B = f^*H$ then B is a non-singular divisor and the assumptions of (5.12,b) hold true. Hence

$$H^b(X, \omega_X \otimes \mathcal{L}) \longrightarrow H^b(X, \omega_X(B) \otimes \mathcal{L})$$

is injective for all b. Since H is in general position we have exact sequences

$$\begin{array}{ccccccc} 0 \longrightarrow & R^bf_*(\mathcal{L}\otimes\omega_X) & \longrightarrow & R^bf_*(\mathcal{L}\otimes\omega_X(B)) & \longrightarrow & R^bf_*(\mathcal{L}\otimes\omega_B) \to 0 \\ & \| & & \| & & \| \\ & R^bf_*(\mathcal{L}\otimes\omega_X) & \longrightarrow & \mathcal{O}_Z(H)\otimes R^bf_*(\mathcal{L}\otimes\omega_X) & \longrightarrow & \mathcal{O}_Z(H)\otimes R^bf_*(\mathcal{L}\otimes\omega_X)\,|_H \end{array}$$

By induction on dim Z we may assume that

$$H^c(H, R^bf_*(\mathcal{L}\otimes\omega_B)) = 0 \ \text{ for } \ c > 0$$

and, if we choose μ large enough, we find by Serre's vanishing theorem

$$H^c(Z, R^bf_*(\mathcal{L}\otimes\omega_X)) = 0 \ \text{ for } \ c \geq 2.$$

In the Leray-spectral sequence (see A.27) all the differentials are zero, since $E_2^{ab} \neq 0$ just for $a = 0$ or $a = 1$, and hence the upper line in the following diagram is exact.

$$\begin{array}{ccccc} 0 \to H^1(Z, R^{b-1}f_*(\mathcal{L}\otimes\omega_X)) & \to & H^b(X, \mathcal{L}\otimes\omega_X) & \to & H^0(Z, R^bf_*(\mathcal{L}\otimes\omega_X)) \\ & & \alpha\downarrow & & \downarrow \\ & & H^b(X, \mathcal{L}\otimes\omega_X(B)) & \to & H^0(Z, R^bf_*(\mathcal{L}\otimes\omega_X(B))) \end{array}$$

Since α is injective and since

$$H^b(X, \mathcal{L}\otimes\omega_X(B)) = H^0(Z, R^bf_*(\mathcal{L}\otimes\omega_X(B)))$$

we find $H^1(Z, R^{b-1}f_*(\mathcal{L}\otimes\omega_X)) = 0$ for all b.

□

§ 7 Some applications of vanishing theorems

The vanishing theorems for integral parts of $\mathbb{Q}$-divisors and for numerically effective sheaves (5.12,c) and (5.12,d), as well as (5.6,a) turned out to be useful for applications in higherdimensional complex projective geometry. We will not be able in these notes to include an outline of the Iitaka-Mori classification of threefolds, and the reader interested in this direction is invited to regard S. Mori's beautiful survey [46].

In this lecture we just want to give a flavour as to how one should try to use vanishing theorems to attack certain types of questions. The choice made is obviously influenced by our personal taste.

We will assume in this lecture:

All varieties are defined over an algebraically closed field k of characteristic zero.

7.1. Example: Surfaces of general type. For a projective surface S' of general type, i.e. for a non-singular S' with $\kappa(\omega_{S'}) = \dim S' = 2$, one can blow down exceptional curves $E \simeq \mathbb{P}^1$ with $E \cdot E = -1$ ([30], p. 414). After finitely many steps one obtains a surface S without any exceptional curve, a minimal model of S' ([30], p. 418). S is characterised by

7.2. Claim. ω_S is *nef*.

PROOF: $\kappa(S) \geq 0$ implies that $\omega_S^N = \mathcal{O}_S(D)$ for D effective. A curve C with $\deg\ (\omega_S\ |_C) < 0$ must be a component of D and $C^2 < 0$. However, the adjunction formula gives

$$-2 \leq 2g(C) - 2 = \deg\ (\omega_S\ |_C) + C^2.$$

Hence the only solution is $C^2 = -1$ and $\deg(\omega_S\ |_C) = -1$, which forces C to be exceptional.

□

D. Mumford in his appendix to [65] used the contraction of (-2) curves to show:

7.3. Theorem. *If S is a minimal model and $\kappa(\omega_S) = 2$, then ω_S is semi-ample.*

X. Benveniste and Y. Kawamata ($\dim X = 3$) and Y. Kawamata and V. Shokurov (see [46] for the references) generalised (7.3) to the higher dimensional case. Their ideas, cut back to the surface case, give a simple proof of (7.3).

PROOF OF 7.3 (FROM THE DIPLOM-THESIS OF T. NAKOVICH, ESSEN):
Step 1.: If $p \in S$ does not lie on any curve C with $\deg\ (\omega_S\ |_C) = 0$, then for some $\nu \gg 0$ there is $s \in H^0(S, \omega_S^\nu)$ with $s(p) \neq 0$.

PROOF: Let $\tau : S' \to S$ be the blowing up of p and E the exceptional curve. One has

$$\deg\ (\tau^*\omega_S^\mu(-E)\ |_{C'}) = \mu \cdot \deg\ (\omega_S\ |_C) - E \cdot C'$$

for curves C' in S' with $C = \tau(C') \neq p$. Hence, for some $\mu \gg 0$ the sheaf $\mathcal{L} = \tau^*\omega_S^\mu(-E)$ will be nef. By (5.12,c) we find

$$H^1(S', \mathcal{L}^{-2}) \cong H^1(S', \omega_{S'} \otimes \mathcal{L}^2) = H^1(S', \tau^*\omega_S^{2\mu+1}(-E)) = 0$$

and hence

$$H^0(S', \tau^*\omega_X^{2\mu+1}) \longrightarrow H^0(E, \mathcal{O}_E)$$

is surjective.

□

Step 2.: For $\nu \gg 0$ let

$$D = \sum_{j=1}^{r} \alpha_j C_j$$

be the base locus of ω_S^ν (i.e. $\omega_S^\nu(-D)$ is generated by $H^0(S, \omega_S^\nu)$ outside of a finite number of points). Then D is a normal crossing divisor, $C_j^2 = -2$ for $j = 1, \ldots, r$ and ω_S^ν is generated by $H^0(S, \omega_S^\nu)$ outside of D.

PROOF: By step 1, if for some $p \in S$ there is no section s of ω_S^ν with $s(p) \neq 0$, then p lies on some curve C with deg $(\omega_S \mid_C) = 0$ and necessarily C is contained in the base locus. We know thereby that deg $(\omega_S \mid_{C_j}) = 0$ for the components C_j of D. By the Hodge-index theorem ([30], p. 364) one finds for any reduced subdivisor C of D that $C \cdot C < 0$. If we take $C = C_j$, then the adjunction formula shows that

$$C \cdot C = -2 \text{ and } C \simeq \mathbb{P}^1.$$

For $C = (C_1 + C_2)$ we get

$$C_1 \cdot C_2 < 2$$

and C_1 and C_2 intersect transversally.

□

Step 3. For D as in Step 2, $\omega_S^\nu(-D)$ is nef, hence $\omega_S^N(-D)$ for $N \geq \nu$ is nef as well. We can choose $N > \alpha_j$ for $j = 1, \ldots r$. For some $i > 0$

$$D' = [\frac{i \cdot D}{N}] = \sum_{j=1}^{r} [\frac{i \cdot \alpha_j}{N}] \cdot C_j$$

will be reduced and non zero. By (5.12,c) again, we have for $\mathcal{L} = \omega_S$

$$H^1(S, \mathcal{L}^{(i)^{-1}}) \cong H^1(S, \omega_S \otimes \mathcal{L}^{(i)}) = H^1(S, \omega_S^{i+1}(-D')) = 0$$

and

$$H^0(S, \omega_S^{i+1}) \longrightarrow H^0(D', \omega_S^{i+1} \mid_{D'})$$

is surjective. Since the right hand side is nontrivial (in fact its dimension is just the number of connected components of D'), for $i + 1$ the base locus does not contain D'. After finitely many steps we are done.

□

The proof of (7.1) is a quite typical example in two respects. First of all, vanishing of H^1 or more general by the surjectivity of the adjunction map in (5.6,a) allows to pull back sections of invertible sheaves on divisors. Secondly it shows again how to play around with integral parts, a method which already appeared in the proof of (5.12).

Corollary (5.12), as stated, has the disadvantage that D has to be a normal crossing divisor. Let us try next to study some weaker conditions.

7.4. Definition. Let X be a normal variety and D be an effective Cartier divisor on X. Let $\tau : X' \to X$ be a blowing up, such that X' is non singular and $D' = \tau^* D$ is a normal crossing divisor. We define:
a)

$$\omega_X\{\frac{-D}{N}\} = \tau_* \omega_{X'}(-[\frac{D'}{N}]).$$

b) $\mathcal{C}_X(D,N) = \text{Coker } (\omega_X\{\frac{-D}{N}\} \longrightarrow \omega_X)$ where ω_X is the reflexive hull of ω_{X_0} for $X_0 = X - \text{Sing } (X)$.
c) (see [23])

$$e(D) = \text{Min } \{N > 0; \mathcal{C}_X(D,N) = 0\}.$$

d) If X is compact and $\mathcal{L}$ invertible, $H^0(X,\mathcal{L}) \neq 0$, then

$$e(\mathcal{L}) = \text{Max } \{e(D);\ D \geq 0 \text{ and } \mathcal{O}_X(D) = \mathcal{L}\}.$$

7.5. Properties (see [23]). Let X and D be as in (7.4).
a) If X has at most rational singularities, then $e(D)$ is finite.
b) If X is non-singular and D a normal crossing divisor then

$$\omega_X\{\frac{-D}{N}\} = \omega_X(-[\frac{D}{N}]).$$

c) $\omega_X\{\frac{-D}{N}\}, \mathcal{C}_X(D,N)$ and $e(D)$ are independent of the blowing up $\tau : X' \to X$ choosen.
d) Let H be a prime Cartier divisor on X, not contained in D, such that H is normal. Then one has a natural inclusion

$$\omega_H\{\frac{-D\mid_H}{N}\} \longrightarrow \omega_X\{\frac{-D}{N}\} \otimes \mathcal{O}_X(H)\mid_H .$$

e) If in d) X and H have rational sigularities, then for $N \geq e(D\mid_H)$, H does not meet the support of $\mathcal{C}_X(D,N)$.

PROOF: a) is obvious since for $N \gg 0$

$$\tau_* \omega_{X'}(-[\frac{D}{N}]) = \tau_* \omega_{X'} = \omega_X.$$

Similar to (5.13), part b) can be deduced from (3.24) and from the fact, that quotient singularities are rational singularities. A more direct argument is as follows. We have an inclusion

$$\omega_X\{\frac{-D}{N}\} \longrightarrow \omega_X(-[\frac{D}{N}])$$

and it is enough to prove b) for some blowing up dominating τ. Hence, it is enough to consider the case that τ is a sequence of blowings with non-singular centers. Let us write

$$\omega_{X'} = \tau^* \omega_X \otimes \mathcal{O}_{X'}(\sum_{i=1}^{t} \alpha_i \cdot E_i).$$

For $m_i = \operatorname{codim}_X(\tau(E_i))$ one has $\alpha_i \geq m_i - 1$.

In fact, assume this to hold true for $\tau_1 : X_1 \longrightarrow X$ and

$$\omega_{X_1} = \tau_1^* \omega_X \otimes \mathcal{O}_{X_1}(\sum_{i=1}^{t-1} \alpha_i \cdot E_i').$$

If $\delta : X' \longrightarrow X_1$ is the blowing up with center S and E_t the exceptional divisor then, for $m = \operatorname{codim}_{X_1}(S)$, one has

$$\omega_{X'} = \delta^* \omega_{X_1} \otimes \mathcal{O}_{X'}((m-1) \cdot E_t)$$

(see [30], p. 188). If $m_t > m$ then S lies on some E_ν' with $m_t - m \leq m_\nu - 1$. Hence

$$\alpha_t \geq m - 1 + \alpha_\nu \geq m_t - 1.$$

On the other hand, assume that $\tau(E_\mu)$ lies on s different components of D, let us say on $D_1, \cdots, D_s$ but not in D_j for $j > s$. Then $m_\mu \geq s$ and, if

$$D = \sum_{j=1}^{r} \alpha_j D_j$$

one has

$$[\sum_{j=1}^{s} \frac{\alpha_j}{N}] \leq \sum_{j=1}^{s} [\frac{\alpha_j}{N}] + s - 1 \leq \sum_{j=1}^{s} [\frac{\alpha_j}{N}] + \alpha_\mu.$$

One obtains

$$[\frac{D'}{N}] \leq \tau^*[\frac{D}{N}] + \sum_{i=1}^{t} \alpha_i \cdot E_i$$

and hence

$$\tau^* \omega_X(-[\frac{D}{N}]) \subset \omega_{X'}(-[\frac{D'}{N}]).$$

c) follows from b). Hence in d) we may assume that D' intersects the proper transform H' on H transversally and, of course, that H' is non-singular. Then

$$[\frac{D'}{N}] \mid_{H'} = [\frac{D' \mid_{H'}}{N}].$$

One has a commutative diagram

$$\begin{array}{ccccc}
\tau_* \omega_{X'}(-[\frac{D'}{N}] + H') & \xrightarrow{\alpha} & \tau_* \omega_{H'}(-[\frac{D'|_{H'}}{N}]) & \longrightarrow & \omega_H \\
\downarrow & & & & \downarrow = \\
\tau_* \omega_{X'}(-[\frac{D'}{N}]) \otimes \mathcal{O}_X(H) & \longrightarrow & \tau_* \omega_{X'}(-[\frac{D'}{N}]) \otimes \mathcal{O}_X(H) \mid_H & \xrightarrow{\gamma} & \omega_H
\end{array}$$

The cokernel of α lies in $R^1\tau_*\omega_{X'}(-[\frac{D'}{N}])$, and (6.11) shows that α is surjective. We obtain therefore a non-trivial morphism

$$\alpha' : \omega_H\{\frac{-D\mid_H}{N}\} \longrightarrow \omega_X\{\frac{-D}{N}\} \otimes \mathcal{O}_X(H)\mid_H .$$

Since $\omega_H\{\frac{-D|_H}{N}\}$ is torsion free d) holds true.
In e) we know that $\omega_H\{\frac{-D|_H}{N}\}$ is isomorphic to ω_H. Hence γ is surjective. Therefore $\omega_X\{\frac{-D}{N}\} \otimes \mathcal{O}_X(H)$ must be isomorphic to $\omega_X \otimes \mathcal{O}_X(H)$ in a neighbourhood of H.

□

7.6. Remark. The diagram used to prove d) gives slightly more. Instead of assuming that $D' = \tau^*D$ it is enough to take any normal crossing divisor D' on X' not containing H'. Then the inclusion

$$\tau_*\omega_{H'}(-[\frac{D'\mid_{H'}}{N}]) \hookrightarrow \tau_*\omega_{X'}(-[\frac{D}{N}]) \otimes \mathcal{O}_X(H)\mid_H$$

exists whenever $H' + D'$ is a normal crossing divisor and $\mathcal{O}_{X'}(-D')$ is τ-numerically effective (see (6.10)).

Up to now, we do not even know that $e(\mathcal{L})$ is finite. This however follows from the first part of the next lemma, since every sheaf $\mathcal{L}$ lies in some ample invertible sheaf.

7.7. Lemma. *Let X be a projective manifold and let $\mathcal{L}$ be an invertible sheaf.*
a) *If $\mathcal{L}$ is very ample and $\nu > 0$, then*

$$e(\mathcal{L}^\nu) \leq \nu \cdot c_1(\mathcal{L})^{\dim X} + 1$$

b) *For $s \in H^0(X, \mathcal{L})$ with zero-locus D assume that for some $p \in X$ the section s has the multiplicity μ i.e.:*

$$s \in m_p^\mu \otimes \mathcal{L} \quad \textit{but} \quad s \notin m_p^{\mu+1} \otimes \mathcal{L}.$$

Then

$$\omega_X\{\frac{-D}{N}\} \longrightarrow \omega_X$$

is an isomorphism in a neighbourhood of p for $N > \mu$.
c) *If under the assumption of* b)

$$\mu' = [\frac{\mu}{N}] - \dim X + 1 \geq 0$$

then $\omega_X\{\frac{-D}{N}\}$ is contained in $m_p^{\mu'} \otimes \omega_X$.

PROOF: a) Let $D \geq 0$ be a divisor, $\mathcal{O}_X(D) = \mathcal{L}^\nu$.

If X is a curve then $[\frac{D}{N}] = 0$ for $N >$ deg $D + 1 = \nu \cdot c_1(\mathcal{L}) + 1$.

In general, let H be the divisor of a general section of $\mathcal{L}$. By induction

$$e(\mathcal{L}^\nu \mid_H) \leq \nu \cdot c_1(\mathcal{L} \mid_H)^{\dim H} + 1 = \nu \cdot c_1(\mathcal{L})^{\dim X} + 1.$$

(7.5,e) tells us that $\mathcal{C}_X(D, N)$ is supported outside of H for

$$N \geq \nu \cdot c_1(\mathcal{L})^{\dim X} + 1$$

and moving H we find $\mathcal{C}_X(D, N) = 0$.

For b) and c) we may assume that the blowing up $\tau : X' \to X$ factors through the blowing up $\varrho : X_p \to X$ of p. For $D_p = \varrho^* D$ and for the exceptional divisor E of ϱ we have

$$\Delta = D_p - \mu \cdot E \geq 0$$

and Δ does not contain E. Assume $N > \mu$. One has

$$\mathcal{O}_E(\Delta \mid_E) = \mathcal{O}_{\mathbf{P}^{n-1}}(\mu)$$

and, by part a), one obtains

$$\omega_E\{\frac{-\Delta \mid_E}{N}\} = \omega_E.$$

From (7.5,e) one knows that

$$\omega_{X_p}\{\frac{-\Delta}{N}\} \longrightarrow \omega_{X_p}$$

is an isomorphism in a neighbourhood of E. Hence

$$\omega_{X_p}(-E) = \omega_{X_p}\{\frac{-\Delta}{N}\} \otimes \mathcal{O}_{X_p}(-E) = \omega_{X_p}\{\frac{-\Delta - N \cdot E}{N}\}$$

is contained in $\omega_{X_p}\{\frac{-D_p}{N}\}$ which implies that $\omega_X = \omega_X\{\frac{-D}{N}\}$ near p.

If $\mu' = [\frac{\mu}{N}] - n + 1 \geq 0$ then

$$\omega_{X_p}\{\frac{-D_p}{N}\} \subset \omega_{X_p}\{\frac{-\mu \cdot E}{N}\} = \omega_{X_p}(-[\frac{\mu}{N}] \cdot E)$$

and

$$\omega_X\{\frac{-D}{N}\} = \varrho_* \omega_{X_p}\{\frac{-D_p}{N}\} \subset \varrho_* \varrho^* \omega_X((n - 1 - [\frac{\mu}{N}]) \cdot E) = m_p^{\mu'} \otimes \omega_X.$$

□

The sheaves $\omega_X\{\frac{-D}{N}\}$ are describing the correction terms needed if one wants to generalize the vanishing theorems (5.12,c) or (5.12,d) to non normal crossing divisors. For example one obtains:

7.8. Proposition. *Let X be a projective manifold, $\mathcal{L}$ be an invertible sheaf and D be a divisor such that $\mathcal{L}^N(-D)$ is numerically effective and*

$$c_1(\mathcal{L}^N(-D))^n > 0$$

for $n = \dim X$. Then

$$H^b(X, \omega_X\{\frac{-D}{N}\} \otimes \mathcal{L}) = 0 \quad \textit{for} \quad b > 0.$$

PROOF: This follows from (5.12,c) and (6.11,b) by using the Leray spectral sequence (A.27).

□

7.9. Remark. Demailly proved in [13] an analytic improvement of Kodaira's vanishing theorem. It would be nice to understand the relation of his positivity condition with the one arising from (7.8), i.e. with the condition that

$$\omega_X\{\frac{-D}{N}\} = \omega_X.$$

One of the reasons for the interest in vanishing theorems as (7.8) is implication that certain sheaves are generated by global sections. For example one has:

7.10. Corollary. *Under the assumptions of (7.8) let $\mathcal{H}$ be a very ample sheaf. Then*

$$\mathcal{H}^{\dim X} \otimes \mathcal{L} \otimes \omega_X\{\frac{-D}{N}\}$$

is generated by global sections.

PROOF.: For

$$\mathcal{F} = \mathcal{L} \otimes \omega_X\{\frac{-D}{N}\} \otimes \omega_X^{-1}$$

we have

$$H^b(X, \mathcal{F} \otimes \mathcal{H}^\nu \otimes \omega_X) = 0 \quad \text{for} \quad b > 0 \quad \text{and} \quad \nu \geq 0.$$

For general sections $H_1, \ldots, H_n$ of $\mathcal{H}$ passing through a given point p and for

$$Y_r = \bigcap_{i=1}^{r} H_i$$

we obtain

$$H^b(Y_r, \mathcal{F} \otimes \mathcal{H}^\nu \otimes \omega_{Y_r}) = 0, \quad \text{for} \quad b > 0 \quad \text{and} \quad \nu \geq 0,$$

by regarding the cohomology sequence given by the short exact sequence

$$0 \longrightarrow \mathcal{F} \otimes \mathcal{H}^\nu \otimes \omega_{Y_r} \longrightarrow \mathcal{F} \otimes \mathcal{H}^{\nu+1} \otimes \omega_{Y_r} \longrightarrow \mathcal{F} \otimes \mathcal{H}^\nu \otimes \omega_{Y_{r+1}} \longrightarrow 0.$$

By induction we may assume that

$$\mathcal{F} \otimes \mathcal{H}^{\dim(Y_{r+1})} \otimes \omega_{Y_{r+1}}$$

is generated by global sections in p, and, using the cohomology sequence again one finds the same for

$$\mathcal{F} \otimes \mathcal{H}^{\dim(Y_r)} \otimes \omega_{Y_r}.$$

□

Let us apply (7.10) for $X = \mathbb{P}^n$ to study the behaviour of zeros of homogeneous polynomials:

7.11. Example: Zeros of polynomials . Let S be a finite set of points in $\mathbb{P}^n$, for $n \geq 2$, and

$$\omega_\mu(S) = \mathrm{Min}\{\ d > 0;\ \text{there exists } s \in H^0(\mathbb{P}^n, \mathcal{O}_{\mathbb{P}^n}(d)) \text{ with multiplicity at least } \mu \text{ in each } p \in S\}.$$

7.12. Claim. For $\mu' < \mu$ one has

$$\frac{\omega_{\mu'}(s)}{\mu' + n - 1} \leq \frac{\omega_\mu(s)}{\mu}.$$

PROOF.: For $d = \omega_\mu(S)$ we have a section $s \in H^0(\mathbb{P}^n, \mathcal{O}_{\mathbb{P}^n}(d))$ with divisor D' such that s has multiplicity at least μ in each $p \in S$. Choose

$$d' = [\frac{d}{\mu}(\mu' + n - 1)].$$

Since d' does not change if we replace d by $\nu \cdot d + 1$ and μ by $\nu \cdot \mu$ for $\nu \gg 0$, we can assume that $D' = D + H$ for a hyperplane H not meeting S.

For $\mathcal{L} = \mathcal{O}_{\mathbb{P}^n}(d' + 1)$, for the divisor $(d' + 1) \cdot D$, and for $N = d$, the assumptions of (7.8) hold true and (7.10) tells us that

$$\mathcal{O}_{\mathbb{P}^n}(n + d' + 1) \otimes \omega_{\mathbb{P}^n}\{\frac{-(d' + 1) \cdot D}{d}\}$$

is globally generated. Since $d' + 1 > \frac{d}{\mu}(\mu' + n - 1)$ and hence $\mu' + n - 1 \leq \frac{(d'+1)\cdot\mu}{d}$ we can apply (7.7,c) and find

$$\mathcal{O}_{\mathbb{P}^n}(n + d' + 1) \otimes \omega_{\mathbb{P}^n}\{\frac{-(d' + 1) \cdot D}{d}\}$$

to be a subsheaf of

$$\mathcal{O}_{\mathbb{P}^n}(d') \otimes \bigotimes_{p \in S} m_p^{\mu'}.$$

□

7.13. Remark. For some generalizations and improvements and for the history of this kind of problem see [18].

Up to this point, the applications discussed are based on the global vanishing theorems for invertible sheaves. J. Kollár's vanishing theorem (5.6,a) for restriction maps or, equivalently, vanishing theorems for the cohomology of higher direct image sheaves ((6.16) and (6.17)) are nice tools to study families of projective varieties:

7.14. Example: Families of varieties over curves .
Let X be a projective manifold, Z a non-singular curve and $f : X \to Z$ a surjective morphism. We call a locally free sheaf $\mathcal{F}$ on Z *semi-positive*, if for some (or equivalently: all) ample invertible sheaf $\mathcal{A}$ on Z and for all $\eta > 0$ the sheaf

$$S^\eta(\mathcal{F}) \otimes \mathcal{A}$$

is ample. One has

7.15. Theorem (Fujita [24]). *For $f : X \to Z$ as above $f_*\omega_{X/Z}$ is semi-positive.*

Here $\omega_{X/Z} = \omega_X \otimes f^*\omega_Z^{-1}$ is the dualizing sheaf of X over Z. In [40] J. Kollár used his vanishing theorem (5.6,a) to give a simple proof of (7.15) and of its generalization to higher dimensional Z, obtained beforehand by Y. Kawamata [35]. As usually, one obtains similar results adding the $\mathcal{L}^{(i)}$. For example, using the notations introduced in (7.4) one has:

7.16. Variant. *Assume in addition that $\mathcal{L}$ is an invertible sheaf, D an effective divisor and that $\mathcal{L}^N(-D)$ is semi-ample. Then one has*
a) *The sheaf $f_*(\mathcal{L} \otimes \omega_{X/Z}\{\frac{-D}{N}\})$ is semi-positive.*
b) *If for a general fibre F of f one has $N \geq e(D\,|_F)$, then $f_*(\mathcal{L} \otimes \omega_{X/Z})$ is semi-positive.*

7.17. Corollary. *Under the assumption of (7.16) assume that D contains a smooth fibre of f. Then, if $N \geq e(D\,|_F)$, the sheaf $f_*(\mathcal{L} \otimes \omega_{X/Z})$ is ample.*

Proof of (7.17): Recall that a vectorbundle $\mathcal{F}$ on Z is ample, if and only if $\tau^*\mathcal{F}$ is ample for some finite cover

$$\tau : Z' \longrightarrow Z.$$

Hence, if $\mathcal{L}'$, D', X' and f' are obtained by pullback from the corresponding objects over Z, it is enough to show that

$$\tau^* f_*(\mathcal{L} \otimes \omega_{X/Z}) = f'_*(\mathcal{L}' \otimes \omega_{X'/Z'})$$

is ample. If, for the ramification locus $\Delta(Z'/Z)$ of Z' over Z, the morphism f is smooth in a neighbourhood of $f^{-1}(\Delta(Z'/Z))$ then $f' : X' \longrightarrow Z'$ satisfies

again the assumption made in (7.14).
Choosing Z' to be ramified of order N over the point $p \in Z$ with $f^{-1}(p) \subseteq D$, we can reduce (7.17) to the case for which D contains the N-th multiple of a fibre, say $N \cdot f^{-1}(p)$. We have

$$f_*(\mathcal{L} \otimes \omega_{X/Z}\{-\frac{D}{N}\}) \subset f_*(\mathcal{L} \otimes \omega_{X/Z}) \otimes \mathcal{O}_Z(-p)$$

and this inclusion is an isomorphism over some open set. In fact, by (7.5,e) the assumption $N \geq e(D|_F)$ implies that $\omega_{X/Z}$ and $\omega_{X/Z}\{-\frac{D}{N}\}$ are the same in a neighbourhood of a general fibre F. Hence

$$f_*(\mathcal{L} \otimes \omega_{X/Z}) \otimes \mathcal{O}_Z(-p)$$

is semi-positive.

□

Proof of (7.16): As in the proof of (7.17) the assumption made in part b) implies that

$$f_*(\mathcal{L} \otimes \omega_{X/Z}\{\frac{-D}{N}\}) \subset f_*(\mathcal{L} \otimes \omega_{X/Z})$$

is an isomorphism over some non-empty open subvariety and b) follows from a).

By definition of $\omega_X\{\frac{-D}{N}\}$ we can assume D to be a normal crossing divisor. Moreover, we can assume that the multiplicities in D are strictly smaller than N and hence $\mathcal{L} = \mathcal{L}(-[\frac{D}{N}])$. Let $p \in Z$ be a point in general position and $F = f^{-1}(p)$. By (5.12,a) applied to the semiample sheaf $\mathcal{L}(F)$ we have a surjection

$$H^0(X, \mathcal{L}(F) \otimes \omega_X(F)) \longrightarrow H^0(F, \mathcal{L}(F) \otimes \omega_F)$$

and $f_*(\mathcal{L} \otimes \omega_X) \otimes \mathcal{O}_Z(2 \cdot p)$ is generatbd by global sections in a neighbourhood of p.

7.18. Claim. $\omega_Z(2 \cdot p) \otimes \bigotimes^\eta f_*(\mathcal{L} \otimes \omega_{X/Z})$ is semi-positive for all $\eta > 0$.

Proof: For $\eta = 1$ (7.18) holds true as we just found a trivial subsheaf $\bigoplus^r \mathcal{O}_Z$ of

$$f_*(\mathcal{L} \otimes \omega_X) \otimes \mathcal{O}_Z(2 \cdot p)$$

of full rank.
In general, let $Y' = X \times_Z \ldots \times_Z X$ (η-times) be the fibre product and $\delta : Y \to Y'$ be a desingularization. The induced morphisms $g' : Y' \to Z$ and $g : Y \to Z$ satisfy:

i) Y' is flat and Gorenstein over Z and

$$\omega_{Y'/Z} = \bigotimes_{j=1}^{\eta} pr_j^* \omega_{X/Z}.$$

ii) For

$$\mathcal{M}' = \bigotimes_{j=1}^{\eta} pr_j^* \mathcal{L} \quad \text{and} \quad \mathcal{M} = \delta^* \mathcal{M}'$$

one has an inclusion, surjective at the general point of Z,

$$g_*(\mathcal{M} \otimes \omega_{Y/Z}) \subset g'_*(\mathcal{M}' \otimes \omega_{Y'/Z}) = \bigotimes^{\eta} f_*(\mathcal{L} \otimes \omega_{X/Z})$$

iii) On a general fibre $\widetilde{F}$ of g the divisor

$$\Delta = \sum_{j=1}^{\eta} \delta^* pr_j^* D$$

has normal crossings, $[\frac{\Delta}{N}]|_{\widetilde{F}} = 0$, and $\mathcal{M}^N(-\Delta)$ is semi-ample.

In fact, i) is the compatibility of relative dualizing sheaves with pullback, ii) follows from flat base change and the inclusion $\delta_* \omega_Y \subset \omega_{Y'}$ and iii) is obvious. Using those three properties, (7.18) follows from the case "$\eta = 1$" applied to $g : Y \to Z$.

□

Since $S^\eta(\ )$ is a quotient of $\bigotimes^\eta(\ )$ and since the quotient of a semi-positive sheaf is again semipositive, one obtains (7.16).

□

7.19. Remarks.
a) If dim $Z > 1$, then the arguments used in the proof of (7.16,a) show that for all $\eta > 0$ and $\mathcal{H}$ very ample the reflexive hull $\mathcal{G}$ of

$$S^\eta(f_*(\mathcal{L} \otimes \omega_{X/Z})) \otimes \mathcal{H}^{\dim(X)+1} \otimes \omega_Z$$

is generated by $H^0(Z, \mathcal{G})$ over some open set. This led the second author to the definition "weakly-positive" (see [64]).
b) One can make (7.17) more explicit and measure the degree of ampleness by giving lower bound for the degree of invertible quotient sheaves of $f_*(\mathcal{L} \otimes \omega_{X/Z})$. Details have been worked out in [23]. These explicit bounds, together with the Kodaira-Spencer map can be used for families of curves over Z to give another proof of the Theorem of Manin saying that the Mordell conjecture holds true for curves over function fields over $\mathbb{C}$.

The vanishing statements for higher direct images (6.11) and its corollaries (7.5,d) and (7.6) are useful to study singularities.

7.20. Example: Deformation of quotient singularities.
Let X be a normal variety and $f : X \to S$ a flat morphism from X to a non-singular curve S.

7.21. Theorem. *Assume that X is normal and that for some $s_0 \in S$ the variety $X_0 = f^{-1}(s_0)$ is a reduced normal surface with quotient singularities. Then the general fibre $f^{-1}(\eta) = X_\eta$ has at most quotient singularities.*

PROOF (SEE [19]):
If Y is a normal surface with rational singularities then Mumford [47] has shown that for $p \in Y$,

$$\text{Spec } (\mathcal{O}_{Y,p}) - p = U$$

has only finitely many non-isomorphic invertible sheaves. Hence for some $N > 0$ one has $\omega_U^N = \mathcal{O}_U$ and for some $N > 0$ the reflexive hull $\omega_X^{[N]}$ of ω_Y^N is invertible. Let

$$\delta : Y' \longrightarrow Y$$

be a desingularization. Since Y has rational singularities we have

$$\delta_* \omega_{Y'} = \omega_Y$$

and

$$\delta^* \omega_Y / torsion \ \subset \omega_{Y'}.$$

We may assume that $\delta^* \omega_Y / torsion = \mathcal{K}$ is invertible and we write

$$\omega_{Y'} = \mathcal{K} \otimes \mathcal{O}_{Y'}(F).$$

With this notation we have for some effective divisor D

$$\delta^* \omega_Y^{[N]} = \mathcal{K}^N \otimes \mathcal{O}_{Y'}(D).$$

The divisors D and F can be used to characterize the quotient singularities among the rational singularities:

7.22. Claim. Y has quotient singularities if and only if $[\frac{D}{N}] \leq F$.

PROOF.: If one replaces D and N by some common multiple, the inequality in (7.22) is not affected. The question being local we may hence assume that N is the smallest integer with $\omega_Y^{[N]}$ invertible and $\omega_Y^{[N]} \simeq \mathcal{O}_Y$.

For $\mathcal{K}^{-1} = \mathcal{L}$ one has $\mathcal{L}^N = \mathcal{O}_{Y'}(D)$ and, as in (3.5), we can consider the cyclic cover Z' obtained by taking the N-th root out of D. Let Z be the normalization of Y in $k(Z')$ and

$$\begin{array}{ccc} Z' & \xrightarrow{\delta'} & Z \\ {\scriptstyle \pi'} \downarrow & & \downarrow {\scriptstyle \pi} \\ Y' & \xrightarrow[\delta]{} & Y \end{array}$$

the induced morphism (Z is usually called *the canonical covering* of Y). One has

$$\pi'_*\omega_{Z'} = \bigoplus_{i=0}^{N-1} \omega_{Y'} \otimes \mathcal{L}^{(i)}.$$

In fact, this follows from (3.11) by duality for finite morphisms (see [30], p. 239) or, since

$$\mathcal{L}^{-i}([\frac{i \cdot D}{N}] + D^{(i)}) = \mathcal{L}^{-i}(D - [\frac{(N-i) \cdot D}{N}]) = \mathcal{L}^{(N-i)}$$

from (3.16,d).
Recall that Z has rational singularities if and only if $\delta'_*\omega_{Z'} = \omega_Z$.
Assume that Y has a quotient singularity in the point p. If $\widetilde{U}$ is the universal cover of $Y - p$ then the normalization $\widetilde{Z}$ of Y in $k(\widetilde{U})$ is non singular and, by construction it dominates Z. Hence Z has quotient singularities and

$$\pi_*\delta'_*\omega_{Z'} = \delta_*\pi'_*\omega_{Z'}$$

is reflexive. In particular $\delta_*\omega_{Y'} \otimes \mathcal{L}^{(1)}$ is reflexive. One has

$$\omega_{Y'} \otimes \mathcal{L}^{(1)} = \mathcal{K} \otimes \mathcal{L} \otimes \mathcal{O}_{Y'}(F - [\frac{D}{N}]) = \mathcal{O}_{Y'}(F - [\frac{D}{N}])$$

and the reflexivity of $\delta_*\omega_{Y'} \otimes \mathcal{L}^{(1)}$ is equivalent to $F \geq [\frac{D}{N}]$.

On the other hand, $F \geq [\frac{D}{N}]$ implies that the summand $\delta_*\mathcal{O}_{Y'}(F - [\frac{D}{N}])$ of $\delta_*\pi'_*\omega_{Z'}$ has one section without zero on Y. Hence $\delta'_*\omega_{Z'}$ has a section without zero on $Z - \pi^{-1}(p)$, which implies that $\delta'_*\omega_{Z'}$ is invertible and coincides with ω_Z. So Z has a rational singularity and is Gorenstein. Those singularities are called rational double points, and they are known to be quotient singularities. Therefore Y has a quotient singularity as well.

□

PROOF OF (7.21): Let $\delta : X' \to X$ be a desingularization. We assume that the proper transform X'_0 of X_0 is non-singular and write $\delta_0 = \delta \mid_{X'_0}$.
By (7.5,d), applied in the case "$D = 0$", we have a natural inclusion

$$\delta_{0*}\omega_{X'_0} \longrightarrow \delta_*\omega_{X'} \otimes \mathcal{O}_X(X_0) \mid_{X_0} .$$

Since X_0 has rational singularities one has

$$\delta_{0*}\omega_{X'_0} = \omega_{X_0} = (\omega_X \otimes \mathcal{O}_X(X_0))|_{X_0}.$$

One obtains $\delta_*\omega_{X'} = \omega_X$, at least if one replaces S by a neighbourhood of s_0. Hence X and X_η have at most rational singularities.

Let us choose $N > 1$, such that both, $\omega_{X_\eta}^{[N]}$ and $\omega_{X_0}^{[N]}$ are invertible (It might happen, nevertheless, that $\omega_X^{[N]}$ is not invertible). We may assume that we have choosen X such that

$$\mathcal{K} = \delta^* \omega_X / torsion$$

is an invertible sheaf. Hence

$$\mathcal{K}_0 = \mathcal{K} \otimes \mathcal{O}_{X'}(\delta^* X_0) \mid_{X'_0}$$

is invertible and generated by global sections. Moreover, one has maps

$$\delta_0^* \omega_{X_0} = \delta^*(\omega_X \otimes \mathcal{O}_X(X_0))|_{X'_0} \longrightarrow \mathcal{K}_0$$

and $\mathcal{K}_0$ contains $\delta_0^* \omega_{X_0} / torsion$. Hence both sheaves must be the same.

Blowing up again, we can assume $\mathcal{M} = \delta^* \omega_X^{[N]} / torsion$ to be locally free and isomorphic to $\mathcal{K}^N(D)$ where D is a divisor in the exceptional locus of δ such that $X'_0 + D$ has at most normal crossings.

We can choose an embedding $\omega_X \hookrightarrow \mathcal{O}_X$ such that the zero-set does not contain X_0. If correspondingly $\mathcal{K} = \mathcal{O}_{X'}(-\Delta)$, for some $\Delta \geq 0$ we can choose the inclusion $\omega_X \hookrightarrow \mathcal{O}_X$ such that

$$D' = N \cdot \Delta - D \geq 0.$$

Blowing up we can assume that $D' + X'_0$ is a normal crossing divisor. By definition

$$\mathcal{O}_X(-D') = \mathcal{O}_X(-N \cdot \Delta + D) = \mathcal{M}$$

and $\mathcal{O}_X(-D')$ is δ-numerically effective.

It is our aim to use (7.6) in order to compare the sheaves $\mathcal{K}_0$ and $\delta_0^* \omega_{X_0}^{[N]}$ with $\mathcal{K}$ and with $\mathcal{M}$. Some unpleasant but elementary calculations will show that the inequality (7.22), applied to X_0, gives a similar inequality for the general fibre X_η.

Let us write

$$\omega_{X'_0} = \mathcal{K}_0(F_0) \text{ and } \delta_0^* \omega_{X_0}^{[N]} = \mathcal{K}_0^N \otimes \mathcal{O}_{X'_0}(D_0).$$

By (7.22) one has

$$F_0 \geq [\frac{1}{N} D_0].$$

Since

$$\mathcal{K}_0^N \otimes \mathcal{O}_{X'_0}(D \mid_{X'_0}) \simeq (\mathcal{K}^N \otimes \mathcal{O}_{X'}(D) \otimes \mathcal{O}_{X'}(+N \cdot \delta^* X_0)) \mid_{X'_0}$$

is a subsheaf of $\delta_0^* \omega_{X_0}^{[N]}$ one obtains that

$$(*) \qquad F_0 \geq [\frac{1}{N} D \mid_{X_0'}] .$$

On the other hand, since

$$\mathcal{K}_0^{N-1} = \mathcal{O}_{X_0'}(-(N-1)\Delta \mid_{X_0'}) \otimes \mathcal{O}_{X'}((N-1)\delta^* X_0) \mid_{X_0'}$$

and

$$[\frac{N-1}{N} D'] = (N-1)\Delta - D + [\frac{1}{N} D]$$

one has

$$\omega_{X_0'}(-[\frac{(N-1)}{N} D' \mid_{X_0'}]) = \mathcal{K}_0(F_0 - (N-1)\Delta \mid_{X_0'} + D \mid_{X_0'} - [\frac{1}{N} D \mid_{X_0'}]) =$$

$$\mathcal{K}_0^N (D \mid_{X_0'} + F_0 - [\frac{1}{N} D \mid_{X_0'}]) \otimes \mathcal{O}_{X'}(-(N-1)\delta^* X_0) \mid_{X_0'} =$$

$$\mathcal{M} \mid_{X_0'} (F_0 - [\frac{1}{N} D \mid_{X_0'}]) \otimes \mathcal{O}_{X'}(\delta^* X_0) \mid_{X_0'} .$$

By the inequality $(*)$ the sheaf

$$\delta_{0*}(\omega_{X_0'}(-[\frac{(N-1)}{N} D' \mid_{X_0'}]) \otimes \mathcal{O}_{X'}(-\delta^* X_0) \mid_{X_0'})$$

contains $\delta_{0*}(\mathcal{M} \mid_{X_0'})$. If F is the divisor with $\omega_{X'} = \mathcal{K} \otimes \mathcal{O}_{X'}(F)$ we get from (7.6) $\delta_{0*}(\mathcal{M} \mid_{X_0'})$ as a subsheaf of

$$\delta_* \omega_X(-[\frac{(N-1)}{N} D']) \mid_{X_0} = \delta_*(\mathcal{K} \otimes \mathcal{O}_{X'}(F) \otimes \mathcal{O}_{X'}(-(N-1)\Delta + D - [\frac{1}{N} D])) \mid_{X_0}$$

$$= \delta_*(\mathcal{K}^N \otimes \mathcal{O}_{X'}(D + F - [\frac{1}{N} D])) \mid_{X_0} = \delta_* \mathcal{M}(F - [\frac{1}{N} D]) \mid_{X_0} .$$

Of course, we have a natural morphism $\delta_* \mathcal{M} \longrightarrow \delta_{0*}(\mathcal{M} \mid_{X_0'})$ and the induced map

$$\delta_* \mathcal{M} \longrightarrow \delta_* \mathcal{M}(F - [\frac{1}{N} D]) \mid_{X_0}$$

is surjective outside of the singular locus of X_0. We have natural maps

$$\omega_X^{[N]} \longrightarrow \delta_* \delta^* \omega_X^{[N]} \longrightarrow \delta_* \mathcal{M} \longrightarrow \delta_* \mathcal{M}(F - [\frac{1}{N} D])|_{X_0} \longrightarrow \omega_X^{[N]}|_{X_0}.$$

The sheaf $\delta_* \mathcal{M}(F - [\frac{1}{N} D])$ is torsionfree and, since X_0 is a Cartier divisor, $\delta_* \mathcal{M}(F - [\frac{1}{N} D])|_{X_0}$ has no torsion as well. Therefore

$$\omega_X^{[N]}|_{X_0} = \delta_* \mathcal{M}(F - [\frac{1}{N} D])|_{X_0}.$$

Since $\mathcal{M} = \delta^* \omega_X^{[N]} / torsion$, this is only possible if $F \geq [\frac{1}{N} \cdot D]$. Hence $F_\eta \geq [\frac{D_\eta}{N}]$ where "η" denotes the restriction to the general fibre and the theorem follows from (7.22).

□

7.23. Example: Adjoint linear systems on surfaces.
Studying adjoint linear systems on higher dimensional manifolds, L. Ein and R. Lazarsfeld [15] realized, that (7.7, b and c) and (7.8) can be used to reprove part of I. Reider's theorem [53] and to obtain similar results for threefolds. We cordially thank them for allowing us to add their argument in the surface case to the final version of these notes.

7.24. Theorem (I. Reider, [53]). *Let S be a non-singular projective surface, defined over an algebraically closed field of characteristic zero, let $p \in S$ be a closed point and let $\mathcal{L}$ be a numerically effective invertible sheaf on S. Assume that $c_1(\mathcal{L})^2 > 4$ and that for all curves C with $p \in C \subset S$ one has $c_1(\mathcal{L}) \cdot C > 1$. Then there is a section $\sigma \in H^0(S, \mathcal{L} \otimes \omega_S)$ with $\sigma(p) \neq 0$.*

PROOF, FOLLOWING §1 OF [15]:
Let $\mathcal{H}$ be an ample invertible sheaf on S and let m_p be the ideal sheaf of p. For $\nu \gg 0$, one has $H^2(S, \mathcal{L}^\nu \otimes \mathcal{H}^{-1}) = 0$ and by the Riemann-Roch formula one finds $a, b \in \mathbb{N}$ with

$$h^0(S, \mathcal{H}^{-1} \otimes \mathcal{L}^\nu \otimes m_p^{2\cdot\nu}) \geq h^0(S, \mathcal{H}^{-1} \otimes \mathcal{L}^\nu) - h^0(S, \mathcal{O}_S/m_p^{2\cdot\nu})$$

$$\geq \frac{1}{2} \cdot c_1(\mathcal{L})^2 \cdot \nu^2 + a \cdot \nu + b - h^0(S, \mathcal{O}_S/m_p^{2\cdot\nu}).$$

Since

$$h^0(S, \mathcal{O}_S/m_p^{2\cdot\nu}) = \frac{1}{2} \cdot (4 \cdot \nu^2 + 2 \cdot \nu)$$

one finds for $\nu \gg 0$ a section s of $\mathcal{H}^{-1} \otimes \mathcal{L}^\nu$ with multiplicity $\mu_p \geq 2 \cdot \nu$ in p (see (7.7,b). Let

$$D = \Delta + \sum_{i=1}^{r} \nu_i \cdot D_i$$

be the zero-divisor of s, where Δ is an effective divisor not containing p and $p \in D_i$ for $i = 1 \cdots r$. If D' is any effective divisor the $\mathcal{H}^\eta(-D')$ will be ample for $\eta \gg 0$. Replacing D by $\eta \cdot D + D'$ and ν by $\nu \cdot \eta$ for a suitably choosen divisor D', we may assume that $r > 1$ and that $\nu_1 > \nu_2 > \cdots > \nu_r$. Of course we can also assume that μ_p is even.

7.25. Claim. If $\mu_p > 2 \cdot \nu_1$ then (7.24) holds true.

PROOF: Let $N = \frac{\mu_p}{2}$ By the choice of s one has $N \geq \nu$ and

$$\mathcal{L}^N(-D) = \mathcal{L}^{N-\nu} \otimes \mathcal{H}$$

is ample. By (7.8) one has

$$H^1(S, \omega_S\{\frac{-D}{N}\} \otimes \mathcal{L}) = 0.$$

By (7.7,b), or just by definition of $\omega_S\{\frac{-D}{N}\}$, one can find some open neighbourhood U of p such that

$$\omega_S\{\frac{-D}{N}\} \longrightarrow \omega_S$$

is an isomorphism on $U - p$. Moreover, by (7.7,c), the inclusion factors like

$$\omega_S\{\frac{-D}{N}\} \longrightarrow \omega_S \otimes m_p \longrightarrow \omega_S.$$

Let $\mathcal{M}$ be the sheaf $j_*j^*(\omega_S\{\frac{-D}{N}\}\otimes\mathcal{L})$ where $j : S-p \longrightarrow S$ is the inclusion. For some nontrivial skyscraper sheaf $\mathcal{C}$ supported in p, one has an exact sequence

$$0 \longrightarrow \omega_S\{\frac{-D}{N}\} \otimes \mathcal{L} \longrightarrow \mathcal{M} \longrightarrow \mathcal{C} \longrightarrow 0.$$

Hence, $\mathcal{M}$ has a section σ with $\sigma(p) \neq 0$.

□

It remains to consider the case where $\mu_p \leq 2 \cdot \nu_1$. If $\mu_p(D_i)$ denotes the multiplicity of D_i in p, then

$$\mu_p = \sum_{i=1}^{r} \nu_i \cdot \mu_p(D_i).$$

Since $r \geq 2$ this implies that $\mu_p(D_1) = 1$.

Let us take $N = \nu_1$. Again, $N \geq \nu$ and by (7.8)

$$H^1(S, \omega_S\{\frac{-D}{N}\} \otimes \mathcal{L}) = 0.$$

One has an inclusion

$$\omega_S\{\frac{-D}{N}\} \otimes \mathcal{L} \longrightarrow \omega_S \otimes \mathcal{L}(-D_1 - [\frac{\Delta}{N}])$$

whose cokernel is a skyscraper sheaf. Hence

$$H^1(S, \omega_S \otimes \mathcal{L}(-D_1 - [\frac{\Delta}{N}])) = 0$$

and the restriction map

$$H^0(S, \omega_S \otimes \mathcal{L}(-[\frac{\Delta}{N}])) \longrightarrow H^0(D_1, \omega_{D_1} \otimes \mathcal{L}(-D_1 - [\frac{\Delta}{N}])|_{D_1})$$

is surjective.

The right hand side contains a section σ with $\sigma(p) \neq 0$, since

$$\deg(\mathcal{L}(-D_1 - [\frac{\Delta}{N}])|_{D_1}) \geq 2.$$

In fact one has:

$$N\cdot\deg(\mathcal{L}(-D_1-[\frac{\Delta}{N}])|_{D_1}) = \deg(\mathcal{L}^N(-D)|_{D_1})+\sum_{i=2}^{r}\nu_i\cdot D_i\cdot D_1+(\Delta-[\frac{\Delta}{N}])\cdot D_1$$

$$\geq (N-\nu)\cdot c_1(\mathcal{L})\cdot D_1 + c_1(\mathcal{H})\cdot D_1 + \sum_{i=2}^{r}\nu_i\cdot D_i\cdot D_1$$

$$> (N-\nu)\cdot c_1(\mathcal{L})\cdot D_1 + \sum_{i=2}^{r}\nu_i\cdot\mu_p(D_i) = (N-\nu)\cdot c_1(\mathcal{L})\cdot D_1 + (\mu_p-\nu_1).$$

Since $c_1(\mathcal{L})\cdot D_1\geq 2$ and $\mu_p\geq 2\cdot\nu$, one obtains

$$N\cdot\deg(\mathcal{L}(-D_1-[\frac{\Delta}{N}])|_{D_1}) > 2\cdot N - 2\cdot\nu + \mu_p - N \geq N.$$

□

7.26. Remark. It is likely that the other parts of I. Reider's theorem [53], i.e. the lower bounds for $c_1(\mathcal{L})^2$ and for $c_1(\mathcal{L})\cdot C$ which imply that $H^0(S,\omega_S\otimes\mathcal{L})$ separates points and tangent directions, can be obtained in a similar way.

§ 8 Characteristic p methods: Lifting of schemes

Up to this point we did not prove the degeneration of the Hodge spectral sequence used in (3.2). Before doing so in Lecture 10 let us first recall what we want to prove.

8.1.

Let X be a proper smooth variety (or a scheme) over a field k. One introduces the *de Rham cohomology*

$$H^b_{DR}(X/k) := \mathbb{H}^b(X,\Omega^\bullet_{X/k})$$

where $\Omega^\bullet_{X/k}$ is the complex of regular differential forms, defined over k, the so called *de Rham complex.*

In order to compute it, one introduces the *"Hodge to de Rham" spectral sequence* associated to the Hodge filtration $\Omega^{\geq a}_{X/k}$ (see (A.25):

$$E_1^{ab} = H^b(X,\Omega^a_{X/k}) \Longrightarrow H^{a+b}_{DR}(X/k).$$

If $k=\mathbb{C}$, the field of complex numbers, the classical Hodge theory tells us that the Hodge spectral sequence

$$E_{1\,an}^{ab} = H^b(X_{an},\Omega^a_{X_{an}}) \Longrightarrow \mathbb{H}^{a+b}(X_{an},\Omega^\bullet_{X_{an}})$$

degenerates in E_1, where $\Omega^\bullet_{X_{an}}$ is the de Rham complex of holomorphic differential forms (see (A.25)).

In fact, one has

$$\mathbb{H}^{a+b}(X_{an}, \Omega^\bullet_{X_{an}}) = H^{a+b}(X_{an}, \mathbb{C})$$

and by Hodge theory

$$\dim\ H^l(X_{an}, \mathbb{C}) = \sum_{a+b=l} \dim\ H^b(X_{an}, \Omega^a_{X_{an}}).$$

As explained in (A.22), this equality is equivalent to the degeneration of $E_{1\ an}$.

As by Serre's GAGA theorems [56],

$$H^b(X_{an}, \Omega^a_{X_{an}}) = H^b(X, \Omega^a_{X/\mathbb{C}}),$$

the Hodge spectral sequence and the "Hodge to de Rham" spectral sequence coincide and therefore the second one degenerates in E_1 as well.

If k is any field of characteristic zero, one obtains the same result by flat base change:

8.2. Theorem. *Let X be a proper smooth variety over a field k of characteristic zero. Then the Hodge to de Rham spectral sequence degenerates in E_1 or, equivalently,*

$$\dim\ H^l_{DR}(X/k) = \sum_{a+b=l} \dim\ H^b(X, \Omega^a_{X/k}).$$

As we have already seen in Lecture 1 and 6, theorem (8.2) implies the Akizuki - Kodaira - Nakano vanishing theorem:

AKNV: *If $\mathcal{L}$ is ample invertible, then*

$$H^b(X, \Omega^a_{X/k} \otimes \mathcal{L}^{-1}) = 0 \quad for \quad a + b < \dim\ X$$

(where, of course, char $k = 0$*).*

Mumford [47] has shown that over a field k of characteristic $p > 0$ the E_1 degeneration for the Hodge to de Rham spectral sequence fails and, finally, Raynaud [52] gave a counterexample to AKNV in characteristic $p > 0$.

The aim of this and of the next three lectures is to present Deligne-Illusie's answer to those counterexamples:

8.3. Theorem (Deligne - Illusie [12]). *Let X be a proper smooth variety over a perfect field k of characteristic $p \geq$ dim X lifting to the ring $W_2(k)$ of the second Witt vectors (see (8.11)). Then both, the E_1-degeneration of the Hodge to de Rham spectral sequence and AKNV hold true.*

Actually they prove a slightly stronger version of (8.3), as will be explained later. Unfortunately one cannot derive from their methods the stronger vanishing theorems mentioned in Lecture 5 such as Grauert-Riemenschneider or Kawamata-Viehweg directly. As indicated, the geometric methods of the first part of these Lecture Notes fail as well. It is still an open problem which of those statements remains true under the assumptions of (8.3).

Finally, by standard techniques of reduction to characteristic $p > 0$, Deligne - Illusie show:

8.4. Proposition. *Theorem (8.2) and AKNV over a field k of characteristic zero are consequences of theorem (8.3).*

In the rest of this lecture, we will try to discuss to some extend elementary properties and examples of liftings to $W_2(k)$.

8.5. Liftings of a scheme.
Let S be a scheme defined over $\mathbb{F}_p$ the field with p elements.

8.6. Definition. A *lifting* of S to $\mathbb{Z}/p^2$ is a scheme $\widetilde{S}$, defined and flat over $\mathbb{Z}/p^2$, such that $S = \widetilde{S} \times_{\mathbb{Z}/p^2} \mathbb{F}_p$.

8.7. Properties.
a) S is defined by a nilpotent ideal sheaf (of square zero) in $\widetilde{S}$. In particular the inclusion $S \subset \widetilde{S}$ or, if one prefers, the projection

$$\mathcal{O}_{\widetilde{S}} \longrightarrow \mathcal{O}_S$$

induces the identity on the *underlying topological spaces* $(S)_{top}$ and $(\widetilde{S})_{top}$.
b) From the exact sequence of $\mathbb{Z}/p^2$-modules

$$0 \longrightarrow p \cdot \mathbb{Z}/p^2 \longrightarrow \mathbb{Z}/p^2 \longrightarrow \mathbb{Z}/p \longrightarrow 0$$

one obtains, since $\mathcal{O}_{\widetilde{S}}$ is flat over $\mathbb{Z}/p^2$, the exact sequence of $\mathcal{O}_{\widetilde{S}}$-modules

$$0 \longrightarrow p \cdot \mathcal{O}_{\widetilde{S}} \longrightarrow \mathcal{O}_{\widetilde{S}} \longrightarrow \mathcal{O}_S \longrightarrow 0$$

and, from the isomorphism of $\mathbb{Z}/p^2$-modules

$$p : \mathbb{Z}/p \longrightarrow p \cdot \mathbb{Z}/p^2,$$

one obtains the isomorphism of $\mathcal{O}_{\widetilde{S}}$-modules

$$p : \mathcal{O}_S \longrightarrow p \cdot \mathcal{O}_{\widetilde{S}}.$$

8.8. Example. Let k be a perfect field of characteristic p and $S = \text{Spec } k$. Then $\widetilde{S}$ exists and is uniquely determined (up to isomorphism) by (8.7,b):

$\widetilde{S} = \text{Spec } W_2(k)$, where $W_2(k)$ is called the *ring of the second Witt vectors* of k.

In concrete terms, $W_2(k) = k \oplus k \cdot p$ as additive group and the multiplication is defined by

$$(x + y \cdot p)(x' + y' \cdot p) = x \cdot x' + (x \cdot y' + x' \cdot y) \cdot p.$$

8.9. Assumptions. Throughout Lectures 8 to 11 S will be a *noetherian scheme over* $\mathbb{F}_p$ *with a lifting* $\widetilde{S}$ *to* $\mathbb{Z}/p^2$.
X will denote a noetherian S-scheme, $D \subset X$ will be a reduced Cartier divisor. X will be supposed to be *smooth over* S, which means that locally X is étale over the affine space $\mathbf{A}^n_S$ over S (here $n = \dim_S X$). *D will be a normal crossing divisor over S*, i.e.:
D is the union of smooth divisors D_i over S and one can choose the previous étale cover such that the coordinates of $\mathbf{A}^n_S$ pull back to a parameter system $(t_1, \ldots, t_n)$ on X, for which D is defined by

$$t_1 \cdot \ldots \cdot t_r, \quad \text{for some} \quad r \leq n.$$

We allow D to be empty.

8.10. Definition. For X smooth and D a normal crossing divisor over S, we define the sheaf $\Omega^1_{X/S}$ (log D) of *one forms with logarithmic poles along D* as the $\mathcal{O}_X$-sheaf generated locally by

$$\frac{dt_i}{t_i}\ , \text{ for } \ i \leq r\ , \text{ and by } \ dt_i\ , \text{ for } \ i > r$$

(where we use the notation from (8.9)). $\Omega^1_{X/S}(\log\ D)$ is locally free of rank n and the definition coincides with the one given in (2.1) for $S = \text{Spec } k$. Finally we define

$$\Omega^a_{X/S}(\log\ D) = \bigwedge^a \Omega^1_{X/S}(\log\ D).$$

8.11. Definition. A *lifting of*

$$D = \sum_{j=1}^{r} D_j \subset X$$

to $\widetilde{S}$ consists of a scheme $\widetilde{X}$ and subschemes $\widetilde{D}_j$ of $\widetilde{X}$, all defined and flat over $\widetilde{S}$ such that $X = \widetilde{X} \times_{\widetilde{S}} S$ and $D_j = \widetilde{D}_j \times_{\widetilde{S}} S$. We write

$$\widetilde{D} = \sum_{j=r}^{r} \widetilde{D}_j.$$

If k is a perfect field of characteristic p and $S = \text{Spec } k$, we say that (X, D) *admits a lifting to* $W_2(k)$ if liftings $\widetilde{X}$ and $\widetilde{D}_j$ exist over $\widetilde{S} = \text{Spec } W_2(k)$.

If $\mathcal{L}$ is an invertible sheaf on X, we say that X and $\mathcal{L}$ *admit a lifting to* $W_2(k)$ if there is a lifting $\widetilde{X}$ of X over $\widetilde{S} = \text{Spec } W_2(k)$ and an invertible sheaf $\widetilde{\mathcal{L}}$ on $\widetilde{X}$ with $\widetilde{\mathcal{L}}|_X = \mathcal{L}$.

8.12. Remark. Of course, $\widetilde{X}$ is also a lifting of the $\mathbb{F}_p$-scheme X to a scheme $\widetilde{X}$ over $\mathbb{Z}/p^2$. In particular, (8.7.a) remains true and we have

$$(\widetilde{X})_{top} = (X)_{top} \quad \text{and} \quad (\widetilde{D})_{top} = (D)_{top}.$$

One can make (8.7,b) more precise:

8.13. Lemma. *Let X be smooth over S and let $\widetilde{X}$ be a scheme over $\widetilde{S}$ with $\widetilde{X} \times_{\widetilde{S}} S = X$. Then the following conditions are equivalent.*

a) *$\widetilde{X}$ is smooth over $\widetilde{S}$.*

b) *$\widetilde{X}$ is a lifting of X to $\widetilde{S}$.*

c) *There is an exact sequence of $\mathcal{O}_{\widetilde{X}}$-modules*

$$0 \longrightarrow p \cdot \mathcal{O}_{\widetilde{X}} \longrightarrow \mathcal{O}_{\widetilde{X}} \xrightarrow{\ r\ } \mathcal{O}_X \longrightarrow 0$$

together with an $\mathcal{O}_{\widetilde{X}}$-isomorphism

$$p : \mathcal{O}_X \longrightarrow p \cdot \mathcal{O}_{\widetilde{X}}$$

satisfying

$$p(x) = p \cdot \widetilde{x} \ , \ \text{for} \ \ \widetilde{x} \in \mathcal{O}_{\widetilde{X}} \ , \ \text{and} \ \ x = r(\widetilde{x}).$$

d) *If $\widetilde{U} \subseteq \widetilde{X}$ is an open subscheme, U its image in X,*

$$\pi : U \longrightarrow \mathbf{A}^n_S = \mathbf{Spec}\ \mathcal{O}_S[t_1, \ldots, t_n]$$

an étale morphism and if $\widetilde{\varphi}_1, \ldots, \widetilde{\varphi}_n \in \mathcal{O}_{\widetilde{U}}$ satisfy $r(\widetilde{\varphi}_i) = \varphi_i = \pi^* t_i$, then π extends to an étale morphism

$$\widetilde{\pi} : \widetilde{U} \longrightarrow \mathbf{A}^n_{\widetilde{S}} = \mathbf{Spec}\ \mathcal{O}_{\widetilde{S}}[t_1, \ldots, t_n]$$

with $\widetilde{\pi}^*(t_i) = \widetilde{\varphi}_i$ for $i = 1, \ldots, n$.

e) *For each $a \geq 0$ one has an exact sequence of $\mathcal{O}_{\widetilde{X}}$-modules*

$$0 \longrightarrow p \cdot \Omega^a_{\widetilde{X}/\widetilde{S}} \longrightarrow \Omega^a_{\widetilde{X}/\widetilde{S}} \xrightarrow{\ r\ } \Omega^a_{X/S} \longrightarrow 0$$

and an $\mathcal{O}_{\widetilde{X}}$*-isomorphism*

$$p : \Omega^a_{X/S} \longrightarrow p \cdot \Omega^a_{\widetilde{X}/\widetilde{S}}$$

satisfying

$$p(\omega) = p \cdot \widetilde{\omega} \text{ , for } \widetilde{\omega} \in \Omega^a_{\widetilde{X}/\widetilde{S}} \text{ , and } \omega = r(\widetilde{\omega}).$$

PROOF: A smooth morphism is flat, and flatness implies c). Obviously d) implies e) and a). Hence the only part to prove is that c) implies d).
Using the notations from d) (for $\widetilde{X} = \widetilde{U}$) we can, of course, define

$$\widetilde{\pi} : \widetilde{X} \longrightarrow \mathbf{A}^n_{\widetilde{S}} \text{ with } \widetilde{\pi}^*(t_i) = \widetilde{\varphi}_i.$$

Given a relation $\sum \widetilde{\lambda}_\nu \widetilde{m}_\nu = 0$ in $\mathcal{O}_{\widetilde{X}}$ between different monomials $\widetilde{m}_\nu$ in $\widetilde{\varphi}_1, \ldots, \widetilde{\varphi}_n$ the exact sequence in c) implies $\widetilde{\lambda}_\nu = p \cdot \widetilde{\mu}_\nu$ for $\widetilde{\mu}_\nu \in \mathcal{O}_{\widetilde{X}}$ and the isomorphism in c) shows that one has

$$\sum \mu_\nu \cdot m_\nu = 0 \text{ for } \mu_\nu = r(\widetilde{\mu}_\nu) \text{ and } m_\nu = r(\widetilde{m}_\nu).$$

Hence $\mu_\nu = 0$ as well as $\widetilde{\mu}_\nu = 0$.

If $g_1, \ldots, g_r$ are locally independent generators of $\mathcal{O}_X$ as a $\mathcal{O}_{\mathbf{A}^n_S}$-module, and if $\widetilde{g}_1, \ldots, \widetilde{g}_r$ are liftings to $\mathcal{O}_{\widetilde{X}}$, then each $\widetilde{x} \in \mathcal{O}_{\widetilde{X}}$ verifies

$$x = r(\widetilde{x}) = \sum_{i=1}^{r} \lambda_i g_i$$

for some $\lambda_i \in \mathcal{O}_{\mathbf{A}^n_S}$. If $\widetilde{\lambda}_1, \ldots, \widetilde{\lambda}_r$ are liftings of $\lambda_1, \ldots, \lambda_r$ to $\mathcal{O}_{\mathbf{A}^n_{\widetilde{S}}}$, then

$$\widetilde{x} - \sum_{i=1}^{r} \widetilde{\lambda}_i \widetilde{g}_i \in p \cdot \mathcal{O}_{\widetilde{X}}$$

and one can find $\widetilde{\mu}_i \in \mathcal{O}_{\mathbf{A}^n_{\widetilde{S}}}$ with

$$\widetilde{x} - \sum_{i=1}^{r} \widetilde{\lambda}_i \widetilde{g}_i = p(\sum_{i=1}^{r} \mu_i g_i) = \sum_{i=1}^{r} p \cdot \widetilde{\mu}_i \widetilde{g}_i,$$

and

$$\widetilde{x} = \sum_{i=1}^{r} (\widetilde{\lambda}_i + p \cdot \widetilde{\mu}_i) \cdot \widetilde{g}_i.$$

In other terms, $\widetilde{g}_1, \ldots \widetilde{g}_r$ are generators of $\mathcal{O}_{\widetilde{X}}$ as a $\mathcal{O}_{\mathbf{A}^n_{\widetilde{S}}}$-module. They are independent by the same argument which gave the independence of the m_μ

above. $\mathcal{O}_{\widetilde{X}}$ as a free $\mathcal{O}_{\mathbf{A}^n_{\widetilde{S}}}$-module is flat.

Finally, (locally in X)

$$\Omega^1_{X/S} = \pi^*\Omega^1_{\mathbf{A}^n_S} = \bigoplus_{1=1}^{n} \mathcal{O}_X d\varphi_i$$

and

$$\widetilde{\pi}^*\Omega^1_{\mathbf{A}^n_{\widetilde{S}}} = \bigoplus_{i=1}^{n} \mathcal{O}_{\widetilde{X}} d\widetilde{\varphi}_i$$

surjects to $\Omega^1_{\widetilde{X}/\widetilde{S}}$. In fact, if $\widetilde{\omega} \in \Omega^1_{\widetilde{X}/\widetilde{S}}$,

$$\widetilde{\omega} - \sum_{i=1}^{n} \widetilde{\lambda}_i d\widetilde{\varphi}_i \in \operatorname{im}(\Omega^1_{X/S} \xrightarrow{\cdot p} \Omega^1_{\widetilde{X}/\widetilde{S}})$$

for some $\widetilde{\lambda}_i \in \mathcal{O}_{\widetilde{X}}$ and, as above, one can modify the $\widetilde{\lambda}_i$ to get

$$\widetilde{\omega} = \sum_{i=1}^{n} (\widetilde{\lambda}_i + p \cdot \widetilde{\mu}_i) d\widetilde{\varphi}_i.$$

As $\widetilde{\pi}^*\Omega^1_{\mathbf{A}^n_{\widetilde{S}}} \longrightarrow \Omega^1_{\widetilde{X}/\widetilde{S}}$ is injective as well, $\widetilde{\pi}$ is étale.

□

8.14. Lemma. *Let X be a smooth S-scheme and*

$$D = \sum_{j=1}^{r} D_j$$

be a normal crossing divisor over S. Let $\widetilde{X}$ be a lifting of X to $\widetilde{S}$ and $\widetilde{D}_j \subseteq \widetilde{X}$ subschemes with

$$\widetilde{D}_j \otimes_{\widetilde{S}} S = D_j$$

for $j = 1, \ldots, r$. Then the following conditions are equivalent:

a)

$$\widetilde{D} = \sum_{j=1}^{r} \widetilde{D}_j \subset \widetilde{X}$$

is a lifting of $D \subset X$ to $\widetilde{S}$.

b) *The components of $\widetilde{D}$ are Cartier divisors in $\widetilde{X}$.*

c) *If in (8.13,d) one knows that $D \mid_U$ is the zero-set of $\varphi_1 \cdot \ldots \cdot \varphi_s$, then one can choose $\widetilde{\pi} : \widetilde{U} \longrightarrow \mathbf{A}^n_{\widetilde{S}}$, such that $\widetilde{D}_j \mid_{\widetilde{U}}$ is the zero set of $\widetilde{\pi}^*(t_j)$.*

Proof: If $\widetilde{D} \subset \widetilde{X}$ is a lifting of $D \subset X$, then the flatness of $\widetilde{X}$ and $\widetilde{D}_j$ over $\widetilde{S}$ implies that the ideal sheaf $J_{\widetilde{D}_j}$ of $\widetilde{D}_j$ is flat over $\widetilde{S}$. We have again an exact sequence

$$0 \longrightarrow p \cdot J_{\widetilde{D}_j} \longrightarrow J_{\widetilde{D}_j} \longrightarrow J_{D_j} \longrightarrow 0$$

where J_{D_j} is the ideal sheaf of D_j, and an isomorphism

$$p : J_{D_j} \longrightarrow p \cdot J_{\widetilde{D}_j}.$$

If $\widetilde{\varphi}_j$ is a lifting of φ_j to $J_{\widetilde{D}_j}$, then for any $\widetilde{g} \in J_{\widetilde{D}_j}$ one has $g = \lambda \cdot \varphi_j$ and

$$\widetilde{g} - \widetilde{\lambda} \cdot \widetilde{\varphi}_j \in p \cdot I_{\widetilde{D}_j}$$

is of the form $p \cdot \widetilde{\mu} \cdot \widetilde{\varphi}_j = p(\mu \cdot \varphi_j)$ for some $\widetilde{\mu} \in \mathcal{O}_{\widetilde{X}}$. Hence $\widetilde{\varphi}_j$ is a defining equation for $\widetilde{D}_j$.
By (8.13,d) b) implies c) and obviously c) implies a).

□

8.15. Definition. Using the notations from (8.14,c) and (8.13,d) we define for a lifting $\widetilde{D} \subset \widetilde{X}$ of $D \subset X$ to $\widetilde{S}$ the sheaf

$$\Omega^1_{\widetilde{X}/\widetilde{S}}(\log\ \widetilde{D})$$

to be the $\mathcal{O}_{\widetilde{X}}$-sheaf generated by

$$\frac{d\widetilde{\varphi}_j}{\widetilde{\varphi}_j} \quad \text{for} \quad j = 1, \ldots, s \quad \text{and} \quad d\widetilde{\varphi}_j \quad \text{for} \quad j = s+1, \ldots, n.$$

8.16. Properties.
a) For all a the sheaves

$$\Omega^a_{\widetilde{X}/\widetilde{S}}(\log\ \widetilde{D}) = \bigwedge^a \Omega^1_{\widetilde{X}/\widetilde{S}}(\log\ \widetilde{D})$$

are locally free over $\mathcal{O}_{\widetilde{X}}$.
b) One has an exact sequence of $\mathcal{O}_{\widetilde{X}}$-modules

$$0 \longrightarrow p \cdot \Omega^a_{\widetilde{X}/\widetilde{S}}(\log\ \widetilde{D}) \longrightarrow \Omega^a_{\widetilde{X}/\widetilde{S}}(\log\ \widetilde{D}) \longrightarrow \Omega^a_{X/S}(\log\ D) \longrightarrow 0$$

and an $\mathcal{O}_{\widetilde{X}}$-isomorphism

$$p : \Omega^a_{X/S}(\log\ D) \longrightarrow p \cdot \Omega^a_{\widetilde{X}/\widetilde{S}}(\log\ \widetilde{D}).$$

8.17. Proposition. *Let X be a smooth S-scheme.*

a) *Locally in the Zariski topology X has a lifting $\widetilde{X}$ to $\widetilde{S}$.*

b) *If $\widetilde{X}$ is a lifting of X to $\widetilde{S}$, if $\widetilde{X}$ is affine and Y a complete intersection in X, then there exists a lifting $\widetilde{Y}$ of Y to $\widetilde{S}$ and an embedding $\widetilde{Y} \subset \widetilde{X}$.*

c) *In particular, if D is a S-normal crossing divisor on X then locally in the Zariski topology $D \subset X$ has a lifting $\widetilde{D} \subset \widetilde{X}$ to $\widetilde{S}$.*

PROOF: Locally X is a complete intersection in an affine space over S. Hence a) follows from b). In b) we may assume that Y is a divisor, let us say the zero set of $\varphi \in \mathcal{O}_X$. We can choose $\widetilde{Y}$ to be the zero set of any lifting $\widetilde{\varphi} \in \mathcal{O}_{\widetilde{X}}$ of φ. In fact, the flatness follows easily from (8.13,c) or from the following argument. Choose

$$\pi : X \longrightarrow \mathbf{A}^n_S = \text{Spec } \mathcal{O}_X[t_1, \ldots, t_n]$$

with $\varphi = \pi^*(t_1)$. By (8.13,d) π extends to an étale map $\widetilde{\pi} : \widetilde{X} \longrightarrow \mathbf{A}^n_{\widetilde{S}}$ with $\widetilde{\varphi} = \widetilde{\pi}^*(t_1)$.

□

8.18. Isomorphisms between liftings

Let in the sequel X be a smooth S-scheme, $D \subset X$ be an S-normal crossing divisor and let, for $i = 1, 2$, $\widetilde{D}^{(i)} \subset \widetilde{X}^{(i)}$ be two liftings of $D \subset X$ to $\widetilde{S}$.

8.19. Notations. A morphism $u : \widetilde{X}^{(1)} \to \widetilde{X}^{(2)}$ is called an *isomorphism of liftings*

$$u : (\widetilde{X}^{(1)}, \widetilde{D}^{(1)}) \longrightarrow (\widetilde{X}^{(2)}, \widetilde{D}^{(2)})$$

if $u \mid_X = id_X$ and if

$$u^*(\mathcal{O}_{\widetilde{X}^{(2)}}(-\widetilde{D}^{(2)})) = \mathcal{O}_{\widetilde{X}^{(1)}}(-\widetilde{D}^{(1)}).$$

8.20. Remark. We have seen in (8.12) that $(\widetilde{X}^{(i)})_{top} = (X)_{top}$ and hence u is the identity on the topological spaces. Henceforth, giving u is the same as giving the morphism

$$u^* : \mathcal{O}_{\widetilde{X}^{(2)}} \longrightarrow \mathcal{O}_{\widetilde{X}^{(1)}}$$

of sheaves of rings on $(X)_{top}$. The assumption $u|_X = id_X$ forces u^* to be an isomorphism.

8.21. Lemma. *Locally in the Zariski topology there exists an isomorphism of liftings*

$$u : (\widetilde{X}^{(1)}, \widetilde{D}^{(1)}) \longrightarrow (\widetilde{X}^{(2)}, \widetilde{D}^{(2)}).$$

PROOF: Locally the diagonal $\Delta \subset X \times X$ is a complete intersection and we can lift it to

$$\widetilde{\Delta} \subset \widetilde{X}^{(1)} \times \widetilde{X}^{(2)}.$$

For example, if $\varphi_1, \ldots, \varphi_n$ are local parameters on X and $\widetilde{\varphi}_1^{(i)}, \ldots, \widetilde{\varphi}_n^{(i)}$ liftings in $\mathcal{O}_{\widetilde{X}^{(i)}}$ such that $D^{(i)}$ is the zero locus of $\varphi_1^{(i)} \cdot \ldots \cdot \varphi_s^{(i)}$, then we can choose $\widetilde{\Delta}$ to be defined by

$$\widetilde{\varphi}_j^{(1)} \otimes 1 - 1 \otimes \widetilde{\varphi}_j^{(2)} \quad \text{for} \quad j = 1, \ldots, n.$$

We have isomorphisms of liftings

$$p_1 : \widetilde{\Delta} \longrightarrow \widetilde{X}^{(1)}, \quad p_2 : \widetilde{\Delta} \longrightarrow \widetilde{X}^{(2)}$$

and $u = p_2 \circ p_1^{-1}$ satisfies

$$u^*(\mathcal{O}_{\widetilde{X}^{(2)}}(-\widetilde{D}^{(2)})) = \mathcal{O}_{\widetilde{X}^{(1)}}(-\widetilde{D}^{(1)}).$$

□

Let

$$u, v : (\widetilde{X}^{(1)}, \widetilde{D}^{(1)}) \longrightarrow (\widetilde{X}^{(2)}, \widetilde{D}^{(2)})$$

be two isomorphisms of liftings. For $\widetilde{x} \in \mathcal{O}_{\widetilde{X}^{(2)}}$ one has

$$(u^* - v^*)(p \cdot \widetilde{x}) = p(u^* - v^*)(\widetilde{x}) = p(id - id)(x) = 0$$

therefore $(u^* - v^*)|_{p \cdot \mathcal{O}_{\widetilde{X}^{(2)}}} = 0$. Of course, the map

$$\mathcal{O}_X = \mathcal{O}_{\widetilde{X}^{(2)}}/p \cdot \mathcal{O}_{\widetilde{X}^{(2)}} \longrightarrow \mathcal{O}_X = \mathcal{O}_{\widetilde{X}^{(1)}}/p \cdot \mathcal{O}_{\widetilde{X}^{(1)}}$$

induced by $(u^* - v^*)$ is zero as well, and $(u^* - v^*)$ factors through

$$(u^* - v^*) : \mathcal{O}_X \longrightarrow p \cdot \mathcal{O}_{\widetilde{X}^{(1)}} = p(\mathcal{O}_X).$$

For $x, y \in \mathcal{O}_X$ with liftings $\widetilde{x}, \widetilde{y} \in \mathcal{O}_{\widetilde{X}^{(2)}}$ one has

$$(u^* - v^*)(\widetilde{x} \cdot \widetilde{y}) = u^*(\widetilde{x}) \cdot u^*(\widetilde{y}) - v^*(\widetilde{x}) \cdot v^*(\widetilde{y}) = x \cdot (u^* - v^*)(\widetilde{y}) + y \cdot (u^* - v^*)(\widetilde{x}).$$

Hence $p^{-1} \circ (u^* - v^*) : \mathcal{O}_X \longrightarrow \mathcal{O}_X$ is a derivation and factors through

$$\mathcal{O}_X \xrightarrow{d} \Omega^1_{X/S} \longrightarrow \mathcal{O}_X$$

where we denote the second morphism by $(u^* - v^*)$ again. If $\widetilde{t}_j^{(i)}$ is a local equation for $\widetilde{D}_j^{(i)}$, $i = 1, 2$, with reduction $t_j \in \mathcal{O}_X$, then

$$u^*(\widetilde{t}_j^{(2)}) = \widetilde{t}_j^{(1)} \cdot (1 + p \cdot \lambda)$$

and

$$v^*(\tilde{t}_j^{(2)}) = \tilde{t}_j^{(1)}(1 + p \cdot \mu).$$

Hence

$$(u^* - v^*)(dt_j) = t_j \cdot (\lambda - \mu) \in t_j \cdot \mathcal{O}_X$$

and $p^{-1} \circ (u^* - v^*)$ even factors through

$$\mathcal{O}_X \xrightarrow{d} \Omega^1_{X/S}(\log\ D) \longrightarrow \mathcal{O}_X.$$

8.22. Proposition. *Keeping the notations from (8.19) let*

$$u : (\tilde{X}^{(1)}, \tilde{D}^{(1)}) \longrightarrow (\tilde{X}^{(2)}, \tilde{D}^{(2)})$$

be an isomorphism of liftings. Then

$$\{v : (\tilde{X}^{(1)}, \tilde{D}^{(1)}) \longrightarrow (\tilde{X}^{(2)}, \tilde{D}^{(2)});\ v \textit{ isomorphism of liftings}\ \}$$

is described by the affine space

$$u^* + \mathrm{Hom}_{\mathcal{O}_X}(\Omega^1_{X/S}(\log\ D), \mathcal{O}_X).$$

PROOF: It just remains to show that for

$$\varphi \in \mathrm{Hom}_{\mathcal{O}_X}(\Omega^1_{X/S}(\log\ D), \mathcal{O}_X)$$

we can find v. Define

$$v^* : \mathcal{O}_{\tilde{X}^{(2)}} \longrightarrow \mathcal{O}_{\tilde{X}^{(1)}}$$

by

$$v^*(\tilde{x}) = u^*(\tilde{x}) - p \cdot \varphi(dx).$$

If $\tilde{t}_j^{(i)}$ is as above an equation of $\tilde{D}_j^{(i)}$,

$$v^*(\tilde{t}_j^{(2)}) = \tilde{t}_j^{(1)} \cdot (1 + p \cdot \lambda) - p \cdot t \cdot \gamma \ \text{ for some } \ \gamma \in \mathcal{O}_X,$$

or

$$v^*(\tilde{t}_j^{(2)}) = \tilde{t}_j^{(1)}(1 + p(\lambda - \gamma))$$

and v^* satisfies the conditions posed in (8.19.).

□

8.23. Proposition. *Let X be an affine scheme, smooth over S and let D be a normal crossing divisor over S. Let $\tilde{D}^{(i)} \subset \tilde{X}^{(i)}$ be two liftings of $D \subset X$ to $\tilde{S}$. Then there exists an isomorphism of liftings*

$$u : (\tilde{X}^{(1)}, \tilde{D}^{(1)}) \longrightarrow (\tilde{X}^{(2)}, \tilde{D}^{(2)}).$$

PROOF: Of course, (8.22) just says that the isomorphisms of liftings over a fixed open set form a "torseur" under the group

$$\mathrm{Hom}_{\mathcal{O}_X}(\Omega^1_{X/S}(\log\ D), \mathcal{O}_X)$$

and, since X is affine and

$$H^1(X, \mathcal{H}om_{\mathcal{O}_X}(\Omega^1_{X/S}(\log\ D), \mathcal{O}_X)) = 0$$

one obtains (8.23). However, to state this in the elementary language used up to now, let us avoid this terminology:

From (8.21) we know that there is an affine open cover $\mathcal{U} = \{X_\alpha\}$ of X and isomorphisms of liftings

$$u_\alpha : (\widetilde{X}^{(1)}_\alpha, \widetilde{D}^{(1)}_\alpha) \longrightarrow (\widetilde{X}^{(2)}_\alpha, \widetilde{D}^{(2)}_\alpha).$$

By (8.22) $p^{-1} \circ (u^*_\alpha - u^*_\beta)$ defines a 1-cocycle with values in the sheaf

$$\mathcal{H}om_{\mathcal{O}_X}(\Omega^1_{X/S}(\log\ D), \mathcal{O}_X).$$

Since

$$H^1(X, \mathcal{H}om_{\mathcal{O}_X}(\Omega^1_{X/S}(\log\ D), \mathcal{O}_X) = 0$$

we find

$$\varphi_\alpha \in \Gamma(X_\alpha, \mathcal{H}om_{\mathcal{O}_X}(\Omega^1_{X/S}(\log\ D), \mathcal{O}_X))$$

in some possibly finer cover $\{X_\alpha\}$ such that

$$p^{-1} \circ (u^*_\alpha - u^*_\beta) = \varphi_\alpha - \varphi_\beta.$$

Hence the isomorphisms of liftings $u^{*'}_\alpha - p \cdot \varphi^*_\alpha$ glue together to $u : \widetilde{X}_1 \to \widetilde{X}_2$. □

§ 9 The Frobenius and its liftings

Everything in this lecture is either elementary or taken from [12].

Let S be a noetherian scheme defined over $\mathbb{F}_p$ and let X be a noetherian S-scheme.

9.1. Definition. The *absolute Frobenius of S* is the endomorphism

$$F_S : S \longrightarrow S$$

defined by the following conditions.
F is the identity on the topological space and

$$F^*_S : \mathcal{O}_S \longrightarrow \mathcal{O}_S$$

is given by $F_S^*(a) = a^p$. In particular for $x \in \mathcal{O}_X$ and $\lambda \in \mathcal{O}_S$ one has

$$F_X^*(\lambda x) = \lambda^p x^p = F_S^*(\lambda) \cdot F_X^*(x),$$

and therefore one has a commutative diagram

$$\begin{array}{ccc} X & \xrightarrow{F_X} & X \\ {\scriptstyle f}\downarrow & & \downarrow{\scriptstyle f} \\ S & \xrightarrow[F_S]{} & S \end{array}$$

For $X' = X \times_{F_S} S$ this allows to factorize F_X:

$$\begin{array}{ccccc} X & \xrightarrow{F} & X' & \xrightarrow{pr_1} & X \\ & {\scriptstyle f}\searrow & {\scriptstyle f'}\downarrow & & \downarrow{\scriptstyle f} \\ & & S & \xrightarrow[F_S]{} & S \end{array}$$

with $F_X = pr_1 \circ F$ and $f' = pr_2$. By abuse of notations we write

$$F_S = pr_1 : X' \longrightarrow X.$$

F is called the *relative Frobenius* (relative to S). For

$$x \otimes \lambda \in \mathcal{O}_{X'} = \mathcal{O}_X \otimes_{F_S} \mathcal{O}_S \quad \text{one has} \quad F^*(x \otimes \lambda) = x^p \cdot \lambda$$

and for

$$x \in \mathcal{O}_X \quad \text{one has} \quad F_S^*(x) = x \otimes 1.$$

9.2. Remark. The absolute Frobenius F_S is a morphism of schemes. In fact, $F_S^* : \mathcal{O}_{S,s} \longrightarrow \mathcal{O}_{S,s}$ satisfies

$$F_S^{*^{-1}}(m_{S,s}) = \{x \in \mathcal{O}_{S,s};\ x^p \in m_{S,s}\} = m_{S,s}$$

for any prime ideal $m_{S,s}$, and hence it is a local homomorphism on the local rings. If $S = \mathrm{Spec}A$, then F_S is induced by the p-th power map $A \to A$.

9.3. Properties.

a) Since $F_S : (S)_{top} \to (S)_{top}$ is the identity,

$$F_S : (X')_{top} \to (X)_{top}$$

is an isomorphism of topological spaces, as well as

$$F : (X)_{top} \to (X')_{top}.$$

b) If $t_1, \ldots, t_m$ are locally on X, generators of $\mathcal{O}_X$, i.e.

$$\mathcal{O}_X = \mathcal{O}_S[t_1, \ldots, t_m]/_{<f_1,\ldots,f_s>}$$

and $f = \sum \lambda_{\underline{i}} \cdot \underline{t}^{\underline{i}}$, for $\underline{t}^{\underline{i}} = t_1^{i_1} \cdot \ldots \cdot t_m^{i_m}$, and $\lambda_{\underline{i}} \in \mathcal{O}_S$, then one has

$$F_S^*(f) = \sum \lambda_{\underline{i}}^p \underline{t}^{\underline{i}} .$$

Hence

$$\mathcal{O}_{X'} = \mathcal{O}_X \otimes_{F_S} \mathcal{O}_S = \mathcal{O}_S[t_1, \ldots, t_m]/_{<F_S^*(f_1),\ldots F_S^*(f_s)>} .$$

For $g = \sum \ \mu_{\underline{i}} \underline{t}^{\underline{i}} \in \mathcal{O}_{X'}$ one has

$$F^*(g) = \sum \ \mu_{\underline{i}} t^{p\cdot\underline{i}} \quad \text{where} \quad t^{p\cdot\underline{i}} = t_1^{p\cdot i_1} \cdot \ldots \cdot t_m^{p\cdot i_m}.$$

c) If X is smooth over S, one has locally étale morphisms $\pi : X \to \mathbf{A}_S^n$, hence a diagram, where the right hand squares are by definition cartesian:

$$\begin{array}{ccccc}
X & \xrightarrow{F} & X' & \xrightarrow{F_S} & X \\
{\scriptstyle\pi}\downarrow & & \downarrow{\scriptstyle\pi'} & & \downarrow{\scriptstyle\pi} \\
\mathbf{A}_S^n & \xrightarrow{F} & (\mathbf{A}_S^n)' & \xrightarrow{F_S} & \mathbf{A}_S^n \\
 & \searrow & \downarrow & & \downarrow \\
 & & S & \xrightarrow{F_S} & S
\end{array}$$

For

$$\mathbf{A}_S^n = \mathrm{Spec}\mathcal{O}_S[t_1 \ldots t_n]$$

we have

$$(\mathbf{A}_S^n)' = \mathrm{Spec}\mathcal{O}_S[t_1 \ldots t_n]$$

and $F^{-1}\mathcal{O}_{(\mathbf{A}_S^n)'}$ is the subsheaf of $\mathcal{O}_{\mathbf{A}_S^n}$ given by $\mathcal{O}_S[t_1^p, \ldots, t_n^p]$. Hence $F_*\mathcal{O}_{\mathbf{A}_S^n}$ is freely generated over $\mathcal{O}_{(\mathbf{A}_S^n)'}$ by

$$F_*(t_1^{a_1} \cdot \ldots \cdot t_n^{a_n}) \ \text{ for } \ 0 \leq a_i < p.$$

We have an isomorphism $f : X \longrightarrow X' \times_{(\mathbf{A}_S^n)'} \mathbf{A}_S^n$ and the left upper square in the diagram is cartesian as well.
In fact, for $x \in X$ we may assume that the maximal ideal $m_{X,x} \subset \mathcal{O}_{X,x}$ is generated by $t_1, \ldots t_n$ and, if $\mathcal{O}_x$ is the local ring of x in $X' \times_{(\mathbf{A}_S^n)'} \mathbf{A}_S^n$, then the maximal ideal m_x of $\mathcal{O}_x$ has the same generators. Hence

$$f^* : \mathcal{O}_x \longrightarrow \mathcal{O}_{X,x}$$

is a local homomorphism inducing a surjection

$$m_x \longrightarrow m_{X,x}/m_{X,x}^2.$$

d) The sheaf $F_*\mathcal{O}_X$ is again a locally free $\mathcal{O}_X$-module. Using the notation from part c) it is generated by

$$F_*\pi^*(t_1^{a_1}\dots t_n^{a_n}) \text{ for } 0 \le a_i < p.$$

Therefore, for any locally free sheaf $\mathcal{F}$ on X, the sheaf $F_*\mathcal{F}$ is locally free over $\mathcal{O}_{X'}$. For example, if D is a normal crossing divisor on X, then $F_*\Omega^a_{X/S}(\log\ D)$ is locally free.

9.4. Definition. When S has a lifting $\widetilde{S}$ to $\mathbb{Z}/p^2$ (see (8.5)), a *lifting* $\widetilde{F}_{\widetilde{S}}$ of F_S is a finite morphism

$$\widetilde{F}_{\widetilde{S}} : \widetilde{S} \longrightarrow \widetilde{S}$$

whose restriction to S is F_S.

Similarly, if X and X' have liftings $\widetilde{X}$ and $\widetilde{X}'$ to $\widetilde{S}$, a *lifting* $\widetilde{F}$ *of the relative Frobenius* F is a finite morphism

$$\widetilde{F} : \widetilde{X} \longrightarrow \widetilde{X}'$$

which restricts to F.

In particular, (8.13,c) gives rise to an exact sequence of $\mathcal{O}_{\widetilde{X}'}$-modules

$$0 \longrightarrow \widetilde{F}_*p\cdot\mathcal{O}_{\widetilde{X}} \longrightarrow \widetilde{F}_*\mathcal{O}_{\widetilde{X}} \longrightarrow F_*\mathcal{O}_X \longrightarrow 0$$

together with an $\mathcal{O}_{\widetilde{X}'}$-isomorphism

$$p : F_*\mathcal{O}_X \longrightarrow \widetilde{F}_*p\cdot\mathcal{O}_{\widetilde{X}} = p\cdot\widetilde{F}_*\mathcal{O}_{\widetilde{X}}.$$

9.5. Assumptions. For the rest of this lecture we assume S to be a scheme over $\mathbb{F}_p$ with a lifting $\widetilde{S}$ to $\mathbb{Z}/p^2$ and a lifting

$$\widetilde{F}_{\widetilde{S}} : \widetilde{S} \longrightarrow \widetilde{S}$$

of the absolute Frobenius.

Moreover we keep the assumptions made in (8.9). Hence X is supposed to be smooth over S and $D \subset X$ is a normal crossing divisor over S. We write $D' = F_S^*(D)$ for $F_S : X' \to X$.

9.6. Example. If k is a perfect field, $S = \text{Spec } k$ and $\widetilde{S} = \text{Spec } W_2(k)$, then one takes

$$\widetilde{F}^*_{\widetilde{S}}(x + y\cdot p) = x^p + y^p\cdot p\ .$$

Furthermore, in this case F_S is an isomorphism of fields and X' is isomorphic to X. In particular, X has a lifting to $\widetilde{S} = \text{Spec } W_2(k)$ if and only if X' does.

9.7. Proposition. *Let $D \subset X$ be an S-normal crossing divisor of the smooth S-scheme X. Let*

$$\widetilde{D}' \subset \widetilde{X}' \quad \textit{be a lifting of} \quad D' \subset X'$$

to $\widetilde{S}$. Then locally in the Zarisky topology

$$D \subset X \quad \textit{has a lifting} \quad \widetilde{D} \subset \widetilde{X}$$

to $\widetilde{S}$ such that F lifts to

$$\widetilde{F} : \widetilde{X} \longrightarrow \widetilde{X}' \quad \textit{with} \quad \widetilde{F}^* \mathcal{O}_{\widetilde{X}'}(-\widetilde{D}') = \mathcal{O}_{\widetilde{X}}(-p \cdot \widetilde{D}).$$

PROOF: By (8.17,c) we know that a lifting $\widetilde{D} \subset \widetilde{X}$ exists locally. Let

$$\pi : X \longrightarrow \mathbf{A}^n_S = \mathbf{Spec}\ \mathcal{O}_S[t_1, \ldots, t_n]$$

be étale and D_j be the zero set of $\varphi_j = \pi^*(t_j)$. By (8.14,c) we can choose liftings of φ_j to $\widetilde{\varphi}_j \in \mathcal{O}_{\widetilde{X}}$ and of

$$\varphi'_j = F^*_S(\varphi_j) = \varphi_j \otimes 1$$

to $\widetilde{\varphi}'_j$ such that $\widetilde{D}_j$ is defined by $\widetilde{\varphi}_j$ and $\widetilde{D}'_j$ by $\widetilde{\varphi}'_j$. We can define

$$\widetilde{F}^* \text{ by } \widetilde{F}^*(\widetilde{\varphi'_j}) = \widetilde{\varphi}^p_j.$$

By the explicit description of F in (9.3,c) $\widetilde{F}$ restricts to F and

$$\widetilde{F}^* \mathcal{O}_{\widetilde{X}'}(-\widetilde{D}') = \mathcal{O}_{\widetilde{X}}(-p \cdot \widetilde{D}).$$

□

9.8. Remark. We have seen in (8.12) that $(\widetilde{X})_{top} = (X)_{top}$ and $(\widetilde{X}')_{top} = (X')_{top}$. By (9.3,a) we have $(X)_{top} \simeq (X')_{top}$ and hence we can regard a lifting $\widetilde{F}$ as a morphism

$$\widetilde{F}^* : \mathcal{O}_{\widetilde{X}'} \longrightarrow \mathcal{O}_{\widetilde{X}}$$

of sheaves of rings over $(X')_{top}$.

Similarly to (8.22) we have

9.9. Proposition. *Keeping notations and assumptions from (9.7) assume that*

$$D \subset X \quad \textit{has a lifting} \quad \widetilde{D} \subset \widetilde{X}$$

to $\widetilde{S}$ and let $\widetilde{F}_0 : \widetilde{X} \to \widetilde{X}'$ be one lifting of F with

$$\widetilde{F}^*_0 \mathcal{O}_{\widetilde{X}'}(-\widetilde{D}') = \mathcal{O}_{\widetilde{X}}(-p \cdot \widetilde{D}).$$

Then

$$\{\widetilde{F}:\widetilde{X}\longrightarrow\widetilde{X}';\ \ \widetilde{F}^*\mathcal{O}_{\widetilde{X}'}(-\widetilde{D}')=\mathcal{O}_{\widetilde{X}}(-p\cdot\widetilde{D}),\ \widetilde{F}\ \textit{lifting of}\ F\}$$

is described by the affine space

$$\widetilde{F}_0^*+\mathrm{Hom}_{\mathcal{O}_{X'}}(\Omega^1_{X'/S}(\log\ D'),F_*\mathcal{O}_X).$$

Proof: As in (8.22) for isomorphisms of liftings one finds that $\widetilde{F}^*-\widetilde{F}_0^*$ is zero on $p\cdot\mathcal{O}_{\widetilde{X}}$ and induces the zero map from $\mathcal{O}_{X'}$ to $\mathcal{O}_X$. Hence $\widetilde{F}^*-\widetilde{F}_0^*$ induces

$$\widetilde{F}^*-\widetilde{F}_0^*:\mathcal{O}_{X'}\longrightarrow p\cdot\mathcal{O}_X.$$

For $\widetilde{x}',\widetilde{y}'\in\mathcal{O}_{\widetilde{X}'}$ one has

$$(\widetilde{F}^*-\widetilde{F}_0^*)(\widetilde{x}'\cdot\widetilde{y}')=F(x')(\widetilde{F}^*-\widetilde{F}_0^*)(\widetilde{y}')+F(y')(\widetilde{F}^*-\widetilde{F}_0^*)(\widetilde{x}')$$

and

$$p^{-1}(\widetilde{F}^*-\widetilde{F}_0^*):\mathcal{O}_{X'}\longrightarrow\mathcal{O}_X$$

factorizes through

$$\mathcal{O}_{X'}\xrightarrow{\ d\ }\Omega^1_{X'/S}\longrightarrow\mathcal{O}_X$$

where the right hand side morphism is $\mathcal{O}_{X'}$-linear and is again denoted by $(\widetilde{F}^*-\widetilde{F}_0^*)$. For $\widetilde{\varphi}_j\in\mathcal{O}_{\widetilde{X}}$ and $\widetilde{\varphi}'_j\in\mathcal{O}_{\widetilde{X}'}$, local parameters for $\widetilde{D}_j$ and $\widetilde{D}'_j$ respectively, which lift φ_j and $\varphi'_j=\varphi_j\otimes 1$, one has

$$\widetilde{F}^*(\widetilde{\varphi}'_j)=\widetilde{\varphi}_j^p\cdot(1+p\cdot\widetilde{\lambda})$$

and

$$\widetilde{F}_0^*(\widetilde{\varphi}'_j)=\widetilde{\varphi}_j^p\cdot(1+p\cdot\widetilde{\lambda}_0)$$

for some $\widetilde{\lambda},\widetilde{\lambda}_0\in\mathcal{O}_{\widetilde{X}}$. Therefore

$$(\widetilde{F}^*-\widetilde{F}_0^*)(\widetilde{\varphi}'_j)=\widetilde{\varphi}_j^p(p(\widetilde{\lambda}-\widetilde{\lambda}_0))$$

and

$$(\widetilde{F}^*-\widetilde{F}_0^*)(d\widetilde{\varphi}'_j)=p^{-1}\circ(\widetilde{F}^*-\widetilde{F}_0^*)(\widetilde{\varphi}'_j)\in\mathcal{O}_X(-p\cdot D_j).$$

Hence

$$(\widetilde{F}^*-\widetilde{F}_0^*):\Omega^1_{X'/S}\longrightarrow\mathcal{O}_X$$

extends to

$$(\widetilde{F}^*-\widetilde{F}_0^*):\Omega^1_{X'/S}(\log\ D')\longrightarrow\mathcal{O}_X.$$

Conversely, for

$$\varphi\in\mathrm{Hom}_{\mathcal{O}_{X'}}(\Omega^1_{X'/S}(\log\ D'),\mathcal{O}_X)$$

we define

$$\widetilde{F}^* = \widetilde{F}_0^* + \varphi^* \text{ by } \widetilde{F}^*(\widetilde{x}) = \widetilde{F}_0^*(\widetilde{x}) - p(\varphi(dx)).$$

We have

$$\mathcal{O}_{X'}(-D') \xrightarrow{d} \Omega^1_{X'/S}(\log\ D')(-D') \xrightarrow{\varphi} \mathcal{O}_X(-p \cdot D)$$

and

$$\widetilde{F}^*(\mathcal{O}_{\widetilde{X}'}(-\widetilde{D}')) = \mathcal{O}_{\widetilde{X}}(-p \cdot \widetilde{D}).$$

□

9.10. Corollary. *Under the assumption of (9.7) assume that X is affine and that*

$$D \subset X \quad \textit{has a lifting} \quad \widetilde{D} \subset \widetilde{X}$$

to $\widetilde{S}$. Then there is a lifting $\widetilde{F} : \widetilde{X} \to \widetilde{X}'$ of F with

$$\widetilde{F}^* \mathcal{O}_{\widetilde{X}'}(-\widetilde{D}') = \mathcal{O}_{\widetilde{X}}(-p \cdot \widetilde{D}).$$

PROOF: One repeats the argument used to prove (8.23), replacing $\widetilde{X}^{(1)}$ by $\widetilde{X}$ and $\widetilde{X}^{(2)}$ by $\widetilde{X}'$ and using (9.7) and (9.9) instead of (8.21) and (8.22).

□

9.11. Remark. Let X be a smooth S-scheme, $\widetilde{X}'$ a lifting of X' and $\widetilde{X}^{(i)}$, for $i = 1, 2$, two liftings of X to $\widetilde{S}$. Assume that we have a lifting

$$\widetilde{F}_2 : \widetilde{X}^{(2)} \longrightarrow \widetilde{X}'$$

and isomorphisms

$$u : \widetilde{X}^{(1)} \longrightarrow \widetilde{X}^{(2)} \quad \text{and} \quad v : \widetilde{X}^{(1)} \longrightarrow \widetilde{X}^{(2)},$$

both lifting the identity. Then, considering again $\mathcal{O}_{\widetilde{X}'}, \mathcal{O}_{\widetilde{X}^{(1)}}$ and $\mathcal{O}_{\widetilde{X}^{(2)}}$ as sheaves of rings on $(X')_{top}$, one has

$$\mathcal{O}_{\widetilde{X}'} \xrightarrow{\widetilde{F}_2^*} \mathcal{O}_{\widetilde{X}^{(2)}} \xrightarrow{(u^* - v^*)} \mathcal{O}_{\widetilde{X}^{(1)}}$$

and

$$(u^* - v^*) \circ \widetilde{F}_2^*(\widetilde{x}') = (u^* - v^*)(d(F^*(x'))) = 0$$

since $F^*(x')$ is a p-th power. We find:

$$(\widetilde{F}_2 \circ u)^* = (\widetilde{F}_2 \circ v)^* : \mathcal{O}_{\widetilde{X}'} \longrightarrow \mathcal{O}_{\widetilde{X}^{(1)}}.$$

In other words, $(\widetilde{F}_2 \circ u)^*$ does not depend on the choice of u.

In (9.10) we used the fact that, for X' affine, the higher cohomology groups of coherent sheaves are zero to obtain the existence of the lifting

$$\widetilde{F} : \widetilde{X} \longrightarrow \widetilde{X}'$$

of F. We have more generally:

9.12. Corollary. *Let X be a smooth scheme and $D \subset X$ be a normal crossing divisor over S. Given liftings*

$$\widetilde{D} \subset \widetilde{X} \quad \text{and} \quad \widetilde{D}' \subset \widetilde{X}' \quad \text{of} \quad D \subset X \quad \text{and} \quad D' \subset X'$$

(respectively) to $\widetilde{S}$, the exact obstruction for lifting F to

$$\widetilde{F} : \widetilde{X} \longrightarrow \widetilde{X}' \quad \text{with} \quad \widetilde{F}^* \mathcal{O}_{\widetilde{X}'}(-\widetilde{D}') = \mathcal{O}_{\widetilde{X}}(-p \cdot \widetilde{D})$$

is a class

$$[F_{\widetilde{X}',\widetilde{D}'}] \in H^1(X', \mathcal{H}om_{\mathcal{O}_{X'}}(\Omega^1_{X'/S}(\log\ D'), F_*\mathcal{O}_X))$$

which does not depend on $(\widetilde{X}, \widetilde{D})$.

Proof: By (9.7) or (9.10) one can cover $\widetilde{X}$ by affine $\widetilde{X}_\alpha$ such that F lifts to $\widetilde{F}_\alpha$ on $\widetilde{X}_\alpha$ with the required property for D. Then by (9.9) $(\widetilde{F}^*_\alpha - \widetilde{F}^*_\beta)$ describes a 1-cocycle with values in

$$\mathcal{H}om_{\mathcal{O}_{X'}}(\Omega^1_{X'/S}(\log\ D'), F_*\mathcal{O}_X).$$

Changing the $\widetilde{F}_\alpha$ corresponds to changing the cocycle by a coboundary. We define $[F_{\widetilde{X}',\widetilde{D}'}]$ to be the cohomology class of this cocycle. If $[F_{\widetilde{X}',\widetilde{D}'}] = 0$ one finds for a possibly finer cover $\{X'_\alpha\}$

$$\varphi_\alpha \in \Gamma(X'_\alpha, \mathcal{H}om_{\mathcal{O}_{X'}}(\Omega^1_{X'/S}(\log\ D'), F_*\mathcal{O}_X))$$

such that the $\widetilde{F}^*_\alpha + \varphi_\alpha$ glue together to give $\widetilde{F} : \widetilde{X} \to \widetilde{X}'$.

If $\widetilde{X}^{(i)}$ are two liftings, $\widetilde{X}^{(i)}_\alpha$ coverings and $\widetilde{F}^{(i)}_\alpha$ liftings of F, for $i = 1, 2$ we can apply (8.21) or (8.23) to get isomorphisms of liftings

$$u_\alpha : \widetilde{X}^{(1)}_\alpha \longrightarrow \widetilde{X}^{(2)}_\alpha.$$

By (9.11) we have on $\widetilde{X}'_\alpha \cap \widetilde{X}'_\beta$

$$(\widetilde{F}^{(2)}_\alpha \circ u_\alpha)^* - (\widetilde{F}^{(2)}_\beta \circ u_\beta)^* = (\widetilde{F}^{(2)}_\alpha \circ u_\alpha)^* - (\widetilde{F}^{(2)}_\beta \circ u_\alpha)^* =$$

$$u^*_\alpha \circ (\widetilde{F}^{(2)*}_\alpha - \widetilde{F}^{(2)*}_\beta) \in \Gamma(X'_\alpha \cap X'_\beta, \mathcal{H}om_{\mathcal{O}_{X'}}(\Omega^1_{X'/S}(\log\ D'), F_*\mathcal{O}_X))$$

As u^*_α is the identity on X'_α the cocycle defined by $\widetilde{F}^{(2)}_\alpha$ and $\widetilde{F}^{(2)}_\alpha \circ u_\alpha$ are the same and $\widetilde{F}^{(2)}_\alpha$ and $\widetilde{F}^{(1)}_\alpha$ define the same cohomology class.

□

9.13. The Cartier operator

Let X be a smooth scheme over S and $F : X \to X'$ be the Frobenius relative to S. The key observation is that the differential d in the de Rham complex $\Omega^\bullet_{X/S}$ is $\mathcal{O}_{X'}$-linear as, using the notations of (9.1),

$$dF^*(x \otimes 1) = dx^p = 0.$$

If D is a normal crossing divisor over S, the homology sheaves

$$\mathcal{H}^a = \mathcal{H}^a(F_*\Omega^\bullet_{X/S}(\log\ D))$$

are $\mathcal{O}_{X'}$-modules computed by the following

9.14. Theorem (Cartier, see [9] [34]).
One has an isomorphism of $\mathcal{O}_{X'}$-modules

$$C^{-1} : \Omega^1_{X'/S}(\log\ D') \longrightarrow \mathcal{H}^1(F_*\Omega^\bullet_{X/S}(\log\ D))$$

such that:
a) For $x \in \mathcal{O}_X$ one has

$$C^{-1}(d(x \otimes 1)) = x^{p-1}dx \quad in \quad \mathcal{H}^1.$$

b) If t is a local parameter defining a component of D, then

$$C^{-1}\left(\frac{d(t \otimes 1)}{t \otimes 1}\right) = \frac{dt}{t} \quad in \quad \mathcal{H}^1.$$

c) C^{-1} is uniquely determined by a) and b).

d) For all $a \geq 0$ one has an isomorphism

$$\bigwedge^a C^{-1} : \Omega^a_{X'/S}(\log\ D') \longrightarrow \mathcal{H}^a(F_*\Omega^\bullet_{X/S}(\log\ D))$$

obtained by wedge product from C^{-1}.

PROOF: c) is obvious since $\Omega^1_{X'/S}(\log\ D')$ is generated by elements of the form

$$d(x \otimes 1) \quad \text{and} \quad \frac{d(t \otimes 1)}{t \otimes 1}.$$

For the existence of C^{-1} let us first assume that $D = \emptyset$. Then

$$(x+y)^{p-1}(dx+dy) - x^{p-1}dx - y^{p-1}dy = df$$

where for

$$\gamma_i \in \mathbb{F}_p \quad \text{with} \quad \gamma_i \equiv \frac{1}{p}\binom{p}{i} \ mod\ p$$

we take
$$f = \sum_{i=1}^{p-1} \gamma_i \cdot x^i \cdot y^{p-i} = \frac{1}{p}[(x+y)^p - x^p - y^p].$$
Moreover, one has
$$(y \cdot x)^{p-1} d(y \cdot x) = x^p \cdot y^{p-1} dy + y^p \cdot x^{p-1} dx$$
and $d(x^{p-1}dx) = 0$. Hence, the property a) defines C^{-1}.

For $D \neq \emptyset$, b) is compatible with the definition of C^{-1} on $\Omega^1_{X'/S}$. In fact,
$$C^{-1}\left(t \otimes 1 \cdot \frac{d(t \otimes 1)}{t \otimes 1}\right) = F^*(t \otimes 1) C^{-1}\left(\frac{d(t \otimes 1)}{t \otimes 1}\right) = t^p \frac{dt}{t}.$$
Having defined C^{-1}, we can define $\bigwedge^a C^{-1}$ as well. As in (9.3,c) we have locally a cartesian square
$$\begin{array}{ccc} X & \xrightarrow{F} & X' \\ {\scriptstyle \pi}\downarrow & & \downarrow{\scriptstyle \pi'} \\ \mathbf{A}^n_S & \xrightarrow{F} & (\mathbf{A}^n_S)' \end{array}$$
with π and π' étale. Hence to show that $\bigwedge^a C^{-1}$ is an isomorphism it is enough to consider the case
$$X = \mathbf{A}^n_S = \mathbf{Spec}\ \mathcal{O}_S[t_1, \ldots t_n]$$
and D to be the zero set of $t_1 \cdot \ldots \cdot t_r$.

If $\mathcal{B}^a$ is the $\mathbb{F}_p$-vector space freely generated by
$$t_1^{i_1} \cdot \ldots \cdot t_n^{i_n} \cdot \omega_{\alpha_1} \wedge \ldots \wedge \omega_{\alpha_a}$$
for
$$0 \le i_\nu < p \quad \text{for} \quad \nu = 1, \ldots, n$$
$$1 \le \alpha_1 < \alpha_2 \ldots < \alpha_a \le n$$
$$\omega_\nu = \begin{cases} \frac{dt_\nu}{t_\nu} & \nu = 1, \ldots, r \\ dt_\nu & \nu = r+1, \ldots, n \end{cases}$$
then $\mathcal{B}^\bullet$, with the usual differential is a subcomplex of $F_*\Omega^\bullet_{X/S}(\log\ D)$. One has $F_*\Omega^\bullet_{X/S}(\log\ D) = \mathcal{O}_{X'} \otimes_{\mathbb{F}_p} \mathcal{B}^\bullet$ and (9.14) follows from the following claim. □

9.15. Claim. One has
i) $H^0(\mathcal{B}^\bullet) = \mathbb{F}_p$.

ii) $H^1(\mathcal{B}^\bullet)$ has the basis $\{\omega_1, \ldots, \omega_r, t_{r+1}^{p-1}\omega_{r+1}, \ldots, t_n^{p-1}\omega_n\}$.

iii) $H^a(\mathcal{B}^\bullet) = \bigwedge^a H^1(\mathcal{B}^\bullet)$.

PROOF: For $n = 1$, this is shown easily:
Obviously ker $(d : \mathcal{B}^0 \longrightarrow \mathcal{B}^1) = \mathbb{F}_p$. For $D = \emptyset$ let us write $\mathcal{K}^\bullet = \mathcal{B}^\bullet$. One has

$$\mathcal{K}^1 = < t^i dt;\ i = 0, \ldots, p-1 >_{\mathbb{F}_p}$$

and

$$d\mathcal{K}^0 = < dt^{i+1} = (i+1) \cdot t^i dt;\ i = 0, \ldots, p-2 >_{\mathbb{F}_p} .$$

For $D \neq \emptyset$ write $\mathcal{L}^\bullet = \mathcal{B}^\bullet$. One has

$$\mathcal{L}^1 = < t^i \frac{dt}{t};\ i = 0, \ldots, p-1 >_{\mathbb{F}_p}$$

and

$$d\mathcal{L}^0 = < dt^i = i \cdot t^i \cdot \frac{dt}{t}; i = 1, \ldots p-1 >_{\mathbb{F}_p} .$$

In both cases (9.15) is obvious. For $n > 1$ one can write

$$\mathcal{B}^\bullet = \underbrace{\mathcal{L} \otimes_{\mathbb{F}_p} \mathcal{L}^\bullet \otimes \ldots \otimes_{\mathbb{F}_p} \mathcal{L}^\bullet}_{r \text{ times}} \otimes_{\mathbb{F}_p} \underbrace{\mathcal{K}^\bullet \otimes \ldots \otimes_{\mathbb{F}_p} \mathcal{K}^\bullet}_{n-r \text{ times.}} .$$

By the Künneth formula (A.8)

$$H^a(\mathcal{B}^\bullet) = \sum_{\sum_{i=1}^n \varepsilon_i = a} = H^{\varepsilon_1}(\mathcal{L}^\bullet) \otimes \ldots \otimes H^{\varepsilon_r}(\mathcal{L}^\bullet) \otimes \ldots \otimes H^{\varepsilon_n}(\mathcal{K}^\bullet).$$

which implies a), b) and c).

□

9.16. Notation. Following Deligne-Illusie, we define

$$\Omega^\bullet_{X/S}(A, B) = \Omega^\bullet_{X/S}(\log\ (A + B))(-A)$$

where $A + B$ is a normal crossing divisor over S.

9.17. Corollary. *The Cartier operator induces an isomorphism*

$$\Omega^a_{X'/S}(A', B') \longrightarrow \mathcal{H}^a(F_* \Omega^\bullet_{X/S}(A, B))$$

PROOF: By (2.7) the residues of

$$d : \mathcal{O}_X(-A) \longrightarrow \Omega^1_{X/S}(\log\ (A+B))(-A) = \Omega^1_{X/S}(A, B)$$

along the components of A are all 1 and by (2.10)

$$\Omega^\bullet_{X/S}(\log\ (A+B))(-p \cdot A) \longrightarrow \Omega^\bullet_{X/S}(A, B)$$

is a quasi isomorphism. Since

$$F_* \Omega^\bullet_{X/S}(\log\ (A+B))(-p \cdot A) = F_* \Omega^\bullet_{X/S}(\log\ (A+B)) \otimes_{\mathcal{O}_{X'}} \mathcal{O}_{X'}(-A')$$

we can apply (9.14).

□

9.18. Duality

Let us keep the notation from (9.16). The wedge product

$$\Omega^{n-i}_{X/S}(\log\ D) \otimes \Omega^{i}_{X/S}(\log\ D) \xrightarrow{\wedge} \Omega^{n}_{X/S} \otimes \mathcal{O}_X(D)$$

is a perfect duality of locally free sheaves. Hence one obtains:

9.19. Lemma. $\Omega^{n-i}_{X/S}(A,B) \otimes \Omega^{i}_{X/S}(B,A) \xrightarrow{\wedge} \Omega^{n}_{X/S}$ *is a perfect duality.*

9.20. Lemma. *One has a perfect duality*

$$F_*\Omega^{n-i}_{X/S}(A,B) \otimes F_*\Omega^{i}_{X/S}(B,A) \longrightarrow \Omega^{n}_{X'/S}$$

given by

$$F_*\Omega^{n-i}_{X/S}(A,B) \otimes F_*\Omega^{i}_{X/S}(B,A) \underset{\wedge}{\longrightarrow} F_*\Omega^{n}_{X/S} \longrightarrow \mathcal{H}^n \underset{C}{\longrightarrow} \Omega^{n}_{X'/S}$$

where C is the Cartier operator.

PROOF: In fact, this is nothing but duality for finite flat morphisms ([30], p 239). One has

$$\begin{aligned} F_*\Omega^{n-i}_{X/S}(A,B) &= F_*\mathcal{H}om_{\mathcal{O}_X}(\Omega^{i}_{X/S}(B,A),\Omega^{n}_{X/S}) \\ &\simeq \mathcal{H}om_{\mathcal{O}_{X'}}(F_*\Omega^{i}_{X/S}(B,A),\Omega^{n}_{X'/S}) \end{aligned}$$

and (9.20) is just saying that $F_*\Omega^{n}_{X/S} \longrightarrow \Omega^{n}_{X'/S}$ is given by the Cartier operator. One can do the calculations by hand.

As in the proof of (9.14) it is enough to consider $X = \mathbf{A}^n_S$, A the zero set of $t_1 \cdot \ldots \cdot t_s$ and B the zero set of $t_{s+1} \ldots t_r$. Define, for $a > 0$, $\mathcal{B}^a(A,B)$ to be the $\mathbb{F}_p$-vector space generated by all

$$\varphi = t_1^{i_1} \cdot \ldots \cdot t_n^{i_n} \cdot \omega_{\alpha_1} \wedge \ldots \wedge \omega_{\alpha_a}$$

with

$$\omega_\nu = \begin{cases} \frac{dt_\nu}{t_\nu} & \text{for} \quad \nu = 1,\ldots s,\ldots,r \\ dt_\nu & \text{for} \quad \nu = r+1,\ldots,n \end{cases}$$

where the indices are given by

$$\begin{array}{lll} 0 < i_\nu \leq p & \text{for} & \nu = 1,\ldots,s \\ 0 \leq i_\nu < p & \text{for} & \nu = s+1,\ldots,r,\ldots,n \\ \text{and by} & & 1 \leq \alpha_1 < \alpha_2 < \ldots < \alpha_a \leq n. \end{array}$$

Similarly we have $\mathcal{B}^a(B,A)$ by taking as index set

$$\begin{array}{lll} 0 \leq i_\nu < p & \text{for} & \nu = 1,\ldots,s \text{ and } \nu = r+1,\ldots,n \\ 0 < i_\nu \leq p & \text{for} & \nu = s+1,\ldots,r \\ \text{and} & & 1 \leq \alpha_1 < \alpha_2 < \ldots < \alpha_a \leq n. \end{array}$$

For $a = n - i$ the only generator δ of $\mathcal{B}^i(B, A)$ with $C(\varphi \wedge \delta) \neq 0$ is

$$\delta = t_1^{j_1} \cdot \ldots \cdot t_n^{j_n} \omega_{\beta_1} \wedge \ldots \wedge \omega_{\beta_i}$$

with

$$\{\beta_1, \ldots, \beta_i\} \cup \{\alpha_1, \ldots, \alpha_{n-i}\} = \{1, \ldots, n\}$$

and

$$i_\nu + j_\nu = \begin{cases} p & \text{for} \quad \nu = 1, \ldots, r \\ p - 1 & \text{for} \quad \nu = r+1, \ldots, n. \end{cases}$$

□

9.21. Remark. For $\varphi \in \mathcal{B}^{n-i-1}(A, B)$ and $\delta \in \mathcal{B}^i(B, A)$ the explicit description of the duality in the proof of (9.20) shows that (up to sign)

$$C(d\varphi \wedge \delta) = C(\varphi \wedge d\delta).$$

Hence we obtain as well:

9.22. Corollary. *Under the duality in (9.20) the transposed of the differential d is again d (up to sign).*

§ 10 The proof of Deligne and Illusie [12]

We keep the assumptions from Lectures 8 and 9. Hence X is supposed to be a smooth noetherian S-scheme, $D \subset X$ a S-normal crossing divisor, and S is a noetherian scheme over $\mathbb{Z}/p$ which admits a lifting $\widetilde{S}$ to $\mathbb{Z}/p^2$ as well as a lifting $\widetilde{F}_{\widetilde{S}} : \widetilde{S} \to \widetilde{S}$ of the absolute Frobenius F_S.

10.1. The two term de Rham complex

is defined as

$$\tau_{\leq 1} \; F_* \Omega^\bullet_{X/S} \; (\log \; D).$$

Hence, as explained in (A.26), it is the complex

$$F_* \mathcal{O}_X \longrightarrow Z^1$$

where

$$Z^1 = Ker \; (F_* \Omega^1_{X/S} \; (\log \; D) \longrightarrow F_* \Omega^2_{X/S} \; (\log \; D)).$$

One has a short exact sequence of complexes

$$0 \longrightarrow \mathcal{H}^0 \longrightarrow \tau_{\leq 1} \; F_* \Omega^\bullet_{X/S} \; (\log \; D) \longrightarrow \mathcal{H}^1[-1] \longrightarrow 0$$

given by

$$\begin{array}{ccccccc} 0 & \longrightarrow & \mathcal{H}^0 & \longrightarrow & F_*\mathcal{O}_X & & \\ & & & & \downarrow & & \\ & & & & Z^1 & \longrightarrow & \mathcal{H}^1 & \longrightarrow & 0 \end{array}$$

where $\mathcal{H}^0 = \mathcal{O}_{X'}$ and $\mathcal{H}^1$ is $\mathcal{O}_{X'}$- isomorphic to $\Omega^1_{X'/S}(\log\ D)$ via the Cartier operator (9.14).

10.2. Definition. A *splitting* of $\tau_{\leq 1}F_*\Omega^\bullet_{X/S}(\log\ D)$ is a diagram

$$\begin{array}{ccc} \tau_{\leq 1}F_*\Omega^\bullet_{X/S}(\log\ D) & \xrightarrow{\sigma} & \mathcal{K}^\bullet \\ & & \uparrow \theta \\ & & \mathcal{H}^0 \oplus \mathcal{H}^1[-1] \end{array}$$

where $\mathcal{K}^\bullet$ is the Čech complex

$$\mathcal{C}^\bullet(\mathcal{U}, \tau_{\leq 1}F_*\Omega^\bullet_{X/S}(\log\ D))$$

associated to some affine open cover $\mathcal{U}$ of X', where σ is the induced morphism, hence a quasi-isomorphism (see (A.6)), and where θ is a quasi-isomorphism. We may assume, of course, that

$$\mathcal{H}^i \xrightarrow{\sigma} \mathcal{H}^i \xrightarrow{\theta^{-1}} \mathcal{H}^i$$

is the identity for $i = 0, 1$.

10.3. Example. Assume that $D \subset X$ and $D' \subset X'$ both lift to

$$\widetilde{D} \subset \widetilde{X} \quad \text{and} \quad \widetilde{D}' \subset \widetilde{X}'$$

on $\widetilde{S}$ and that F lifts to $\widetilde{F} : \widetilde{X} \to \widetilde{X}'$ in such a way that

$$\widetilde{F}^*\mathcal{O}_{\widetilde{X}'}(-\widetilde{D}') = \mathcal{O}_{\widetilde{X}}(-p \cdot \widetilde{D}).$$

For example, if $S = \operatorname{Spec} k$ for a perfect field k and if $D \subset X$ has a lifting $\widetilde{D} \subset \widetilde{X}$ then as we have seen in (9.6) $D' \subset X'$ has a lifting as well. By (9.12) the existence of $\widetilde{F}$ is equivalent to

$$[F_{\widetilde{X}',\widetilde{D}'}] = 0 \ \text{ in } \ H^1(X', \mathcal{H}om_{\mathcal{O}_{X'}}(\Omega^1_{X'/S}(\log\ D'), F_*\mathcal{O}_X)).$$

For example it automatically exists if this group vanishes.

Anyway, if the liftings $\widetilde{D}, \widetilde{X}, \widetilde{D}', \widetilde{X}'$ and $\widetilde{F}$ exist, the morphism

$$\widetilde{F}^* : \Omega^1_{\widetilde{X}'/\widetilde{S}}(\log\ \widetilde{D}') \longrightarrow \Omega^1_{\widetilde{X}/\widetilde{S}}(\log\ \widetilde{D})$$

verifies

$$\widetilde{F}^*|_{p\cdot\Omega^1_{\tilde{X}'/\tilde{S}}(\log\ \tilde{D}')} = 0.$$

In fact,

$$F^* : \Omega^1_{X'/S}(\log\ D') \longrightarrow \Omega^1_{X/S}(\log\ D)$$

is given by $F^*(d(t \otimes 1)) = d(t^p)$ and hence it is the zero map. We have a commutative diagram

$$\begin{array}{ccc} \Omega^1_{X'/S}(\log\ D') & \xrightarrow[\simeq]{p} & p\cdot\Omega^1_{\widetilde{X}'/\widetilde{S}}(\log\ \widetilde{D}') \\ \downarrow{\scriptstyle F^*} & & \downarrow{\scriptstyle \widetilde{F}^*} \\ \Omega^1_{X/S}(\log\ D) & \xrightarrow[\simeq]{p} & p\cdot\Omega^1_{\widetilde{X}/\widetilde{S}}(\log\ \widetilde{D}) \end{array}$$

and hence the vertical morphisms are both zero. The same argument shows that the factorization

$$\widetilde{F}^* : \Omega^1_{X'/S}(\log\ D') \longrightarrow \Omega^1_{\widetilde{X}/\widetilde{S}}(\log\ \widetilde{D})$$

takes values in

$$p\cdot\Omega^1_{\widetilde{X}/\widetilde{S}}(\log\ \widetilde{D}) = p\cdot\Omega^1_{X/S}(\log\ D).$$

The induced map

$$p^{-1}\circ\widetilde{F}^* : \Omega^1_{X'/S}(\log\ D') \longrightarrow \Omega^1_{X/S}(\log\ D)$$

can be written in coordinates as follows. For $x \in \mathcal{O}_X$ let $\widetilde{x} \in \mathcal{O}_{\widetilde{X}}$ be a lifting of x and let $\widetilde{x}' \in \mathcal{O}_{\widetilde{X}'}$ be a lifting of $x' = x \otimes 1$. One writes

$$\widetilde{F}^*(\widetilde{x}') = \widetilde{x}^p + p(u(\widetilde{x},\widetilde{x}'))$$

for some $u(\widetilde{x},\widetilde{x}') \in \mathcal{O}_X$. Then

$$p^{-1}\circ\widetilde{F}^*(d\widetilde{x}') = x^{p-1}dx + du(\widetilde{x},\widetilde{x}').$$

In particular the image of $p^{-1}\circ\widetilde{F}^*$ lies in

$$Z^1 \subset F_*\Omega^1_{X/S}(\log\ D)$$

and the composition with $Z^1 \longrightarrow \mathcal{H}^1$ gives back the Cartier operator.

In this example, i.e. if the liftings $\widetilde{D}, \widetilde{X}, \widetilde{D}', \widetilde{X}'$ and $\widetilde{F}$ all exist, we can take $\mathcal{U} = \{X'\}$ and define

$$\theta : \mathcal{O}_{X'} \oplus \Omega^1_{X'/S}(\log\ D')[-1] \longrightarrow \tau_{\leq 1}F_*\Omega^\bullet_{X/S}(\log\ D)$$

by

$$\begin{array}{ccc} \mathcal{O}_{X'} & \longrightarrow & F_*\mathcal{O}_X \\ \downarrow{\scriptstyle 0} & & \downarrow{\scriptstyle d} \\ \Omega^1_{X'/S} & \xrightarrow[p^{-1}\circ\tilde{F}^*]{} & Z^1 \end{array}$$

and, by (9.14), θ is a quasi-isomorphism.

10.4. Notation. We call a cohomology class

$$\varphi \in \mathbb{H}^1(X', \mathcal{H}om_{\mathcal{O}_{X'}}(\mathcal{H}^1, F_*\mathcal{O}_X) \to \mathcal{H}om_{\mathcal{O}_{X'}}(\mathcal{H}^1, Z^1))$$

a *splitting cohomology class*, if φ maps to the identity in

$$H^0(X', \mathcal{H}om_{\mathcal{O}_{X'}}(\mathcal{H}^1, \mathcal{H}^1)) = \mathbb{H}^1(X', \mathcal{H}om_{\mathcal{O}_{X'}}(\mathcal{H}^1, \mathcal{H}^1)[-1]).$$

10.5. Proposition. *The splittings of*

$$\tau_{\leq 1} F_*\Omega^\bullet_{X/S}(\log\ D)$$

are in one to one correspondence with the splitting cohomology classes

$$\varphi \in \mathbb{H}^1(X', \mathcal{H}om_{\mathcal{O}_{X'}}(\mathcal{H}^1, F_*\mathcal{O}_X) \longrightarrow \mathcal{H}om_{\mathcal{O}_{X'}}(\mathcal{H}^1, Z^1)).$$

PROOF: Let φ be a splitting cohomology class, realized as cocycle

$$\varphi_{\alpha\beta} \in \Gamma(X'_{\alpha\beta}, \mathcal{H}om_{\mathcal{O}_{X'}}(\mathcal{H}^1, F_*\mathcal{O}_X))$$

and

$$\psi_\alpha \in \Gamma(X'_\alpha, \mathcal{H}om_{\mathcal{O}_{X'}}(\mathcal{H}^1, Z^1))$$

for some affine open cover $\mathcal{U} = \{X'_\alpha\}$ of X'. Hence, using the notations from (A.6) for the differential in the Čech complex, $\delta\varphi = 0$ and $d\varphi - \delta\psi = 0$. By assumption ψ_α induces the identity in

$$\Gamma(X'_\alpha, \mathcal{H}om_{\mathcal{O}_{X'}}(\mathcal{H}^1, \mathcal{H}^1)).$$

Then $\theta = (id, (\varphi_{\alpha\beta}, \psi_\alpha))$ is the map wanted, i.e.

$$\begin{array}{ccc} \mathcal{O}_{X'} & \xrightarrow{id} & \bigoplus_{\mathcal{U}} \varrho_{\alpha *} F_*\mathcal{O}_X|_{X'_\alpha} \\ {\scriptstyle 0}\downarrow & & \downarrow{\scriptstyle \delta\oplus d} \\ \mathcal{H}^1 & \xrightarrow{(\varphi_{\alpha\beta}, \psi_\alpha)} & \bigoplus_{\mathcal{U}} \varrho_{\alpha\beta *} F_*\mathcal{O}_X|_{X'_{\alpha\beta}} \oplus \bigoplus_{\mathcal{U}} \varrho_{\alpha *}\ Z^1|_{X'_\alpha} \\ \downarrow & & \downarrow{\scriptstyle (-\delta\oplus d,\ 0\oplus -\delta)} \\ 0 & \longrightarrow & \bigoplus_{\mathcal{U}} \varrho_{\alpha\beta\gamma *} F_*\mathcal{O}_X|_{X'_{\alpha\beta\gamma}} \oplus \bigoplus_{\mathcal{U}} \varrho_{\alpha\beta *}\ Z^1|_{X'_{\alpha\beta}}. \end{array}$$

where

$$\varrho_{\alpha_1,\cdots\alpha_r} : X'_{\alpha_1,\cdots\alpha_r} \longrightarrow X'$$

denotes the embedding, and where

$$(-\delta \oplus d,\ 0 \oplus -\delta)(x_{\alpha\beta}, z_\alpha) = (-\delta(x), d(x) - \delta(z)).$$

Conversely, let for some $\mathcal{U}$

$$\tau_{\leq 1}(F_*\Omega^\bullet_{X/S}(\log\ D)) \xrightarrow{\sigma} \mathcal{K}^\bullet \xleftarrow{\theta} \mathcal{H}^0 \oplus \mathcal{H}^1[-1]$$

be a splitting. As $\mathcal{H}^1$ is $\mathcal{O}_{X'}$-locally free, $\sigma \otimes id$ and $\theta \otimes id$ are quasi-isomorphisms of the corresponding complexes tensored with $\mathcal{H}om_{\mathcal{O}_{X'}}(\mathcal{H}^1, \mathcal{O}_{X'})$. We obtain therefore maps

$$H^0\Big(X', \mathcal{H}om_{\mathcal{O}_{X'}}(\mathcal{H}^1, \mathcal{H}^1)\Big) = \mathbb{H}^1\Big(X', \mathcal{H}^1 \otimes \mathcal{H}om_{\mathcal{O}_{X'}}(\mathcal{H}^1, \mathcal{O}_{X'})[-1]\Big)$$

$$\big\downarrow$$

$$\mathbb{H}^1\Big(X', \Big[\mathcal{H}^0 \oplus \mathcal{H}^1[-1]\Big] \otimes \mathcal{H}om_{\mathcal{O}_{X'}}(\mathcal{H}^1, \mathcal{O}_{X'})\Big)$$

$$\big\downarrow \simeq$$

$$\mathbb{H}^1\Big(X', \tau_{\leq 1} F_*\Omega^\bullet_{X/S}(\log\ D) \otimes \mathcal{H}om_{\mathcal{O}_{X'}}(\mathcal{H}^1, \mathcal{O}_{X'})\Big)$$

$$\big\downarrow =$$

$$\mathbb{H}^1(X', \mathcal{H}om_{\mathcal{O}_{X'}}(\mathcal{H}^1, F_*\mathcal{O}_X) \to \mathcal{H}om_{\mathcal{O}_{X'}}(\mathcal{H}^1, Z^1))$$

$$\big\downarrow$$

$$H^0\Big(X', \mathcal{H}om_{\mathcal{O}_{X'}}(\mathcal{H}^1, \mathcal{H}^1)\Big)$$

where the last map comes from the short exact sequence in (10.1). By definition of a splitting, the composed map is the identity. Therefore, the image of

$$id_{\mathcal{H}^1} \in H^0(X', \mathcal{H}om_{\mathcal{O}_{X'}}(\mathcal{H}^1, \mathcal{H}^1))$$

is a splitting cohomology class φ. Obviously, the both constructions are inverse to each other.

□

10.6. Remark. In (10.2) one could have replaced in the definition of a splitting the Čech complex by any complex $\mathcal{K}^\bullet$ bounded below and quasi-isomorphic to $\tau_{\leq 1} F_*\Omega^\bullet_{X/S}(log\ D)$. Then we would have proven as in (10.5), that a splitting cohomology class defines a splitting. However, to get the converse, we need that σ and θ define a map from

$$\mathbb{H}^0(X', \mathcal{H}om_{\mathcal{O}_{X'}}(\mathcal{H}^1, \mathcal{O}_{X'}) \otimes \mathcal{K}^\bullet)$$

to
$$\mathbb{H}^1(X', \mathcal{H}om_{\mathcal{O}_{X'}}(\mathcal{H}^1, \mathcal{O}_{X'}) \otimes \tau_{\leq 1} F_* \Omega^1_{X/S}(\log\ D)).$$
This is of course the case when $\mathcal{K}^\bullet$ is a complex of $\mathcal{O}_{X'}$-modules, but not in general. One needs a bit more knowledge on the derived category; in particular one needs the global Hom (,) in this category.

10.7. Main theorem. *Let X be a smooth scheme over S and $D \subset X$ be a S-normal crossing divisor. Then*

a) *A lifting $\widetilde{D}' \subset \widetilde{X}'$ of $D' \subset X'$ to $\widetilde{S}$ defines a splitting cohomology class*
$$\varphi = \varphi_{(\widetilde{X}', \widetilde{D}')}.$$

b) *Every splitting cohomology class φ is of the shape $\varphi = \varphi_{(\widetilde{X}', \widetilde{D}')}$ for some lifting $\widetilde{D}' \subset \widetilde{X}'$ of $D' \subset X'$ to $\widetilde{S}$.*

We will only need part a) in the proof of Theorem (8.3). Even if it might be more elegant to use more formal arguments we will give the necessary calculations in an explicit way for cycles in the Čech-cohomology.

PROOF: a) Let $\mathcal{U} = \{X_\alpha\}$ be an affine cover of X, such that the $X_{\alpha\beta}$ are affine, and such that (8.17) and (9.10) give liftings
$$\widetilde{D}_\alpha \subset \widetilde{X}_\alpha \quad \text{of} \quad D_\alpha \subset X_\alpha$$
to $\widetilde{S}$ and liftings
$$\widetilde{F}_\alpha : \widetilde{X}_\alpha \longrightarrow \widetilde{X}' \quad \text{of} \quad F : X \longrightarrow X'$$
satisfying
$$\widetilde{F}_\alpha^* \mathcal{O}_{\widetilde{X}'}(-\widetilde{D}') = \mathcal{O}_{\widetilde{X}_\alpha}(-p \cdot \widetilde{D}_\alpha).$$
For $\widetilde{X}_{\alpha\beta} = \widetilde{X}_\alpha|_{X_{\alpha\beta}}$ (8.23) implies the existence of isomorphisms of liftings
$$u_{\alpha\beta} : \widetilde{X}_{\alpha\beta} \longrightarrow \widetilde{X}_{\beta\alpha}$$
As in the proof of (9.12) one uses (9.9) to define
$$\varphi_{\alpha\beta} = p^{-1} \circ (\widetilde{F}_\alpha^* - (\widetilde{F}_\beta \circ u_{\alpha\beta})^*) \in \Gamma(X'_{\alpha\beta}, \mathcal{H}om_{\mathcal{O}_{X'}}(\Omega^1_{X'/S}(\log\ D'), F_*\mathcal{O}_X))$$
and by (9.11) $\varphi_{\alpha\beta}$ is a cocycle. On $\widetilde{X}_\alpha$ we obtained in (10.3) a map
$$\psi_\alpha = p^{-1} \circ \widetilde{F}_\alpha^* \in \Gamma(X'_\alpha, \mathcal{H}om_{\mathcal{O}_{X'}}(\Omega^1_{X'/S}(\log\ D'), Z^1))$$
lifting the Cartier operator C^{-1}.

□

10.8. Claim. One has

$$p^{-1} \circ \widetilde{F}^*_\beta|_{X'_{\beta\alpha}} = p^{-1} \circ (\widetilde{F}_\beta \circ u_{\alpha\beta})^*$$

in

$$\Gamma(X'_{\beta\alpha}, \mathcal{H}om_{\mathcal{O}_{X'}}(\Omega^1_{X'/S}(\log\ D'), Z')).$$

PROOF: For $\widetilde{x}' \in \mathcal{O}_{\widetilde{X}'}$ and $\widetilde{x}_\beta \in \mathcal{O}_{\widetilde{X}_\beta}$ we can write

$$\widetilde{F}^*_\beta(\widetilde{x}') = \widetilde{x}^p_\beta + p \cdot u(\widetilde{x}_\beta, \widetilde{x}').$$

Since $u^*_{\alpha\beta}|_{p\cdot\mathcal{O}_{X_{\alpha\beta}}}$ is the identity, one obtains

$$u^*_{\alpha\beta}\widetilde{F}^*_\beta(\widetilde{x}') = u^*_{\alpha\beta}(\widetilde{x}_\beta)^p + p \cdot u(\widetilde{x}_\beta, \widetilde{x}').$$

By (10.3) we have

$$p^{-1} \circ \widetilde{F}^*_\beta(d\widetilde{x}') = x^{p-1}dx + du(\widetilde{x}_\beta, \widetilde{x}') = p^{-1} \circ (\widetilde{F}_\beta \circ u_{\alpha\beta})^*.$$

□

Now (10.8) is just saying that $\delta\psi = d\varphi$ and therefore $(\varphi_{\alpha\beta}, \psi_\beta)$ defines a cohomology class $\varphi_{(\widetilde{X}',\widetilde{D}')}$ in

$$\mathbb{H}^1(X', \mathcal{H}om_{\mathcal{O}_{X'}}(\Omega^1_{X'/S}(\log\ D'), F_*\mathcal{O}_X) \longrightarrow \mathcal{H}om_{\mathcal{O}_{X'}}(\Omega^1_{X'/S}(\log\ D'), Z^1)).$$

By construction its image in

$$H^0(X', \mathcal{H}om_{\mathcal{O}_{X'}}(\Omega^1_{X'/S}(\log\ D'), \mathcal{H}^1))$$

is given by (ψ_β) and hence it is the Cartier operator.

b) Conversely, let $(\varphi_{\alpha\beta}, \psi_\alpha)$ be the cocycle giving the splitting cohomology class φ for some affine covering $\mathcal{U}' = \{X'_\alpha\}$. First we want to add some coboundary to get a new representative of φ.

By (8.17), (9.10) and (8.21) we can assume that we have:

i) *Liftings to* $\widetilde{S}$:

$$\widetilde{D}_\alpha \subset \widetilde{X}_\alpha \text{ of } D_\alpha = D|_{X_\alpha} \subset X_\alpha = X|_{X'_\alpha},$$

$$\widetilde{D}'_\alpha \subset \widetilde{X}'_\alpha \text{ of } D'_\alpha = D'|_{X'_\alpha} \subset X'_\alpha$$

and

$$\widetilde{F}_\alpha : \widetilde{X}_\alpha \longrightarrow \widetilde{X}'_\alpha \text{ of } F_\alpha = F|_{X_\alpha} : X_\alpha \longrightarrow X'_\alpha$$

with

$$\widetilde{F}^*_\alpha : \mathcal{O}_{\widetilde{X}'_\alpha}(-\widetilde{D}'_\alpha) = \mathcal{O}_{\widetilde{X}_\alpha}(-p \cdot \widetilde{D}_\alpha).$$

ii) *Isomorphisms of liftings*:

$$u'_{\alpha\beta} : \widetilde{X}'_{\alpha\beta} \longrightarrow \widetilde{X}'_{\beta\alpha} \quad \text{and} \quad u_{\alpha\beta} : \widetilde{X}_{\alpha\beta} \longrightarrow \widetilde{X}_{\beta\alpha}$$

were we keep the notation $\widetilde{X}'_{\alpha\beta} = \widetilde{X}'_{\alpha}|_{X_{\alpha\beta}}$.

For $\widetilde{x}' \in \mathcal{O}_{\widetilde{X}'_{\alpha\beta}}$ we can write $u'^{*}_{\alpha\beta}(\widetilde{x}') = \widetilde{x}' + p \cdot \lambda_{\alpha\beta}(\widetilde{x}')$. Then

$$\widetilde{F}^{*}_{\alpha} u'^{*}_{\alpha\beta}(\widetilde{x}') = \widetilde{F}^{*}_{\alpha}(\widetilde{x}') + p \cdot F^{*}\lambda_{\alpha\beta}(\widetilde{x}') = \widetilde{F}^{*}_{\alpha}(\widetilde{x}') + p \cdot \lambda_{\alpha\beta}(\widetilde{x}')^{p}.$$

Since $d(\lambda_{\alpha\beta}(\widetilde{x}')^{p}) = 0$ the explicit description of $p^{-1} \circ \widetilde{F}^{*}_{\alpha}$ in (10.3) gives

10.9. Claim. One has $p^{-1} \circ \widetilde{F}^{*}_{\alpha}|_{X'_{\alpha\beta}} = p^{-1} \circ (u'_{\alpha\beta} \circ \widetilde{F}_{\alpha})^{*}$ in

$$\Gamma(X'_{\alpha\beta}, \mathcal{H}om_{\mathcal{O}_{X'}}(\Omega^{1}_{X'/S}(\log\ D'),\ Z^{1})).$$

Define

$$\theta_{\alpha} = p^{-1} \circ \widetilde{F}^{*}_{\alpha} \in \Gamma(X'_{\alpha}, \mathcal{H}om_{\mathcal{O}_{X'}}(\Omega^{1}_{X'/S}(\log\ D'),\ F_{*}\mathcal{O}_{X})).$$

Replacing $\mathcal{U}'$ by some finer cover if necessary we find

$$f_{\alpha} \in \Gamma(X'_{\alpha}, \mathcal{H}om_{\mathcal{O}_{X'}}(\Omega^{1}_{X'/S}(\log\ D'),\ F_{*}\mathcal{O}_{X}))$$

such that $df_{\alpha} = \theta_{\alpha} - \psi_{\alpha}$.

10.10. Claim. For $\sigma'_{\alpha\beta} = \varphi_{\alpha\beta} + \delta f_{\alpha}$ the cohomology class φ is represented by the cocycle $(\sigma'_{\alpha\beta}, \theta_{\alpha})$.

PROOF: This is obvious since

$$(\sigma'_{\alpha\beta}, \theta_{\alpha}) = (\varphi_{\alpha\beta} + \delta f_{\alpha}, \psi_{\alpha} + df_{\alpha}).$$

□

The main advantage of $\sigma'_{\alpha\beta}$ is that it comes in geometric terms:
Write, using (9.9)

$$\sigma_{\alpha\beta} = p^{-1} \circ \Big((u'_{\alpha\beta} \circ \widetilde{F}_{\alpha})^{*} - (\widetilde{F}_{\beta} \circ u_{\alpha\beta})^{*} \Big)$$

in

$$\Gamma\Big(X'_{\alpha\beta}, \mathcal{H}om_{\mathcal{O}_{X'}}(\Omega^{1}_{X'/S}(\log\ D')\ , F_{*}\mathcal{O}_{X})\Big).$$

Applying (10.9) to the first and (10.8) to the second summand one gets

$$d\sigma_{\alpha\beta} = \theta_{\alpha} - \theta_{\beta} = df_{\alpha} - df_{\beta} + \psi_{\alpha} - \psi_{\beta} = d(f_{\alpha} - f_{\beta}) + d\varphi_{\alpha\beta}.$$

Hence $g_{\alpha\beta} = \sigma_{\alpha\beta} - \varphi_{\alpha\beta} - (f_\alpha - f_\beta)$ is closed and lives in

$$\Gamma\Big(X'_{\alpha\beta}, \mathcal{H}om_{\mathcal{O}_{X'}}(\Omega^1_{X'/S}(\log\ D'), \mathcal{O}_{X'})\Big).$$

By (8.22) $u'^*_{\alpha\beta} - g_{\alpha\beta}$ defines a new isomorphism of liftings

$$v'_{\alpha\beta} : \widetilde{X}'_{\alpha\beta} \longrightarrow \widetilde{X}'_{\beta\alpha}.$$

As $\widetilde{F}^*_\alpha = pF^*$ on $p \cdot \mathcal{O}_{X'_\alpha}$, one has $p^{-1} \circ \widetilde{F}^*_\alpha \circ g_{\alpha\beta} = g_{\alpha\beta}$ and

$$\begin{aligned} \sigma'_{\alpha\beta} = \quad & \varphi_{\alpha\beta} + (f_\alpha - f_\beta) = \sigma_{\alpha\beta} - g_{\alpha\beta} = \\ & p^{-1} \circ \Big[(u'_{\alpha\beta} \circ \widetilde{F}_\alpha)^* - (\widetilde{F}_\beta \circ u_{\alpha\beta})^* - \widetilde{F}^*_\alpha \cdot g_{\alpha\beta}\Big] \,. \end{aligned}$$

One obtains

10.11. Claim. $\sigma'_{\alpha\beta} = p^{-1} \circ \Big[(v'_{\alpha\beta} \circ \widetilde{F}_\alpha)^* - (\widetilde{F}_\beta \circ u_{\alpha\beta})^*\Big]$.

The proof of (10.7) ends with

10.12. Claim. The cocycle condition for $\sigma'_{\alpha\beta}$ allows the glueing of $\widetilde{X}'_\alpha$ to $\widetilde{X}'$ using $v'_{\alpha\beta}$.

PROOF: One has to show that

$$v'_{\alpha\gamma} = v'_{\beta\gamma} \circ v'_{\alpha\beta}$$

or, by (8.22), that the homomorphism defined there verifies

$$v'^*_{\alpha\beta} \circ v'^*_{\beta\gamma} - v'^*_{\alpha\gamma} = 0.$$

Since $\widetilde{F}^*_\alpha$ is injective, it is enough to show that

$$p^{-1} \circ \widetilde{F}^*_\alpha \circ [v'^*_{\alpha\beta} \circ v'^*_{\beta\gamma} - v'^*_{\alpha\gamma}] = 0$$

as homomorphism in

$$\mathcal{H}om_{\mathcal{O}_{X'}}(\Omega^1_{X'/S}(\log\ D'), F_*\mathcal{O}_X).$$

The cocycle condition for σ' is

$$p^{-1} \circ [\widetilde{F}^*_\alpha \circ v'^*_{\alpha\beta} - u^*_{\alpha\beta} \circ \widetilde{F}^*_\beta - \widetilde{F}^*_\alpha \circ v'^*_{\alpha\gamma} + u^*_{\alpha\gamma} \circ \widetilde{F}^*_\gamma + \widetilde{F}^*_\beta \circ v'^*_{\beta\gamma} - u^*_{\beta\gamma} \circ \widetilde{F}^*_\gamma] = 0.$$

Since

$$\widetilde{F}^*_\alpha \circ v'^*_{\alpha\beta} - u^*_{\alpha\beta} \circ \widetilde{F}^*_\beta$$

is a homomorphism from $\mathcal{O}_{X'}$ to $p \cdot \mathcal{O}_X$, we can replace it by

$$[\widetilde{F}^*_\alpha \circ v'^*_{\alpha\beta} - u^*_{\alpha\beta} \circ \widetilde{F}^*_\beta] \circ v^*_{\beta\gamma}$$

as $v^*_{\beta\gamma}$ is the identity on $\mathcal{O}_{X'_{\beta\gamma}}$. Similarly, we can add some $u^*_{\alpha\beta}$ at the right hand side (see (9.11)) and get

$$0 = p^{-1} \circ [\widetilde{F}^*_\alpha \circ v'^*_{\alpha\beta} \circ v'^*_{\beta\gamma} - u^*_{\alpha\beta} \circ \widetilde{F}^*_\beta \circ v'^*_{\beta\gamma} - \widetilde{F}^*_\alpha \circ v'^*_{\alpha\gamma} +$$

$$+ u^*_{\alpha\gamma} \circ \widetilde{F}^*_\gamma - u^*_{\alpha\beta} \circ \widetilde{F}^*_\beta \circ v'^*_{\beta\gamma} - u^*_{\alpha\beta} u^*_{\beta\gamma} \circ \widetilde{F}^*_\gamma]$$

where all the summands are morphisms from $\mathcal{O}_{\widetilde{X}'_\gamma} \longrightarrow \mathcal{O}_{\widetilde{X}_\alpha}$. This is the same as

$$0 = p^{-1} \circ \widetilde{F}^*_\alpha \circ (v'^*_{\alpha\beta} \circ v'^*_{\beta\gamma} - v'^*_{\alpha\gamma}) + p^{-1} \circ (u^*_{\alpha\gamma} - u^*_{\alpha\beta} u^*_{\beta\gamma}) \widetilde{F}^*_\gamma.$$

By (9.11) the term on the right is zero and

$$0 = p^{-1} \circ \widetilde{F}^*_\alpha \circ (v'^*_{\alpha\beta} \circ v'^*_{\beta\gamma} - v'^*_{\alpha\gamma}).$$

□

10.13. Splittings of the de Rham complex.

Let us generalize (10.2) to $\tau_{\leq i} F_* \Omega^\bullet_{X/S}(\log\ D)$ for $i > 1$. As remarked in (10.6) one can, using the derived category, replace the complex $\mathcal{K}^\bullet$ in the following definition by any complex $\mathcal{K}^\bullet$ bounded below and quasi-isomorphic to $\tau_{\leq i} F_* \Omega^\bullet_{X/S}(\log\ D)$.

10.14. Definition. A *splitting* of $\tau_{\leq i} F_* \Omega^\bullet_{X/S}(\log\ D)$ is a diagram

$$\begin{array}{ccc} \tau_{\leq i} F_* \Omega^\bullet_{X/S}(\log\ D) & \xrightarrow{\ \sigma\ } & \mathcal{K}^\bullet \\ & & \uparrow \theta \\ & & \bigoplus_{j \leq i} \mathcal{H}^j[-j] \end{array}$$

where $\mathcal{K}^\bullet$ is the Čech complex

$$\mathcal{C}^\bullet(\mathcal{U}, \tau_{\leq i} F_* \Omega^\bullet_{X/S}(\log\ D))$$

associated to some affine cover $\mathcal{U}$ of X (and hence σ a quasi-isomorphism) and where θ is a quasi-isormorphism. Here again,

$$\bigoplus_{j \leq i} \mathcal{H}^j[-j]$$

is the complex with zero differential and with $\mathcal{H}^j$ in degree j and $\tau_{\leq i}$ is the filtration explained in (A.26).

10.15. Example. Let us return to the assumptions made in (10.3), i.e. that the liftings $\widetilde{D}, \widetilde{X}, \widetilde{D}', \widetilde{X}'$ and especially $\widetilde{F}$ exist. We had defined there a morphism

$$\psi = p^{-1} \circ \widetilde{F}^* : \Omega^1_{X'/S}(\log\ D') \longrightarrow Z^1$$

which was a lifting of C^{-1}.

We define

$$\psi^j(\omega_1 \wedge \ldots \wedge \omega_j) = \psi(\omega_1) \wedge \psi(\omega_2) \wedge \ldots \wedge \psi(\omega_j)$$

where $\omega_l \in \Omega^1_{X'/S}(\log\ D')$. Since $\psi(\omega_l)$ is closed, the image of ψ^j lies in Z^j and, since the Cartier operator was defined as $\bigwedge^j C^{-1}$ the map ψ^j induces the Cartier operator on in

$$\mathcal{H}om_{\mathcal{O}_{X'}}(\Omega^j_{X'/S}(\log\ D'), \mathcal{H}^j).$$

10.16. Theorem. *Let X be a smooth S-scheme, $D \subset X$ be a normal crossing divisor over S and let*

$$\widetilde{D}' \subset \widetilde{X}' \text{ be a lifting of } D' \subset X'$$

to $\widetilde{S}$. Then the splitting cohomology class $\varphi_{(\widetilde{X}', \widetilde{D}')}$ of (10.7,a) induces a splitting of

$$\tau_{\leq i} F_* \Omega^\bullet_{X/S}(\log\ D) \text{ for } i < p = \mathrm{char}\,(S).$$

In particular, if $p > \dim_S X$, it induces a splitting of the whole de Rham complex

$$F_* \Omega^\bullet_{X/S}(\log\ D).$$

Proof: Let $(\varphi_{\alpha\beta}, \psi_\alpha)$ be a Čech cocycle for $(\varphi_{\widetilde{X}', \widetilde{D}'})$ where we regard $(\varphi_{\alpha\beta}, \psi_\alpha)$ as an $\mathcal{O}_{X'}$-homomorphism:

$$(\varphi, \psi) : \Omega^1_{X'/S}(\log\ D') \longrightarrow \mathcal{C}^1(F_* \mathcal{O}_X) \oplus \mathcal{C}^0(Z^1).$$

We define

$$(\varphi, \psi)^{\otimes j}(\omega_1, \otimes \cdots \otimes \omega_j) \in \mathcal{C}^j(\tau_{\leq i} F_* \Omega^\bullet_{X/S}(\log\ D))$$

for all $0 < j \leq i$ and all

$$\omega_1 \otimes \cdots \otimes \omega_j \in \bigotimes_1^j \Omega^1_{X'/S}(\log\ D')$$

by the following inductive formula:
For any cocycle

$$\underline{b} := (b_j, \ldots, b_0), \quad b_l \in \mathcal{C}^l(F_* \Omega^{j-l}_{X/S}(\log\ D)),$$

with

$$db_{j-s} + (-1)^{j-s} \delta b_{j-s-1} = 0 \text{ for all } 0 \leq s \leq j,$$

we define

$$\underline{b} \otimes (\varphi, \psi)(\omega_{j+1}) = (a_{j+1}, \ldots, a_0) =: \underline{a},$$

with

$$a_l \in \mathcal{C}^l(F_*\Omega^{j+1-l}_{X/s}(\log\ D)),$$

by the rule

$$a_{j+1-s} := (-1)^s b_{j-s} \cup \varphi + b_{j+1-s} \cup \psi$$

where

$$(b_{j-s} \cup \varphi)_{\alpha_0 \ldots \alpha_{j+1-s}} := (b_{j-s})_{\alpha_0 \ldots \alpha_{j-s}} \cdot \varphi_{\alpha_{j-s}, \alpha_{j+1-s}}(\omega_{j+1})$$

and

$$(b_{j+1-s} \cup \psi)_{\alpha_0 \ldots \alpha_{j+1-s}} := (b_{j+1-s})_{\alpha_0 \ldots \alpha_{j+1-s}} \cdot \psi_{\alpha_{j+1-s}}(\omega_{j+1}).$$

One has

$$da_{j+1-s} + (-1)^{j+1-s}\delta a_{j-s} = 0$$

and therefore $\underline{a}$ is a Čech cocycle.

We have, for $j \leq i$, a diagram

$$\begin{array}{ccc}
(\Omega^1_{X'/S}(\log\ D'))^{\otimes\ j} & \xrightarrow{(\varphi,\psi)^{\otimes\ j}} & \mathcal{C}^j(\tau_{\leq i} F_*\Omega^1_{X/S}(\log\ D))_{\text{closed}} \\
\downarrow \pi^j & & \downarrow \\
\Omega^j_{X'/S}(\log\ D') & \xrightarrow{\bigwedge^j C^{-1}} & \mathcal{H}^j
\end{array}$$

and any section δ^j of π^j allows to define

$$\theta_j = (\bigwedge^j C^{-1})^{-1} \circ \delta^j \circ (\varphi, \psi)^{\otimes j}.$$

The splitting

$$\theta : \bigoplus_{j \leq i} \mathcal{H}^j[-j] \longrightarrow \mathcal{K}^\bullet \quad \text{is} \quad \theta = \bigoplus_{j \leq i} \theta_j[-j].$$

Such sections δ^j exist for $j \leq i < \text{char}\ (S)$:

$$\delta^j(\omega_1 \otimes \ldots \otimes \omega_j) = \frac{1}{j!} \sum_{s \in \Sigma_j} sign\ (s) \cdot \omega_{s(1)} \wedge \ldots \wedge \omega_{s(j)},$$

where Σ_j denotes the symmetric group.

□

10.17. Corollary. *Let $D = A + B$ in (10.16). Then for $i < p$ the splitting cohomology class $\varphi_{(\tilde{X},\tilde{D})}$ induces a splitting of*

$$\tau_{\leq i} F_*\Omega^\bullet_{X/S}(A, B)$$

as well, i.e. a quasi-isomorphism

$$\theta : \bigoplus_{j\leq i} \Omega^j_{X'/S}(A', B')[-j] \longrightarrow \mathcal{K}^\bullet(A, B)$$

where $\mathcal{K}^\bullet(A, B)$ *is the Čech complex of*

$$\tau_{\leq i} F_* \Omega^\bullet_{X/S}(A, B).$$

PROOF: As in (9.17) one obtains from (2.7) and (2.10) a quasi-isomorphism

$$\begin{array}{c} F_*\Omega^1_{X/S}(\log\ (A + B)) \otimes_{\mathcal{O}_{X'}} \mathcal{O}_{X'}(-A') \\ \| \\ F_*(\Omega^\bullet_{X/S}(\log\ (A + B))(-p \cdot A)) \\ \downarrow \\ F_*\Omega^\bullet_{X/S}(A, B)\ . \end{array}$$

For $\mathcal{K}^\bullet$, the Čech complex of $\tau_{\leq i} F_* \Omega^\bullet_{X/S}(\log\ D)$, we have a quasi-isomorphism

$$\mathcal{K}^\bullet \otimes_{\mathcal{O}_{X'}} \mathcal{O}_{X'}(-A) \longrightarrow \mathcal{K}^\bullet(A, B)$$

and the existence of θ follows from (10.16).

□

10.18. Remark. In (10.3) and (10.15) we have seen that if both $D' \subset X'$ and $D \subset X$ lift to $\widetilde{D}' \subset \widetilde{X}'$ and $\widetilde{D} \subset \widetilde{X}$, and if F lifts to: $\widetilde{F} : \widetilde{X} \to \widetilde{X}'$ with

$$\widetilde{F}^* \mathcal{O}_{\widetilde{X}'}(-\widetilde{D}') = \mathcal{O}_{\widetilde{X}}(-p \cdot \widetilde{D}),$$

then

$$\psi = p^{-1} \circ \widetilde{F}^* : \Omega^1_{X'/S}(\log\ D') \longrightarrow Z^1 \subset \Omega^1_{X/S}(\log\ D)$$

lifts the Cartier operator, and gives an especially nice splitting of

$$\tau_{\leq 1} F_* \Omega^\bullet_{X/S}(\log\ D).$$

This defines

$$\bigwedge^l \psi : \Omega^l_{X'/S}(\log\ D') \longrightarrow Z^l \subset \Omega^l_{X/S}(\log\ D)$$

lifting the Cartier operator, and therefore one obtains a quasi-isomorphism:

$$\bigoplus_j \Omega^j_{X'/S}(\log\ D')[-j] \xrightarrow{\ \bigwedge^\bullet \psi\ } F_*\Omega^\bullet_{X/S}(\log\ D),$$

which gives via (10.17) a quasi-isomorphism

$$\bigoplus_j \Omega^j_{X'/S}(A',B')[-j] \xrightarrow{\bigwedge^\bullet \psi} F_*\Omega^\bullet_{X/S}(A,B)$$

if $D = A+B$. In particular, there is here no restriction on $\dim_S X$ in this case. In general one has

10.19. Proposition. *Let X, A and B be as in (10.17). Then the splitting cohomology class $\varphi_{(\tilde{X}',\tilde{D}')}$ induces a splitting of*

$$F_*\Omega^\bullet_{X/S}(A,B)$$

when $\dim_S X \leq p$ and S is affine.

PROOF: Of course this is nothing but (10.17) if $\dim_S X < p$.
For $\dim_S X \geq p$, we observe first that whenever $j : U \to X$ is the embedding of an open set such that $(X-U)$ is a divisor, then for coherent sheaves $\mathcal{F}$ on U and $\mathcal{G}$ on X one has

$$\mathcal{H}om_{\mathcal{O}_X}(j_*\mathcal{F},\mathcal{G})_x = \begin{cases} \mathcal{H}om_{\mathcal{O}_X}(\mathcal{F},\mathcal{G}|_U)_x & \text{for } x \in U \\ 0 & \text{for } x \in (X-U), \end{cases}$$

that is

$$\mathcal{H}om_{\mathcal{O}_X}(j_*\mathcal{F},\mathcal{G}) = j_!\mathcal{H}om_{\mathcal{O}_X}(\mathcal{F},\mathcal{G}|_U).$$

Let us consider the $\mathcal{O}_{X'}$-maps defined in (10.17) for $0 \leq l \leq p-1$:

$$\Omega^l_{X'/S}(B',A') \longrightarrow \mathcal{C}^l(F_*\mathcal{O}_X) + \cdots + \mathcal{C}^0(Z^l)$$

$$(\varphi_{\alpha_0\ldots\alpha_l}, \ \cdots \ , \varphi_{\alpha_0})$$

with the cocycle condition

$$d\varphi_{\alpha_0\ldots\alpha_k} + (-1)^l\delta\varphi_{\alpha_0\ldots\alpha_{k-1}} = 0.$$

The composite map

$$\Omega^l_{X'/S}(B',A') \longrightarrow \mathcal{C}^0(Z^l) \longrightarrow \mathcal{C}^0(\mathcal{H}^l)$$

is just $\bigwedge^l C^{-1}$.
Applying for $1 \leq k \leq l \leq p-1$ the functor $\mathcal{H}om_{\mathcal{O}_{X'}}(-,\Omega^n_{X'/S})$, where $n = \dim_S X$, one obtains $\mathcal{O}_{X'}$-linear maps (see (9.19) and (9.20))

$$(j_{\alpha_0\ldots\alpha_k})_!F_*\Omega^{n-(l-k)}_{X/S}(A,B) \xrightarrow{\varphi^\vee_{\alpha_0\ldots\alpha_k}} \Omega^{n-l}_{X'/S}(A',B')$$

and therefore $\mathcal{O}_{X'}$-maps

$$F_*\Omega_{X/S}^{n-(l-k)}(A,B) \xrightarrow{\varphi^{\vee}_{\alpha_0\dots\alpha_k}} \mathcal{C}^k(\Omega^{n-l}_{X'/S}(A',B')).$$

For $k=0$, one has an exact sequence

$$0 \longrightarrow Z^l \longrightarrow F_*\Omega^l_{X/S}(B,A) \longrightarrow F_*\Omega^{l+1}_{X/S}(B,A),$$

and applying (9.20) again, one obtains that

$$\mathcal{H}om_{\mathcal{O}_{X'}}(Z^l, \Omega^n_{X'/S}) = \frac{F_*\Omega^{n-l}_{X/S}(A,B)}{dF_*\Omega^{n-l-1}_{X/S}(A,B)}$$

which gives similarly a $\mathcal{O}_{X'}$-linear map

$$\frac{F_*\Omega^{n-l}_{X/S}(A,B)}{dF_*\Omega^{n-l-1}_{X/S}(A,B)} \xrightarrow{\varphi^{\vee}_{\alpha_0}} \mathcal{C}^0(\Omega^{n-l}_{X'/S}(A',B')).$$

The cocycle condition tells us that

$$\varphi^{\vee}_{\alpha_0\dots\alpha_k} \circ d + (-1)^l \delta\varphi^{\vee}_{\alpha_0\dots\alpha_{k-1}} = 0.$$

This means that $\varphi^{\vee}$ defines a map of complexes

$$\tau_{\geq n-l} F_*\Omega^{\bullet}_{X/S}(A,B)[(n-l)] \longrightarrow \tau_{\leq l}\mathcal{C}^{\bullet}(\Omega^{n-l}_{X'/S}(A',B')),$$

where

$$\tau_{\geq n-l} F_*\Omega^{\bullet}_{X/S}(A,B)[(n-l)] :=$$

$$\frac{F_*\Omega^{n-l}_{X/S}(A,B)}{dF_*\Omega^{n-l}_{X/S}(A,B)} \longrightarrow F_*\Omega^{n-l+1}_{X/S}(A,B) \longrightarrow \cdots \longrightarrow F_*\Omega^n_{X/S}(A,B).$$

The composite map

$$\mathcal{H}^{n-l} \longrightarrow \tau_{\geq n-l} F_*\Omega^{\bullet}_{X/S}(A,B)[(n-l)] \longrightarrow \mathcal{C}^{\bullet}(\Omega^{n-l}_{X'/S}(A',B'))$$

is given by $\bigwedge^{n-l} C$. As $\tau_{\geq n-l}$ maps to $\tau_{\geq n-l+1}$, we find in this way a $\mathcal{O}_{X'}$-map

$$\tau_{\geq n-p+1} F_*\Omega^{\bullet}_{X/S}(A,B) \xrightarrow{\varphi^{\vee}} \bigoplus_{i=n-p+1}^{n} \mathcal{C}^{\bullet}(\Omega^i_{X'/S}(A',B'))[-i]$$

which is a quasi-isomorphism. In particular, for any open set $U' \subset X'$ and any $\mathcal{O}_{X'}$-sheaf $\mathcal{F}'$ one has

$$\mathbb{H}^l(U', \tau_{\geq n-p+1} F_*\Omega^{\bullet}_{X/S}(A,B) \otimes \mathcal{F}') = \bigoplus_{i=n-p+1}^{n} H^{l-i}(U', \Omega^i_{X'/S}(A',B') \otimes \mathcal{F}').$$

Consider now $n = p$ as needed to finish the proof of (10.19). From the exact sequence

$$0 \longrightarrow \mathcal{H}^0 \longrightarrow F_*\Omega^\bullet_{X/S}(A,B) \longrightarrow \tau_{\geq 1} F_*\Omega^\bullet_{X/S}(A,B) \longrightarrow 0$$

we obtain an exact sequence of hypercohomology groups

$$\mathbb{H}^p(F_*\Omega^\bullet_{X/S}(A,B) \otimes \mathcal{H}^{p^\vee}) \longrightarrow \bigoplus_{a=1}^{p} H^{p-a}(\Omega^a_{X'/S}(A',B') \otimes \mathcal{H}^{p^\vee}) \longrightarrow$$

$$\longrightarrow H^{p+1}(\mathcal{H}^0 \otimes \mathcal{H}^{p^\vee})$$

where $\mathcal{H}^{p^\vee} := \mathcal{H}om_{\mathcal{O}_{X'}}(\mathcal{H}^p, \mathcal{O}_{X'})$.

As $\dim_S X = p$ and S is affine, one has $H^{p+1}(\mathcal{H}^0 \otimes \mathcal{H}^{p^\vee}) = 0$, and therefore

$$\bigwedge^p C \in H^0(\Omega^p_{X'/S}(A',B') \otimes \mathcal{H}^{p^\vee})$$

lifts to some

$$\Gamma \in \mathbb{H}^p(F_*\Omega^\bullet_{X/S}(A,B) \otimes \mathcal{H}^{p^\vee}).$$

Representing Γ by a Čech cocycle

$$\mathcal{H}^p \xrightarrow{[\Gamma]} \mathcal{C}^p(F_*\mathcal{O}_X) + \cdots + \mathcal{C}^0(F_*\Omega^p_{X/S}(A,B)),$$

and taking a common refinement $\mathcal{U}$ of the Čech covers defining φ and Γ, one obtains altogether a quasi-isomorphism

$$(\varphi, [\Gamma] \circ \bigwedge^p C^{-1}) : \bigoplus_j \Omega^j_{X'/S}(A',B')[-j] \longrightarrow \mathcal{C}^\bullet(F_*\Omega^\bullet_{X/S}(A,B)).$$

□

10.20. Remark. If $\dim_S X = n > p$, and S is affine, one considers the exact sequence

$$0 \longrightarrow \tau_{\leq n-p} F_*\Omega^\bullet_{X/S}(A,B) \longrightarrow F_*\Omega^\bullet_{X/S}(A,B) \longrightarrow \tau_{\geq n-p+1}\Omega^\bullet_{X/S}(A,B) \longrightarrow 0$$

giving the short exact sequences

$$\mathbb{H}^q(F_*\Omega^\bullet_{X/S}(A,B) \otimes \mathcal{H}^{q^\vee}) \longrightarrow \bigoplus_{a=n-p+1}^{n} H^{q-a}(\Omega^a_{X'/S}(A',B') \otimes \mathcal{H}^{q^\vee}) \longrightarrow$$

$$\longrightarrow H^{q+1}(\tau_{\leq n-p} F_*\Omega^\bullet_{X/S}(A,B) \otimes \mathcal{H}^{q^\vee})$$

for all $p \leq q \leq n$. If $n - p \leq p - 1$, then

$$H^{q+1}(\tau_{\leq n-p} F_* \Omega^\bullet_{X/S}(A,B) \otimes \mathcal{H}^{q^\vee}) = \bigoplus_{a=0}^{n-p} H^{q+1-a}(\Omega^a_{X'/S}(A',B') \otimes \mathcal{H}^{q^\vee}).$$

If those groups are vanishing, then the same argument as above shows that one obtains a splitting of $F_*\Omega^\bullet_{X/S}(A,B)$.

For example, take $D = \emptyset$. For $n = p+1$, one requires the vanishing of

$$H^{p+1}(\mathcal{H}^{p^\vee}),\ H^p(\Omega^1_{X'/S} \otimes \mathcal{H}^{p^\vee}) \text{ and of } H^{p+1}(\Omega^1_{X'/S} \otimes \mathcal{H}^{p+1^\vee}),$$

that is, via duality, the vanishing of

$$H^0(\Omega^p_{X'/S} \otimes \Omega^{p+1}_{X'/S}) \quad \text{and} \quad H^1(\Omega^{p\otimes 2}_{X'/S}).$$

Using (10.19) it is now quite easy to prove theorem (8.3) and some generalizations.

10.21. Theorem. *Let $f : X \to S$ be a smooth proper S-scheme,* $\dim_S X \leq p$, *and let $D \subset X$ be a S-normal crossing divisor. Assume that there exists a lifting*

$$\widetilde{D}' \subset \widetilde{X}' \text{ of } D' \subset X'$$

to $\widetilde{S}$. Let $D = A + B$. Then one has:

a) *The $\mathcal{O}_S$-sheaves*

$$E_1^{ab} = R^b f_* \Omega^a_{X/S}(A,B)$$

are locally free and compatible with arbitrary base change.

b) *The Hodge to de Rham spectral sequence*

$$E_1^{ab} \Longrightarrow \mathbb{R}^{a+b} f_* \Omega^\bullet_{X/S}(A,B)$$

degenerates in E_1.

PROOF: Assuming a), part b) follows if one knows that $\mathbb{R}^l f_* \Omega^\bullet_{X/S}(A,B)$ is a locally free $\mathcal{O}_S$-module of rank

$$\sum_{a+b=l} \operatorname{rank}_{\mathcal{O}_S}(E_1^{ab}).$$

Hence for a) and b) we can assume S to be affine.

(10.17), for $i = \dim_S X < p$, or (10.19) for $\dim_S X = p$ imply that

$$\mathbb{R}^l f_* \Omega^\bullet_{X/S}(A,B) = \mathbb{R}^l f'_*(F_* \Omega^\bullet_{X/S}(A,B)) = \bigoplus_a R^{l-a} f'_* \Omega^a_{X'/S}(A',B').$$

Hence, if a) holds true for $f : X \to S$, then

$$\mathbb{R}^l f_* \Omega^\bullet_{X/S}(A,B) = \bigoplus_a F_S^* R^{l-a} f_* \Omega^a_{X/S}(A,B)$$

is locally free of the right rank and one obtains b).

By "cohomology and base change" ([50], II §5, for example), there exist for $a = 0, \ldots, l$ bounded complexes $\mathcal{E}^\bullet_a$ of vector bundles on S, such that

$$\mathcal{H}^l(\mathcal{E}^\bullet_a) = R^{l-a} f_* \Omega^a_{X/S}(A,B)$$

and, for any affine map $\varphi : T \longrightarrow S$,

$$\mathcal{H}^l(\varphi^* \mathcal{E}^\bullet_a) = R^{l-a} f_{T*} \Omega^a_{X_T/T}(A_T, B_T)$$

where X_T, A_T, B_T, X'_T, $f_T : X_T \to T$ and $f'_T : X'_T \to T$ are obtained by pullback from X, A, B, X', f and $f' : X' \to S$.

For example, $R^{l-a} f'_* \Omega^a_{X'/S}(A', B')$ is given by $\mathcal{H}^l(F_S^* \mathcal{E}^\bullet_a)$ and hence $\mathbb{R}^l f_* \Omega^\bullet_{X/S}(A,B)$ by the l-th homology of the complex

$$F_S^* \mathcal{E}^\bullet \quad \text{for} \quad \mathcal{E}^\bullet = \bigoplus_a \mathcal{E}^\bullet_a.$$

To prove a), we have to show that for all l, the sheaf

$$\mathcal{H}^l(\mathcal{E}^\bullet) = \bigoplus_a \mathcal{H}^l(\mathcal{E}^\bullet_a)$$

is locally free. If this is wrong, then we take l_0 to be the maximal l with $\mathcal{H}^l(\mathcal{E}^\bullet)$ not locally free. Hence, if $\partial_\bullet$ denotes the differential in $\mathcal{E}^\bullet$, ker ∂_{l_0} is a vector bundle, let us say of rank r, but the image of

$$\partial_{l_0-1} : \mathcal{E}^{l_0-1} \longrightarrow \ker \partial_{l_0}$$

is not a subbundle.

For some closed point $s \in S$ one finds an infinitesimal neighbourhood $\hat{S}$, for example one of the form

$$\hat{S} = \mathrm{Spec}(\mathcal{O}_{S,s}/m^\mu_{S,s}) \quad \text{for} \quad \mu >> 0,$$

such that $\mathrm{im}(\partial_{l_0-1}|_{\hat{S}})$ is not a subbundle of $\ker(\partial_{l_0}|_{\hat{S}})$. Let us write $\hat{S} = \mathrm{Spec}R$ where R is an Artin ring. For any R-module M let $\mathrm{lg}(M)$ denote the lenght.

∂_{l_0-1} is represented by a matrix Δ_{l_0-1}. For the point $s \in \hat{S}$ one has

$$h = \mathrm{rank}_{k(s)} \mathcal{H}^{l_0}(\mathcal{E}^\bullet \otimes k(s)) > \frac{\mathrm{lg}(\mathcal{H}^{l_0}(\mathcal{E}^\bullet \otimes R))}{\mathrm{lg}(R)}.$$

Let $n_0 \subset R$ be the ideal generated by the $(r-h+1)$ minors of Δ_{l_0-1}. Then

$$\mathcal{H}^{l_0}(\mathcal{E}^\bullet \otimes R/n_0)$$

is free of rank h as an R/n_0-module. If $n_0' \subset n_0 \subset R$ is another ideal with $n_0' \neq n_0$, then

$$\lg(\mathcal{H}^{l_0}(\mathcal{E}^\bullet \otimes R/n_0')) < h \cdot \lg(R/n_0').$$

Repeating this construction, starting with R/n_0 instead of R we find after finitly many steps some ideal n such that the R/n modules

$$\mathcal{H}^l(\mathcal{E}^\bullet \otimes R/n)$$

are free of rank $h(l)$ over R/n for all l, but for some l' and for all ideals $n' \subset n$ with $n' \neq n$ one has

$$\lg(\mathcal{H}^{l'}(\mathcal{E}^\bullet \otimes R/n')) < h(l') \cdot \lg(R/n') \ .$$

In particular, this holds true for the ideal n' of R generated by $F_{\hat{S}}^* n$. Let us write $T' = \operatorname{Spec}(R/n')$ and $T = \operatorname{Spec}(R/n)$. We have affine morphisms

$$\begin{array}{ccccc} T' & \xrightarrow{\delta} & T & \xrightarrow{j} & T' \\ \downarrow & & \downarrow & & \downarrow \\ S & \xrightarrow{F_S} & S & \xrightarrow{=} & S \ . \end{array}$$

$\mathcal{H}^l(\mathcal{E}^\bullet|_T)$ is free of rank $h(l)$ over R/n for all l and hence

$$\mathcal{H}^l(\delta^* j^* \mathcal{E}^\bullet|_{T'}) = \mathcal{H}^l((F_S^* \mathcal{E}^\bullet)|_{T'})$$

is free of rank $h(l)$ over R/n'. The Hodge to de Rham spectral sequence implies that

$$\lg(\mathcal{H}^l(\mathcal{E}^\bullet|_{T'})) = \sum_a \lg(R^{l-a} f_{T'*} \Omega^a_{X_{T'}/T'}(A_{T'}, B_{T'})) \geq$$

$$\geq \lg(\mathbb{R}^l f_{T'*} \Omega^\bullet_{X_{T'}/T'}(A_{T'}, B_{T'})).$$

On the other hand,

$$\mathbb{R}^l f_{T'*} \Omega^\bullet_{X_{T'}/T'}(A_{T'}, B_{T'}) = \mathbb{R}^l f'_*(F_* \Omega^\bullet_{X/S}(A,B) \otimes f'^*(R/n')) =$$

$$= \bigoplus_a R^{l-a} f'_*(\Omega^a_{X'/S}(A', B') \otimes f'^*(R/n')).$$

We can apply base change and find the latter to be

$$\bigoplus_a \mathcal{H}^l(\delta^* \mathcal{E}^\bullet_a|_T) = \mathcal{H}^l(\delta^* \mathcal{E}^\bullet|_T) \ .$$

Since the sheaves $\mathcal{H}^l(\mathcal{E}^\bullet|_T)$ are locally free, we have altogether

$$\lg(\mathcal{H}^l(\mathcal{E}^\bullet|_{T'})) \geq \lg(\mathcal{H}^l(\delta^*\mathcal{E}^\bullet|_T)) =$$
$$= \lg(\delta^*\mathcal{H}^l(\mathcal{E}^\bullet|_T)) = h(l) \cdot \lg(R/n') .$$

For $l = l'$ this contradicts the choice of n and n'. Hence $\mathcal{H}^l(\mathcal{E}^\bullet)$ is locally free. □

10.22. Remark. As shown in [12], 4.1.2. it is enough to assume that for all a and b the dimension of $H^b(X_s, \Omega^a_{X_s}(A_s, B_s))$ is finite for closed points $s \in S$ and that the conjugate spectral sequence

$$_cE_2^{ij} = R^i f'_{T*}\mathcal{H}^j(F_*\Omega^\bullet_{X_T/T}(A_T, B_T)) \Longrightarrow \mathbb{R}^{i+j} f_{T*}\Omega^\bullet_{X_T/T}(A_T, B_T)$$

satisfies

$$_cE_2^{ij} = {}_cE_\infty^{ij} \text{ for } i + j = l$$

and for all T, in order to obtain (10.21).

10.23. Corollary. *Let K be a field of characteristic zero, X be a smooth proper scheme over K and $D = A + B$ be a reduced normal crossing divisor defined over K. Then the Hodge to de Rham spectral sequence*

$$E_1^{ab} = H^b(X, \Omega^a_X(A, B)) \Longrightarrow \mathbb{H}^{a+b}(X, \Omega^\bullet_X(A, B))$$

degenerates in E_1.

PROOF: By flat base change we may assume that K is of finite type over $\mathbb{Q}$. Hence we find a ring R, of finite type over $\mathbb{Z}$, such that K is the quotient field over R. Let $f : \mathcal{X} \to \operatorname{Spec} R$ be a proper morphism with $X = \mathcal{X} \times_R K$, and $\mathcal{A}$ and $\mathcal{B}$ divisors on $\mathcal{X}$ with $A = \mathcal{A}|_X$ and $B = \mathcal{B}|_X$.
Replacing $\operatorname{Spec} R$ by some open affine subscheme, we may assume that f is smooth, that $\mathcal{D} = \mathcal{A} + \mathcal{B}$ is a normal crossing divisor over $\operatorname{Spec} R$ and that

$$\mathbb{R}^l f_*\Omega^\bullet_{\mathcal{X}/\operatorname{Spec} R}(\mathcal{A}, \mathcal{B}) \text{ and } R^b f_*\Omega^a_{\mathcal{X}/\operatorname{Spec} R}(\mathcal{A}, \mathcal{B})$$

are locally free for all l, a and b. Take a closed point $s \in \operatorname{Spec} R$ with

$$\operatorname{char} k(s) = p > \dim_S X.$$

Hence

$$X_s = \mathcal{X} \times_R k(s) \longrightarrow S = k(s) , \quad A_s = \mathcal{A}|_{X_s} \text{ and } B_s = \mathcal{B}|_{X_s}$$

satisfy the assumptions made in (10.21) and

$$\sum_{a+b=l} \operatorname{rank} R^b f_*\Omega^a_{\mathcal{X}/\operatorname{Spec} R}(\mathcal{A}, \mathcal{B}) = \operatorname{rank} \mathbb{R}^l f_*\Omega^\bullet_{\mathcal{X}/\operatorname{Spec} R}(\mathcal{A}, \mathcal{B}).$$

□

As mentioned in (3.18) the corollary (10.23) finally ends the proof of (3.2) in characteristic zero, and hence of the different vanishing theorems and applications discussed in Lectures 5 - 7. A slightly different argument, avoiding the use of (3.19) or (3.22) can be found at the end of this lecture.

As promised we are now able to prove (3.2,b and c) for fields of characteristic $p \neq 0$ as well.

PROOF OF (3.2,B) AND C) IN CHARACTERISTIC $p \neq 0$: Recall, that on the projective manifold X we considered the invertible sheaves

$$\mathcal{L}^{(i)} = \mathcal{L}^{(i,D)} = \mathcal{L}^{i}(-[\frac{i \cdot D}{N}]) \text{ where}$$

$$D = \sum_{j=1}^{r} \alpha_i D_j$$

is a normal crossing divisor and $\mathcal{L}$ an invertible sheaf with $\mathcal{L}^N = \mathcal{O}_X(D)$. For N prime to char k, we constructed in §3, for $i = 0, \ldots, N-1$, integrable logarithmic connections

$$\nabla_{(i)} : \mathcal{L}^{(i)^{-1}} \longrightarrow \Omega^1_X(\log\ D^{(i)}) \otimes \mathcal{L}^{(i)^{-1}}$$

with poles along

$$D^{(i)} = \sum_{\substack{j=1 \\ \frac{i \cdot \alpha_j}{N} \notin \mathbb{Z}}}^{r} D_j.$$

The residue of $\nabla_{(i)}$ along $D_j \subset D^{(i)}$ is given by multiplication with

$$(i \cdot \alpha_j - N \cdot [\frac{i \cdot \alpha_j}{N}]) \cdot N^{-1}.$$

If A and B are reduced divisors such that A, B and $D^{(i)}$ have pairwise no common component then we want to prove:

□

10.24. Claim. Let k be a perfect field, N prime to char $k = p$ and assume that $p \geq \dim\ X$. If X, D, A and B admit a lifting to $W_2(k)$, then the spectral sequence

$$E_1^{ab} = H^b(X, \Omega^a_X(\log\ (A + B + D^{(i)}))(-B) \otimes \mathcal{L}^{(i)^{-1}}) \Longrightarrow$$

$$\mathbb{H}^{a+b}(X, \Omega^\bullet_X(\log\ (A + B + D^{(i)}))(-B) \otimes \mathcal{L}^{(i)^{-1}})$$

degenerates in E_1.

PROOF: Let $F : X \to X'$ be the relative Frobenius morphism and $\mathcal{L}', D', A'$ and B' be the sheaf and the divisors on X' obtained by field extensions. Then

$$F^*\mathcal{L}'^{(i,D')} = F^*\mathcal{L}'^{(i)} = \mathcal{L}^{p\cdot i}(-p\cdot[\frac{i\cdot D}{N}])$$

contains

$$\mathcal{L}^{(p\cdot i)} = \mathcal{L}^{p\cdot i}(-[\frac{p\cdot i\cdot D}{N}]).$$

Since p is prime to N one has $D^{(i)} = D^{(p\cdot i)}$. The connection

$$\nabla_{(p\cdot i)^{-1}} : \mathcal{O}_X(-B)\otimes\mathcal{L}^{(p\cdot i)^{-1}} \longrightarrow \Omega^1_X(\log\,(A+B+D^{(i)}))(-B)\otimes\mathcal{L}^{(p\cdot i)^{-1}}$$

induces a connection with the same poles on

$$\mathcal{O}_X(-B)\otimes F^*\mathcal{L}'^{(i)^{-1}}$$

whose residues along $D_j \subset D^{(i)}$ are given by multiplication with

$$N^{-1}\cdot p\cdot i\cdot\alpha_j - [\frac{p\cdot i\cdot\alpha_j}{N}] + ([\frac{p\cdot i\cdot\alpha_j}{N}] - p[\frac{i\cdot\alpha_j}{N}]) \,.$$

Obviously this number is zero modulo p. Since

$$[\frac{p\cdot i\cdot\alpha_j}{N}] \le \frac{p\cdot i\cdot\alpha_j}{N} < p\cdot([\frac{i\cdot\alpha_j}{N}]+1) = p\cdot[\frac{i\cdot\alpha_j}{N}]+p$$

(2.10) implies that the complexes

$$\Omega^\bullet_X\ (\log\,(A+B+D^{(i)}))(-B)\otimes F^*\mathcal{L}'^{(i)^{-1}}$$

and

$$\Omega^\bullet_X(\log\,(A+B+D^{(i)}))(-B)\otimes\mathcal{L}^{(p\cdot i)^{-1}}$$

are quasi-isomorphic.

10.25. Claim. *The complex*

$$F_*(\Omega^\bullet_X(\log\,(A+B+D^{(i)}))(-B)\otimes F^*\mathcal{L}'^{(i)^{-1}})$$

is isomorphic to the complex

$$F_*(\Omega^\bullet_X(\log\,(A+B+D^{(i)}))(-B))$$

tensorized with $\mathcal{L}'^{(i)^{-1}}$.

PROOF: Let $\pi : Y \to X$ be the cyclic cover obtained by taking the N-th root out of D, let $F : Y \to Y'$ be the relative Frobenius of Y. We have the induced diagram

$$\begin{array}{ccc} Y & \xrightarrow{F} & Y' \\ {\scriptstyle\pi}\downarrow & & \downarrow{\scriptstyle\pi'} \\ X & \xrightarrow{F} & X' \end{array}$$

and, on $X' - \mathrm{Sing}(D_{red})$,

$$\pi'_*\mathcal{O}_{Y'} = Ker(d : \pi'_*F_*\mathcal{O}_Y \longrightarrow \pi'_*F_*\Omega^1_Y) =$$

$$Ker(d : F_*(\bigoplus_{j=0}^{N-1} \mathcal{L}^{(j)^{-1}}) \longrightarrow F_*(\bigoplus_{j=0}^{N-1} \Omega^1_X(\log\ D^{(j)}) \otimes \mathcal{L}^{(j)^{-1}})).$$

Since F^* of the i-th eigenspace $\mathcal{L}'^{(i)}$ of $\pi'_*\mathcal{O}_{Y'}$ lies in the $p \cdot i$-th eigenspace $\mathcal{L}^{(i \cdot p)}$ of $\pi_*\mathcal{O}_Y$ one has

$$\begin{aligned} \mathcal{L}'^{(i)^{-1}} &= Ker(\nabla_{(p \cdot i)} : F_*\mathcal{L}^{(p \cdot i)^{-1}} \longrightarrow F_*(\Omega^1_X(\log\ D^{(i)}) \otimes \mathcal{L}^{(p \cdot i)^{-1}})) \\ &= Ker(\nabla_{(p \cdot i)} : (F_*\mathcal{O}_X) \otimes \mathcal{L}'^{(i)^{-1}} \longrightarrow (F_*\Omega^1_X(\log\ D^{(i)})) \otimes \mathcal{L}'^{(i)^{-1}}) \end{aligned}$$

By the Leibniz rule $\nabla_{(p \cdot i)}$ restricted to $(F_*\mathcal{O}_X) \otimes \mathcal{L}'^{(i)^{-1}}$ is nothing but $d \otimes id$ as claimed.

□

From (10.19) and by base change

$$\dim \mathbb{H}^l(X, \Omega^\bullet_X(\log\ (A + B + D^{(i)}))(-B) \otimes \mathcal{L}^{(p \cdot i)^{-1}}) =$$

$$\dim \mathbb{H}^l(X', F_*(\Omega^\bullet_X(\log\ (A + B + D^{(i)}))(-B) \otimes \mathcal{L}^{(p \cdot i)^{-1}})) =$$

$$\sum_{a+b=l} \dim H^b(X', \omega^a_{X'}(\log\ (A' + B' + D'^{(i)}))(-B') \otimes \mathcal{L}'^{(i)^{-1}} =$$

$$\sum_{a+b=l} \dim H^b(X, \omega^a_X(\log\ (A + B + D^{(i)}))(-B) \otimes \mathcal{L}^{(i)^{-1}} \geq$$

$$\dim \mathbb{H}^l(X, \Omega^\bullet_X(\log\ (A + B + D^{(i)}))(-B) \otimes \mathcal{L}^{(i)^{-1}}).$$

For some $\nu > 0$ one has $p^\nu \equiv 1\ mod\ N$. Repeating the argument $\nu - 1$ times one finds

$$\dim \mathbb{H}^l(X, \Omega^\bullet_X(\log\ (A + B + D^{(i)}))(-B) \otimes \mathcal{L}^{(i)^{-1}}) \geq$$

$$\dim \mathbb{H}^l(X, \Omega^\bullet_X(\log\ (A + B + D^{(i)}))(-B) \otimes \mathcal{L}^{(p^{\nu-1} \cdot i)^{-1}}) \geq \cdots$$

$$\cdots \geq \sum_{a+b=l} \dim H^b(X, \Omega^a_X(\log\ (A + B + D^{(i)}))(-B) \otimes \mathcal{L}^{(i)^{-1}}) \geq$$

$$\dim \mathbb{H}^l(X, \Omega^\bullet_X(\log\ (A + B + D^{(i)}))(-B) \otimes \mathcal{L}^{(i)^{-1}}) .$$

Hence all the inequalities must be equalities and one obtains (10.24).

□

2. PROOF OF (3.2,B) AND C) IN CHARACTERISTIC 0:
The arguments used in (10.23) to reduce (10.23) to (10.21) show as well that 3.2,b and c in characteristic 0 follow from (10.24).

□

§ 11 Vanishing theorems in characteristic p.

In this lecture we start with the elegant proof of the Akizuki-Kodaira-Nakano vanishing theorem, due to Deligne, Illusie and Raynaud [12]. Then we will discuss some generalizations. However they only seem to be of interest if one assumes that one has embedded resolutions of singularities in characteristic p.

11.1. Lemma. *Let k be a perfect field, let X be a proper smooth k-scheme and D a normal crossing divisor, both admitting a lifting $\widetilde{D} \subset \widetilde{X}$ to $W_2(k)$. Let $\mathcal{M}$ be a locally free $\mathcal{O}_X$-module. Then, for $l <$ char(k) one has*

$$\sum_{a+b=l} \dim\ H^b(X, \Omega^a_X(\log\ D) \otimes \mathcal{M}) \leq \sum_{a+b=l} \dim\ H^b(X, \Omega^a_X(\log\ D) \otimes F^*_X\mathcal{M}).$$

PROOF: By (10.16) we have

$$\dim\ \mathbb{H}^l(X, \Omega^\bullet_X(\log\ D) \otimes F^*_X\mathcal{M}) = \sum_{a+b=l} \dim\ H^b(X', \Omega^a_{X'}(\log\ D') \otimes \mathcal{M}')$$

for the sheaf $\mathcal{M}' = {pr_1}^*\mathcal{M}$ on $X' = X \times_{F_S} S$ and $D' = D \times_{F_S} S$. By base change the right hand side is

$$\sum_{a+b=l} \dim\ H^b(X, \Omega^a_X(\log\ D) \otimes \mathcal{M}).$$

The Hodge to de Rham spectral sequence implies that the left hand side is smaller than or equal to

$$\sum_{a+b=l} \dim\ H^b(X, \Omega^a_X(\log\ D) \otimes F^*_X\mathcal{M}).$$

□

11.2. Corollary. *Under the assumptions of (11.1) assume that $\mathcal{M}$ is invertible. Then*

$$\sum_{a+b=l} \dim\ H^b(X, \Omega^a_X(\log\ D) \otimes \mathcal{M}) \leq \sum_{a+b=l} \dim\ H^b(X, \Omega^a_X(\log\ D) \otimes \mathcal{M}^p).$$

11.3. Corollary (Deligne, Illusie, Raynaud, see [12]).
For $a+b < \mathrm{Min}\ \{\mathrm{char}(k),\ \dim\ X\}$ and $\mathcal{L}$ ample and invertible, one has under the assumptions of (11.1.)

$$H^b(X, \Omega^a_X(\log\ D) \otimes \mathcal{L}^{-1}) = 0.$$

PROOF: For ν large enough, and $\mathcal{L}^{-1} = \mathcal{M}$ one has

$$H^b(X, \Omega^a_X(\log\ D) \otimes \mathcal{M}^{p^\nu}) = 0$$

for $b < \dim\ X$. By (11.2) one has

$$H^b(X, \Omega^a_X(\log\ D) \otimes \mathcal{M}^{p^{\nu-1}}) = 0$$

and after finitely many steps one obtains (11.3)

□

The following corollary is, as well known in characteristic zero, a direct application of (11.3) for $a = 0$. It will be needed in our discussion of possible generalizations of (11.3).

11.4. Corollary. *Let k be a perfect field, let X be a proper smooth k-scheme with $\dim X \leq \mathrm{char}\ k$ and let $\mathcal{L}$ be a numerically effective sheaf (see (5.5)). Assume that $\mathcal{A}$ is a very ample sheaf, such that X and $\mathcal{A}$ lift to $\widetilde{X}$ and $\widetilde{\mathcal{A}}$ over $W_2(k)$, with $\widetilde{\mathcal{A}}$ very ample over $\widetilde{S}$. Then one has:*

a) $\mathcal{A}^{\dim X+1} \otimes \mathcal{L} \otimes \omega_X$ *is generated by global sections.*

b) $\mathcal{A}^{\dim X+2} \otimes \mathcal{L} \otimes \omega_X$ *is very ample.*

PROOF: By (11.3) and Serre duality $H^1(X, \mathcal{A}^{\dim X} \otimes \mathcal{L} \otimes \omega_X) = 0$. Hence one has a surjection

$$H^0(X, \mathcal{A}^{\dim X+1} \otimes \mathcal{L} \otimes \omega_X) \longrightarrow H^0(H, \mathcal{A}^{\dim X} \otimes \mathcal{L} \otimes \omega_H)$$

where H is a smooth zero divisor of a general section of $\mathcal{A}$. Since $\mathcal{A}$ lifts to $W_2(k)$, we can choose H such that it lifts to $W_2(k)$ as well. By induction on $\dim\ X$ we can assume that

$$\mathcal{A}^{\dim X} \otimes \mathcal{L} \otimes \omega_H$$

is generated by global sections and, moving H, we obtain a).
Part b) follows directly from a) (see [30], II. Ex.7.5).

□

11.5. Proposition. *Let k be a perfect field of characteristic $p > 0$, let X be a proper smooth k-variety, let D be an effective normal crossing divisor and $\mathcal{L}$ be an invertible sheaf on X. Assume that (X, D) and $\mathcal{L}$ admit liftings to $W_2(k)$*

and that one has:

(∗) *For some* $\nu_0 \in \mathbb{N}$ *and all* $\nu \geq 0$ *the sheaf* $\mathcal{L}^{\nu_0+\nu} \otimes \mathcal{O}_X(-D)$ *is ample.*

Then, for $a + b < \dim X \leq \operatorname{char} k$ *one has*

$$H^b(X, \Omega_X^a(\log D) \otimes \mathcal{L}^{-1}) = 0.$$

PROOF: By (5.7), the assumption (∗) implies that $\mathcal{L}$ is numerically effective. Let us choose μ_0 such that $\mathcal{L}^{\mu_0 \cdot \nu_0}(-\mu_0 \cdot D)$ is very ample and

$$\mathcal{L}^{\mu_0 \cdot \nu_0}(-\mu_0 \cdot D) \otimes \omega_X^{-1} \quad \text{and} \quad \mathcal{L}^{\mu_0 \cdot \nu_0}(-\mu_0 \cdot D + D_{red}) \otimes \omega_X^{-1}$$

are both ample. From (11.4) we find for $n = \dim X$ that both,

$$\mathcal{L}^{\mu_0 \cdot \nu_0 (n+3)+\nu}(-\mu_0(n+3) \cdot D) \quad \text{and} \quad \mathcal{L}^{\mu_0 \cdot \nu_0 (n+3)+\nu}(-\mu_0(n+3) \cdot D + D_{red})$$

are very ample for all $\nu \geq 0$. Hence the assumption (∗) in (11.5) can be replaced by

(∗∗) *For some* $\nu_0 \in \mathbb{N}$ *and all* $\nu \geq 0$ *the sheaves* $\mathcal{L}^{\nu_0+\nu} \otimes \mathcal{O}_X(-D)$ *and* $\mathcal{L}^{\nu_0+\nu} \otimes \mathcal{O}_X(-D + D_{red})$ *are very ample.*

Choose $\eta \in \mathbb{N} - \{0\}$ such that $N = p^\eta + 1 > \nu_0$ and $[\frac{D}{N}] = 0$. Let H be the zero set of a general section of $\mathcal{L}^N \otimes \mathcal{O}_X(-D)$. In §3 we constructed an integrable logarithmic connection $\nabla_{(i)}$ on the sheaf

$$\mathcal{L}^{(i)^{-1}} = \mathcal{L}^{-i}([\frac{i \cdot (D + H)}{N}]) .$$

Let $F : X \to X'$ be the relative Frobenius morphism and $\mathcal{L}', D', H'$ the sheaf and the divisors on X', obtained by field extension via $F_{\operatorname{Spec} k} : k \to k$ from $\mathcal{L}, D$ and H.

As we have seen in the proof of (10.24) and in (10.25) one has an inclusion of complexes

$$(F_* \Omega_X^\bullet(\log (D + H))) \otimes \mathcal{L}'^{(i)^{-1}} \longrightarrow F_*(\Omega_X^\bullet(\log (D + H)) \otimes \mathcal{L}^{(p \cdot i)^{-1}}).$$

This inclusion is a quasi-isomorphism and as in (10.24) one obtains from (10.19) and base change the inequalities

$$\dim \mathbb{H}^l(X, \Omega_X^\bullet(\log (D + H)) \otimes \mathcal{L}^{(1)^{-1}}) \leq$$

$$\sum_{a+b=l} \dim H^b(X, \Omega_X^a(\log (D + H)) \otimes \mathcal{L}^{(1)^{-1}}) =$$

$$\sum_{a+b=l} \dim H^b(X', \Omega^a_{X'}(\log\, (D'+H')) \otimes \mathcal{L}'^{(1)^{-1}}) =$$

$$\dim \mathbb{H}^l(X, \Omega^\bullet_X(\log\, (D+H)) \otimes \mathcal{L}^{(p)^{-1}}) \le \cdots$$

$$\dim \mathbb{H}^l(X, \Omega^\bullet_X(\log\, (D+H)) \otimes \mathcal{L}^{(p^\gamma)^{-1}}) \le$$

$$\sum_{a+b=l} H^b(X, \Omega^a_X(\log\, (D+H)) \otimes \mathcal{L}^{(p^\gamma)^{-1}})$$

for all $\gamma > 0$. For $\gamma = \eta$ we have $p^\eta = N-1$ and

$$\mathcal{L}^{(p^\eta)} = \mathcal{L}^{(N-1)} = \mathcal{L}^{N-1}(-[\tfrac{(N-1)\cdot(D+H)}{N}]) =$$

$$= \mathcal{L}^{N-1}(-D - [\tfrac{-D}{N} + \tfrac{(N-1)\cdot H}{N}]) = \mathcal{L}^{N-1}(-D + D_{red}).$$

Hence $\mathcal{L}^{(p^\eta)}$ is ample and from (11.3) we obtain, for

$$l < \dim\ X \le \text{char}\ k,$$

that

$$\sum_{a+b=l} H^b(X, \Omega^a_X(\log\, (D+H)) \otimes \mathcal{L}^{(p^\eta)^{-1}}) = 0.$$

Since $\mathcal{L}^{(1)} = \mathcal{L}$ we obtain for $a+b < \dim\ X \le \text{char}\ k$

$$H^b(X, \Omega^a_X(\log\, (D+H)) \otimes \mathcal{L}^{-1}) = 0.$$

Finally, since $\mathcal{L}^N \otimes \mathcal{O}_X(-D)$ lifts to $W_2(k)$, we can choose H such that H and $D|_H$ both lift to $W_2(k)$ and the exact sequence

$$H^{b-1}(H, \Omega^{a-1}_H(\log\ D)|_H) \otimes \mathcal{L}^{-1}) \longrightarrow H^b(X, \Omega^a_X(\log\ D) \otimes \mathcal{L}^{-1}) \longrightarrow$$

$$\longrightarrow H^b(X, \Omega^a_X(\log\, (D+H)) \otimes \mathcal{L}^{-1}).$$

allows to prove (11.5) by induction.

□

11.6. Remarks..
a) By (5.7) and (5.4,d) the assumption (∗) in (11.5) implies that $\mathcal{L}$ is numerically effective and of maximal Iitaka dimension.

b) If, on the other hand, $\mathcal{L}$ is numerically effective and of maximal Iitaka dimension, then there exists some effective divisor D such that the sheaf

$$\mathcal{L}^{\nu_0+\nu} \otimes \mathcal{O}_X(-D)$$

is ample for some $\nu_0 \in \mathbb{N}$ and all $\nu \ge 0$. However, in general, this divisor is not a normal crossing divisor and henceforth (11.5) is of no use.

If one assumes that the embedded resolution of singularities holds true over k and even over $W_2(k)$, (11.5) would give an affirmative answer to

11.7. Problem. *Let k be a perfect field of characteristic $p > 0$, let X be a proper smooth k-variety and $\mathcal{L}$ an invertible sheaf. Assume that X and $\mathcal{L}$ admit liftings to $W_2(k)$, that $\mathcal{L}$ is numerically effective and that $\kappa(\mathcal{L}) = \dim X$.*

Does this imply that

$$H^b(X, \mathcal{L}^{-1}) = 0 \quad for \quad b < \dim X \leq \text{char } k \ ?$$

11.8. Remark.
a) If $\dim X = 2$ then (11.5) gives an affirmative answer to the problem (11.7), since we have imbedded resolution of singularities for curves on surfaces. In the surface case however, [12], Cor. 2.8, gives the vanishing of $H^b(X, \mathcal{L}^{-1})$, for $b < 2$, without assuming that $\mathcal{L}$ lifts to $W_2(k)$.
b) As mentioned in Lecture 1 and 8, even if we restrict ourselves to the case where $\mathcal{L}$ is semi-ample and of maximal Iitaka dimension, we do not know the answer to problem (11.7) for higherdimensional X.

§ 12 Deformation theory for cohomology groups

In this lecture we will recall D. Mumford's description of higher direct image sheaves, already used in (10.21), and their base change properties and, following Green and Lazarsfeld [26], deduce the deformation theory for cohomology groups.

12.1. Theorem (Mumford). *Let $g : Z \longrightarrow Y$ be a projective flat morphism of noetherian schemes, let $Y_0 \subset Y$ be an affine open subscheme,*

$$Z_0 = g^{-1}(Y_0) \quad , \quad g_0 = g|_{Z_0}$$

and let $\mathcal{B}$ be a locally free sheaf on Z. Then there exists a bounded complex $(\mathcal{E}^\bullet, \delta_\bullet)$ of locally free $\mathcal{O}_{Y_0}$ modules of finite rank such that

$$\mathcal{H}^b(\mathcal{E}^\bullet \otimes \mathcal{F}) = R^b g_{0*}(\mathcal{B}|_{Z_0} \otimes g_0^* \mathcal{F})$$

for all coherent sheaves $\mathcal{F}$ on Y_0.

In order to construct $\mathcal{E}^\bullet$, D. Mumford uses in [50], II, §5, the description of higher direct images by Čech complexes. The " Coherence Theorem" of Grauert-Grothendieck allows to realize $(\mathcal{E}^\bullet)$ as a complex of locally free sheaves of finite rank. The proof of (12.1) can be found as well in [30], III, §12.

From (12.1) one obtains easily the base change theorems of Grauert and Grothendieck, as well as the ones used at the end of Lecture 10.

12.2. Example. Let $y \in Y_0$ be a point and $\mathcal{F} = k(y)$. For $Z_y = g^{-1}(y)$ one obtains

$$\begin{array}{ccc} \mathcal{H}^b(\mathcal{E}^\bullet \otimes k(y)) & \xrightarrow{=} & H^b(Z_y, B|_{Z_y}) \\ {\scriptstyle\tau}\uparrow & & {\scriptstyle\eta}\uparrow \\ \mathcal{H}^b(\mathcal{E}^\bullet) \otimes k(y) & \xrightarrow{=} & R^b g_*(\mathcal{B}) \otimes k(y), \end{array}$$

where η is the base change morphism ([30], III, 9.3.1). In general, due to the fact that the images of

$$\delta_{b-1} : \mathcal{E}^{b-1} \longrightarrow \mathcal{E}^b \quad \text{and} \quad \delta_b : \mathcal{E}^b \longrightarrow \mathcal{E}^{b+1}$$

are not subbundles of $\mathcal{E}^b$ and $\mathcal{E}^{b+1}$, τ and hence η will be neither injective nor surjective.

12.3. Example. Let X be a projective manifold, defined over an algebraically closed field k and let $Y \subset \mathrm{Pic}^0(X)$ be a closed subscheme,

$$Z = X \times Y \quad \text{and} \quad g = pr_2 : Z \longrightarrow Y.$$

Recall that on Z we have a Poincaré bundle $\mathcal{P}$ (see for example [50]), i.e. an invertible sheaf $\mathcal{P}$ such that

$$\mathcal{P}|_{g^{-1}(y)} \simeq \mathcal{N}_y,$$

if $\mathcal{N}_y$ is the linebundle on X corresponding to

$$y \in Y \subset \mathrm{Pic}^0(X).$$

In more fancy terms, the functor

$$T \longmapsto \mathcal{P}ic(X \times T/T)$$

is represented by a locally noetherian group-scheme Pic(X), whose connected component containing zero is $\mathrm{Pic}^0(X)$. The invertible sheaf $\mathcal{P}$ is the restriction of the universal bundle on $X \times \mathrm{Pic}(X)$ to

$$X \times Y \subset X \times \mathrm{Pic}^0(X)$$

(see [28]). For $y \in Y$ let $T_{y,Y} = (m_{y,Y}/m_{y,Y}^2)^*$ be the Zariski tangent space. We have an exact sequence

$$0 \longrightarrow T_{y,Y}^* \longrightarrow \mathcal{O}_{y,Y}/m_{y,Y}^2 \longrightarrow k(y) \longrightarrow 0.$$

Since

$$g^*(T^*_{y,Y}) \otimes_{\mathcal{O}_Z} \mathcal{P} = g^*(T^*_{y,Y}) \otimes \mathcal{N}_y$$

one obtains the exact sequence

$$0 \longrightarrow g^*(T^*_{y,Y}) \otimes \mathcal{N}_y \longrightarrow g^*(\mathcal{O}_{y,Y}/m^2_{y,Y}) \otimes \mathcal{P} \longrightarrow \mathcal{N}_y \longrightarrow 0$$

on $X \simeq g^{-1}(y)$. If, identifying X with $g^{-1}(y)$,

$$\zeta \in H^1(X, g^*(T^1_{y,Y})) = H^1(X, \mathcal{O}_X) \otimes T^*_{y,Y}$$

is the extension class of this sequence, the induced edge morphism

$$H^b(X, \mathcal{N}_y) \longrightarrow H^{b+1}(X, g^*(T^*_{y,Y}) \otimes \mathcal{N}_y) = H^{b+1}(X, \mathcal{N}_y) \otimes_{k(y)} T^*_{y,Y}$$

is the cup-product with ζ.

12.4.

Keeping the notations from (12.3), let $\mathcal{M}$ be a locally free sheaf on X and $\mathcal{B} = \mathcal{P} \otimes pr_1^* \mathcal{M}$. Since the exact sequence

$$0 \longrightarrow g^*(T^*_{y,Y}) \otimes \mathcal{N}_y \otimes \mathcal{M} \longrightarrow g^*(\mathcal{O}_{y,Y}/m^2_{y,Y}) \otimes \mathcal{B} \longrightarrow \mathcal{N}_y \otimes \mathcal{M} \longrightarrow 0$$

is obtained from

$$0 \longrightarrow g^*(T^*_{y,Y}) \otimes \mathcal{N}_y \longrightarrow g^*(\mathcal{O}_{y,Y}/m^2_{y,Y}) \otimes \mathcal{P} \longrightarrow \mathcal{N}_y \longrightarrow 0$$

by tensorproduct with $\mathcal{M}$, the induced edge morphism

$$H^b(X, \mathcal{M} \otimes \mathcal{N}_y) \longrightarrow H^{b+1}(X, \mathcal{M} \otimes \mathcal{N}_y) \otimes_{k(y)} T^*_{y,Y}$$

is again the cup-product with ζ. Let us finally remark that ζ induces a morphism

$$T_{y,Y} \xrightarrow{\ \zeta\ } H^1(X, \mathcal{O}_X)$$

which, due to the universal property of $\mathcal{P}$ is injective. In fact, if we represent $\tau \in T_{y,Y}$ by a morphism

$$\zeta' : D = \text{Spec } k[\epsilon] \longrightarrow Y \text{ with } \zeta'(<\epsilon>) = y \ ,$$

where $k[\epsilon] = k[t]/t^2$ is the ring of dual numbers, then for $\tau \neq 0$ the pullback of $\mathcal{P}$ to $X \times D$ is non trivial and hence the extension class $\zeta(\tau)$ of

$$0 \longrightarrow \mathcal{N}_y \xrightarrow{\ \cdot\epsilon\ } \mathcal{P}|X \times D \longrightarrow \mathcal{N}_y \longrightarrow 0$$

is non zero. If $Y = \text{Pic}^0(X)$ one has $\dim T_{y,Y} = \dim H^1(X, \mathcal{O}_X)$ and ζ is surjective as well.

12.5. Notations. For $X, \mathcal{M}$ as above let us write

$$S^b(X, \mathcal{M}) = \{y \in \text{Pic}^0(X); H^b(X, \mathcal{N}_y \otimes \mathcal{M}) \neq 0\}.$$

The first part of the following lemma is well known and an easy consequence of the semicontinuity of the dimensions of cohomology groups. To fix notations we will prove it nevertheless.

12.6. Lemma.
a) $S^b(X,\mathcal{M})$ *is a closed subvariety of* $\mathrm{Pic}^0(X)$.

b) If $Y \subset S^b(X,\mathcal{M})$ *is an ireducible component and*

$$m = \mathrm{Min}\{\dim H^b(X,\mathcal{N}_y \otimes \mathcal{M});\ y \in Y\},$$

then the set

$$U = \{y \in Y;\ \dim H^b(X,\mathcal{N}_y \otimes \mathcal{M}) = m\}$$

is open and dense in Y.

c) For $y \in U$ *and* $\zeta : T_{y,Y} \hookrightarrow H^1(X,\mathcal{O}_X)$ *as in (12.4) the cup-products*

$$\zeta(T_{y,Y}) \otimes H^{b-1}(X,\mathcal{N}_y \otimes \mathcal{M}) \longrightarrow H^b(X,\mathcal{N}_y \otimes \mathcal{M})$$

and

$$\zeta(T_{y,Y}) \otimes H^b(X,\mathcal{N}_y \otimes \mathcal{M}) \longrightarrow H^{b+1}(X,\mathcal{N}_y \otimes \mathcal{M})$$

are both zero.

PROOF: For any open affine $P_0 \subset \mathrm{Pic}^0(X)$ let $\mathcal{E}^\bullet$ be the complex from (12.1) describing the higher direct images of

$$\mathcal{B}|_{X\times P_o} = pr_1^*\mathcal{M} \otimes \mathcal{P}|_{X\times P_o}$$

and there base change. If we write

$$\mathcal{W}_b = \mathrm{Coker}(\delta_{b-1} : \mathcal{E}^{b-1} \longrightarrow \mathcal{E}^b),$$

then for $\mathcal{F}$ coherent on P_0 we have

$$\mathcal{W}_b \otimes \mathcal{F} = \mathrm{Coker}(\delta_{b-1} : \mathcal{E}^{b-1} \otimes \mathcal{F} \longrightarrow \mathcal{E}^b \otimes \mathcal{F})$$

and an exact sequence

$$0 \longrightarrow \mathcal{H}^b(\mathcal{E}^\bullet \otimes \mathcal{F}) \longrightarrow \mathcal{W}_b \otimes \mathcal{F} \longrightarrow \mathcal{E}^{b+1} \otimes \mathcal{F} \longrightarrow \mathcal{W}_{b+1} \otimes \mathcal{F} \longrightarrow 0.$$

If $S^b(X,\mathcal{M}) \neq \mathrm{Pic}^0(X)$, then $S^b(X,\mathcal{M})$ is just the locus where $\mathcal{W}_b \to \mathcal{E}^{b+1}$ is not a subbundle. Obviously this condition defines a closed subscheme of P_0. For $y \in Y \cap P_0 = Y_0$ we have

$$\dim \mathcal{H}^b(\mathcal{E}^\bullet \otimes k(y)) = -\mathrm{rank}(\mathcal{E}^{b+1}) + \dim(\mathcal{W}_b \otimes k(y)) + \dim(\mathcal{W}_{b+1} \otimes k(y))$$

and the open set U in b) is nothing but the locus where both,

$$\mathcal{W}_b \otimes \mathcal{O}_{Y_0} \quad \text{and} \quad \mathcal{W}_{b+1} \otimes \mathcal{O}_{Y_0}$$

are locally free $\mathcal{O}_{Y_0}$ modules. On U the sequence

$$0 \longrightarrow \mathcal{H}^b(\mathcal{E}^\bullet|_U) \longrightarrow \mathcal{W}_b|_U \longrightarrow \mathcal{E}^{b+1}|_U \longrightarrow \mathcal{W}_{b+1}|_U \longrightarrow 0$$

is an exact sequence of vector bundles and

$$\mathcal{H}^b(\mathcal{E}^\bullet \otimes \mathcal{F}) = \mathcal{H}^b(\mathcal{E}|_U) \otimes \mathcal{F}$$

for all coherent $\mathcal{O}_U$ modules $\mathcal{F}$. In particular for $y \in U$ the exact sequence

$$0 \longrightarrow \mathcal{E}^\bullet \otimes T^*_{y,Y} \longrightarrow \mathcal{E}^\bullet \otimes \mathcal{O}_{y,Y}/m^2_{y,Y} \longrightarrow \mathcal{E}^\bullet \otimes k(y) \longrightarrow 0$$

induces

$$\begin{array}{ccccc} \mathcal{H}^b(\mathcal{E}^\bullet \otimes T^*_{y,Y}) & \longrightarrow & \mathcal{H}^b(\mathcal{E}^\bullet \otimes \mathcal{O}_{y,Y}/m^2_{y,Y}) & \longrightarrow & \mathcal{H}^b(\mathcal{E}^\bullet \otimes k(y)) \\ {\scriptstyle =}\downarrow & & {\scriptstyle =}\downarrow & & {\scriptstyle =}\downarrow \\ \mathcal{H}^b(\mathcal{E}^\bullet) \otimes T^*_{y,Y} & \longrightarrow & \mathcal{H}^b(\mathcal{E}^\bullet) \otimes \mathcal{O}_{y,Y}/m^2_{y,Y} & \longrightarrow & \mathcal{H}^b(\mathcal{E}^\bullet) \otimes k(y) \end{array}$$

and the edge morphisms

$$\mathcal{H}^i(\mathcal{E}^\bullet \otimes k(y)) \longrightarrow \mathcal{H}^{i+1}(\mathcal{E}^\bullet \otimes T^*_{y,Y})$$

are zero for $i = b$ and $b-1$. Using (12.1) we have identified in (12.4) this edge morphism with the cup-product

$$H^i(X, \mathcal{M} \otimes \mathcal{N}_y) \longrightarrow H^{i+1}(X, \mathcal{M} \otimes \mathcal{N}_y) \otimes_{k(y)} T^*_{y,Y}$$

with the extension class $\zeta \in H^1(X, \mathcal{O}_X) \otimes T^*_{y,Y}$.

□

12.7. Corollary(Green, Lazarsfeld [26]). *If $Y \subset S^b(X, \mathcal{M})$ is an irreducible component and $y \in Y$ is a point in general position then*

$$\mathrm{codim}_{\mathrm{Pic}^0(X)}(Y) \geq \mathrm{codim}(\Gamma \subset H^1(X, \mathcal{O}_X))$$

where

$$\Gamma = \{\varphi \in H^1(X, \mathcal{O}_X);\ \alpha \cup \varphi = 0 \ \textit{ and } \ \beta \cup \varphi = 0 \ \textit{ for all}$$
$$\alpha \in H^{b-1}(X, \mathcal{N}_y \otimes \mathcal{M}) \ \textit{ and } \ \beta \in H^b(X, \mathcal{N}_y \otimes \mathcal{M})\}.$$

12.8. Remark. Even if one seems to loose some information, in the applications of (12.7) in Lecture 13 we will replace Γ by the larger space

$$\{\varphi \in H^1(X, \mathcal{O}_X);\ \beta \cup \varphi = 0 \ \text{ for all } \ \beta \in H^b(X, \mathcal{N}_y \otimes \mathcal{M})\}$$

in order to obtain lower bounds for $\mathrm{codim}_{\mathrm{Pic}^0(X)}(S^b(X, \mathcal{M}))$ for certain invertible sheaves $\mathcal{M}$.

§ 13 Generic vanishing theorems [26], [14]

In this section we want to use (12.7) and Hodge-duality to prove some bounds for

$$\mathrm{codim}_{\mathrm{Pic}^0(X)}(S^b(X,\mathcal{M}))$$

for the subschemes $S^b(X,\mathcal{M})$ introduced in §12. In particular, we lose a little bit the spirit of the previous lectures, where we tried to underline as much as possible the *algebraic* aspects of vanishing theorems. Everything contained in this lecture is either due Green-Lazarsfeld [26] or to H. Dunio [14]. The use of Hodge duality will force us to assume that X is a complex manifold. Without mentioning it we will switch from the algebraic to the analytic language and use the comparison theorem of [56] whenever needed.

13.1. Notations. Let X be a projective complex manifold. The Picard group $\mathrm{Pic}(X)$ is $H^1(X,\mathcal{O}_X^*)$ and, using the exponential sequence, $\mathrm{Pic}^0(X)$ is identified with

$$H^1(X,\mathcal{O}_X)/H^1(X,\mathbb{Z}).$$

Let $\mathcal{P}$ be the Poincaré bundle on $X \times \mathrm{Pic}^0(X)$, and $g = pr_2$. If

$$\zeta : T_{y,\mathrm{Pic}^0(X)} \longrightarrow H^1(X,\mathcal{O}_X)$$

is the extension class of

$$0 \longrightarrow g^*T_{y,\mathrm{Pic}^0(X)} \otimes \mathcal{N}_y \longrightarrow g^*(\mathcal{O}_{y,\mathrm{Pic}^0}/m^2_{y,\mathrm{Pic}^0}) \otimes \mathcal{P} \longrightarrow \mathcal{N}_y \longrightarrow 0$$

then ζ is the identity. Let

$$\mathrm{Alb}(X) = H^0(X,\Omega^1_X)^*/H_1(X,\mathbb{Z})$$

be the Albanese variety of X and

$$\alpha : X \longrightarrow \mathrm{Alb}(X)$$

be the Albanese map. The morphism α induces an isomorphism

$$\alpha^* : H^0(\mathrm{Alb}(X),\Omega^1_{\mathrm{Alb}(X)}) \longrightarrow H^0(X,\Omega^1_X).$$

In particular,

$$\dim \alpha(X) = \mathrm{rank}_{\mathcal{O}_X}(\mathrm{im}(H^0(X,\Omega^1_X) \otimes_{\mathbb{C}} \mathcal{O}_X \longrightarrow \Omega^1_X)) .$$

13.2. Theorem (Green-Lazarsfeld [26]). *Let X be a complex projective manifold and*

$$S^b(X) = \{y \in \mathrm{Pic}^0(X);\ H^b(X,\mathcal{N}_y) \neq 0\}.$$

Then

$$\mathrm{codim}_{\mathrm{Pic}^0(X)}(S^b(X)) \geq \dim(\alpha(X)) - b.$$

In particular, if $\mathcal{N} \in \mathrm{Pic}^0(X)$ is a generic line bundle, then $H^b(X,\mathcal{N}) = 0$ for $b < \dim \alpha(X)$.

PROOF: Using (12.7) or (12.8) one obtains (13.2) from

□

13.3. Claim. *Assume that* $H^b(X, \mathcal{N}_y) \neq 0$ *then for*

$$\Gamma = \{\varphi \in H^1(X, \mathcal{O}_X);\ \beta \cup \varphi = 0 \ \ \textit{for all} \ \ \beta \in H^b(X, \mathcal{N}_y)\}$$

one has

$$\operatorname{codim}(\Gamma \subset H^1(X, \mathcal{O}_X)) \geq \dim(\alpha(X)) - b.$$

PROOF:
Step 1: $\mathcal{N}_y$ is a flat unitary bundle on X, obtained from a unitary representation of the fundamental group. In particular the conjugation of harmonic forms with values in $\mathcal{N}_y$ gives a complex antilinear isomorphism, the so called Hodge duality,

$$\iota : H^b(X, \Omega^a_X \otimes \mathcal{N}_y) \longrightarrow H^a(X, \Omega^b_X \otimes \mathcal{N}_y^{-1})$$

(see (13.5) and (13.6) for generalizations). Moreover, if $\varphi \in H^1(X, \mathcal{O}_X)$ and if $\omega = \bar{\varphi} \in H^0(X, \Omega^1_X)$ is the Hodge-dual of φ, then for $\beta \in H^b(X, \Omega^a_X \otimes \mathcal{N}_y)$ one has

$$\iota(\beta \cup \varphi) = \iota(\beta) \wedge \omega \in H^a(X, \Omega^{b+1}_X \otimes \mathcal{N}_y^{-1}).$$

Hence

$$\iota(\Gamma) = \bar{\Gamma} \subset H^0(X, \Omega^1_X)$$

is the subspace of forms $\omega \in H^0(X, \Omega^1_X)$ such that

$$\beta \wedge \omega = 0 \ \ \text{for all} \ \ \beta \in H^0(X, \Omega^b_X \otimes \mathcal{N}_y^{-1}).$$

Step 2: Consider the natural map

$$\gamma : H^0(X, \Omega^1_X) \otimes \mathcal{O}_X \longrightarrow \Omega^1_X.$$

Since all one-forms are pullback of one-forms on $\alpha(X) \subset \operatorname{Alb}(X)$, the subsheaf $\operatorname{im}(\gamma)$ of Ω^1_X is of rank $\dim \alpha(X)$ and

$$r = \operatorname{rank}\gamma(\overline{\Gamma} \otimes \mathcal{O}_X) = \dim \overline{\Gamma} - \operatorname{rank}(\ker(\gamma) \cap \overline{\Gamma} \otimes \mathcal{O}_X) \geq$$

$$\dim \overline{\Gamma} - \dim \ker(\gamma) = \dim \overline{\Gamma} - (\dim H^0(X, \Omega^1_X) - \dim \alpha(X))$$

$$= \dim \alpha(X) - \operatorname{codim}(\Gamma \subset H^1(X, \mathcal{O}_X)) \ .$$

We assumed that $H^0(X, \Omega^b_X \otimes \mathcal{N}_y^{-1}) \neq 0$. Hence we have at least one element

$$\beta \in H^0(X, \Omega^b_X \otimes \mathcal{N}_y^{-1}) \ \ \text{and} \ \ \beta \wedge \gamma(\overline{\Gamma} \otimes \mathcal{O}_X) = 0.$$

Since

$$(\wedge^b \Omega^1_X) \otimes (\wedge^{n-b} \Omega^1_X) \longrightarrow \Omega^n_X$$

is a nondegenerate pairing, for $n = \dim X$, we find some meromorphic differential form

$$\delta \in \Omega_X^{n-b} \otimes \mathbb{C}(X) \quad \text{with} \quad \delta \wedge \beta \neq 0.$$

Hence δ lies in

$$\Omega_X^{n-b} \otimes \mathbb{C}(X) - \{\gamma(\overline{\Gamma} \otimes \mathcal{O}_X) \wedge \Omega_X^{n-b-1} \otimes \mathbb{C}(X)\}.$$

This however is only possible if $n - b \leq n - r$ or $b \geq r$. Altogether we find

$$b \geq \dim \alpha(X) - \operatorname{codim}(\Gamma \subset H^1(X, \mathcal{O}_X)).$$

□

13.4.

If one tries to use the same methods for $S^b(X, \mathcal{M})$ one has to make sure that $H^b(X, \mathcal{M} \otimes \mathcal{N}_y)$ is in Hodge duality with $H^0(X, \Omega_X^b \otimes \mathcal{M}' \otimes \mathcal{N}_y^*)$ for some sheaf $\mathcal{M}'$. As shown in (3.23) this holds true for the sheaves $\mathcal{L}^{(i)}$ arising from cyclic coverings, at least if one considers $\Omega_X^b(\log\ D)$ instead of Ω_X^b. More generally one has:

13.5. Theorem (K. Timmerscheidt [59]). *Let D be a normal crossing divisor on X, let $\mathcal{V}$ be a locally free sheaf and*

$$\nabla : \mathcal{V} \longrightarrow \Omega_X^1(\log\ D) \otimes \mathcal{V}$$

an integrable logarithmic connection. Assume that for all components D_i of D, the real part of all eigenvalues of $\operatorname{res}_{D_i}(\nabla)$ lies in $(0,1)$
(which implies that conditions (*) and (!) of (2.8) are satisfied, and that $\mathcal{V}$ is the canonical extension defined by Deligne [10]).
Assume moreover that the local constant system

$$V = \ker(\nabla : \mathcal{V}|_U \longrightarrow \Omega_U^1 \otimes \mathcal{V}|_U)$$

is unitary for $U = X - D$. Then one has:
a) The Hodge to de Rham spectral sequence

$$E_1^{ab} = H^b(X, \Omega_X^a(\log\ D) \otimes \mathcal{V}) \longrightarrow \mathbb{H}^{a+b}(X, \Omega_X^\bullet(\log\ D) \otimes \mathcal{V})$$

degenerates at E_1.
b) There exists a $\mathbb{C}$ - antilinear isomorphism

$$\iota : H^b(X, \Omega_X^a(\log\ D) \otimes \mathcal{V}) \longrightarrow H^a(X, \Omega_X^b(\log\ D) \otimes \mathcal{V}^*(-D_{\text{red}}))$$

such that for $\varphi \in H^1(X, \mathcal{O}_X)$ and $\omega = \overline{\varphi} \in H^0(X, \Omega_X^1)$ the diagram

$$\begin{array}{ccc} H^b(X, \Omega_X^a(\log\ D) \otimes \mathcal{V}) & \xrightarrow{\iota} & H^a(X, \Omega^b(\log\ D) \otimes \mathcal{V}^*(-D_{\text{red}})) \\ \downarrow{\scriptstyle \cup\varphi} & & \downarrow{\scriptstyle \wedge\omega} \\ H^{b+1}(X, \Omega_X^a(\log\ D) \otimes \mathcal{V}) & \xrightarrow{\iota} & H^a(X, \Omega_X^{b+1}(\log\ D) \otimes \mathcal{V}^*(-D_{\text{red}})) \end{array}$$

commutes.

13.6. Examples.
a) If $D = 0$ and if $\mathcal{N}_y$ is the invertible sheaf corresponding to $y \in \mathrm{Pic}^0(X)$, then $\mathcal{N}_y$ has an integrable connection ∇, whose kernel is a unitary rank one local constant system. In this case (13.5) is wellknown and proven by the usual arguments from classical Hodge-theory, applied to $\mathcal{N}_y \otimes \mathcal{N}_y^*$ see [11].
b) If the sheaf $\mathcal{V}$ in (13.5) is of the form $\mathcal{V} = \mathcal{L}^{(i)^{-1}}$ for

$$D = \sum_{j=1}^{r} \alpha_j D_j \quad \text{and} \quad \mathcal{L}^N = \mathcal{O}_X(D),$$

then the assumptions made in (13.5) are satisfied whenever

$$\frac{i \cdot \alpha_j}{N} \notin \mathbb{Z} \ \text{ for } \ j = 1, \ldots, r.$$

Hence, using the notations from (3.2), one has $D^{(i)} = D^{(N-i)} = D_{red}$ and

$$\mathcal{L}^{(i)}(-D_{red}) = \mathcal{L}^i(-[\frac{i \cdot D}{N}] - D_{red}) = \mathcal{L}^i(-\{\frac{i \cdot D}{N}\}) =$$

$$\mathcal{L}^i([\frac{-i \cdot D}{N}]) = \mathcal{L}^{i-N}([\frac{(N-i) \cdot D}{N}]) = \mathcal{L}^{(N-i)^{-1}}$$

Hence (3.2) and (3.23) imply (13.5) for $\mathcal{M} = \mathcal{L}^{(i)^{-1}}$.
c) Finally, for $\mathcal{M} = \mathcal{L}^{(i)^{-1}} \otimes \mathcal{N}_y$ one can use (13.6,a) on the finite covering Y of X obtained by taking the N-th root out of D and the arguments used to prove (3.23) imply (13.5)in that case.

13.7. Corollary (H. Dunio [14]). *Keeping the assumptions made in (13.5) and the notations introduced in (13.1) and (12.5) one has*

$$\mathrm{codim}_{\mathrm{Pic}_0(X)}(S^b(X, \mathcal{V})) \geq \dim(\alpha(X)) - b.$$

PROOF: Again, it is sufficient to give a lower bound for

$$\mathrm{codim}(\Gamma \subset H^1(X, \mathcal{O}_X))$$

where

$$\Gamma = \{\varphi \in H^1(X, \mathcal{O}_X); \ \beta \cup \varphi = 0 \ \text{ for all } \ \beta \in H^b(X, \mathcal{N}_y \otimes \mathcal{V})\},$$

or using (13.5,b), for

$$\mathrm{codim}(\bar{\Gamma} \subset H^0(X, \Omega_X^1))$$

where

$$\bar{\Gamma} = \{\omega \in H^0(X, \Omega_X^1); \beta \wedge \omega = 0 \ \text{ for all}$$
$$\beta \in H^0(X, \Omega_X^b(\log D) \otimes \mathcal{N}_y^* \otimes \mathcal{V}^*(-D_{red}))\}.$$

As in (13.3), if

$$\gamma : H^0(X, \Omega^1_X) \otimes \mathcal{O}_X \longrightarrow \Omega^1_X$$

is the natural map, one has

$$r = \operatorname{rank}(\gamma(\bar{\Gamma} \otimes \mathcal{O}_X)) \geq \dim \alpha(X) - \operatorname{codim}(\bar{\Gamma} \subset H^0(X, \Omega^1_X)).$$

Assume that one has some

$$0 \neq \beta \in H^0(X, \Omega^b_X(\log\ D) \otimes \mathcal{N}^*_y \otimes \mathcal{V}^*(-D_{red})).$$

Let $v_1, \ldots., v_s$ be a basis of

$$\mathcal{N}^*_y \otimes \mathcal{V}^*(-D_{red}) \otimes \mathbb{C}(X),$$

then one has $\beta = \sum_{i=1}^{s} \beta_i v_i$ for some $\beta_i \in \Omega^b_X \otimes \mathbb{C}(X)$ and $\beta_i \wedge \omega = 0$ for all i and all

$$\omega \in \gamma(\bar{\Gamma} \otimes \mathcal{O}_X) \otimes \mathbb{C}(X).$$

As in (13.3) this is only possible if $b \geq r$.

□

13.8. Example. If $\mathcal{L}$ is an invertible sheaf on X and if D is a normal crossing divisor let $N \in \mathbb{N}$ be larger than the multiplicities of the components of D. If $\mathcal{L}^N = \mathcal{O}_X(D)$ then $\mathcal{V} = \mathcal{L}^{-1}$ satisfies the assumptions made in (13.7) and

$$H^b(X, \mathcal{L}^{-1} \otimes \mathcal{N}_y) = 0$$

for $y \in \operatorname{Pic}^0(X)$ in general position and $b < \dim \alpha(X)$.
On the other hand, (5.12,e) tells us, that

$$H^b(X, \mathcal{L}^{-1} \otimes \mathcal{N}_y) = 0$$

for all $y \in \operatorname{Pic}^0(X)$ and $b < \kappa(\mathcal{L})$. Hence (13.7) is only of interest if

$$\dim \alpha(X) - \kappa(\mathcal{L}) > 0.$$

In this situation the bounds given in (13.7) can be improved. The generic vanishing theorem remains true for

$$b < \dim \alpha(X) + \kappa(\mathcal{L}) - \dim \alpha'(Z)$$

where Z is a desingularization of the image of the rational map

$$\Phi_\nu : X \longrightarrow \mathbb{P}(H^0(X, \mathcal{L}^\nu))$$

for ν sufficiently large (see (5.3)) and where

$$\alpha' : Z \longrightarrow \operatorname{Alb}(Z)$$

is the Albanese map of Z. To be more precise:

13.9. Assumptions and Notations. Let X be a complex projective manifold, let $\mathcal{L}$ be an invertible sheaf on X, let

$$D = \sum_{j=1}^{r} \alpha_j D_j$$

be a normal crossing divisor and let N be a natural number with

$$0 < \alpha_j < N \quad \text{for} \quad j = 1, ..., r.$$

Assume that either $\mathcal{L}^N(-D)$ is semi-ample or that (more generally) $\mathcal{L}^N(-D)$ is numerically effective and

$$\kappa(\mathcal{L}^N(-D)) = \nu(\mathcal{L}^N(-D))$$

(see (5.9) and (5.11)). For some $\mu > 0$ the rational map (see (5.3))

$$\Phi_\mu : X \longrightarrow \Phi_\mu(X) \subset \mathbb{P}(H^0(X, \mathcal{L}^\mu))$$

has an irreducible general fibre and $\dim(\Phi_\mu(X)) = \kappa(\mathcal{L})$. For such a μ let Z be a desingularization of $\Phi_\mu(X)$ and X' a blowing up of X such that the induced rational map

$$\Phi' : X' \longrightarrow Z$$

is a morphism of manifolds. Φ'^* defines a morphism

$$\Phi^* : \mathrm{Pic}^0(Z) \longrightarrow \mathrm{Pic}^0(X') = \mathrm{Pic}^0(X)$$

and $\Phi^*(\mathrm{Pic}^0(Z))$ is an abelian subvariety of $\mathrm{Pic}^0(X)$ independent of the desingularization choosen. Let

$$\alpha : X \longrightarrow \mathrm{Alb}(X) \quad \text{and} \quad \alpha' : Z \longrightarrow \mathrm{Alb}(Z)$$

be the Albanese maps.

13.10. Theorem (H. Dunio [14]). *Under the assumptions made in (13.9) one has:*

a) $S^b(X, \mathcal{L}^{-1}) = 0$ *for* $b < \kappa(\mathcal{L})$.

b) $S^b(X, \mathcal{L}^{-1})$ *lies in the subgroup of* $\mathrm{Pic}^0(X)$, *which is generated by torsion elements and by* $\Phi^*(\mathrm{Pic}^0(Z))$, *for* $b = \kappa(\mathcal{L})$.

c) $\mathrm{codim}_{\mathrm{Pic}^0(X)}(S^b(X, \mathcal{L}^{-1})) \geq \dim\ \alpha(X) - \dim\ \alpha'(Z) + \kappa(\mathcal{L}) - b$.

PROOF: a) is nothing but (5.12,e) and it has already be shown twice in these notes. Nevertheless, when we prove (13.10,c) it will come out again.

First of all, since $S^b(X, \mathcal{L}^{-1})$ is compatible with blowing ups of X, we may assume that the rational map $\Phi : X \to Z$ is a morphism. Moreover, as in the proof of (5.12), we can assume that $\mathcal{L}^N(-D)$ is semi-ample or even, replacing N and D by some common multiple, that

$$\mathcal{L}^N(-D) = \mathcal{O}_X(H)$$

where H is a non singular divisor and $D + H$ a normal crossing divisor. Since

$$\mathcal{L} = \mathcal{L}^{(1,D)} = \mathcal{L}^{(1,D+H)}$$

we can as well assume that $\mathcal{L}^N = \mathcal{O}_X(D)$. By (13.5,b) or (13.6,c) the space

$$H^b(X, \mathcal{L}^{-1} \otimes \mathcal{N}_y)$$

is Hodge dual to

$$H^0(X, \Omega^b_X(\log\ D) \otimes \mathcal{L}^{(N-1)^{-1}} \otimes \mathcal{N}_y^{-1}).$$

Let $\mathcal{G}^b_\Phi \hookrightarrow \Omega^b_X(\log\ D)$ be the largest subsheaf which over some open non empty subvariety of X coincides with

$$\Phi^*\Omega^\kappa_Z \wedge \Omega^{b-\kappa}_X(\log\ D).$$

Of course $\mathcal{G}^b_\Phi = 0$ for $b < \kappa = \kappa(\mathcal{L})$ and $\delta \in \mathcal{G}^b_\Phi \otimes \mathbb{C}(X)$ if and only if δ is a meromorphic b-form with $\delta \wedge \Phi^*\Omega^1_Z = 0$.

Since $\mathcal{L}^{(N-1)} \subseteq \mathcal{L}^{N-1}$ and since

$$(\mathcal{L}^{(N-1)})^N = \mathcal{O}_X(N \cdot D_{red} - D),$$

we have $\kappa(\mathcal{L}^{(N-1)}) = \kappa(\mathcal{L})$ and $(\mathcal{L}^{(N-1)})^\mu$ contains $\mathcal{L}^N$ for some $\mu > 0$.

13.11. Claim. If $\mathcal{M}$ is an invertible sheaf such that $\mathcal{M}^\mu$ contains $\mathcal{L}^N$ for some $\mu > 0$, then

$$H^0(X, \mathcal{G}^b_\Phi \otimes \mathcal{M}^{-1} \otimes \mathcal{N}_y^{-1}) = H^0(X, \Omega^b_X(\log\ D) \otimes \mathcal{M}^{-1} \otimes \mathcal{N}_y^{-1}).$$

PROOF: The methods used to prove (13.11) are due to F. Bogomolov [6]. A section $\beta \in H^0(X, \Omega^b_X(\log\ D) \otimes \mathcal{M}^{-1} \otimes \mathcal{N}_y^{-1})$ gives an inclusion

$$\beta : \mathcal{M} \longrightarrow \Omega^b_X(\log\ D) \otimes \mathcal{N}_y^{-1}$$

and we have to show that $\Phi^*\Omega^1_Z \wedge \beta(\mathcal{M}) = 0$.

If $\tau : X' \longrightarrow X$ is generically finite and $D' = \tau^* D$ a normal crossing divisor, then $\tau^*\Omega^b_X(\log\ D)$ is a subsheaf of $\Omega^b_{X'}(\log\ D')$. In fact, if (locally) D' is the zero set of $x'_1, ..., x'_r$ and if x is a local parameter on X defining one component of D, then

$$\tau^* x = \prod_{j=1}^{r} x'^{\nu_j}_j \quad \text{and} \quad \tau^* \frac{dx}{x} = \sum_{j=1}^{r} \nu_j \frac{dx'_j}{x'_j} \in \Omega^b_{X'}(\log\ D').$$

Hence β induces

$$\beta' : \tau^*\mathcal{M} \longrightarrow \Omega^b_{X'}(\log\ D') \otimes \tau^*(\mathcal{N}_y^{-1}).$$

Since

$$\beta'(\tau^*\mathcal{M}) \wedge \tau^*\Phi^*\Omega^1_Z = 0$$

implies that

$$\beta(\mathcal{M}) \wedge \Phi^*\Omega^1_Z = 0,$$

we can replace X by X' whenever we like. For example, if

$$s_0, ..., s_\kappa \in H^0(X, \mathcal{L}^\nu) \subset H^0(X, \mathcal{M}^\mu)$$

are choosen such that the functions

$$\frac{s_1}{s_0}, ..., \frac{s_\kappa}{s_0}$$

are algebraic independent, we can take X' as a desingularization of the covering obtained by taking the μ-th root out of $s_0, s_1, ..., s_\kappa$. Hence to prove (13.11) we may assume that $\mathcal{M}$ itself has sections

$$s_0, ..., s_\kappa \ \text{ with } \ f_1 = \frac{s_1}{s_0}, ..., f_\kappa = \frac{s_\kappa}{s_0}$$

algebraic independent. From (13.5) we know that $d(\beta(s_i)) = 0$, which by the Leibniz rule implies

$$0 = d(\beta(s_i)) = d(f_i \cdot \beta(s_0)) = d(f_i) \wedge \beta(s_0).$$

However, $d(f_1), ..., d(f_\kappa)$ are generators of $\Phi^*\Omega^1_Z$ over some non empty open subset.

□

Part a) of (13.10) follows from (13.11) since for $b < \kappa$ the sheaf $\mathcal{G}^b_\Phi = 0$.

If $b = \kappa$ then

$$\mathcal{G}^b_\Phi = \Phi^*\omega_Z \otimes \mathcal{O}_X(\Delta)$$

for some effective divisor Δ on X, not meeting the general fibre F of Φ. Hence, for $y \in S^\kappa(X, \mathcal{L}^{-1})$ (13.11) implies that $\mathcal{L}^{(N-1)^{-1}} \otimes \mathcal{N}_y^{-1}|_F$ has a non trivial section and therefore $\mathcal{N}_y^{-1}|_F = \mathcal{L}^{(N-1)}|_F$. The divisor $D + H$ does not meet the general fibre F and, as we claimed in (13.10,b), $N \cdot y \in \Phi^*(\mathrm{Pic}^0(Z))$.

13.12. Remark. If $\mathcal{L}$ is semi-ample and $b > \kappa$, then a similar argument shows that $\mathcal{L}^{(N-1)^{-1}} \otimes \mathcal{N}_y^{-1}|_F \otimes \Omega_F^{b-\kappa}$ has a non trivial section. This implies, as we have seen in the proof of (13.2), that those $\mathcal{N}_y|_F$ are corresponding to points y in a subvariety of $\mathrm{Pic}^0(F)$ of codimension larger than or equal to

$$\dim(\alpha(F)) - b + \kappa = \dim(\alpha(X)) - \dim(\alpha(Z)) - b + \kappa,$$

which gives (13.10,c).

We instead generalise the argument used in step 2 of the proof of (13.3): If

$$\Gamma = \{\varphi \in H^1(X, \mathcal{O}_X);\ \beta \cup \varphi = 0 \ \text{ for all } \ \beta \in H^b(X, \mathcal{L}^{-1} \otimes \mathcal{N}_y)\}$$

then the Hodge dual of Γ is

$$\bar{\Gamma} = \{\omega \in H^0(X, \Omega_X^1);\ \beta \cup \omega = 0 \ \text{ for all } \ \beta \in H^0(X, \mathcal{G}_\Phi^b \otimes \mathcal{L}^{(N-1)^{-1}} \otimes \mathcal{N}_y^{-1})\}.$$

If γ is the composed map

$$H^0(X, \Omega_X^1) \otimes \mathcal{O}_X \longrightarrow \Omega_X^1 \longrightarrow \Omega_{X/Z}^1$$

then

$$H^0(X, \mathcal{G}_\Phi^b \otimes \mathcal{L}^{(N-1)^{-1}} \otimes \mathcal{N}_y^{-1}) \neq 0$$

implies that

$$\gamma(\bar{\Gamma} \otimes \mathcal{O}_X) \wedge \bar{\beta} = 0 \ \text{ for some } \ \bar{\beta} \in \Omega_{X/Z}^{b-\kappa} \otimes \mathbb{C}(X)$$

or, in other terms, that

$$\Omega_{X/Z}^{n-b} \otimes \mathbb{C}(X) \neq \gamma(\bar{\Gamma} \otimes \mathcal{O}_X) \wedge \Omega_{X/Z}^{n-b-1} \otimes \mathbb{C}(X).$$

Again this is only possible for

$$n - b \leq n - \kappa - \mathrm{rank}(\gamma(\bar{\Gamma} \otimes \mathcal{O}_X))$$

or

$$b - \kappa \geq \mathrm{rank}(\gamma(\bar{\Gamma} \otimes \mathcal{O}_X)).$$

However,

$$\mathrm{rank}(\gamma(\bar{\Gamma} \otimes \mathcal{O}_X)) \geq \dim(\alpha(X)) - \dim(\alpha'(Z)) - \mathrm{codim}(\bar{\Gamma} \subset H^0(X, \Omega_X^1))$$

and (13.10,c) follows from (12.7)

□

13.13. Remarks.
a) If $Z(\varphi)$ denotes the zero locus of a global one-form φ and

$$w(X) = \mathrm{Max}\{\mathrm{codim}_X Z(\varphi); \varphi \in H^0(X, \Omega_X^1)\},$$

then a second result of Green and Lazarsfeld [26] says that, for a generic line bundle

$$\mathcal{N} \in \mathrm{Pic}^0(X) \quad \text{and} \quad a + b < w(X),$$

one has

$$H^b(X, \Omega_X^a \otimes \mathcal{N}) = 0.$$

b) In [27] Green and Lazarsfeld obtain moreover a more explicit description of the subvarieties $S^b(X)$ of $\mathrm{Pic}^0(X)$. They show that the irreducible components of $S^b(X)$ are translates of subtori of $\mathrm{Pic}^0(X)$. This description generalizes results due to A. Beauville [5], who studied $S^1(X)$ and showed the same result in this case.

c) Finally, C. Simpson recently gave in [58] a complete description of the $S^b(X)$ and similar "degeneration loci". In particular he showed that the components of $S^b(X)$ are even translates of subtori of $\mathrm{Pic}^0(X)$ by points of finite order, a result conjectured and proved for $b = 1$ by A. Beauville.

d) Writing these notes we would have liked to prove the generic vanishing theorems for invertible sheaves in the algebraic language used in the first part. However, we were not able to replace the use of Hodge duality by some algebraic argument.

APPENDIX: Hypercohomology and spectral sequences

1.

In this appendix, we list some formal properties of cohomology of complexes that we are using throughout these notes. However we do not pretend making a complete account on this topic. In particular, we avoid the use of the derived category, which is treated to a broad extend in the literature (see [60], [29], [7], [8], [33], [31]).

2.

Through this section X is a variety over a commutative ring k.

3.

We consider *complexes* $\mathcal{F}^\bullet$ *of sheaves* of $\mathcal{O}$-modules, where $\mathcal{O}$ is a sheaf of commutative rings. For example

$$\mathcal{O} = \mathbb{Z}\ ,\ \mathcal{O} = k \ \text{ or } \ \mathcal{O} = \text{ structure sheaf of } X.$$

Any map of $\mathcal{O}$-modules

$$\sigma : \mathcal{F}^\bullet \longrightarrow \mathcal{G}^\bullet$$

between two such complexes induces a map of *cohomology sheaves*:

$$\mathcal{H}^i(\sigma) : \mathcal{H}^i(\mathcal{F}^\bullet) \longrightarrow \mathcal{H}^i(\mathcal{G}^\bullet)$$

where $\mathcal{H}^i(\mathcal{F}^\bullet)$ is the sheaf associated to the presheaf

$$U \mapsto \frac{\ker\ \Gamma(U, \mathcal{F}^i) \to \Gamma(U, \mathcal{F}^{i+1})}{\text{im}\ \Gamma(U, \mathcal{F}^{i-1}) \to \Gamma(U, \mathcal{F}^i)}$$

in the given topology.
One says that σ is a *quasi-isomorphism* if $\mathcal{H}^i(\sigma)$ is an isomorphism for all i.

4.

We will only consider complexes $\mathcal{F}^\bullet$ which are *bounded below*, that means $\mathcal{F}^i = 0$ for i sufficiently negative.

5. Example: the analytic de Rham complex.

X is a complex manifold. Then the standard map

$$\mathbb{C} \longrightarrow \ (\ \mathcal{O}_X \to \Omega^1_X \to \Omega^2_X \to \Omega^3_X \to \cdots\ )$$

from the constant sheaf $\mathbb{C}$ to the analytic de Rham complex $\Omega^\bullet_X$ is a quasi-isomorphism as by the so called "Poincaré lemma"

$$\mathcal{H}^i(\Omega^\bullet_X) = 0 \text{ for } i > 0, \text{ and } \mathcal{H}^0(\Omega^\bullet_X) = \mathbb{C}.$$

6. Example: the Čech complex.

Let $\mathcal{U} = \{U_\alpha; \alpha \in A\}$, for $A \subset \mathbb{N}$, be some open covering of the variety X defined over k. To a bounded below complex $\mathcal{F}^\bullet$ one associates its Čech complex $\mathcal{G}^\bullet$ defined as follows.

$$\mathcal{G}^i := \bigoplus_{a \geq 0} \mathcal{C}^a(\mathcal{U}, \mathcal{F}^{i-a})$$

where

$$\mathcal{C}^a(\mathcal{U}, \mathcal{F}^{i-a}) = \prod_{\alpha_0 < \alpha_1 < \ldots < \alpha_a} \varrho_* \mathcal{F}^{i-a}|_{U_{\alpha_0 \ldots \alpha_a}}.$$

Here, for any

$$U_{\alpha_0 \ldots \alpha_a} := U_{\alpha_0} \cap \ldots \cap U_{\alpha_a}$$

ϱ denotes the open embedding

$$\varrho = \varrho_{\alpha_0 \ldots \alpha_a} : U_{\alpha_0 \ldots \alpha_a} \longrightarrow X,$$

and for any sheaf $\mathcal{F}$, one writes $\varrho_* \mathcal{F}$ for the sheaf associated to the presheaf

$$U \mapsto \Gamma(U_{\alpha_0 \ldots \alpha_a} \cap U, \mathcal{F}).$$

As $\mathcal{F}^i = 0$ for $i << 0$, the direct sum in the definition of $\mathcal{G}$ has finitely many summands.

The differential Δ of $\mathcal{G}^\bullet$ is defined by

$$\Delta(s) = (-1)^i \delta s + d_{\mathcal{F}^\bullet} s \text{ for } s \in \mathcal{C}^a(\mathcal{U}, \mathcal{F}^{i-a}),$$

where δ is the Čech differential defined by

$$(\delta s)_{\alpha_0 \ldots \alpha_{a+1}} = \sum_0^{a+1} (-1)^l s_{\alpha_0 \ldots \hat{\alpha}_l \ldots \alpha_{a+1}}|_{U_{\alpha_0 \ldots \alpha_{a+1}}}$$

and $d_{\mathcal{F}^\bullet}$ is the differential of $\mathcal{F}^\bullet$. Then the natural map

$$\sigma : \mathcal{F}^\bullet \longrightarrow \mathcal{G}^\bullet$$

defined by

$$\mathcal{F}^i \xrightarrow{\varrho} \prod_{\alpha \in A} \varrho_* \mathcal{F}^i|_{U_\alpha} = \mathcal{C}^0(\mathcal{U}, \mathcal{F}^i)$$

is a quasi-isomorphism.

To show this one considers first a single sheaf $\mathcal{F}$ (see [30], III 4.2) and then one computes that, whenever one has a double complex

$$\begin{array}{ccccccccc}
 & \uparrow & & \uparrow & & \uparrow & & & \\
\longrightarrow & \mathcal{K}^{1,i-1} & \longrightarrow & \mathcal{K}^{1,i} & \longrightarrow & \mathcal{K}^{1,i+1} & \longrightarrow & & \\
 & \uparrow & & \uparrow & & \uparrow & & \uparrow d_{vert} & \\
\longrightarrow & \mathcal{K}^{0,i-1} & \longrightarrow & \mathcal{K}^{0,i} & \longrightarrow & \mathcal{K}^{0,i+1} & \longrightarrow & & \\
 & \uparrow & & \uparrow & & \uparrow & & & \\
\longrightarrow & \mathcal{F}^{i-1} & \longrightarrow & \mathcal{F}^{i} & \longrightarrow & \mathcal{F}^{i+1} & \longrightarrow & \cdots & \underset{d_{hor}}{\longrightarrow}
\end{array}$$

such that $\mathcal{F}^i \longrightarrow \mathcal{K}^{\bullet,i}$ is a quasi-isomorphism for all i, then $\mathcal{F}^\bullet \longrightarrow \mathcal{D}^\bullet$ is a quasi-isomorphism as well, where $\mathcal{D}^\bullet$ is the associated *double complex*:

$$\mathcal{D}^i := \bigoplus_a \mathcal{K}^{a,i-a} \longrightarrow \mathcal{D}^{i+1} := \bigoplus_a \mathcal{K}^{a,i+1-a}$$

with differential $(-1)^i d_{vert} + d_{hor}$.

7.

If $\mathcal{F}^\bullet$ and $\mathcal{G}^\bullet$ are two complexes of $\mathcal{O}$-modules bounded below one defines the *tensor product* $\mathcal{F}^\bullet \otimes \mathcal{G}^\bullet$ by

$$(\mathcal{F}^\bullet \otimes \mathcal{G}^\bullet)^i := \bigoplus_a \mathcal{F}^a \otimes \mathcal{G}^{i-a}$$

with differential from

$$\mathcal{F}^a \otimes \mathcal{G}^{i-a} \text{ to } \mathcal{F}^{a+1} \otimes \mathcal{G}^{i-a} \oplus \mathcal{F}^a \otimes \mathcal{G}^{i+1-a}. \text{ given by}$$

$$d(f_a \otimes g_{i-a}) = df_a \otimes g_{i-a} + (-1)^a f_a \otimes dg_{i-a}.$$

As both $\mathcal{F}^\bullet$ and $\mathcal{G}^\bullet$ are bounded below, a takes finitely many values, and $(\mathcal{F}^\bullet \otimes \mathcal{G}^\bullet)$ is a complex of $\mathcal{O}$-modules bounded below.

8.

If in 7 we assume moreover that *locally* the $\mathcal{O}$-modules $\mathcal{F}^a$ and $\mathcal{G}^b$ are *free* and $d\mathcal{F}^{a-1}$ as well as $d\mathcal{G}^{b-1}$ are subbundles for all a and b, then locally one has some decomposition

$$\begin{array}{ccccccc}
\mathcal{F}^a & = & d\mathcal{F}^{a-1} & \oplus & \mathcal{H}^a(\mathcal{F}^\bullet) & \oplus & \mathcal{F}'^a \\
\mathcal{G}^b & = & d\mathcal{G}^{b-1} & \oplus & \mathcal{H}^b(\mathcal{G}^\bullet) & \oplus & \mathcal{G}'^b
\end{array}$$

where

$$\begin{array}{cccc}
d: & \mathcal{F}'^a & \longrightarrow & d\mathcal{F}^a \\
d: & \mathcal{G}'^b & \longrightarrow & d\mathcal{G}^b
\end{array}$$

are isomorphisms. In particular,

$$\begin{aligned}\mathcal{F}^a \otimes \mathcal{G}^b &= d(\mathcal{F}^{a-1} \otimes d\mathcal{G}^{b-1} + \mathcal{F}^{a-1} \otimes \mathcal{H}^b + \mathcal{H}^a \otimes \mathcal{G}^{b-1}) \\ &\oplus \mathcal{H}^a(\mathcal{F}^\bullet) \otimes \mathcal{H}^b(\mathcal{G}^\bullet) \\ &\oplus (d\mathcal{F}^{a-1} + \mathcal{H}^a(\mathcal{F}^\bullet)) \otimes \mathcal{G}'^b \\ &\oplus \mathcal{F}'^a \otimes (d\mathcal{G}^{b-1} + \mathcal{H}^b(\mathcal{G}^\bullet)) \\ &\oplus \mathcal{F}'^a \otimes \mathcal{G}'^b\end{aligned}$$

and therefore one has the Künneth decomposition

$$\mathcal{H}^i(\mathcal{F}^\bullet \otimes \mathcal{G}^\bullet) = \bigoplus_a \mathcal{H}^a(\mathcal{F}^\bullet) \otimes \mathcal{H}^{i-a}(\mathcal{G}^\bullet).$$

9.

The map $\sigma : \mathcal{F}^\bullet \longrightarrow \mathcal{I}^\bullet$ is called an *injective resolution of* $\mathcal{F}^\bullet$ if $\mathcal{I}^\bullet$ is a complex of $\mathcal{O}$-modules bounded below, σ is a quasi-isomorphism, and the sheaves $\mathcal{I}^i$ are injective for all i, i.e.:

$$\mathcal{H}om_{\mathcal{O}}(\mathcal{B}, \mathcal{I}^i) \longrightarrow \mathcal{H}om_{\mathcal{O}}(\mathcal{A}, \mathcal{I}^i)$$

is surjective for any injective map $\mathcal{A} \longrightarrow \mathcal{B}$ of sheaves of $\mathcal{O}$-modules. It is an easy fact that if $\mathcal{O}$ is a constant commutative ring, for example $\mathcal{O} = \mathbb{Z}$, $\mathcal{O} = k$, then every complex of $\mathcal{O}$-modules which is bounded below admits an injective resolution (see [33], (6.1)).

10.

From now on we assume that $\mathcal{O}$ is a constant commutative ring. Let $\mathcal{F}^\bullet$ be a complex of $\mathcal{O}$-modules, bounded below. One defines the *hypercohomology group* $\mathbb{H}^a(X, \mathcal{F}^\bullet)$, to be the $\mathcal{O}$-module

$$\mathbb{H}^a(X, \mathcal{F}^\bullet) := \frac{\ker\ \Gamma(X, \mathcal{I}^a) \to \Gamma(X, \mathcal{I}^{a+1})}{\operatorname{im}\ \Gamma(X, \mathcal{I}^{a-1}) \to \Gamma(X, \mathcal{I}^a)}.$$

One verifies that this definition does not depend on the injective resolution choosen (see [30] III, 1.0.8).

In particular, if $\sigma : \mathcal{F}^\bullet \longrightarrow \mathcal{G}^\bullet$ is a *quasi-isomorphism*, then σ induces an *isomorphism of the hypercohomology groups*:

$$\mathbb{H}^a(X, \mathcal{F}^\bullet) \xrightarrow{\sim} \mathbb{H}^a(X, \mathcal{G}^\bullet)$$

(by taking an injective resolution $\mathcal{I}^\bullet$ of $\mathcal{G}^\bullet$ which is also an injective resolution of $\mathcal{F}^\bullet$).

11.

By definition, $H^a(X, \mathcal{I}) = 0$ for $a > 0$ if $\mathcal{I}$ is an injective sheaf. We will verify in (A.28) that if $\sigma : \mathcal{F}^\bullet \longrightarrow \mathcal{G}^\bullet$ is a quasi-isomorphism and if $H^a(X, \mathcal{G}^i) = 0$

for all $a > 0$ and all i, then

$$\mathbb{H}^a(X, \mathcal{F}^\bullet) = \frac{\ker\ \Gamma(X, \mathcal{G}^a) \to \Gamma(X, \mathcal{G}^{a+1})}{\mathrm{im}\ \Gamma(X, \mathcal{G}^{a-1}) \to \Gamma(X, \mathcal{G}^a)}.$$

We call $(\mathcal{G}^\bullet, \sigma)$ an *acyclic resolution* of $\mathcal{F}^\bullet$ in this case.

12.

$\mathbb{H}$ transforms *short exact sequences*

$$0 \longrightarrow \mathcal{A}^\bullet \longrightarrow \mathcal{B}^\bullet \longrightarrow \mathcal{C}^\bullet \longrightarrow 0$$

of complexes of $\mathcal{O}$-modules which are bounded below into *long exact sequences*

$$\cdots \longrightarrow \mathbb{H}^i(\mathcal{A}^\bullet) \longrightarrow \mathbb{H}^i(\mathcal{B}^\bullet) \longrightarrow \mathbb{H}^i(\mathcal{C}^\bullet) \longrightarrow \mathbb{H}^{i+1}(\mathcal{A}^\bullet) \longrightarrow \ldots$$

of $\mathcal{O}$-modules.

13.

We assume now that the complex $\mathcal{F}^\bullet$ of $\mathcal{O}$-modules, bounded below, has subcomplexes

$$\ldots \subset Filt_{i-1} \ \subset Filt_i \subset \ldots \subset \mathcal{F}^\bullet$$

$$(\text{or} \quad \ldots \subset Filt^i \subset Filt^{i-1} \subset \ldots \subset \mathcal{F}^\bullet)$$

such that

$$\bigcup_i \ Filt_i = \mathcal{F}^\bullet$$

$$(\text{or} \quad \bigcup_i \ Filt^i = \mathcal{F}^\bullet)$$

and such that

$$Filt_i = 0 \ \text{ for } \ i << 0$$

$$(\text{or} \quad Filt^i = 0 \ \text{ for } \ i \gg 0).$$

One says that $\mathcal{F}^\bullet$ is *filtered* by the subcomplexes $Filt_i$ (or $Filt^i$). Via (A.12), this filtration defines a *filtration on the hypercohomology groups*:

$$Filt_i \mathbb{H}^a(X, \mathcal{F}^\bullet) := \ \mathrm{im}(\mathbb{H}^a(X, Filt_i) \longrightarrow \mathbb{H}^a(X, \mathcal{F}^\bullet))$$

$$(\text{or} \ \ Filt^i \mathbb{H}^a(X, \mathcal{F}^\bullet) := \ \mathrm{im}(\mathbb{H}^a(X, Filt^i) \longrightarrow \mathbb{H}^a(X, \mathcal{F}^\bullet))).$$

This just means that the group $\mathbb{H}^a(X, \mathcal{F}^\bullet)$ has subgroups

$$\ldots \subset Filt_{i-1}\mathbb{H}^a(X, \mathcal{F}^\bullet) \subset Filt_i \mathbb{H}^a(X, \mathcal{F}^\bullet) \subset \ldots \subset \mathbb{H}^a(X, \mathcal{F}^\bullet)$$

$$(\text{or} \quad \ldots \subset Filt^i \mathbb{H}^a(X, \mathcal{F}^\bullet) \subset Filt^{i-1}\mathbb{H}^a(X, \mathcal{F}^\bullet) \subset \ldots \subset \mathbb{H}^a(X, \mathcal{F}^\bullet)).$$

We pass from *increasing* to *decreasing* filtrations by setting

$$Filt_i := Filt^{-i}.$$

14.

We define

$$Gr_i := Filt_i / Filt_{i-1}.$$

One has a diagram of exact sequences

$$\begin{array}{ccccccccc}
 & & & & 0 & & 0 & & \\
 & & & & \downarrow & & \downarrow & & \\
 & & 0 & \longrightarrow & Filt_{i-1} & \longrightarrow & Filt_i & \longrightarrow & Gr_i \longrightarrow 0 \\
 & & & & \downarrow & & \downarrow & & \\
 & & & & \mathcal{F}^\bullet & \xrightarrow{\text{identity}} & \mathcal{F}^\bullet & & \\
 & & & & \downarrow & & \downarrow & & \\
0 & \longrightarrow & Gr_i & \longrightarrow & \mathcal{F}^\bullet / Filt_{i-1} & \longrightarrow & \mathcal{F}^\bullet / Filt_i & \longrightarrow & 0 \\
 & & & & \downarrow & & \downarrow & & \\
 & & & & 0 & & 0 & &
\end{array}$$

which gives via (A.12):

$$\begin{aligned}
Gr_i(Filt_\bullet \mathbb{H}^a(\mathcal{F}^\bullet)) &:= Filt_i \mathbb{H}^a(\mathcal{F}^\bullet) / Filt_{i-1} \mathbb{H}^a(\mathcal{F}^\bullet) \\
&= \frac{\mathbf{H}^a(Filt_i)}{\mathbf{H}^a(Filt_{i-1}) + \mathbf{H}^{a-1}(\mathcal{F}^\bullet / Filt_i)} \\
&= \frac{\ker(\mathbf{H}^a(Gr_i) \to \mathbf{H}^{a+1}(Filt_{i-1}))}{\ker(\mathbf{H}^a(Gr_i) \to \mathbf{H}^a(\mathcal{F}^\bullet / Filt_{i-1}))}.
\end{aligned}$$

We define

$$E_\infty^{a-i,i} := Gr_i(Filt_\bullet \mathbb{H}^a(\mathcal{F}^\bullet)) \text{ and } E_2^{a-i,i} := \mathbb{H}^a(Gr_i).$$

Obviously $E_\infty^{a-i,i}$ is a *subquotient* of $E_2^{a-i,i}$.

15.

The formations of *spectral sequences* has the aim to compute $E_\infty^{a-i,i}$ only in terms of $E_2^{s,t}$, by filtering the terms $\mathbb{H}^{a+1}(Filt_{i-1})$ and $\mathbb{H}^a(\mathcal{F}^\bullet / Filt_{i-1})$ appearing in the description (A.14) by the induced filtrations:

$$Filt_l \; \mathbb{H}^{a+1}(Filt_{i-1}) := \mathrm{im}(\mathbb{H}^{a+1}(Filt_l) \longrightarrow \mathbb{H}^{a+1}(Filt_{i-1}))$$

for $l \leq i-1$ and

$$Filt_l\ \mathbb{H}^a(\mathcal{F}^\bullet/Filt_{i-1}) := \mathrm{im}(\mathbb{H}^a(Filt_l/Filt_{i-1}) \longrightarrow \mathbb{H}^a(\mathcal{F}^\bullet/Filt_{i-1}))$$

for $l \geq i-1$, and by computing the corresponding graded quotients.

16.

If we assume that for some a, the filtration $Filt_i\mathbb{H}^a(\mathcal{F}^\bullet)$ is *exhausting*, that is:

$$\bigcup_i Filt_i\mathbb{H}^a(\mathcal{F}^\bullet) = \mathbb{H}^a(\mathcal{F}^\bullet),$$

then

$$Gr\ \mathbb{H}^a(\mathcal{F}^\bullet) = \bigoplus_i E_\infty^{a-i,i}$$

is the corresponding graded group. In particular, assume that $\mathbb{H}^a(\mathcal{F}^\bullet)$ and $E_\infty^{a-i,i}$ are *free $\mathcal{O}$-modules* (where $\mathcal{O}$ is $\mathbb{Z}$ or k), and that $\mathbb{H}^a(\mathcal{F}^\bullet)$ is of finite rank. Then

$$\mathrm{rank}_{\mathcal{O}}\mathbb{H}^a(\mathcal{F}^\bullet) = \sum_i \mathrm{rank}_{\mathcal{O}} E_\infty^{a-i,i}.$$

If $E_2^{a-i,i}$ is also free, then $\mathrm{rank}_{\mathcal{O}}\mathbb{H}^a(\mathcal{F}^\bullet) \leq \sum_i \mathrm{rank}_{\mathcal{O}} E_2^{a-i,i}$ and one has equality if and only if

$$E_\infty^{a-i,i} = E_2^{a-i,i} \quad \text{for all } i.$$

17. Example: One step filtration:

$$0 = Filt_{s-1} \subset Filt_s = \mathcal{F}^\bullet.$$

Then $\mathbb{H}^a(\mathcal{F}^\bullet) = \mathbb{H}^a(Gr_s) = E_\infty^{a-s,s} = E_2^{a-s,s}$.

18. Example: Two steps filtration:

$$0 = Filt_{s-2} \subset Filt_{s-1} \subset Filt_s = \mathcal{F}^\bullet$$

Then one has the exact sequence

$$0 \longrightarrow Gr_{s-1} \longrightarrow \mathcal{F}^\bullet \longrightarrow Gr_s \longrightarrow 0$$

and

$$0 \longrightarrow E_\infty^{a-(s-1),(s-1)} \longrightarrow \mathbb{H}^a(\mathcal{F}^\bullet) \longrightarrow E_\infty^{a-s,s} \longrightarrow 0$$

with

$$E_\infty^{a-(s-1),(s-1)} = \mathbb{H}^a(Gr_{s-1})/\mathbb{H}^{a-1}(Gr_s)$$

$$E_\infty^{a-s,s} = \ker\ \mathbb{H}^a(Gr_s) \longrightarrow \mathbb{H}^{a+1}(Gr_{s-1})$$

19.

Define the differential d_2 as the connecting morphism of the vertical exact sequence on the right hand side (or of the above horizontal exact sequence) of the diagram

$$\begin{array}{ccccccccc}
 & & & & 0 & & 0 & & \\
 & & & & \downarrow & & \downarrow & & \\
0 & \longrightarrow & Gr_{i-1} & \longrightarrow & Filt_i/Filt_{i-2} & \longrightarrow & Gr_i & \longrightarrow & 0 \\
 & & \| & & \downarrow & & \downarrow & & \\
0 & \longrightarrow & Gr_{i-1} & \longrightarrow & Filt_{i+1}/Filt_{i-2} & \longrightarrow & Filt_{i+1}/Filt_{i-1} & \longrightarrow & 0 \\
 & & & & \downarrow & & \downarrow & & \\
 & & & & Gr_{i+1} & \xrightarrow{\text{identity}} & Gr_{i+1} & & \\
 & & & & \downarrow & & \downarrow & & \\
 & & & & 0 & & 0 & &
\end{array}$$

and the $\mathcal{O}$-module $E_3^{a-i,i}$ by

$$E_3^{a-i,i} = \frac{\ker\ E_2^{a-i,i} \xrightarrow{d_2} E_2^{a-i+2,i-1}}{\mathrm{im}\ E_2^{a-i-2,i+1} \xrightarrow{d_2} E_2^{a-i,i}}.$$

For $\varepsilon = 1, 2$ and the ε-steps filtration one has

$$E_\infty^{a-i,i} = E_{\varepsilon+1}^{a-i,i},$$

as we saw in (A.17) and (A.18), and the filtration on $\mathbb{H}^a(\mathcal{F}^\bullet)$ is exhausting.

20.

The right vertical exact sequence of the diagram (A.19) gives a surjection

$$\mathbb{H}^a(Filt_{i+1}/Filt_{i-1}) \xrightarrow{\gamma} \ker(\mathbb{H}^a(Gr_{i+1}) \to \mathbb{H}^{a+1}(Gr_i))$$

and the middle horizontal sequence induces a morphism

$$\ker(\mathbb{H}^a(Gr_{i+1}) \to \mathbb{H}^{a+1}(Gr_i)) \xrightarrow{d_3'} \mathbb{H}^{a+1}(Gr_{i-1})/\mathrm{im}\ \mathbb{H}^a(Gr_i)$$

Replacing i by $i-1$, one obtains maps

$$\mathbb{H}^a(Filt_i/Filt_{i-2}) \xrightarrow{\gamma} \ker(\mathbb{H}^a(Gr_i) \to \mathbb{H}^{a+1}(Gr_{i-1})) \xrightarrow{d_3'} \frac{\mathbb{H}^{a+1}(Gr_{i-2})}{\mathrm{im}\ \mathbb{H}^a(Gr_{i-1})}$$

where γ is surjective. As the extension

$$0 \longrightarrow Gr_{i-2} \longrightarrow Filt_i/Filt_{i-3} \longrightarrow Filt_i/Filt_{i-2} \longrightarrow 0$$

lifts to an extension

$$0 \longrightarrow Filt_{i-2}/Filt_{i-4} \longrightarrow Filt_i/Filt_{i-4} \longrightarrow Filt_i/Filt_{i-2} \longrightarrow 0$$

the image of $d_3' \circ \gamma$ lies in fact in

$$\frac{\ker\ \mathbb{H}^{a+1}(Gr_{i-2}) \to \mathbb{H}^{a+2}(Gr_{i-3})}{\operatorname{im}\ \mathbb{H}^{a}(Gr_{i-1}) \to \mathbb{H}^{a+1}(Gr_{i-2})} = E_3^{a-i+3,i-2} ,$$

as well as the image of d_3'. The map d_3' factorizes through $E_3^{a-i,i}$. The resulting map is written as

$$d_3 : E_3^{a-i,i} \longrightarrow E_3^{a-i+3,i-2}.$$

One defines

$$E_4^{a-i,i} = \frac{\ker\ E_3^{a-i,i} \xrightarrow{d_3} E_3^{a-i+3,i-2}}{\operatorname{im}\ E_3^{a-i-3,i+2} \xrightarrow{d_3} E_3^{a-i,i}} .$$

By construction a class in $E_3^{a-i,i}$ may be represented by a class in

$$\mathbb{H}^a(Filt_i/Filt_{i-2})$$

and similarly a class in $E_4^{a-i,i}$ may be represented by a class in

$$\mathbb{H}^a(Filt_i/Filt_{i-3}).$$

More generally, one defines inductively in the same vein differentials

$$E_r^{a-i,i} \xrightarrow{d_r} E_r^{a-i+r,i-r+1},$$

and $\mathcal{O}$-modules:

$$E_{r+1}^{a-i,i} := \frac{\ker\ E_r^{a-i,i} \xrightarrow{d_r} E_r^{a-i+r,i-r+1}}{\operatorname{im}\ E_r^{a-i-r,i+r-1} \xrightarrow{d_r} E_r^{a-i,i}}$$

which are subquotients of E_r. A class in $E_{r+1}^{a-i,i}$ is represented by a lifting to $\mathbb{H}^a(Filt_i/Filt_{i-r})$.

When $Filt_{\sigma-1} = 0$, and $Filt_\sigma \neq 0$, one defines the induced filtration on $Filt_{\sigma+\rho}$ by

$$Filt_l\ Filt_{\sigma+\rho} = \begin{cases} Filt_l & \text{if } l \leq \sigma+\rho \\ Filt_{\sigma+\rho} & \text{if } l \geq \sigma+\rho\ . \end{cases}$$

Then one has:

$$E_\infty^{a-i,i}(\mathbb{H}^a(Filt_{\sigma+\varrho})) = E_{\varrho+2}^{a-i,i}$$

and the filtration on $\mathbb{H}^a(Filt_{\sigma+\varrho})$ is exhausting.

21.

This shows that in general, for going from $\mathbb{H}^a(Filt_{\sigma+\varrho})$ to $\mathbb{H}^a(\mathcal{F}^\bullet)$ one has to introduce infinitely many r and groups $E_r^{a-i,i}$, a reason for the notation $E_\infty^{a-i,i}$.

One says that the *spectral sequence* with E_2 term $E_2^{a-i,i}$ and d_2 differential d_2 (simply noted $(E_2^{a-i,i}, d_2)$) *degenerates* in E_r if:

The filtration on $\mathbb{H}^a(\mathcal{F}^\bullet)$ is exhausting and $E_\infty^{a-i,i} = E_r^{a-i,i}$ for all i, or equivalently $d_{r+l} = 0$ for all $l \geq 0$.

With this terminology, we have seen in (A.20) that a $(\varrho + 1)$-steps filtration defines an E_2 spectral sequence which degenerates in $E_{\varrho+2}$.

22.

Under the assumptions of (A.16), assume moreover that $E_r^{a-i,i}$ is also a free $\mathcal{O}$-module (for example if $\mathcal{O}$ is a field). Then one has:

$$\operatorname{rank}_{\mathcal{O}} \mathbb{H}^a(\mathcal{F}^\bullet) \leq \sum_i \operatorname{rank}_{\mathcal{O}} E_r^{a-i,i}$$

for all r, and the spectral sequence degenerates in E_r if and only if this is an equality.

23.

By (A.19), to say that the spectral *sequence degenerates* in E_2 means that for all i, one has an exact sequence

$$0 \longrightarrow \mathbb{H}^a(Filt_{i-1}) \longrightarrow \mathbb{H}^a(Filt_i) \longrightarrow \mathbb{H}^a(Gr_i) \longrightarrow 0,$$

and of course that the filtration is exhausting.

24.

One says that the spectral sequence (E_2, d_2) *converges* to $\mathbb{H}^a(\mathcal{F}^\bullet)$ if it degenerates in E_r for some r. We sometimes write

$$(E_2^{a-i,i}, d_2) \Longrightarrow \mathbb{H}^a(\mathcal{F}^\bullet)$$

instead of "(E_2, d_2) converges to $\mathbb{H}^a(\mathcal{F}^\bullet)$".

25. The Hodge to de Rham spectral sequence.

On $\mathcal{F}^\bullet$, a complex of $\mathcal{O}$-modules, bounded below, we define the *Hodge filtration* (often called the stupid filtration) by:

$$Filt^i\mathcal{F}^\bullet = \mathcal{F}^{\geq i} = Filt_{-i}\mathcal{F}^\bullet$$

where

$$(\mathcal{F}^{\geq i})^l = \begin{cases} 0 & \text{if } l < i \\ \mathcal{F}^l & \text{if } l \geq i. \end{cases}$$

Then $Gr_{-i} = Filt_{-i}/Filt_{-i-1} = \mathcal{F}^i[-i]$ where $[\alpha]$ means:

$$(\mathcal{F}^\bullet[\alpha])^l = \mathcal{F}^{l+a}.$$

This is the so called *shift* by α to the right. The E_2 spectral sequence reads:

$$E_2^{a-i,i} = \mathbb{H}^a(Gr_i) = \mathbb{H}^a(\mathcal{F}^{-i}[i]) = H^{a+i}(\mathcal{F}^{-i})$$

where the differential d_2 goes to

$$\mathbb{H}^{a+1}(Gr_{i-1}) = \mathbb{H}^{a+1}(\mathcal{F}^{-(i-1)}[i-1]) = H^{a+i}(\mathcal{F}^{-i+1})$$

and is just induced by the differential in the complex.

This spectral sequence is usually rewritten as an E_1 spectral sequence by setting

$$E_1^{-i,a+i} = E_2^{a-i,i} \quad \text{or} \quad E_1^{\alpha,\beta} = E_2^{\beta+2\alpha,-\alpha}$$

with differentials:

$$\begin{array}{ccc} E_2^{\beta+2\alpha,-\alpha} & \xrightarrow{d_2} & E_2^{\beta+2\alpha+2,-\alpha-1} \\ \| & & \| \\ E_1^{\alpha,\beta} & \xrightarrow{d_1} & E_1^{\alpha+1,\beta} \end{array}$$

The E_1 spectral sequence obtained is called the Hodge to de Rham spectral sequence, at least when $\mathcal{F}^\bullet$ is some de Rham complex on X, possibly with some poles, possibly with non-trivial coefficients . . .

For a given a, one has

$$\mathbb{H}^a(\mathcal{F}^\bullet) = \mathbb{H}^a(\mathcal{F}^{\leq a+1})$$

where

$$(\mathcal{F}^{\leq i})^l = \begin{cases} \mathcal{F}^l & \text{if } l \leq i \\ 0 & \text{if } l > i. \end{cases}$$

In particular this is a finite complex, on which the Hodge filtration induces a finite step filtration. Therefore the E_1 Hodge to de Rham spectral sequence always converges.

To say that it degenerates in E_1 means that for any i one has exact sequences

$$0 \longrightarrow \mathbb{H}^a(\mathcal{F}^{\geq i+1}) \longrightarrow \mathbb{H}^a(\mathcal{F}^{\geq i}) \longrightarrow H^{a-i}(\mathcal{F}^i) \longrightarrow 0.$$

Putting those sequences together, one obtains exact sequences

$$0 \longrightarrow \mathbb{H}^a(\mathcal{F}^{\geq i+j}) \longrightarrow \mathbb{H}^a(\mathcal{F}^{\geq i}) \longrightarrow \mathbb{H}^a(\mathcal{F}^{[i,i+j)}) \longrightarrow 0$$

where

$$(\mathcal{F}^{[i,i+j)})^l = \begin{cases} \mathcal{F}^l & \text{if } l \in [i, i+j) \\ 0 & \text{if not.} \end{cases}$$

If moreover, one knows that $\mathbb{H}^a(\mathcal{F}^\bullet)$ is a free $\mathcal{O}$-module of finite rank, as well as $H^{a-i}(\mathcal{F}^i)$, then the E_1 Hodge to de Rham spectral sequence degenerates in E_1 if and only if

$$\mathrm{rank}_{\mathcal{O}}\ \mathbb{H}^a(\mathcal{F}^\bullet) = \sum_i \mathrm{rank}_{\mathcal{O}}\ H^{a-i}(\mathcal{F}^i).$$

26. The conjugate spectral sequence.

On $\mathcal{F}^\bullet$, a complex of $\mathcal{O}$-modules bounded below, we defined the τ-filtration:

$$(\tau_{\leq i}\mathcal{F}^\bullet)^l = \begin{cases} \mathcal{F}^l & \text{for } l < i \\ \ker d & \text{for } l = i \\ 0 & \text{otherwise.} \end{cases}$$

Then $Gr_i = \mathcal{H}^i[-i]$ is the cohomology sheaf in degree i. The E_2 spectral sequence reads

$$E_2^{a-i,i} = \mathbb{H}^a(Gr_i) = H^{a-i}(\mathcal{H}^i)$$

where the d_2 differential goes to

$$\mathbb{H}^{a+1}(Gr_{i-1}) = H^{a-i+2}(\mathcal{H}^{i-1}).$$

It is called the conjugate spectral sequence. For a given a, one has

$$\mathbb{H}^a(\mathcal{F}^\bullet) = \mathbb{H}^a(\tau_{\leq a+1}\mathcal{F}^\bullet).$$

However $\tau_{\leq a+1}\mathcal{F}^\bullet$ is a finite complex on which the τ-filtration induces a finite step filtration. Therefore the E_2-conjugate spectral sequence always converges. Furthermore, if $\sigma : \mathcal{F}^\bullet \longrightarrow \mathcal{G}^\bullet$ is a *quasi-isomorphism* then σ induces quasi-isomorphisms

$$\tau_{\leq i}\mathcal{F}^\bullet \longrightarrow \tau_{\leq i}\mathcal{G}^\bullet$$

for all i, and therefore σ induces an isomorphism of the conjugate spectral sequences.

To say that the conjugate spectral sequence degenerates in E_2 means that one has exact sequences

$$0 \longrightarrow \mathbb{H}^a(\tau_{\leq i-1}) \longrightarrow \mathbb{H}^a(\tau_{\leq i}) \longrightarrow H^{a-i}(\mathcal{H}^i) \longrightarrow 0$$

for all i. If $\mathbb{H}^a(\mathcal{F}^\bullet)$ is a free $\mathcal{O}$-module of finite rank, as well as $\mathbb{H}^{a-i}(\mathcal{H}^i)$, then the degeneration in E_2 is equivalent to

$$\mathrm{rank}_{\mathcal{O}}\mathbb{H}^a(\mathcal{F}^\bullet) = \sum_i \mathrm{rank}_{\mathcal{O}}\ H^{a-i}(\mathcal{H}^i).$$

27. The Leray spectral sequence.

Let $f : X \longrightarrow Y$ be a morphism between two k-varieties, and let $\mathcal{F}^\bullet$ be a complex of $\mathcal{O}$-modules on X, bounded below.

Let $\mathcal{F}^\bullet \longrightarrow \mathcal{I}^\bullet$ be an injective resolution. We consider the *direct image functor* (already used and defined but not named in 6):

$$(f_*\mathcal{K})_x = \lim_{\substack{\longrightarrow \\ x\in U}} H^0(f^{-1}(U), \mathcal{K})$$

for any $\mathcal{O}$-sheaf $\mathcal{K}$ on X and any open set U in Y. In particular, by definition $H^0(X, \mathcal{K}) = H^0(Y, f_*\mathcal{K})$. Therefore one has

$$\mathbb{H}^a(X, \mathcal{F}^\bullet) = \frac{\ker \ H^0(Y, f_*\mathcal{I}^a) \longrightarrow H^0(Y, f_*\mathcal{I}^{a+1})}{\mathrm{im}\ H^0(Y, f_*\mathcal{I}^{a-1}) \longrightarrow H^0(Y, f_*\mathcal{I}^a)}.$$

One verifies immediately that, by definition, $f_*\mathcal{I}^i$ is an injective sheaf as well, which allows to write (A.10):

$$\mathbb{H}^a(X, \mathcal{F}^\bullet) = \mathbb{H}^a(Y, (f_*\mathcal{I}^\bullet)),$$

where $f_*\mathcal{I}^\bullet$ is the complex

$$(f_*\mathcal{I}^\bullet)^l = f_*\mathcal{I}^l.$$

One considers the conjugate spectral sequence for $(f_*\mathcal{I}^\bullet)$:

$$E_2^{a-i,i} = H^{a-i}(Y, \mathcal{H}^i(f_*\mathcal{I}^\bullet)).$$

One defines

$$R^if_*\mathcal{F}^\bullet := \mathcal{H}^i(f_*\mathcal{I}^\bullet).$$

By definition

$$(R^if_*\mathcal{F}^\bullet)_x := \lim_{\substack{\longrightarrow \\ x\in U}} \frac{\ker\ H^0(f^{-1}(U),\mathcal{I}^i)\rightarrow H^0(f^{-1}(U),\mathcal{I}^{i+1})}{\mathrm{im}\ H^0(f^{-1}(U),\mathcal{I}^{i-1})\rightarrow H^0(f^{-1}(U),\mathcal{I}^i)}$$

$$= \lim_{\substack{\longrightarrow \\ x\in U}} \mathbb{H}^i(f^{-1}(U), \mathcal{F}^\bullet)$$

In particular, $R^if_*\mathcal{F}^\bullet$ does not depend on the injective resolution chosen.

The E_2 spectral sequence reads

$$E_2^{a-i,i} = H^{a-i}(Y, R^if_*\mathcal{F}^\bullet),$$

with d_2 differential to $H^{a-i+2}(Y, R^{i-1}f_*\mathcal{F}^\bullet)$, and is called the *Leray spectral sequence for f*.

As the conjugate spectral sequence for $(f_*\mathcal{I}^\bullet)$ converges to $\mathbb{H}^a(f_*\mathcal{I}^\bullet)$ (A.26), the *Leray spectral sequence for f always converges to* $\mathbb{H}^a(\mathcal{F}^\bullet)$.
In particular, if $\mathcal{F}$ is just a sheaf for which $R^if_*\mathcal{F} = 0$ for $i > 0$, one has:

$$H^a(X, \mathcal{F}) = H^a(Y, f_*\mathcal{F}) \quad \text{for all} \quad a.$$

28.

Let $\mathcal{G}^\bullet$ be a complex of $\mathcal{O}$-modules, bounded below, such that $H^i(\mathcal{G}^j) = 0$ for $i > 0$ and all j. Then the E_1 Hodge to de Rham spectral sequence $E_1^{ij} = H^j(\mathcal{G}^i)$ degenerates in E_2 and one has

$$\begin{aligned} E_\infty^{i,0} = E_2^{i,0} &= \frac{\ker \ H^0(\mathcal{G}^i) \to H^0(\mathcal{G}^{i+1})}{\operatorname{im} \ H^0(\mathcal{G}^{i-1}) \to H^0(\mathcal{G}^i)} \\ &= \mathbb{H}^i(\mathcal{G}^\bullet) \\ &= \mathbb{H}^i(\mathcal{F}^\bullet) \ \text{ for any quasi-isomorphism } \quad \mathcal{F}^\bullet \longrightarrow \mathcal{G}^\bullet. \end{aligned}$$

29.

Take for $\mathcal{F}^\bullet$ a complex of quasi-coherent sheaves (for example some de Rham complex). We consider a collection of very ample Cartier divisors D_α with empty intersection, such that the open covering of X defined by $U_\alpha := X - D_\alpha$ consists of affine varieties. Then one has:

$$H^a(X, \varrho_*\mathcal{F}^j|_{U_{\alpha_0 \dots \alpha_i}}) = H^a(U_{\alpha_0 \dots \alpha_i}, \mathcal{F}^j|_{U_{\alpha_0 \dots \alpha_i}})$$

for all a where $\varrho : U_{\alpha_0 \dots \alpha_i} \longrightarrow X$ is the natural embedding of the affine set $U_{\alpha_0 \dots \alpha_i}$. In fact, one has

$$\begin{aligned} (R^i\varrho_*\mathcal{F}^j)_x &= \varinjlim_{x \in V} \ H^i(V \cap U_{\alpha_0 \cdots \alpha_i}, \mathcal{F}^j) \\ &= 0 \ \text{ for } \ i > 0 \end{aligned}$$

and one applies (A.27). By (A.28) one obtains:

$$\mathbb{H}^a(X, \mathcal{F}^\bullet) = \frac{\ker \ \oplus C^i(\mathcal{U}, \mathcal{F}^{a-i}) \to \oplus \ C^{i'}(\mathcal{U}, \mathcal{F}^{a+1-i'})}{\operatorname{im} \ \oplus C^{i'}(\mathcal{U}, \mathcal{F}^{a-1-i'}) \to \oplus \ C^i(\mathcal{U}, \mathcal{F}^{a-i})}$$

where

$$\begin{aligned} C^i(\mathcal{U}, \mathcal{F}^j) &= H^0(X, \mathcal{C}^i(\mathcal{U}, \mathcal{F}^j)) \\ &= \bigoplus\nolimits_{\alpha_0 < \dots < \alpha_i} H^0(U_{\alpha_0 \dots \alpha_i}, \mathcal{F}^j). \end{aligned}$$

References

[1] Y. Akizuki, S. Nakano: Note on Kodaira-Spencer's proof of Lefschetz's theorem. Proc. Jap. Acad., Ser A **30** (1954), 266 - 272

[2] D. Arapuro: A note on Kollár's theorem. Duke Math. Journ. **53** (1986), 1125 - 1130

[3] D. Arapuro and D.B. Jalte: On Kodaira vanishing for singular varieties. Proc. Amer. Math. Soc. **105** (1989), 911 - 916

[4] I. Bauer, S. Kosarew: On the Hodge spectral sequences for some classes of non-complete algebraic manifolds. Math. Ann. **284** (1989), 577 - 593

[5] A. Beauville: Annulation du H^1 et systèmes paracanoniques sur les surfaces. Journ. reine angew. Math. **388** (1988), 149 - 157 & 219 - 220

[6] F. Bogomolov: Unstable vector bundles and curves on surfaces. Proc. Intern. Congress of Math., Helsinki 1978, 517 - 524

[7] A. Borel et al.: Intersection Cohomology. Progress in Math. **50** (1984) Birkhäuser

[8] A. Borel et al.: Algebraic D-Modules. Perspectives in Math. **2** (1987) Acadamic Press

[9] P. Cartier: Une nouvelle opération sur les formes différentielles. C. R. Acad. Sci., Paris, **244** (1957), 426 - 428

[10] P. Deligne: Equations différentielles à points singuliers réguliers. Springer Lect. Notes Math. **163** (1970)

[11] P. Deligne: Théorie de Hodge II. Publ. Math., Inst. Hautes Etud. Sci. **40** (1972), 5 - 57

[12] P. Deligne, L. Illusie: Relèvements modulo p^2 et décomposition du complexe de de Rham. Inventiones math. **89** (1987), 247 - 270

[13] J.-P. Demailly: Une généralisation du théorème d'annulation de Kawamata - Viehweg. C. R. Acad. Sci. Paris. **309** (1989), 123 - 126

[14] H. Dunio: Über generische Verschwindungssätze. Diplomarbeit, Essen 1991

[15] L. Ein, R. Lazarsfeld: Global generation of pluricanonical and adjoint linear series on smooth projective threefolds. manuscript

[16] H. Esnault: Fibre de Milnor d'un cône sur une courbe plane singulière. Inventiones math. **68** (1982), 477 - 496

[17] H. Esnault, E. Viehweg: Revêtement cycliques. Algebraic Threefolds, Proc. Varenna 1981. Springer Lect. Notes in Math. **947** (1982), 241 - 250

[18] H. Esnault, E. Viehweg: Sur une minoration du degré d'hypersurfaces s'annulant en certains points. Math. Ann. **263** (1983), 75 - 86

[19] H. Esnault, E. Viehweg: Two-dimensional quotient singularities deform to quotient singularities. Math. Ann. **271** (1985), 439 - 449

[20] H. Esnault, E. Viehweg: Logarithmic De Rham complexes and vanishing theorems. Inventiones math. **86** (1986), 161 - 194

[21] H. Esnault, E. Viehweg: Revêtements cycliques II, Proc. Algebraic Geometry, La Rabida 1984, Hermann, Travaux en cours **23** (1987), 81 - 96

[22] H. Esnault, E. Viehweg: Vanishing and non-vanishing theorems. Proc. Hodge Theory, Luminy 1987 (Ed.: Barlet, Elzein, Esnault, Verdier, Viehweg). Astérisque **179 - 180** (1989), 97 - 112

[23] H. Esnault, E. Viehweg: Effective bounds for semipositive sheaves and the height of points on curves over complex function fields. Compositio math. **76** (1990), 69 - 85

[24] T. Fujita: On Kähler fibre spaces over curves. J. Math. Soc. Japan **30** (1978), 779 - 794

[25] H. Grauert, O. Riemenschneider: Verschwindungssätze für analytische Kohomologiegruppen auf komplexen Räumen. Inventiones math. **11** (1970), 263 - 292

[26] M. Green, R. Lazarsfeld: Deformation theory, generic vanishing theorems, and some conjectures of Enriques, Catanese and Beauville. Inventiones math. **90** (1987), 389 - 407

[27] M. Green, R. Lazarsfeld: Higher obstructions to deforming cohomology groups of line bundles. Journal of the AMS. **4** (1991), 87 - 103

[28] A. Grothendieck: Fondements de la Géométrie Algébrique. Séminaire Bourbaki 1957 - 62

[29] R. Hartshorne: Residues and Duality. Springer Lect. Notes in Math. **20** (1966)

[30] R. Hartshorne: Algebraic Geometry. Graduate Texts in Mathematics 52. Springer Verlag, New York, 1977

[31] L. Illusie: Catégories dérivées et dualité. Travaux de J.-L. Verdier. L'Enseignement Math. **36** (1990), 369 - 391

[32] L. Illusie: Réduction semi-stable et décomposition de complexes de de Rham à coefficients. Duke Math. Journ. **60** (1990), 139 - 185

[33] B. Iversen: Cohomology of Sheaves. Universitext, Springer-Verlag (1986)

[34] N. Katz: Nilpotent Connections and the Monodromy Theorem. Publ. Math., Inst. Hautes Etud. Sci. **39** (1970), 175 - 232

[35] Y. Kawamata: Characterization of abelian varieties. Compositio math. **43** (1981), 253 - 276

[36] Y. Kawamata: A generalization of Kodaira-Ramanujam's vanishing theorem. Math. Ann. **261** (1982), 43 - 46

[37] Y. Kawamata: Pluricanonical systems on minimal algebraic varieties. Inventiones math. **79** (1985), 567 - 588

[38] Y. Kawamata, K. Matsuda, M. Matsuki: Introduction to the mimimal model problem. Algebraic Geometry, Sendai 1985. Adv. Stud. Pure Math. **10** (1987), 283 - 360

[39] K. Kodaira: On a differential-geometric method in the theory of analytic stacks. Proc. Natl. Acad. Sci. USA **39** (1953), 1268 - 1273

[40] J. Kollár: Higher direct images of dualizing sheaves I. Ann. Math. **123** (1986), 11 - 42

[41] J. Kollár: Higher direct images of dualizing sheaves II. Ann. Math. **124** (1986), 171 - 202

[42] I. Kosarew, S. Kosarew: Kodaira vanishing theorems on non-complete algebraic manifolds. Math. Z. **205** (1990), 223 - 231

[43] K. Maehara: Vanishing theorems. manuscript

[44] K. Maehara: Kawamata covering and logarithmic de Rham complexes. manuscript

[45] Y. Miyaoka: On the Mumford-Ramanujam vanishing theorem on a surface. Proc. of the conference "Géométrie alg. d'Angers 1979", Oslo 1979, 239 - 248

[46] S. Mori: Classification of higher-dimensional varieties. Algebraic Geometry. Bowdoin 1985. Proc. of Symp. in Pure Math. **46** (1987), 269 - 331

[47] D. Mumford: Pathologies of modular surfaces. Am. J. Math. **83** (1961), 339 - 342

[48] D. Mumford: The topology of normal singularities of an algebraic surface and a criterion for simplicity. Publ. Math. IHES, **9** (1961), 229 - 246

[49] D. Mumford: Pathologies, III. Amer. Journ. Math. **89** (1967), 94 - 104

[50] D. Mumford: Abelian Varieties. Oxford University Press (1970)

[51] C.P. Ramanujam: Remarks on the Kodaira vanishing theorem. Journ. Indian Math. Soc. **36** (1972), 41 - 51

[52] M. Raynaud: Contre-exemple au "Vanishing Theorem" en caractéristique $p > 0$. C. P. Ramanujam - A tribute. Studies in Mathematics 8, Tata Institute of Fundamental Research, Bombay (1978), 273 - 278

[53] I. Reider: Vector bundles of rank 2 and linear systems on algebraic surfaces. Ann. of Math. **127** (1988), 309 - 316

[54] M. Saito: Introduction to mixed Hodge Modules. Proc. Hodge Theory, Luminy 1987 (Ed.: Barlet, Elzein, Esnault, Verdier, Viehweg). Astérisque **179 - 180** (1989), 145 - 162

[55] J.-P. Serre: Faisceaux algébriques cohérents. Ann. of Math. **61** (1955), 197 - 278

[56] J.-P. Serre: Géométrie algébrique et géométrie analytique. Ann. Inst. Fourier **6** (1956), 1 - 42

[57] B. Shiffman, A.J. Sommese: Vanishing Theorems on Complex Manifolds. Progress in Math. 56, Birkhäuser (1985)

[58] C. Simpson: Subspaces of moduli spaces of rank one local systems. manuscript

[59] K. Timmerscheidt: Mixed Hodge theory for unitary local systems. Journ. reine angew. Math. **379** (1987), 152 - 171

[60] J.L. Verdier: Catégories dérivées, état 0. SGA 4 1/2 (Ed.: Deligne) Springer Lect. Notes in Math. **569** (1977), 262 - 311

[61] J.L. Verdier: Classe d'homologie associée à un cycle. Séminaire de Géometrie Analytique, Astérisque **36/37** (1976), 101-151

[62] E. Viehweg: Rational singularities of higher dimensional schemes. Proc. AMS **63** (1977), 6 - 8

[63] E. Viehweg: Vanishing theorems. Journ. reine angew. Math. **335** (1982), 1 - 8

[64] E. Viehweg: Vanishing theorems and positivity in algebraic fibre spaces. Proc. ICM, Berkeley 1986, 682 - 688

[65] O. Zariski: The theorem of Riemann-Roch for high multiples of an effective divisor on an algebraic surface. Ann. of Math. **76** (1962), 560 - 615

Hélène Esnault and Eckart Viehweg
Fachbereich 6, Mathematik
Universität GH Essen
Universitätsstr. 3
D-W-4300 Essen1
Germany

Previously published in the series DMV Seminar:

Volume 1: **Manfred Knebusch/Winfried Scharlau, Algebraic Theory of Quadratic Forms.**
1980, 48 pages, softcover, ISBN 3-7643-1206-8.

Volume 2: **Klas Diederich/IngoLieb, Konvexitaet in der Komplexen Analysis.**
1980, 150 pages, softcover, ISBN 3-7643-1207-6.

Volume 3: **S. Kobayashi/H. Wu/C. Horst, Complex Differential Geometry.**
2nd edition 1987, 160 pages, softcover, ISBN 3-7643-1494-X.

Volume 4: **R. Lazarsfeld/A. van de Ven, Topics in the Geometry of Projective Space.**
1984, 52 pages, softcover, ISBN 3-7643-1660-8.

Volume 5: **Wolfgang Schmidt, Analytische Methoden für Diophantische Gleichungen.**
1984, 132 pages, softcover, ISBN 3-7643-1661-6.

Volume 6: **A. Delgado/D. Goldschmidt/B. Stellmacher, Groups and Graphs: New Results and Methods.**
1985, 244 pages, softcover, ISBN 3-7643-1736-1.

Volume 7: **R. Hardt/L. Simon, Seminar on Geometric Measure Theory.**
1986, 118 pages, softcover, ISBN 3-7643-1815-5.

Volume 8: **Yum-Tong Siu, Lectures on Hermitian-Einstein Metrics for Stable Bundles and Kaehler-Einstein Metrics.**
1987, 172 pages, softcover, ISBN 3-7643-1931-3.

Volume 9: **Peter Gaenssler/Winfried Stute, Seminar on Empirical Processes.**
1987, 114 pages, softcover, ISBN 3-7643-1921-6.

Volume 10: **Jürgen Jost, Nonlinear Methods in Riemannian and Kaehlerian Geometry.**
2nd edition1991, 154 pages, softcover, ISBN 3-7643-2685-9.

Volume 11: **Tammo tom Dieck/Ian Hambleton, Surgery Theory and Geometry of Representations.**
1988, 122 pages, softcover, ISBN 3-7643-2204-7.

Volume 12: **Jacobus van Lint/Gerard van der Geer, Introduction to Coding Theory and Algebraic Geometry.**
1988, 83 pages, softcover, ISBN 3-7643-2230-6.

Volume 13: **Hanspeter Kraft/Peter Slodowy/Tonny A. Springer (Eds.), Algebraic Transformation Groups and Invariant Theory.**
1989, 220 pages, softcover, ISBN 3-7643-2284-5.

Volume 14: **Rabi Bhattacharya/Manfred Denker, Asymptotic Statistics.**
1990, 122 pages, softcover, ISBN 3-7643-2282-9.

Volume 15: **Alfred H. Schatz/Vidar Thomée/Wolfgang L. Wendland, Mathematical Theory of Finite and Boundary Element Methods.**
1990, 280 pages, softcover, ISBN 3-7643-2211-X.

Volume 16: **Joseph Mecke/Rolf Schneider/Dietrich Stoyan/Wolfgang Weil, Stochastische Geometrie.**
1990, 216 pages, softcover, ISBN 3-7643-2543-7.

Volume 17: **Lennart Ljung/Georg Pflug/Harro Walk, Stochastic Approximation and Optimization of Random Processes.**
1992, 120 pages, softcover, ISBN 3-7643-2733-2.

Volume 18: **Klaus Roggenkamp/Martin Taylor, Group Rings and Class Groups.**
1992, 208 pages, softcover, ISBN 3-7643-2734-0.

Volume 19: **Piet Groeneboom/Jon A. Wellner, Information Bounds and Nonparametric Maximum Likelihood Estimation.**
1992, 134 pages, softcover, ISBN 3-7643-2794-4.

GPSR Compliance

The European Union's (EU) General Product Safety Regulation (GPSR) is a set of rules that requires consumer products to be safe and our obligations to ensure this.

If you have any concerns about our products, you can contact us on ProductSafety@springernature.com

In case Publisher is established outside the EU, the EU authorized representative is:

Springer Nature Customer Service Center GmbH
Europaplatz 3
69115 Heidelberg, Germany

Batch number: 10584556

Printed by Printforce, the Netherlands

Pathogene Mikroorganismen im Grund- und Trinkwasser

Transport – Nachweismethoden – Wassermanagement

Adrian Auckenthaler
Peter Huggenberger
(Hrsg.)

Springer Basel AG

Herausgeber

Adrian Auckenthaler
Kantonales Laboratorium BL
Trinkwasserinspektorat
Liestal, Schweiz

PD Dr. Peter Huggenberger
Angewandte und Umweltgeologie
Departement Erdwissenschaften der
Universität Basel, Schweiz

Bibliografische Information der Deutschen Bibliothek
Die Deutsche Bibliothek verzeichnet diese Publikation in der Deutschen Nationalbiografie; detaillierte bibliografische Daten sind im Internet über http://dnb.ddb.de abrufbar.

ISBN 978-3-7643-6950-7 ISBN 978-3-0348-8009-1 (eBook)
DOI 10.1007/978-3-0348-8009-1

Ursprünglich erschienen bei Birkhäuser Verlag 2003

Computer-to-plate Vorlage durch die Herausgeber erstellt
Gedruckt auf säurefreiem Papier, hergestellt aus chlorfrei gebleichtem Zellstoff TCF ∞
Umschlaggestaltung: Micha Lotrovsky, CH-4106 Therwil, Schweiz
Bildmotiv *E. coli*: Tamara Fretz, Universität Basel, Zentrum für Mikroskopie ZMB Bio-Pharmazentrum

ISBN 978-3-7643-6950-7

9 8 7 6 5 4 3 2 1

www.birkhasuer-science.com

Inhaltsverzeichnis

Vorwort

Routinekontrollen des Trinkwassers zeigen zeitweise mikrobiologische Belastungen mit fäkalen Indikatorkeimen. Die Bedeutung dieser positiven Befunde im Grund- und Trinkwasser ist noch weitgehend unklar, da die Herkunft und der Transport der Mikroorganismen meist unbekannt sind und grössere Krankheitsausbrüche in der Schweiz und in Deutschland nur selten auftreten oder nicht erkannt werden. In Studien mit neuen Analyseverfahren in der Mikrobiologie konnten jedoch neben den Indikatorkeimen auch pathogene Mikroorganismen wie Cryptosporidien und Viren festgestellt werden. Es ist deshalb davon auszugehen, dass das Vorhandensein von fäkalen Indikatorkeimen eine gesundheitliche Gefährdung bedeuten kann.

Die Autoren des Buches, ein Umweltnatur- und ein Erdwissenschaftler sowie zwei Mikrobiologen forschen in den Bereichen Grund- und Trinkwasser. Sie haben zudem in verschiedenen praxisbezogenen Projekten für kantonale Fachstellen oder Gemeinden spezifische Grundlagen für den Grund- und Trinkwasserschutz erarbeitet. Es zeigt sich dabei, dass vertiefte Kenntnisse der jeweiligen Prozesse und der lokalen hydrogeologische Gegebenheiten die Grundvoraussetzung für einen wirkungsvollen Gewässerschutz und eine sichere Trinkwassergewinnung ist.

Das vorliegende Buch ist eine umfassende Zusammenstellung der Literatur zum Transport, zum analytischen Nachweis und zur epidemiologischen Bedeutung von pathogenen Mikroorganismen im Grund- und Trinkwasser. Es zeigt den aktuellen Kenntnisstand der Forschung auf diesem interdisziplinären Gebiet. Beispiele von neueren Feldversuchen sollen die Theorie illustrieren. Neben der rein naturwissenschaftlichen Sicht des Partikeltransportes im Untergrund wird dem Aspekt der Nutzung der Ressource Grundwasser für die Trinkwassergewinnung ein besonderer Stellenwert eingeräumt und die Probleme der Wasserversorger angesprochen. Mögliche Lösungsstrategien auf lokaler und regionaler Ebene werden diskutiert und konkrete Lösungsansätze für Wasserversorger aufgezeigt. Eine effektive und nachhaltige Sicherung der Trinkwasserqualität verlangt je nach Gefährdungssituation kurz-, mittel- und langfristige Strategien. Diese setzen unter anderem das Erkennen von bisher wenig beachteten, bei Niederschlagsereignissen und Hochwässern auftretenden Belastungsspitzen voraus.

Wir möchten allen danken, die zum Gelingen dieses Buches beigetragen haben. Unser besonderer Dank gilt Prof. Dr. Hauke Harms und Dr. Antonis Chatzinotas für die Verfassung des Kapitels „Nachweismethoden von Mikroorganismen". Dr. Daniel Hartmann und Dr. Benjamin Meylan vom Bundesamt für Umwelt, Wald und Landschaft sei für die fachliche Begleitung und die finanzielle Unterstützung bestens gedankt. Für die kritische Durchsicht, Änderungsvorschläge und Anregungen danken wir Dr. Tom Bosma, Dr. Urs von Gunten, Dr. Mario Snozzi, Dr. Paul Svoboda, Rainer Fretz und Christian Regli. Für sprachliche und stilistische Korrekturen danken wir Jutta Röthlisberger und für die grosse Hilfe bei der Gestaltung der Figuren und die Formatierung des Textes Andrea Voegelin. Den Mitarbeitern des Birkhäuser Verlages, insbesondere Dr. Detlef Klüber, danken wir für die gute Zusammenarbeit.

Mit diesem Buch möchten wir nicht nur Fachwissen vermitteln, sondern auch Anregungen für die Umsetzung der dargelegten Ideen und Konzepte geben und damit zu einer steten Verbesserung der Trinkwasserqualität beitragen.

Basel, Oktober 2002

Adrian Auckenthaler und Peter Huggenberger

Editoren

Adrian Auckenthaler hat 1995 sein Studium in Umweltnaturwissenschaften an der Eidgenössischen Technischen Hochschule in Zürich (ETHZ) abgeschlossen. Im selben Jahr übernahm er die Leitung des Trinkwasserinspektorates des Kantonalen Laboratoriums Basel-Landschaft. Eine seiner Hauptaufgaben ist die Umsetzung neuer Erkenntnisse zum Schutz von Trinkwasser. Seit rund vier Jahren arbeitet er im Rahmen eines Projekts zum Partikeltransport im Karst zusammen mit dem Geologischen Institut der Universität Basel und dem Schweizerischen Tropeninstitut. Die neusten Ergebnisse und Erkenntnisse aus diesem Projekt sind in das Buch mit eingeflossen.

Peter Huggenberger studierte Erdwissenschaften an der UNI/ETH Zürich und promovierte 1985 im Bereich Strukturgeologie. Nach längeren Forschungstätigkeiten an der Eidgenössischen Anstalt für Wasserversorgung, Abwasserreinigung und Gewässerschutz (EAWAG) und an der Universität Genf wurde er 1997 als Leiter der Forschungsgruppe Angewandte und Umweltgeologie sowie als Kantonsgeologe von Basel-Stadt an die Universität Basel berufen. Seine Forschungsschwerpunkte sind die Hydrogeologie komplex aufgebauter regionale Grundwassersysteme, der Transport von Schadstoffen und Mikroorganismen in heterogenen Grundwasserleitern sowie die Fluss-Grundwasserinteraktionen. Neben der reinen Forschungstätigkeit ist er auch an der Durchführung von praxisbezogenen Projekten wie beispielsweise dem Grundwassermanagement von Grossbaustellen beteiligt.

Autoren

Adrian Auckenthaler

Kantonales Laboratorium BL,
Trinkwasserinspektorat,
Liestal, Schweiz

adrian.auckenthaler@vsd.bl.ch

PD Dr. Peter Huggenberger

Angewandte und Umweltgeologie
Departement Erdwissenschaften der
Universität Basel, Schweiz

peter.huggenberger@unibas.ch

Prof. Dr. Hauke Harms

ENAC-ISTE-LPE
Bâtiment GR
École Polytechnique Fédérale de Lausanne
(EPFL), Schweiz

hauke.harms@epfl.ch

Dr. Antonis Chatzinotas

ENAC-ISTE-LPE
Bâtiment GR
École Polytechnique Fédérale de Lausanne
(EPFL), Lausanne, Schweiz

antonis.chatzinotas@epfl.ch

Einleitung 1

Adrian Auckenthaler und Peter Huggenberger

Trinkwasser ist ein Naturprodukt und wird aus Poren-, Kluft- oder Karstgrundwasserleitern, Oberflächengewässern oder aus Uferfiltrat gewonnen. In all diesen Wässern kommen natürlicherweise Mikroorganismen vor. Kritisch wird es dann, wenn pathogene Keime ins Trinkwasser gelangen oder wenn sie in grosser Anzahl im Rohwasser (Wasser vor der Aufbereitung zu Trinkwasser) auftreten. Die mikrobiologische Belastung von Grundwasser in Flussnähe oder von Karstquellen kann grossen zeitlichen Schwankungen unterliegen, da der Transport der Mikroorganismen in diesen Systemen von Hochwässern und Niederschlagsereignissen beeinflusst ist. Für die Wasserversorgungen stellen die mikrobiologischen Qualitätsschwankungen in der Rohwasserqualität deshalb eines der Hauptprobleme dar.

Die Beurteilung der mikrobiologischen Trinkwasserqualität basiert auf Indikatorkeimen, deren Herkunft menschliche oder tierische Fäkalien sind. Die Nachweismethoden für diese Keime gehen auf Analyseprinzipien zurück, die vor rund 100 Jahren entwickelt wurden. Das Konzept der Indikatorkeime hat viele Vorteile, wie z.B. der einfache und zuverlässige Nachweis, ist letztlich aber für eine "end of pipe" Kontrolle des Trinkwassers ungeeignet. Denn bei der Aufbereitung von Trinkwasser werden Mikroorganismen selektiv abgetötet, allen voran die fäkalen Indikatorkeime *E. coli*. So kommt es, dass Trinkwasser, das frei von Indikatorkeimen ist, trotzdem pathogene Mikroorganismen enthält, wie der bisher grösste Ausbruch einer Cryptosporidienepidemie in Milwaukee, USA 1994 mit 400'000 Erkrankten gezeigt hat (MacKenzie et al., 1994; Karansis et al., 1996). In den USA und England sind seit 1984 dank gutem Überwachungs- und Meldesystem mehrere wasserbedingte Cryptosporidienausbrüche dokumentiert. Auch in der Schweiz und in Deutschland sind in den letzten Jahren wasserbedingte Erkrankungsfälle durch Giardia und Norwalk-like Viren (NLV) aufgetreten (Häfliger et al., 2000; Maurer und Stürchler, 2000; Gornik et al., 2000). Protozoen und Viren werden weltweit im Grund- und teilweise im Trinkwasser gefunden (Metzler et al., 1996; Svoboda et al., 1999). Die Infektionsdosis dieser Pathogene ist sehr gering. Bei Cryptosporidien muss bereits ab einer Konzentration von 10 bis 30 Oozysten pro 100 L mit einem Epidemieausbruch gerechnet werden. Bei Rotaviren genügte bereits ein Viruspartikel, um 27% gesunder Erwachsener zu infizieren (Gale, 1996; Haas und Rose, 1996). Aufgrund dieser sehr geringen Infektionsdosis und den festgestellten Epidemien sind die Protozoen Cryptosporidium und Giardia und die Viren als Problemkeime der Wasserversorgungen sowohl in Industriestaaten als auch in Entwicklungsländern erkannt worden.

Die Wasserversorgungen in der Schweiz und in Deutschland sind abgesehen von den grossen Städten, meist dezentral organisiert. Viele Gemeinden beziehen ihr Trinkwasser aus eigenen Grundwasserbrunnen oder Quellwasservorkommen. Bei einer mikrobiologischen Belastung des Trinkwassers, festgestellt durch die Analyse von fäkalen Indikatorkeimen, wird häufig eine einstufige Aufbereitung in Form einer Desinfektion mit UV (Ultraviolett Bestrahlung) oder Chlor gewählt. Diese kostengünstigen Anlagen sind einfach und schnell zu installieren und bringen vordergründig gute Resultate. Weisen Rohwässer aber bei Regenereignissen oder Hochwässern mikrobiologische Qualitätsschwankungen mit pathogenen Keimen und erhöhte Trübungen auf, stossen einstufige Aufbereitungsverfahren rasch an ihre Grenzen. Bei Karstquellwässern, Grundwasserfassungen nahe bei Flüssen oder bei Seewasserwerken muss deshalb in der Regel mehrstufig aufbereitet werden. Die Rohwasseraufbereitung mit Flockungsmitteln und einer Schnellsandfiltration kann jedoch die Clusterbildung von Mikroorganismen verstärken, was die individuelle Gefährdung der Konsumenten erhöht (Gale et al., 1997).

Die Entwicklung und Validierung von molekularen Methoden für die Trinkwasseranalyse wie zum Beispiel die PCR (Polymerase Chain Reaction) und der Vergleich ihrer Aussagekraft mit klassischen Methoden wird heute weltweit vorangetrieben (Köster et al., 2002). Durch diese neuen Nachweismethoden wird es möglich sein, ein klareres Bild über die mikrobiologische Gefährdung des Trinkwassers zu zeichnen. Der Nachweis von Protozoen oder Viren im Trinkwasser wird vor allem kleinere Wasserversorgungen vor eine grosse Herausforderung stellen. Es gibt mehrere Ansätze, um diese Aufgabe anzugehen.

Die Einzugsgebiete der Wasserfassungen müssen nach hydrogeologischen Gesichtspunkten und nach den mikrobiologischen Belastungsmustern erfasst werden. Für die Ausscheidung von Schutzzonen in Karstgebieten wurde in der Schweiz die Multikriterienmethode EPIK entwickelt, die auf einer differenzierten hydrogeologischen Betrachtung des Einzugsgebietes der Fassung beruht. In ähnlicher Weise sollten auch Schutzzonen für Porengrundwasserleiter ausgeschieden werden.

Das Management der Wasserentnahmen ist ein weiteres Element für eine sichere Trinkwassergewinnung. Unter Management wird die online Überwachung von Systemparametern wie Niederschlag, Schüttung bei Quellen oder Pegelstand bei Lockergesteinsgrundwasserleitern und weiteren Parametern verstanden. Bei einer Veränderung dieser Parameter gegenüber einem Grundzustand (Basisabfluss) ist mit einer Verunreinigung durch Mikroorganismen zu rechnen. Das Management soll aufzeigen, ab welchem Zeitpunkt in Phasen mit erhöhtem Risiko das gewonnene Wasser aufzubereiten oder zu verwerfen ist, das heisst, die Trinkwasserversorgung anderweitig sichergestellt werden muss.

Die letzte Stufe zur Sicherung von qualitativ hochstehendem Trinkwasser ist die Aufbereitung. Eine Trinkwasseraufbereitung kann nur dann richtig dimensioniert werden und entsprechend funktionieren, wenn Art und Ausmass der mikrobiologischen, chemischen und physikalischen Verunreinigung bekannt sind. Für die Dimensionierung einer Aufbereitungsanlage und die optimale Wahl des Verfahrens sind deshalb die Kenntnisse aus den hydrogeologischen Abklärungen zu den Grundwasserschutzzonen sowie die online Überwachung nötig. Diese naturwissenschaftlichen Erkenntnisse dienen ebenfalls dem Aufbau eines optimalen Qualitätssicherungssystems und dem Verständnis des gesamten Prozesses der lokalen Trinkwassergewinnung.

Eine Strategie für die nachhaltige Qualitätssicherung der Trinkwassergewinnung führt zur Formulierung von kurz-, mittel- und langfristigen Zielen auf lokaler und regionaler Ebene und verlangt ein interdisziplinäres Vorgehen von Wasserversorgern, Behörden auf nationaler, kantonaler und kommunaler Ebene, Geologie- und Ingenieurbüros sowie Forschungsinstitutionen.

Pathogene Mikroorganismen im Trinkwasser 2

Adrian Auckenthaler

Einleitung

Wasserbedingte Infektionen durch hygienisch nicht einwandfreies Trinkwasser haben weltweit eine grosse Bedeutung; sie verursachen jährlich rund eine Milliarde Erkrankungen und drei Millionen Tote (Ford, 1999). Die Erkrankungen treten vor allem bei schlechter Sanitärhygiene auf. In Europa ist die Inzidenz der wasserbedingten Erkrankungen seit dem Beginn des 20. Jahrhunderts drastisch zurückgegangen, namentlich als Folge der Trinkwasseraufbereitung und –desinfektion, der verbesserten Sanitärhygiene (Abwasserentsorgung), der Einführung von Massnahmen zum Schutz der Trinkwasserressourcen und der Wasserversorgungsinfrastruktur. Die mikrobiologische Trinkwasserqualität ist jedoch heute wieder vermehrt gefährdet durch: 1) zu hohe Tierdichten auf Landwirtschaftsbetrieben, 2) grössere Mengen an flüssigem Hofdünger aufgrund der Umstellung landwirtschaftlicher Betriebe auf Schwemmentmistung, 3) Einsatz von Hofdüngern in der Nähe von Trinkwasserfassungen mit zu klein dimensionierten oder falsch ausgeschiedenen Schutzzonen, 4) veränderte Siedlungsstrukturen und zunehmende Nutzungskonflikte in dicht besiedelten Gebieten, 5) teilweise nicht den Risiken angepasste oder veraltete Trinkwasseraufbereitung und 6) vermehrtes Auftreten pathogener Keime, die eine hohe Infektiosität aufweisen können.

Durch die Landwirtschaft werden Hofdünger mit zoonotischen Erregern wie Cryptosporidien ausgebracht. Zusätzlich werden in der Schweiz jährlich 80'000 Tonnen Klärschlamm als Dünger eingesetzt, wovon 20'000 bis 30'000 Tonnen unhygienisiert sind. Durch diesen Eintrag werden wahrscheinlich erhebliche Mengen humanpathogener Viren in die Ackerbaugebiete verfrachtet (Baumgartner, 2001). Über Drainagen in landwirtschaftlich genutzten Gebieten und Regenwasserentlastungen bei Kläranlagen werden zudem bei Niederschlagsereignissen innerhalb kurzer Zeit grosse Mengen an Mikroorganismen in die Oberflächengewässer eingetragen. Diese Einträge spielen vor allem bei schlechten Boden- und Untergrundfiltern eine Rolle, wie sie in Karstgebieten oder bei der Infiltration von Flusswasser ins Grundwasser vorkommen. Routineuntersuchungen des Kantonalen Laboratoriums Basel-Landschaft (KLBL) zeigen pro Jahr in rund 20% der Trinkwasserversorgungen, die Karstquellwasser mittels einstufiger Verfahren aufbereiten, den Fäkalindikatorkeim *E. coli* (KLBL, 2001). Detaillierte Abklärungen bei einer Karstquelle in der Nordwestschweiz haben ergeben, dass die pathogenen Mikroorganismen *Cryptosporidium* spp., verotoxinbildende *E. coli* (VTEC) und *Campylobacter jejuni* konzentriert bei Regenereignissen auftreten (Auckenthaler et al., 2002). Im selben Studiengebiet konnten in einer Quelle auch Norwalk-like Viren gefunden werden (Hobbins et al., 2001).

Die epidemiologische Bedeutung dieser Befunde lässt sich heute nur schwer abschätzen. Die erforderliche Datengrundlage fehlt noch weitgehend, weil in der Schweiz keine Meldepflicht für bedeutende durch Wasser übertragbare Krankheitserreger besteht. In Deutschland ist die Situation besser, seit Januar 2001 sind verschiedene pathogene Mikroorganismen zusätzlich meldepflichtig geworden (Ammon und Bräunig, 2002). Englische Studien über gastrointestinale virale Erkrankungen haben jedoch ergeben, dass das Verhältnis der Erkrankten in der Bevölkerung zu Erkrankten, die einen Arzt aufsuchen 1 zu 1562 beträgt. Eine Person, die einen Arzt aufsucht, repräsentiert somit rund 1500 Fälle in der Bevölkerung (Wheeler, 1999).

In der Schweiz ist die Bevölkerung in den letzten Jahren in den ländlichen Gebieten um die Ballungszentren durch Neuzuzüge stark gewachsen. Dieser Bevölkerungszuwachs stellt höhere Anforderungen an die Wasserversorgungen, da sie eine grössere Menge Trinkwasser zur Verfügung stellen müssen. Das Risiko von Trinkwasser bedingten Krankheitsausbrüchen nimmt zu, da eine grössere Gruppe von Personen versorgt wird und die Überwachungs- und Aufbereitungssysteme diesem Risiko oft nicht angepasst werden. Muss die Aufbereitung aufgrund schlechter Rohwasserqualität zwingend verbessert werden, kommen hochtechnische Verfahren zum Einsatz, die für grosse Wasserversorgungen konzipiert sind und für die in den kleineren Wasserversorgungen oft kein qualifiziertes Personal vorhanden ist (Tibbetts, 2000).

In Zukunft ist wegen der demographischen Entwicklung mit immer mehr älteren Menschen zu rechnen – in den USA wird geschätzt, dass sich die Anzahl der über 65 Jährigen von 1980 bis 2020 verdoppelt, was die Grösse der Risikogruppe für Erkrankungen durch enterische Mikroorganismen erhöht. Zu dieser Gruppe gehören nebst älteren Menschen auch Kleinkinder, schwangere Frauen und Immunsupprimierte. Dieses Segment der Bevölkerung macht in den USA heute 20% aus (Gerba et al., 1996). Für die Schweiz und Deutschland ist mit einer ähnlichen Zahl zu rechnen (Ammon und Bräunig, 2002). Durch die verbesserte Analytik sind in letzter Zeit bisher unbekannte pathogene Keime wie *Cryptosporidium* spp., Norwalk-like Viren und verotoxinbildende *E. coli* (VTEC) nachweisbar geworden, die für die Wasserversorgungen von Bedeutung sind. Dass auch in Zukunft mit neuen Krankheitserregern gerechnet werden muss, zeigt sich daran, dass seit 1973 weltweit insgesamt 21 bisher unbekannte humanpathogene Erreger entdeckt wurden (Sonntag, 1997). Diese über das Wasser übertragbaren Erreger zeigen teilweise eine hohe Infektiosität. Die geringen Infektionsdosen von lediglich 1 bis 10 Partikeln stellen eine zusätzliche Herausforderung für die Wasseraufbereitung dar. Als wichtigstes Lebensmittel und Ausgangsprodukt für andere Lebensmittel sollte Wasser deshalb möglichst frei von pathogenen Keimen sein.

Tab. 2.1: Kurze Charakterisierung von Bakterien, Viren und Protozoen

Viren	Bakterien	Protozoen
Viren sind 10 bis 300 nm gross, ihr genetisches Material ist in Proteinhüllen gepackt. Sie können den biochemischen Apparat einer geeigneten (meist sehr spezifischen) Wirtszelle auf Produktion neuer Viren der eigenen Art umprogrammieren. Viren haben selbst keinen Stoffwechsel und sind auf chemische Bausteine, Energie und Enzyme lebender Zellen angewiesen. Heute sind rund 1500 Viren bekannt.	Einzellige Mikroorganismen, die den Pflanzen und Tieren als selbständige systematische Einheit gegenüberstehen. Das Zellinnere der 0,5 bis 5 µm grossen Bakterien weist nur eine geringe Differenzierung auf und enthält keinen Zellkern. Die Vermehrung erfolgt durch Querteilung. Bei genügend Nährstoffen kann eine Generationszeit bei einigen Arten weniger als 20 min dauern. Es sind über 5000 Arten bekannt.	Rund 40'000 bekannte Arten von 1 µm bis 2 mm grossen Einzellern. Das genetische Material ist in einem Zellkern zusammengefasst. Die Fortpflanzung findet geschlechtlich oder ungeschlechtlich statt. Protozoen können umweltresistente Dauerstadien bilden (Zysten, Oozysten) und sich mit Hilfe von Scheinfüsschen, Geisseln oder Wimpern fortbewegen.

Allerdings ist die Forderung nach einer Nulltoleranz, wie sie in gewissen Risikoabschätzungen gestellt wird – mehrere 1000 L Trinkwasser müssten von Viren frei sein, um ein Infektionsrisiko auszuschliessen – unrealistisch (Botzenhardt, 2000a). Es ist aber zu beachten, dass bei wasserbedingten Ausbrüchen meist mit Erkrankungsraten von mehreren hundert bis tausend Personen zu rechnen ist, wie die beiden Fälle La Neuveville und Küblis in der Schweiz gezeigt haben (Maurer und Schtürchler, 2000; Lüthi und Beuret, 2001).

In diesem Kapitel sind Mikroorganismen beschrieben, die direkt Krankheitsausbrüche verursacht haben oder bei denen aufgrund epidemiologischer Abklärungen vermutet wird, dass sie mit wasserbedingten Ausbrüchen im Zusammenhang stehen.

Bakterien

Im 18. und am Anfang des 19. Jahrhunderts haben Seuchen wie Pest und Cholera in Europa Angst und Schrecken verbreitet. Die beiden Erreger *Yersinia pestis* und *Vibrio cholerae* wurden erst durch die Arbeiten von Louis Pasteur (1822 – 1895) und Robert Koch (1843 – 1910) als bakterielle Pathogene erkannt. Nach der ersten grossen Pest-Pandemie im 14. Jahrhundert warf der „schwarze Tod" über Jahrhunderte hinweg seinen Schatten über Europa. Die Cholera trat im 19. Jahrhundert von China her kommend in Europa auf und verursachte bis zum ersten Weltkrieg fünf Pandemien (Jütte, 1997). Schlechte Sanitärhygiene oder der Verzicht auf die Trinkwasseraufbereitung kann noch heute zu wasserbedingten Choleraausbrüchen führen, wie die Epidemie in Peru 1991 zeigte (Daniel, 1998).

Abb. 2.1: „Cholera“ gezeichnet als geflügelter Drache vom Schweizer Maler Arnold Böcklin (1827 – 1901) (mit freundlicher Genehmigung, Archiv für Kunst und Geschichte, Berlin)

In den USA waren in den Jahren 1989 bis 1997 23,5% der wasserbedingten Gastroenteritiden auf bakterielle Erreger zurückzuführen. In 3 von 4 Fällen sind die „neuen“ pathogenen *E. coli* (VTEC) als verantwortlich für Ausbrüche erkannt worden (MMWR, 2000). Pathogene *E. coli* waren ebenfalls in Kanada im Jahr 2000 die Ursache für wasserbedingte Erkrankungen in einer öffentlichen Wasserversorgung (CCDR, 2000). Es ist zu erwarten, dass auch in Zukunft immer wieder neue, bisher unbekannte pathogene Keime im Wasser auftreten werden.

Bakterien werden im Gegensatz zu Viren und Protozoen gegenwärtig am häufigsten im Grund- und Trinkwasser untersucht. Dieses Teilkapitel beschreibt eine Auswahl der wichtigsten Vertreter unter besonderer Berücksichtigung der Indikatorkeime, die zur Trinkwasserüberwachung genutzt werden. Das Vorhandensein von Indikatorkeimen soll das Vorhandensein pathogener Keime, meist fäkalen Ursprungs, aufzeigen. Neben den bekannten werden auch neu erkannte pathogene Keime beschrieben. Am Schluss wird kurz auf Biofilme und Filterverkeimungen eingegangen, die hauptsächlich in technischen Systemen zur Trinkwasseraufbereitung und zum Trinkwassertransport auftreten. Die Tabellen 2.2, 2.3, 2.4 und 2.6 geben einen Überblick über die beschriebenen Bakterien, Viren und Protozoen.

Tab. 2.2: Im Grund- und Trinkwasser vorkommende pathogene Bakterien und ihre Infektionsdosis

Bakterien	Infektionsdosis [Anzahl Keime][1]	Vorkommen im Grund- und Trinkwasser		Quelle
		weltweit	Schweiz / Deutschland	
Campylobacter jejuni	10-500	Grund- und Trinkwasser, Ausbrüche	Grund- und Trinkwasser	*AWWA, 1999c; Tschäppe, 2000; Raso, 2001*
Shigella spp.	10-100	Grund- und Trinkwasser, Ausbrüche	Grund- und Trinkwasser, Ausbrüche	*Barwick et al., 2000; Maurer und Stürchler, 2000*
Salmonella sp. typhoide Salmonellen enteritische Salmonellen	10^1-10^5 10^2-10^3 10^5-10^6	Grund- und Trinkwasser, Ausbrüche	unbekannt	*Covert, 1999; Tschäppe, 2000*
Yersinia enterocolitica	10^4-10^9	Grund- und Trinkwasser, Ausbrüche	möglich	*Schiemann, 1990; Fricker, 1999b; Tschäppe, 2000*
Staphylococcus aureus	10^6/g LM	Grund- und Trinkwasser	unbekannt	*Geldreich, 1999*
Vibrio cholerae	10^6-10^8	im Grund- und Trinkwasser nachgewiesen	unbekannt	*Daniel 1998; Ford, 1999*
Pathogene *E. coli*	10-100	Grund- und Trinkwasser, Ausbrüche	Grundwasser	*Rice, 1999; Tschäppe, 2000; Auckenthaler et al., 2002*
Helicobacter pylori	unbekannt	Vorkommen im Wasser unbekannt	unbekannt	*AWWA, 1999c*
Mycobacterium avium	unbekannt	Trinkwassernetz, dokumentierte Ausbrüche in Nord Amerika	unbekannt	*AWWA, 1999c*
Legionella pneumophila	>10	technische Wassersysteme	technische Wassersysteme	*Hall, 1999*
Pseudomonas aeruginosa	grosse Anzahl	ubiquitär Wasser, Boden, Gemüse	ubiquitär Wasser, Boden, Gemüse	*Geldreich, 1999*

([1]Zahlen Infektionsdosis aus Beck, 1997 und Ford, 1999)

Indikatorkeime

Das Konzept der Indikatorkeime ist fast hundert Jahre alt und wird weltweit immer noch angewendet. Die Idee ist, dass man mit der Untersuchung einiger ausgewählter Bakterien die mikrobiologische Trinkwasserqualität beurteilen kann. Die Abwesenheit von Indikatorkeimen in einer Wasserprobe wird dabei als Indiz für die Abwesenheit von pathogenen Mikroorganismen gewertet. Damit dieses Konzept funktioniert, muss ein optimaler Indikator verschiedene Kriterien erfüllen. Diese sind (verändert nach Bonde, 1962):

- Anwesenheit, wenn die Krankheitserreger anwesend sind, oder wenn deren Anwesenheit möglich ist
- Keine Vermehrung in der Umwelt aber genügend lange Überlebensdauer der einzelnen Bakterien
- Vorkommen in grösserer Anzahl als die Krankheitserreger
- Grössere Resistenz in der Umwelt und gegen die Wasseraufbereitung als die Krankheitserreger
- Schnelles, selektives Wachstum auf einfachen Nährböden und möglichst eindeutige Identifikation
- Möglichst homogene Verteilung in der zu untersuchenden Probe

Ein Problem der meisten heutigen Indikatorkeime zur Beurteilung der Trinkwasserqualität ist ihre geringe Resistenz gegen die Wasseraufbereitung. *E. coli* beispielsweise stirbt bereits bei sehr geringen Desinfektionsmittelkonzentrationen ab. In der Literatur wird oft bemängelt, dass die Indikatorkeime nicht mit den pathogenen Keimen im Wasser korrelieren. Dabei wird allerdings kaum zwischen Rohwasser und aufbereitetem Trinkwasser unterschieden. Im Rohwasser ist eine Korrelation von Indikatorkeimen und pathogenen Mikroorganismen sehr viel wahrscheinlicher als nach der Aufbereitung (Auckenthaler et al., 2002). Die Wirkung eines Desinfektionsmittels variiert von Mikroorganismus zu Mikroorganismus. Die Filtration ist ebenfalls von verschiedenen Faktoren wie z.B. der Oberfläche und der Grösse der Partikel abhängig (siehe dazu Kap. Trinkwasseraufbereitung). Es erstaunt deshalb wenig, wenn nach der Aufbereitung keine Korrelation der Indikatorkeime mit den pathogenen Mikroorganismen besteht.

Eine weitere Lücke im Konzept der Indikatorkeime ist die hohe Infektiosität gewisser Viren. Teilweise genügen bereits 10 bis 100 Viruspartikel, um eine Infektion auszulösen. Das akzeptable Risiko einer Virusinfektion durch das Trinkwasser ist in den USA von der Environmental Protection Agency (EPA) mit einer Infektion pro 10'000 exponierter Personen und Jahr festgesetzt worden (Regli et al., 1991; Botzenhardt, 1994). Zur Einhaltung dieser Forderung dürfte demnach nur eine infektiöse Einheit pro 140 m^3 Trinkwasser enthalten sein (Walter, 1999). Mit der Abwesenheit von *E. coli* in 100 mL untersuchten Wassers kann deshalb keine Beurteilung auf Virusabwesenheit gemacht werden. In diesem Fall muss die Trinkwasseraufbereitung die geforderten Reduktionsraten garantieren können.

Tab. 2.3: Indikatorkeime für Grund- und Trinkwasseruntersuchungen

Bakterien	Infektionsdosis [Anzahl Keime][1]	Vorkommen im Grund- und Trinkwasser		Quelle
		weltweit	Schweiz / Deutschland	
aerobe mesophile Keime	$>10^9$/ g LM[2]	Indikatorkeim Grund- und Trinkwasser	Indikatorkeim Grund- und Trinkwasser in der Schweiz	*HyV, 1995*
E. coli	$>10^6$/ g LM[2]	Indikatorkeim Grund- und Trinkwasser	Indikatorkeim Grund- und Trinkwasser	*HyV, 1995; EC, 1998*
Enterokokken	$>10^6$/ g LM[2]	eher total Coliforme untersucht	Indikatorkeim Grund- und Trinkwasser	*HyV, 1995; EC, 1998*
Clostridium perfringens (anaerober Sporenbildner)	10^6-10^8/ g LM	in der EU als Indikatorkeim aufgenommen	Grundwasser	*EC, 1998; Brockmann und Botzenhart, 2000; Raso, 2001*

([1]Zahlen Infektionsdosis aus Beck, 1997 und Ford, 1999; [2]Toleranzwerte in Lebensmitteln gemäss Hygieneverordnung (HyV, 1995))

Aerobe mesophile Keime, *E. coli*, Enterokokken

In der Schweiz und in Deutschland werden zur Trinkwasserbeurteilung hauptsächlich die drei Indikatorkeime aerobe mesophile Keime (AMK), *E. coli* und Enterokokken verwendet. In Deutschland kommt mit der neuen Trinkwasserverordnung ab dem 1. Januar 2003 nun auch *Clostridium perfringens* hinzu (siehe dazu nächsten Abschnitt).

Die AMK bilden eine grosse Gruppe von Bakterien, Hefen und Pilzen, die unter normalem Luftsauerstoff auf einem unspezifischen Nährmedium bei mittlerer Temperatur (30°C) wachsen. Im Trinkwassernetz dürfen in der Schweiz maximal 300 AMK/mL nachweisbar sein, unbehandeltes Grundwasser darf höchstens 100 Keime/mL und behandeltes lediglich 20 Keime/mL enthalten (Anh. 2b, HyV). Die unterschiedlichen Zahlen kommen daher, dass sich die Zahl der AMK im Verteilnetz durch eine Wiederverkeimung erhöhen kann. Diese kann auf lange Standzeiten des Wassers im Verteilnetz, erhöhte Konzentration von wachstumslimitierenden Substanzen (z.B. Phosphat, organisches Material) oder auf Biofilme in den Rohrleitungen zurück geführt werden. Durch Wiederverkeimung kann die Anzahl AMK auf mehrere 1000/mL anwachsen. Eine akute Gesundheitsgefährdung ist dabei nicht zu erwarten. Es sind jedoch Massnahmen, wie Netzspülungen, Chlorung des Trinkwassers und allenfalls Ersatz alter Leitungen sowie, bei wiederholtem Auftreten, Verminderung des

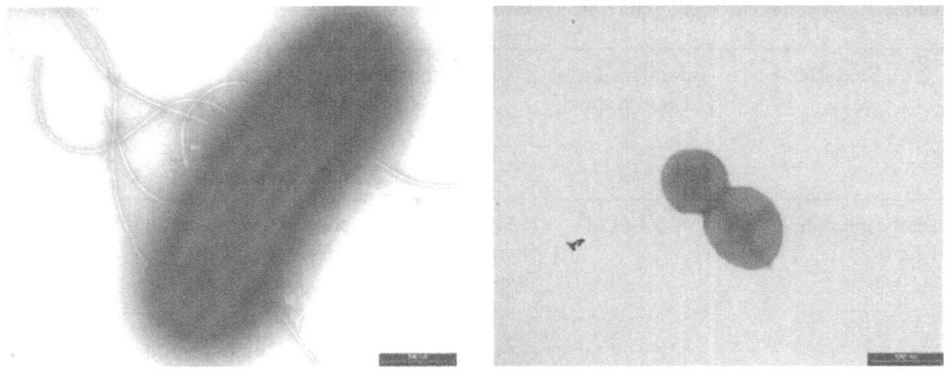

Abb. 2.2: *E. coli* (links) und Enterokokken (rechts), elektronenmikroskopische Aufnahmen (mit freundlicher Genehmigung, Tamara Fretz, Zentrum für Mikroskopie ZMB Bio-Pharmazentrum, Universität Basel)

Eintrags von organischer Substanz und anderen wachstumslimitierenden Substanzen ins Leitungsnetz zu treffen Laut der neuen Trinkwasserverordnung in Deutschland wird die Koloniezahl, welche den aeroben mesophilen Keimen entspricht, bei 22°C bestimmt. Als Bewertungskriterium darf keine anormale Veränderung der Koloniezahl bei mehrmaligen Untersuchungen des Wassers auftreten, eine Richtzahl oder zulässige Höchstkonzentration ist aber nicht bestimmt worden (Oehmichen et al., 2001).

E. coli und Enterokokken gelten als klassische, fäkale Indikatorkeime und haben eine Grösse von rund 0,5 bis 2 µm. Sie dürfen in 100 mL Trinkwasser nicht nachweisbar sein (Anh. 2b, HyV; Anh. 1 Teil A, EC, 1998). Bei diesem Wert handelt es sich in der Schweiz um einen Toleranzwert und in Deutschland um einen Parameterwert. Bei der Überschreitung des Wertes sind entsprechende Massnahmen zur Wiederherstellung des gesetzeskonformen Zustandes einzuleiten. *E. coli* und Enterokokken stammen aus dem Darm von Warmblütern, wobei letztere auch aus Insekten stammen können. Bei der Untersuchung der Enterokokken werden *Streptococcus faecialis* und *Streptococcus faecium* nachgewiesen. Die Enterokokken sollen gegenüber Umwelteinflüssen und Trinkwasseraufbereitung resistenter sein als *E. coli*. Die Untersuchungen der Kantonalen Laboratorien zeigen aber, dass beide Keime ins Trinkwassernetz gelangen können, und dass *E. coli* zahlenmässig oft stärker vertreten ist. Beide Keime galten bisher als harmlos und ihr Auftreten im Trinkwasser führt zwar zur Einleitung von Massnahmen für die Eliminierung der Keime, hat aber meist keine grösseren Konsequenzen für die Wasserversorgungen.

Nun jedoch wurde erkannt, dass es neben den harmlosen auch gefährliche, verotoxinbildende *E. coli* (VTEC) gibt, und dass bei den Enterokokken antibiotikaresistente Keime auftreten können. Auf die Problematik der VTEC wird im untenstehenden Abschnitt näher eingegangen. Die Situation der Antibiotikaresistenzen im Zusammenhang mit der Wasserversorgung ist noch kaum geklärt. Es gibt erst wenige Untersuchungen, doch ist zu erwarten, dass in Zukunft vermehrt antibiotikaresistente Stämme im Wasser nachgewiesen werden können.

Clostridium perfringens

In die neue EG-Trinkwasserrichtlinie 98/83/EG über die Qualität von Wasser für den menschlichen Gebrauch wurde *C. perfringens* als Indikator für die Beurteilung der mikrobiologischen Qualität aufgenommen (EC, 1998). Das Bakterium darf in 100 mL Wasser nicht nachweisbar sein. In Deutschland wird mit in Kraft treten der neuen Trinkwasserverordnung im Januar 2003 die EG-Trinkwasserrichtlinie umgesetzt.

C. perfringens produziert eine Reihe von hoch aktiven Enzymen und Toxinen. Die wichtigste durch *C. perfringens* verursachte Erkrankung ist der Gasbrand, eine ohne Therapie tödlich verlaufende Wundinfektion. In ca. 80% der Gasbranderkrankungen wird der Erreger nachgewiesen.

Die mit dem Wasser aufgenommenen Clostridien stellen keine direkte gesundheitliche Gefahr für den Menschen dar. Die Verwendung von verunreinigtem Trinkwasser für die Zubereitung von Nahrungsmitteln und deren unsachgemässe Lagerung kann hingegen zu einer Vermehrung der Clostridien und in der Folge zu ernsthaften Erkrankungen führen (Brockmann und Botzenhardt, 2000).

C. perfringens ist ubiquitär vorhanden und konnte unter anderem in Luft, Staub, Erde, Wasser und Nahrungsmitteln nachgewiesen werden. In menschlichen Fäzies liegt die Konzentration bei 10^1 bis 10^8/g.

Als Indikatorkeim wurde *C. perfringens* deshalb vorgeschlagen, weil das Bakterium umweltresistente Sporen bildet, die gegenüber der Aufbereitung von Trinkwasser eine ähnliche Resistenz wie Oozysten von Cryptosporidium oder Zysten von Giardia aufweisen. Aufgrund von Untersuchungen zur Überwachung von Flusswasseraufbereitungen schlagen Payment und Franco, 1993 die routinemässige Erfassung von *C. perfringens* im filtrierten und aufbereiteten Wasser als ergänzende Messung neben den Indikatorkeimen vor. *E. coli* ist zur Überwachung eines Aufbereitungsprozesses ungeeignet, da das Bakterium bereits bei geringen Desinfektionsmittelkonzentrationen abstirbt (siehe dazu Kap. Trinkwasseraufbereitung).

Bei Karstwasseruntersuchungen in der Nordwestschweiz konnte *C. perfringens* im Rohwasser festgestellt werden. Die Konzentrationen lagen allerdings stets um 1 bis 2 Grössenordnungen tiefer als diejenigen von *E. coli* (Raso, 2001). Für die Überwachung von Multibarrierensystemen könnte die Konzentration von *C. perfringens* deshalb zu gering sein, bei reiner Desinfektion des Wassers hingegen wäre *C. perfringens* eine gute Ergänzung zu den heutigen Indikatorkeimen.

Bekannte pathogene Bakterien

Die nachfolgend aufgeführten bekannten pathogenen Bakterien werden in der Schweiz und in Deutschland normalerweise in Grund- und Trinkwasseruntersuchungen nicht analysiert. Ihr Vorkommen im Grundwasser ist deshalb kaum bekannt, was eine breitere Abschätzung der Gefährdung für die TrinkwasserkonsumentInnen erschwert. In detaillierten Karstwasseruntersuchungen in der Nordwestschweiz konnten Campylobacter während Regenereignissen nachgewiesen werden (Raso, 2001), was lokal sicher zu einer Gefährdungszunahme führen kann. Epidemiologische Zahlen aus den USA zeigen immer wieder wasserbedingte Ausbrüche durch bekannte pathogene Bakterien. In der Schweiz wurden bei der Trinkwasserverunreinigung in La Neuveville 1998 bei den epidemiologischen Abklärungen unter anderem *Shigella sonnei* und *Campylobacter jejuni* als Ursache eruiert (Maurer und Stürchler, 2000). Dies zeigt, dass pathogene Bakterien sehr wohl bis ins Trinkwassernetz gelangen können, sobald Trink- und Grundwasserschutzmassnahmen vernachlässigt und Hygienestandards nicht erfüllt werden oder technische Massnahmen wie z.B. die Trinkwasseraufbereitung versagen.

Campylobacter spp.

Campylobacteriosen können beim Menschen zu akuter Enteritis mit Durchfall, Bauchschmerzen, Fieber und Erbrechen und in sehr seltenen Fällen zur Hospitalisierung der Betroffenen führen (Krämer, 1992). 99% der Erkrankungen werden durch *C. jejuni* verursacht (AWWA, 1999). *C. jejuni* ist resistent gegenüber einer Reihe von Antibiotika, jedoch nicht gegen Erythromycin und Fluoroquinolone (Murphy et al., 1996).

Bei Campylobacter Enteritis handelt es sich um eine Zoonose. Das Reservoir der Bakterien sind verschiedene Haus- und Wildtiere, insbesondere Vögel (Fricker, 1999a). Die Übertragung erfolgt fäkal-oral über verunreinigte Nahrungsmittel oder Wasser, wobei unsorgfältig zubereitetes oder zu wenig erhitztes Pouletfleisch die häufigste Ursache von Campylobacteriosen ist (Hopkins und Scott, 1983). Die Infektionsdosis wird auf lediglich 500 Keime geschätzt (RKI, 2000). Verschiedene komparative Studien von Spitälern über Prävalenzraten in Stuhlproben haben ergeben, dass Campylobacter doppelt so häufig vorhanden sind wie Salmonellen und 4,5 bis 7 Mal häufiger als Shigellen (AWWA, 1999a). In der Schweiz haben die gemeldeten Fälle von Campylobacteriosen neuerdings die gemeldeten Fälle von Salmonellosen übertroffen (BAG, 2001). In Deutschland sind die Campylobacter-Infektionen von 1992 bis 2000 stark angestiegen, ihre Anzahl liegt jedoch noch unter den rückläufigen Salmonellen-Infektionen (Ammon und Bräunig, 2002).

In der Literatur sind einige wasserbedingte Campylobacter-Ausbrüche mit hunderten bis tausenden von Erkrankten beschrieben. So sind in den USA in der Zeit von 1978 bis 1986 11 Fälle mit durchschnittlich 453 Erkrankten aufgetreten (AWWA, 1999a). In Schweden wurden zwischen 1980 und 1995 ebenfalls 11 Fälle mit maximal 3000 Erkrankten registriert (Andersson et al., 1997). Bei den Ausbrüchen konnte der verantwortliche Organismus nur selten im Trinkwasser nachgewiesen werden. Der Grund dafür könnte im sporadischen Auftreten des Bakteriums im Wasser, in lückenhaften Analyseverfahren oder der in Präsenz von lebenden aber nicht kultivierbaren Bakterien liegen (Fricker, 1999a). In der Schweiz konnten Campylobacter im Rohwasser einer Karstquelle (Raso, 2001) und in einer Quelle aus einem Porengrundwasserleiter (Schaffter und Parriaux, 2002) nachgewiesen werden.

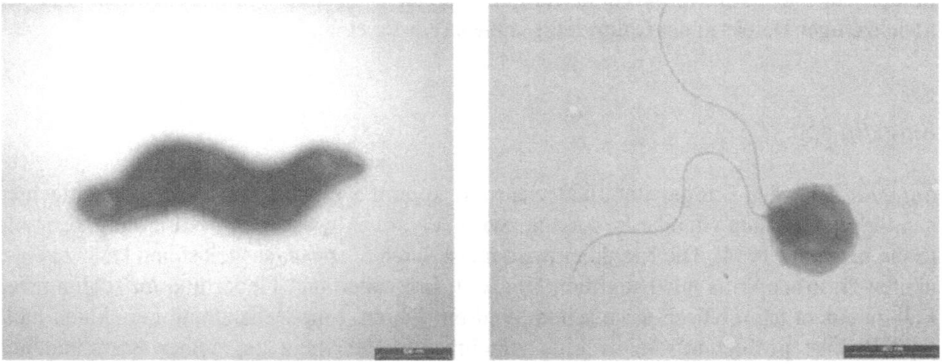

Abb. 2.3 *Campylobacter jejuni*, elektronenmikroskopische Aufnahmen (mit freundlicher Genehmigung, Tamara Fretz, Zentrum für Mikroskopie ZMB Bio-Pharmazentrum, Universität Basel)

Salmonellen

Die Gattung Salmonella gehört zur Familie der Enterobacteriaceae. Diese bildet eine grosse und vielfältige Gruppe von mehr als 2000 Serotypen, die im Boden, Wasser, in Pflanzen, Fäkalien und in der natürlichen Darmflora von Tieren vorkommen. Die Bakterien haben eine Länge von rund 2 bis 5 µm und einen Durchmesser von 0,8 bis 1,5 µm. Viele der Salmonellen sind pathogen für den Menschen oder für Tiere (Covert, 1999). Salmonellose ist weltweit eines der grössten Gesundheitsprobleme. Pro Jahr erkranken rund 16 Millionen Menschen an *S. typhi*, dem Erreger von Typhus, wobei die Krankheit in 600'000 Fällen tödlich verläuft (Ford, 1999). In Europa war Typhus vor allem im 19. und am Anfang des 20. Jahrhunderts ein Problem bis das Trinkwasser desinfiziert wurde. Die enterischen Salmonellen jedoch nehmen seit dem 2. Weltkrieg an Bedeutung zu und kommen heute in unserer Bevölkerung endemisch und epidemisch vor (Tauxe, 1997).

Für die typhösen Salmonellen ist der Mensch das einzige Reservoir. Für die enterischen Salmonellen dienen Wild- und Haustiere, so zum Beispiel Hühner, Schweine, Kühe, Vögel, Nagetiere, Hunde und Katzen als Wirte (Covert, 1999). Während bei den typhösen Salmonellen die Infektionsdosis 10^2 bis 10^3 Bakterien beträgt, übersteigt sie bei den enterischen Salmonellen in der Regel 10^5 bis 10^6 Bakterien. Die Bakterien können sich jedoch in Nahrungsmitteln vermehren, so dass ausgehend von einer geringen Anzahl Bakterien, die Infektionsdosis leicht erreicht werden kann.

In den USA wurden in der Zeit von 1976 bis 1980 insgesamt 223 meist kleinere Fälle von wasserbedingten Salmonellosen registriert. In Riverside, Kalifornien wurde 1965 ein grosser Ausbruch mit 18'000 Erkrankten aufgrund epidemiologischer Abklärungen dem Wasser zugeordnet, die Ursache der Salmonellose konnte aber nicht mit Sicherheit eruiert werden (Covert, 1999).

In der Schweiz ist die letzte registrierte wasserbedingte Epidemie von *S. typhi* 1963 in Zermatt mit 437 erkrankten Personen aufgetreten. Ursache waren ungeklärte Abwässer, die in den für die Trinkwasserversorgung genutzten Zmuttbach geleitet wurden und ein gleichzeitiger Defekt in der Chlorierungsanlage (Köster et al., 2002).

Shigella spp.

Shigella spp., der Erreger der Bakterienruhr, kommt weltweit vor allem bei schlechten sanitärhygienischen Standards vor und wird oft mit Abwasser in Verbindung gebracht (Samonis et al., 1994). Die Krankheit wird meist durch *S. sonnei* ausgelöst und kann zu einer akuten Gastroenteritis mit Durchfall, Übelkeit, Erbrechen und Fieber und im schlimmsten Fall zu einem hämolytisch-urämischen Syndrom führen. Eine Behandlung der akuten Fälle ist teilweise problematisch, da eine zunehmende Resistenz gegenüber verschiedenen Antibiotika zu verzeichnen ist (Meyers et al., 1998). Als Reservoir für die stäbchenförmigen Shigellen mit einem Durchmesser von rund 0,3 bis 1 µm und einer Länge von 1 bis 6 µm dienen infizierte Menschen. Die Übertragung erfolgt auf der fäkal-oralen Route über Wasser, Nahrungsmittel und Milch. Die Infektionsdosis ist mit 10 bis 100 Keimen gering (Beck, 1997). Wasserbedingte Epidemien sind wie bei Yersinia meist auf ungenügend geschütztes und unbehandeltes Trinkwasser zurückzuführen, da Shigella in gechlortem Wasser relativ schnell abstirbt. In den USA sind zwischen 1985 und 1994 insgesamt 17 Ausbrüche dokumentiert (CDC, 1996). Für die Jahre 1997 und 1998 wurde ein Fall mit 83 erkrankten und 4 hospitalisierten Personen beschrieben (Barwick et al., 2000).

In der Schweiz sind 1998 in La Neuveville 1600 Personen an akutem Brechdurchfall erkrankt. Bei den Patienten konnten *Shigella sonnei*, *Campylobacter jejuni* und zur Hauptsache SRSV (small round structured viruses) isoliert werden (Maurer und Stürchler, 2000; Häfliger et al., 2000). Die Ursache war eine Verunreinigung des Trinkwassers durch eine defekte Abwasseranlage in der engeren Schutzzone S2. Solche Anlagen sind allerdings in der engeren Schutzzone weder in der Schweiz noch in Deutschland zulässig.

Staphylococcus

S. aureus, *S. epidermis* und *S. saprophyticus* sind opportunistische Pathogene, die Infektionen der Haut auslösen können. Das Enterotoxin von *S. aureus* kann bei verunreinigten Nahrungsmitteln auch zu Durchfall und Erbrechen führen. Das Reservoir der Staphylokokken bilden Warmblüter und Abwasser. *S. aureus* kommt in der natürlichen Flora der menschlichen Haut und in Fäkalien vor. Die Übertragung der Keime erfolgt vor allem beim Baden, wo sie in Hautwunden, Augen und Ohren gelangen können. Verunreinigtes Wasser kann bei der Zubereitung von Nahrungsmitteln eine Bedeutung haben, da sich die Keime vermehren, Enterotoxine bilden und dadurch Erkrankungen auslösen können.

Bisher wurden keine wasserbedingten Fälle von Staphylokokken beschrieben. Die Keime haben aber in Spitälern durch die Antibiotikaresistenz an Bedeutung gewonnen und das Wasser könnte eine Übertragung solcher Stämme begünstigen (Geldreich, 1999).

Vibrio cholerae

Die Cholera wird durch das Bakterium *Vibrio cholerae* verursacht, das in verunreinigten Nahrungsmitteln und verschmutztem Trinkwasser vorkommt. Die Folge der Infektion ist eine Entzündung des Darmes mit starkem Brechdurchfall. Der stetige Flüssigkeitsverlust verursacht eine Austrocknung des Körpers und einen Mangel an Mineralien. Eine zu starke Dehydrierung kann zum Tod führen (Jütte, 1997; Daniel, 1998).

Die Cholera trat erst im 19. Jahrhundert von China her kommend in Europa auf. Die Sterblichkeitsrate lag damals bei 60 bis 80%. Bis zum Ersten Weltkrieg zählte man fünf Cholera-Pandemien in Europa (1830 bis 1838, 1847 bis 1858, 1865 bis 1875, 1884 bis 1896 und 1902 bis 1912) (Jütte, 1997). Cholera ist noch heute eine häufige infektiöse Erkrankung in sich entwickelnden Ländern und weltweit erkranken laut WHO ca. 581'000 Menschen pro Jahr wobei 18'000 an dieser Krankheit sterben (Daniel, 1998). Die Bedeutung des Trinkwassers als Überträger der Krankheit wurde bereits 1854 vom englischen Armenarzt John Snow schlüssig nachgewiesen. Schlechte Sanitärhygiene oder der Verzicht auf die Trinkwasseraufbereitung kann zu wasserbedingten Choleraausbrüchen führen, wie die Epidemie in Peru 1991 mit rund einer Million Erkrankten und 10'000 Toten aufgezeigt hat. Die Behörden hatten wegen der Angst vor Desinfektionsmittelnebenprodukten die Chlorung des Trinkwassers ausgeschaltet (Ford, 1999; Daniel, 1998).

Yersinia

Y. enterocolitica wurde erstmals 1939 isoliert und ist heute als Verursacher gastrointestinaler Erkrankungen beim Menschen bekannt. Betroffen sind vor allem Kinder unter 7 Jahren. Die Erkrankung tritt vorwiegend im Herbst und Winter auf, verursacht Fieber, Durchfall und Bauchkrämpfe und dauert rund zwei Wochen. Als Reservoir für *Y. enterocolitica* dienen Boden, Wasser und Tiere, bei denen das Bakterium zur natürlichen Darmflora gehören kann. Die Bakterien mit einem Durchmesser von 0,5 bis 0,8 μm und einer Länge von 1 bis 3 μm überleben im Wasser bis zu Temperaturen von minimal 0°C über einen längeren Zeitraum (Highsmith und Abshire, 1975; Schillinger und McFeters, 1978). Die Übertragung findet auf fäkal-oralem Weg statt. Verschiedene Nahrungsmittel wie Milch, Tofu, Sojasprossen und Schweinefleisch dienen als Vektoren (Andersen et al., 1991; Simonet und Catteau, 1997). In Deutschland sind humanpathogene Stämme von *Y. enterocolitica* die dritthäufigsten bakteriellen Erreger von Nahrungsmittelinfektionen (Ammon und Bräunig, 2002).

Wasserbedingte Epidemien sind relativ selten und dann eher bei unbehandeltem Trinkwasser aufgetreten, da *Y. enterocolitica* in chloriertem Wasser ähnlich schnell abstirbt wie *E. coli*. Der grösste mit Yersinia assoziierte Fall trat 1977 in Montana (USA) auf. Bei diesem Ausbruch konnte zwar kein klinischer Nachweis des Erregers in Stuhlproben erbracht werden. Die Analysen des unbehandelten Trinkwassers waren jedoch positiv für verschiedene Stämme von Yersinia (Fricker, 1999b).

Basierend auf einer serologischen und biochemischen Charakterisierung kommt eine Studie zum Schluss, dass Isolate von Yersinia aus Trinkwasseraufbereitungsanlagen keine gesundheitliche Signifikanz haben (Aleksic und Bockemühl, 1988). Im Gegensatz dazu konnte Trinkwasser in verschiedenen Fällen als Ursache für eine Erkrankung durch Yersinia beschrieben werden (Schiemann, 1990). Die Wahrscheinlichkeit jedoch, kleinere Ausbrüche von Yersinia zu entdecken, ist gering, da der Nachweis des Erregers relativ schwierig ist. Zudem gibt es keine Standards für die Detektion von Yersinia, weshalb diese fälschlicherweise auch als Coliforme beurteilt werden können (Fricker, 1999b).

Neu erkannte pathogene Bakterien

Pathogene *E. coli* (VTEC), *Mycobacterium avium* Complex und *Helicobacter pylori* wurden von der American Water Works Association als „emerging pathogens" benannt. Von Seiten der Lebensmittelkontrollen sind die VTEC vor allem durch Erkrankungen nach dem Konsum von „Hamburgern" bekannt. Im Rohwasser einer Karstquelle in der Nordwestschweiz konnten VTEC in Abhängigkeit der *E. coli* Konzentration nachgewiesen werden (Auckenthaler et al., 2002), womit wasserbedingte Ausbrüche auch in der Schweiz denkbar sind. *Mycobacterium avium* Complex und *Helicobacter pylori* konnten zwar im Wasser nachgewiesen werden, die Bedeutung dieser Befunde ist allerdings noch nicht geklärt. Die hier vorgestellten pathogenen Bakterien bilden nur eine kleine Auswahl. Dass in Zukunft noch weitere Keime an Bedeutung gewinnen könnten, zeigt eine Studie von Payment et al., 1991. In Haushalten, die Trinkwasser ohne auffällige mikrobiologische Verunreinigung (weder Indikatorkeime noch Pathogene) erhielten, wurden Umkehrosmosefilter eingebaut. Nach der Installation gingen die gastrointestinalen Erkrankungen gegenüber Kontrollhaushalten ohne Filter um 25% zurück.

Pathogene *E. coli*

Neben den harmlosen treten pathogene *E. coli* Stämme auf, die ernsthafte Infekte verursachen können. Diese Gruppe umfasst enteropathogene *E. coli* (EPEC), enterotoxische *E. coli* (ETEC), enteroinvasive *E. coli* (EIEC), enterohämorrhagische *E. coli* (EHEC), enteroaggregative *E. coli* (EAggEC) und diffuse adherierende *E. coli* (DAEC) (Burnens, 2001). Bei den letzten beiden Gruppen besteht heute kaum Information über deren Transmission (AWWA, 1999c), weshalb sie hier nicht weiter diskutiert werden. Den ersten vier Gruppen ist gemeinsam, dass sie alle Durchfall auslösen können. ETEC ist eine häufige Ursache der Reisediarrhoe. Die hitzestabilen Enterotoxine sind dem von *Vibrio cholera* gebildeten Toxin sehr ähnlich. EIEC verursacht eine ähnliche Erkrankung wie *Shigella*. EHEC gehört zu der Gruppe der verotoxinbildenden *E. coli* (VTEC), die seit 1977 bekannt sind. Die von ihnen produzierten Verotoxine Typus I und II haben eine Ähnlichkeit mit dem von *Shigella dysenteriae* produzieren Shiga-Toxin (Beutin, 1996).

Die Infektionsdosis von VTEC wird auf 10 bis 100 Keime geschätzt (Tschäppe, 2000; Willshaw et al., 1994). Rund 10% der Hauptrisikogruppen (ältere Menschen und Kleinkinder) entwickeln eine Woche nach dem Einsetzen der Durchfälle ein hämolytisch-urämisches Syndrom (HUS), das zu Nierenversagen führen kann (Su und Brandt, 1997). In 50% der HUS-Fälle wird eine Dialyse erforderlich, ca. 10 bis 30% der Erkrankungen enden mit einer terminalen Niereninsuffizienz (lebenslange Dialysepflicht, beziehungsweise vorzeitiger Tod der Patienten) (Beutin, 1996). Die Therapie der durch VTEC verursachten Durchfälle und des HUS ist rein symptomatisch. Antibiotika sind wirkungslos, da das Verotoxin zum Zeitpunkt der Erkrankung bereits in den Zellen gebunden ist (Meyers et al., 1998).

Als Reservoir der VTEC gelten Kälber und Rinder (Wilson et al., 1992) mit einer altersabhängigen Prävalenz von 23,7% bei Kälbern und 2,3% bei Kühen (Stephan et al., 2000). Die Überlebensfähigkeit der VTEC in Kälberfaecies kann mehr als 50 Tage, im Boden mehr als 130 Tage betragen (Maule, 2000). Die Übertragung findet über Fleischprodukte, unpasteurisierte Milch, rohes Gemüse und Wasser statt (Tschäppe, 1997). Wasserbedingte Epidemien durch pathogene *E. coli* wurden aus den USA, Japan und Ungarn beschrieben (Rice, 1999). Mitte 2000 erkrankten in Walkerton (Ontario, Kanada) 2000 Personen, deren Trinkwasser mit VTEC und Campylobacter verunreinigt war. Sechs Menschen starben im Spital an den Folgen der Erkrankung (CCDR, 2000). In der Schweiz konnten VTEC im Rohwasser einer Karstquelle, die zu Trinkwasserzwecken genutzt wird, nachgewiesen werden. Ab einer Konzentration von 300 *E. coli* pro 100 mL konnten praktisch in allen Proben VTEC nachgewiesen werden. (Auckenthaler et al., 2002, Raso, 2001). Die Desinfektion des Wassers mit Chlor ist bei pathogenen *E. coli* ebenso wirksam wie bei den harmlosen *E. coli* (Rice, 1999).

Helicobacter pylori

H. pylori ist ein bakterisches Pathogen, das den Magen und den oberen Gastrointestinaltrakt befällt und als wichtiges Agens für Gastritis genannt wird. Das Bakterium wurde erst 1983 erstmals bei Patienten entdeckt und als mögliche Ursache für Magengeschwüre erkannt. Die Übertragungswege sind noch nicht endgültig geklärt, doch scheint die fäkal-orale Route und die Übertragung durch das Wasser sehr wahrscheinlich. In einer Studie von 1998 in den USA wurden in 40% von 42 Oberflächengewässerproben und in 65% von 20 Grundwasserbrunnen mit geringer vertikaler Filterstrecke *H. pylori* gefunden. In einem kleinen Cluster von 10 Personen konnte auch ein Zusammenhang von *H. pylori* Infektionen und kontaminiertem Trinkwasser festgestellt werden (AWWA, 1999a). Neuere Untersuchungen zeigen, dass *H. pylori* durch Chlor rasch inaktiviert wird (Johnson et al., 1997). Weitere Untersuchungen über die Prävalenz in unchloriertem Wasser und die Übertragungswege sind notwendig, um die gesundheitliche Bedeutung von *H. pylori* und die Rolle des Wassers als Vektor abklären zu können (Ford, 1999; AWWA, 1999a).

Mycobacterium avium Complex (MAC)

Mycobacterium avium Complex (MAC; z.B. *M. avium* und *M. intracellulare*) ist ein opportunistisches humanes Pathogen, das die Lunge infiziert und Husten, Müdigkeit, Gewichtsverlust und Nachtschweiss, ähnlich wie *M. tuberculosis* verursacht (LeChevallier, 1999). MAC sind ubiquitär in der Umwelt vorhanden und konnten im Boden, Hausstaub, Wasser (Abwasser, Oberflächengewässer, Grund- und Trinkwasser) sowie in Tieren nachgewiesen werden. Das Bakterium wächst in wenig sauerstoffhaltigem Wasser bei hohem Anteil an organischer Substanz und Zink. Die hydrophobe Zellwand begünstigt die Kolonisierung von

Oberflächen und die Bildung von Biofilmen (AWWA 1999a). Die Übertragung über das Trinkwasser ist noch nicht ganz geklärt, jedoch deuten gastrointestinale Erkrankungen bei AIDS Patienten darauf hin, dass Nahrungsmittel oder das Wasser die Hauptübertragungswege sind (AWWA, 1999a). Der analytische Nachweis von MAC im Wasser ist schwierig und die Anzahl detektierter Keime abhängig von der verwendeten Methode. In Trinkwasserverteilnetzen konnten zwischen 0,08 und 0,14 CFU/100 mL (Haas et al., 1983) beziehungsweise 10 und 45'000 CFU/100 mL (Fischeder et al., 1991) gefunden werden. In der Schweiz konnten im Trinkwasser ebenfalls MAC nachgewiesen werden (pers. Komm. Ewald, 2001). Die meisten Spezies von Mycobacterium überleben eine Chlorkonzentration von 1 mg/L (LeChevallier, 1999), was eine Inaktivierung der Keime äusserst schwierig macht. *M. avium* und *M. intracellulare* wurden von der EPA (United States Environmental Protection Agency) in die Liste der möglichen Verunreinigungs-Kandidaten für eine gesetzliche Regelung aufgenommen (USEPA, 1998).

Biofilme und Filterverkeimung

Biofilme können sich unter anderem an den Innenwänden von Trinkwasserleitungen bilden. Sie entstehen durch die Kolonisierung der Oberflächen durch verschiedene Mikroorganismen, können eine Dicke von mehr als 100 µm erreichen und bilden eine schützende Umgebung unter anderem für pathogene Keime wie Legionellen und Pseudomonaden. Die Biofilmbildung wird erleichtert durch anorganisches und organisches Material, rauhe Oberflächen und Totenden in den Leitungsnetzen sowie durch erhöhte Temperaturen. Beim organischen Material ist insbesondere der AOC (assimilierbarer organischer Kohlenstoff) von Bedeutung, der teilweise bei der Aufbereitung von Trinkwasser mittels Ozon oder Chlor entstehen kann.

Aus dem Biofilm werden Bakterien ins Trinkwasser ausgetragen. Mit Desinfektionsmitteln lassen sich Biofilme nur schwer bekämpfen, da sie eine gallertartige äussere Schicht besitzen, die reaktive Produkte rasch absorbiert. Die Desinfektionsmittelresistenz im Biofilm nimmt gegenüber frei lebenden Mikroorganismen um das 150- bis 3000-fache zu (Fox und Reasoner, 1999). Für die Bekämpfung von Biofilmen ist Chloramin besser geeignet als Chlor, da es weiter in die Matrix des Biofilmes eindringen kann (vgl. Kap. Trinkwasseraufbereitung).

Legionella pneumophila

Legionellen kommen in der aquatischen Umwelt ubiquitär vor und besiedeln sowohl Oberflächengewässer und Grundwasser als auch Trinkwasserleitungsnetze, Hausinstallationen und Kühlwassersysteme. Die optimale Wachstumstemperatur liegt zwischen 35 und 45°C. Die von Legionellen verursachten Erkrankungen sind die Legionärskrankheit und das Pontiac Fieber. Die Legionärskrankheit löst in 1 bis 6% der Infektionen schwere Lungenentzündungen aus und kann zum Tod führen. Beim Pontiac Fieber ist die Erkrankungsrate nahezu 100%, die Krankheit ist jedoch selbstlimitierend, Todesfälle sind keine bekannt. Übertragen werden die beiden Krankheiten durch wasserhaltige Aerosole.

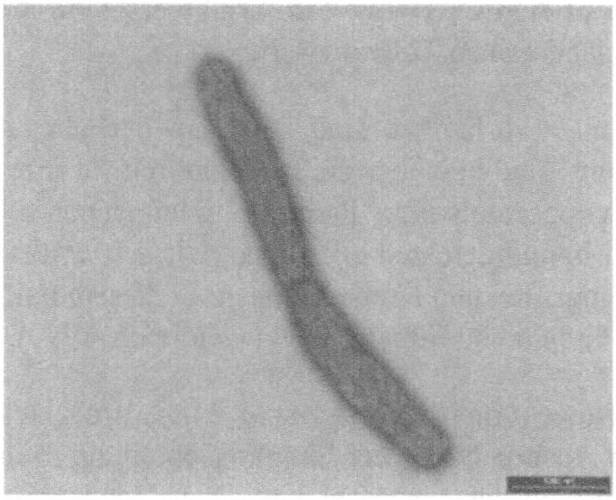

Abb. 2.4: *Legionella pneumophila,* elektronenmikroskopische Aufnahme (mit freundlicher Genehmigung, Tamara Fretz, Zentrum für Mikroskopie ZMB Bio-Pharmazentrum, Universität Basel)

Weltweit gibt es immer wieder Ausbrüche von Legionellenerkrankungen. In den USA wird die Zahl Erkrankter auf jährlich 17'000 bis 23'000 Fälle geschätzt (Hall, 1999). Untersuchungen in der Schweiz haben vor allem in Alters- und Pflegeheimen sowie in Spitälern teilweise hohe Belastungen des Trinkwassers mit Legionellen aufgezeigt. Erkrankungen durch Legionellen sind in der Schweiz jedoch kaum erfasst und bekannt.

Legionellen können in Amöben überleben und sich in ihnen vermehren (Hall, 1999). Die Bekämpfung dieser Bakterien mittels Desinfektionsmitteln wie Chlor ist deshalb nur bedingt möglich und erst ab Konzentrationen von 1 bis 2 mg/L freiem Chlor gegeben. Wirkungsvoller ist die thermische Behandlung durch Erhitzen des Boilers über 60°C und Spülen der Hausinstallationen mit heissem Wasser (über 60°C). Sanitärinstallationen mit stagnierendem Wasser sollten vermieden werden. Luftbefeuchter und Klimaanlagen (insbesondere in Alters-/Pflegeheimen, Spitälern etc.) bei denen Wasser versprüht wird, müssen regelmässig und fachlich korrekt gereinigt und desinfiziert werden. Nach Möglichkeit sind sie durch Systeme ohne Aerosolbildung zu ersetzen

Pseudomonas aeruginosa

P. aeruginosa kommt ubiquitär vor und besiedelt unter anderem Oberflächengewässer, Grundwasser, Trinkwasser, in Behältnisse abgefülltes Wasser, Böden und Gemüse. Favorisierte Habitate sind Wasser-Luft Grenzflächen, wie sie beispielsweise in Sand- und Aktivkohlefiltern, Duschköpfen und Befeuchtern vorkommen. Konzentrationen von *P. aeruginosa* wurden im Abwasser mit 4200 CFU/mL (Highsmith und Abshire, 1975) im abgefüllten Wasser mit 0 bis 10 CFU/mL bei 2% positiven Befunden (Edberg et al., 1996) und beim Gemüse mit 10 bis 600 CFU/mL (Kominos et al., 1972; Correa et al., 1991) angegeben. Im Trinkwasser ist lediglich eine geringe Anzahl dieser Bakterien zu erwarten. Allerdings kann eine Vermehrung stattfinden (Hardalo und Edberg, 1997). Die optimale Wachstumstemperatur beträgt 15 bis 30°C (Geldreich, 1999). In Behältnissen abgefülltes Trinkwasser darf in der Schweiz in 100 mL keine *P. aeruginosa* aufweisen (Anh. 2b, Nr. 31, HyV) Die Trinkwasserrichtlinie in der EG schreibt vor, dass 250 mL frei von *P. aeruginosa* sein müssen (Anh. 1 Teil A, EC, 1998). Für Trinkwasser im Verteilnetz gibt es in der Schweiz und in Deutschland keinen Toleranzwert.

Da *P. aeruginosa* ubiquitär vorhanden sind und man deshalb täglich mit einer grossen Anzahl in Kontakt kommt, sind keine speziellen Vektoren für ihre Verbreitung notwendig. Erkrankungen durch das opportunistische Pathogen treten jedoch selten auf, da praktisch nur Personen mit speziellen Immundefekten erkranken, wie z.B. Patienten mit Verbrennungen, grossen chirurgischen Eingriffen und Krebspatienten. In diesem Fall kann die Infektion rasch und tödlich verlaufen (Hardalo und Edberg, 1997; Geldreich, 1999).

Wasserbedingte Ausbrüche von *P. aeruginosa* sind praktisch keine bekannt. Ein Ausnahmefall stellt der Ausbruch in einer Säuglingsabteilung in einem Spital in den USA dar. Das Trinkwasser war durch Abwasser und Oberflächenwasser verunreinigt worden (Geldreich, 1999).

Viren

Bis vor 30 Jahren waren die Ursachen der nicht-bakteriellen Gastroenteritiden unbekannt, obwohl der amerikanische Kinderarzt, John Zahorsky bereits 1929 auf ein Syndrom hingewiesen hatte, das er "winter vomiting disease“ nannte (Zahorsky, 1929). Die Ursache des Syndroms wurde erst bei der Abklärung eines Falles von 1969 in Norwalk (Ohio, USA) als virales Agens entdeckt und ist heute unter dem Begriff Norwalk-like Virus bekannt (Adler und Zickel, 1969; Kapikian et al., 1972).

Lebensmittelbedingte, virale Gastroenteritiden sind von zunehmender Bedeutung (Sockett et al., 1993) und mehrere Viren wurden als etiologisches Agens identifiziert. Nach WHO-Definition liegt ein lebensmittelbedingter Ausbruch vor, wenn zwei oder mehr Personen eine ähnliche Krankheit haben, nachdem sie das selbe Lebensmittel konsumierten oder aus dem Lebensmittel das krankmachende Agens isoliert wurde (Gerigk, 1993). Eine Übertragung über Nahrungsmittel und Wasser wurde dokumentiert für Enteroviren, Rotaviren, Astroviren und Norwalk-like Viren (Hedberg und Osterholm, 1993).

Die biologischen Erfordernisse des Eindringens der Viren in eine lebende Wirtszelle und die dortige Vermehrung macht sie zu obligaten Zellparasiten und Krankheitserregern. Das Auftreten von Viren im Trinkwasser ist deshalb stets mit einer Infektionsgefahr verbunden. Die Festsetzung eines Grenzwertes für Viren im Wasser ist daher problematisch, sie dürfen im untersuchten Wasservolumen überhaupt nicht nachweisbar sein (Botzenhardt, 2000). Die WHO schreibt denn in ihren 1993 überarbeiteten Trinkwasserrichtlinien für die Trinkwasserqualität auch vor, dass Rohwasser keine pathogenen Viren enthalten darf. Einfache Verfahren zur Aufbereitung von Trinkwasser, wie beispielsweise eine alleinige Desinfektion, können Viren nicht vollständig eliminieren.

Tab. 2.4: Über Wasser übertragbare Viren und ihr Vorkommen im Grund- und Trinkwasser

Viren	**Infektionsdosis [Anzahl Partikel]**	**Vorkommen im Grund- und Trinkwasser**		**Quelle**
		weltweit	**Schweiz/ Deutschland**	
Astrovirus	1-10	Grund- und Trinkwasser, Ausbrüche	unbekannt	*Hermann, 1999; Walter, 2000; Botzenhart, 2000, Ford, 1999*
Norwalk-like Virus (Synonyme: Human Calicivirus, Norwalk Virus, SRSV)	1	Grund- und Trinkwasser, Ausbrüche	Grund- und Trinkwasser, Ausbrüche	*Hobbins, 1999; Gerba und Rose, 1990; Kukkula et al., 1999*
Rotavirus	10-100	Grund- und Trinkwasser, Ausbrüche	Grund- und Trinkwasser	*Metzler et al., 1996; Botzenhardt, 2000; Abbaszadegan, 1999*
Enterovirus: Poliovirus, Coxackie A und B, Echovirus, Enterovirus	5-10	Grund- und Trinkwasser	Grund- und Trinkwasser	*Metzler et al., 1996; Walter, 2000, Gerba, 1999*
Hepatitis-A-Virus	1-10	Grund- und Trinkwasser, Ausbrüche	k.A.	*Walter, 2000; Sobsey, 1999; Ford, 1999*
Hepatitis-E-Virus	1-10	Grund- und Trinkwasser, Ausbrüche	k.A.	*Walter, 2000; Sobsey, 1999; Ford, 1999*

(k.A.: keine Angaben)

Bisher sind in der Schweiz und in Deutschland erst wenige Grund- und Trinkwasseruntersuchungen auf Viren durchgeführt worden, da die Analytik neue, relativ aufwändige Verfahren bedingt und das Bewusstsein für die Übertragung dieser Krankheitserreger über das Wasser zum Teil noch wenig vorhanden ist. Die wenigen vorliegenden Untersuchungen zeigen aber verschiedene humanpathogene Viren im Abwasser, Oberflächengewässer und Grundwasser (Metzler et al., 1996; Hobbins et al., 2001; Häfliger et al., 2000). Die in diesem Teilkapitel beschriebenen Viren wurden nach ihrer Rolle als Krankheitserreger zusammengefasst. Die Tabelle 2.4 gibt einen Überblick über mögliche wasserübertragbare Viren.

Erkrankungen des Verdauungstraktes

Viral bedingte gastrointestinale Erkrankungen, ausgelöst durch den Konsum von Nahrungsmitteln und Trinkwasser, sind weltweit von steigender Bedeutung. Obwohl Gastroenteritis in den entwickelten Ländern meist keine dramatischen Krankheitsverläufe zeigt, hat sie doch eine grosse ökonomische Relevanz. Jährlich werden Millionen von Arbeitsausfalltagen durch Befindlichkeitsstörungen in Folge dieser Infektionen verursacht. Astro-, Rota- und Norwalk-like Viren (NLV) zählen zu den wichtigsten Erregern von fiebrigen Magen-Darmerkrankungen im Kindes- und Jugendalter (Walter, 2000), wobei während epidemischen Ausbrüchen wie beispielsweise in La Neuveville alle Altersgruppen betroffen sind (Häfliger et al., 2000). Die WHO schätzt die Zahl der jährlichen Todesfälle durch Darminfektionen auf 5 bis 18 Millionen weltweit. Bei rund 60% der Durchfälle, bei denen keine Bakterien oder Protozoen nachgewiesen werden können, werden virale Erreger vermutet.

Astrovirus

Die Astroviren gehören zur Familie der Astroviridae. Sie haben eine Grösse von 27 bis 32 nm und sind unter dem Elektronenmikroskop als sternförmige Gebilde erkennbar, was ihnen ihren Namen gegeben hat (Walter, 2000). Astroviren gehören zu den weltweit wichtigsten Durchfallerregern bei Kindern (Lüthi, 1997). Eine Untersuchung der Antikörper Prävalenz bei Kindern in Oxfordshire (UK) zeigte eine Zunahme von 4% in einem Alter von 6 bis 12 Monaten über 64% bei 3 bis 4-jährigen bis 87% bei 5 bis 10-jährigen (Kurtz und Lee, 1978). Die Symptome sind ähnlich wie bei Infektionen durch Rotaviren, nur verläuft die Krankheit leichter und seltener mit Erbrechen. Die Erkrankungen treten ganzjährig auf, jedoch mit Spitzen während der kalten Jahreszeit.

Die Übertragung erfolgt fäkal-oral direkt oder indirekt via Erbrochenem, Nahrungsmitteln oder Wasser. Die Übertragung über das Wasser ist uneindeutig. Wasserbedingte Ausbrüche sind nicht dokumentiert (Hermann, 1999), jedoch wurde kontaminiertes Wasser als Ursache für Ausbrüche vermutet (Kurtz und Lee, 1987). Aufgrund der epidemiologischen Charakteristik kann davon ausgegangen werden, dass die Astroviren eine hohe Umweltresistenz besitzen. Mit den Fäkalien werden mindestens 10^6 Viruspartikel/g ausgeschieden. Damit ist die Virusübertragung durch Wasser und andere Umweltkompartimente wahrscheinlich (Walter, 2000).

Norwalk-like Virus (NLV)

Die Systematik der NLV ist teilweise recht uneinheitlich. Sie werden als small round structured viruses (SRSV), Human Calicivirus (HuCV) und Norwalk-like Viren bezeichnet. Folgt man der Klassifikation des internationalen Komitees zur Taxierung der Viren, werden Norwalk Viren und Norwalk-like Viren in der Familie der Caliciviridae im Genus Calicivirus und der Spezies Human Calicivirus eingeordnet (Murphy et al., 1995). Die NLVs sind weltweit verbreitet und haben in den letzten Jahren an Bedeutung gewonnen. In den USA werden 66,6% aller lebensmittelbürtigen Erkrankungen allein auf NLV zurückgeführt (Mead et al., 1999). In England liegt die jährliche Inzidenz der Infektionen bei 12,5 von 1000 Personen (Wheeler et al., 1999) und in der Schweiz wird die Zahl der Infizierten pro Jahr auf 410'000 geschätzt (Baumgartner, 2001). Der Mensch ist neben Rindern und Schweinen das wichtigste Reservoir für NLV (Schwab, 1999). Die Erkrankung führt bei Kindern typischerweise zu Erbrechen, bei Erwachsenen eher zu Durchfall (Hurst, 1999). Die Ausbrüche finden das ganze Jahr über statt, akzentuiert ist die Erkrankung allerdings in den Wintermonaten (AWWA, 1999b). Die Übertragung erfolgt fäkal-oral über Nahrungsmittel, Wasser und Erbrochenes (Aerosol) (KLBL, 1999). Die geringe Infektionsdosis von 10 bis 100 Viruspartikel und die hohe Infektiosität führen zu hohen Erkrankungsraten und sekundären Fällen bei Kontakt mit erkrankten Personen (Parashar, 2001).

Im Stuhl der infizierten und erkrankten Personen werden pro Gramm bis zu 10^{12} Viruspartikel mit einer Grösse von 30 bis 35 nm ausgeschieden (Walter, 2000). In der Umwelt sind die Viren für einige Tage stabil (Lüthi, 1997). In einer Studie in der Nordwestschweiz wurden NLV im geklärten Abwasser in 44%, im Oberflächenwasser in 48% und im Karstquellwasser in 4% der untersuchten Proben gefunden (Hobbins et al., 2001). Wasserbedingte Ausbrüche wurden in der Schweiz ebenfalls beschrieben. So konnten bei der Trinkwasserverunreinigung in La Neuveville mit rund 1600 Erkrankten NLV nachgewiesen werden (Häfliger et al., 2000). Ein zweiter Fall ist aus Küblis bekannt, als an der Jugendolympiade verunreinigtes, nicht aufbereitets Grundwasser zur Herstellung isotonischer Getränke benutzt wurde. Hier erkrankten 1400 Personen. Die NLV wurden sowohl im Stuhl der Erkrankten als auch im Trinkwasser festgestellt (Lüthi und Beuret, 2001, Baumgartner, 2001). Eine weitere NLV-Epidemie mit rund 650 Erkrankten ist 2001 in zwei Gemeinden des Kantons Zürich, vermutlich durch kontaminiertes Trinkwasser, aufgetreten. Die NLV konnten zwar lediglich in den Stuhlproben der Patienten, nicht aber im Trinkwasser nachgewiesen werden. Der Ausbruch korreliert zeitlich allerdings mit dem Ausfall der Desinfektion bei der Trinkwasseraufbereitung (Fretz, 2001). In den USA hat die CDC (Center of Disease Control) zwischen 1996 und 2000 3% aller NLV Fälle auf Wasser zurückführen können (Parashar, 2001).

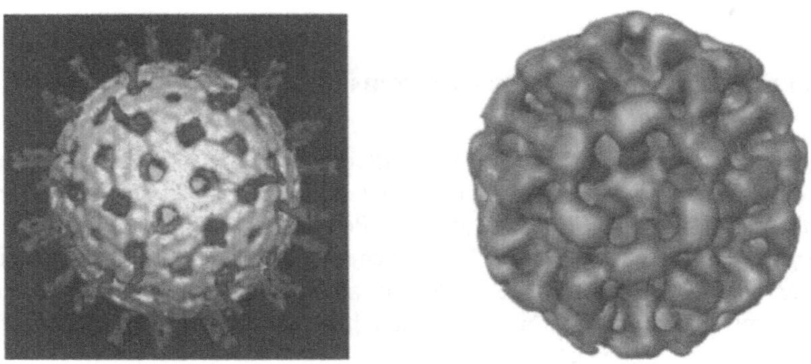

Abb. 2.5: Rotavirus (links), Norwalk like Virus (rechts), elektronenmikroskopische Aufnahme (mit freundlicher Genehmigung von Dr. B.V.V. Prasad, Ph. D. Baylor Collage of Medicine, Houston, TX 77030)

Rotaviren

Die Rotaviren gehören zur Familie der Reoviridae. Es werden sechs Gruppen A bis F unterschieden, wobei nur die Gruppen A, B und C für den Menschen infektiös sind. In Europa und den USA werden die meisten Rotavirus Infektionen durch die Gruppe A verursacht. Die Viren haben eine Grösse von 65 bis 75 nm, sind resistent gegen tiefe und hohe pH-Werte und können für mehrere Wochen im Wasser persistieren (Abbaszadegan, 1999; Nadal, 2001).

Rotaviren verursachen weltweit am häufigsten schwere Durchfallerkrankungen und werden jährlich für eine Million Todesfälle verantwortlich gemacht (Walter, 2000). Praktisch jedes Kind bis zum Alter von vier Jahren erkrankt einmal an dem Virus. Wie Studien in der Schweiz zeigen, werden bei 22 bis 43% der wegen (Brech-)Durchfall hospitalisierten Kinder Rotaviren festgestellt. Rotaviren sind in Deutschland die häufigsten viralen Erreger. Im Jahr 2001 wurden 47'485 Gastroenteritiserkrankungen registriert (Ammon und Bräunig, 2002). In den USA wird die Zahl der jährlich erkrankten Kinder auf 3,5 Millionen geschätzt. Die Infektionen treten in den Wintermonaten gehäuft auf. Die Übertragung erfolgt fäkal-oral (Abbaszadegan, 1999; Nadal, 2001).

Die Infektionsdosis ist sehr tief, ein einziges Viruspartikel kann bei einem empfänglichen Individuum eine Infektion auslösen. Dies führt unter anderem dazu, dass ausgehend von einer betroffenen Person, die Erkrankung oft auf andere Familienmitglieder übertragen wird (Lüthi, 1997).

Die infizierten Personen scheiden pro Gramm Stuhl 10^{10} bis 10^{12} infektiöse Viruspartikel aus (Walter, 2000). Im Abwasser und im belastetem Oberflächengewässer können die Viren deshalb auch häufig nachgewiesen werden (Botzenhardt, 2000). In der Schweiz konnten im Abwasser, in Flüssen und Seen und im Grundwasser Rotaviren in 68% der Proben festgestellt werden (Metzler et al., 1996). In den USA wurden zudem verschiedene wasserbedingte Ausbrüche beschrieben mit Erkrankungsraten unter Erwachsenen von über 43% (Abbaszadegan, 1999).

Erkrankungen des Zentralnervensystems

Poliomyelitis (Kinderlähmung) ist seit dem Altertum bekannt. Die Krankheit gewann zu Beginn des 20. Jahrhunderts an Bedeutung, als sie jeweils im Sommer und Herbst epidemisch auftrat. Nach der Entwicklung der Poliovirus-Vakzine und grossangelegten Impfprogrammen ist die Krankheit in der westlichen Hemisphäre seit fünf Jahren nicht mehr aufgetreten (AWWA, 1999b). Heute spielen vor allem die anderen Enteroviren eine Rolle, die Symptome über den Gastrointestinaltrakt hinaus auslösen und Erkrankungen des Zentralnervensystems verursachen können.

Abb. 2.6: Poliomyelitis: Die Muskulatur des rechten Beines des Mannes auf der Darstellung der ägyptischen Stele ist zurückgebildet (mit freundlicher Genehmigung, Oliver Ocasek Regional Medical Library Northern Ohio Universities, College of Medicine)

Enteroviren

Die Enteroviren gehören zur Familie der Picornaviridae. Diese kleinsten animalen Viren mit einer Grösse von lediglich 22 bis 30 nm haben eine hohe Persistenz und können auch bei ausserordentlichen Umweltbedingungen über einen längeren Zeitraum infektiös bleiben (AWWA, 1999b). Im Grundwasser wurden Überlebenszeiten bis zu 81 Tagen gemessen (Botzenhardt, 2000). Die Enteroviren schliessen die Typenspezies Poliovirus 1, 2, 3, Coxakievirus A1-A 22+24, Coxackievirus B1-6 Echovirus 1-9, 11-27, 29-34 und Enterovirus 68-71 ein. Sie haben weltweite Verbreitung (Walter, 2000). Die verursachten Erkrankungen können von einer gewöhnlichen Grippe über Meningitis, Enzephalitis bis zu Schäden des zentralen Nervensystems (Poliomyelitis) reichen. Ergebnisse aus verschiedenen Studien zeigen, dass die minimale infektiöse Einheit bei Enteroviren bei 5 bis 10 Viruspartikeln liegt (Walter, 2000).

Der Mensch bildet das einzige bekannte Reservoir für Enteroviren. Die Übertragung erfolgt fäkal-oral oder respiratorisch (Gerba, 1999). Die Enteroviren vermehren sich im Darmepithel, wie der Name besagt, und sind daher im Stuhl und im Abwasser nachweisbar (Botzenhardt, 2000). Die Konzentration der Viren ist in den gemässigten Breiten im Spätsommer und Herbst im Abwasser erhöht. In den USA wurde die durchschnittliche Konzentration im Abwasser mit 7000 Viruspartikeln/L angegeben. Die Coxackieviren sind die am häufigsten im Wasser und Trinkwasser gefundenen Enteroviren (Gerba, 1999). In der Schweiz konnten im Abwasser, in Flüssen und Seen und im Grundwasser Enteroviren in über 50% der Proben nachgewiesen werden (Metzler et al., 1996).

Tab. 2.5: Krankheitsbilder der Enteroviren (nach Gerba, 1999; Walter, 2000; Botzenhardt, 2000)

Spezies	Anzahl Serotypen	Krankheitsbilder
Poliovirus, Typen 1-3	3	Poliomyelitis (spinale Kinderlähmung), aseptische Meningitis, Fieber
Coxackievirus, Gruppe A, Typen 1-24	24	aseptische Meningitis, Herpangina, Hand-, Fuss- und Mundkrankheit, Atemwegserkrankungen „gewöhnliche Grippe“, akute hämorrhagische Konjunktivitis
Coxackievirus, Gruppe B, Typen 1-6	6	aseptische Meningitis, fatale Neugeborenenkrankheit, Pleurodynie oder Bornholmerkrankheit (Teufelsgriff), Myocarditis (Herzmuskelentzündung), infektiöse Perikarditis (Herzbeutelentzündung)
Echovirus, Typen 1-34	34	aseptische Meningitis, fieberhafte Erkrankungen, Hautausschläge, Atemwegserkrankungen
Enterovirus 68-71	4	Paralysen, aseptische Meningitis, Hand-, Fuss- und Mundkrankheit, akute hämorrhagische Konjunktivitis

Wasserbedingte Krankheitsausbrüche sind nicht schlüssig dokumentiert. Einige trinkwasserbedingte Ausbrüche wurden mit Enteroviren in Zusammenhang gebracht, leider fehlten aber die epidemiologischen Abklärungen (Gerba, 1999).

Infektiöse Leberentzündungen (Hepatitis)

Die Hepatitis ist eine Entzündung der Leber, die durch verschiedene Viren hervorgerufen werden kann. Die Gelbsucht ist bereits seit Hippokrates, 700 Jahre vor Christus bekannt. Seit den 40er Jahren wurden mehrere Viren als Erreger von Leberentzündungen nachgewiesen. Zwei davon, Hepatitis-A-Virus und Hepatitis-E-Virus, können über das Wasser übertragen werden.

Hepatitis-A-Virus

Der Hepatitis-A-Virus (HAV) gehört zur Familie der Picornaviridae und hat eine weltweite Verbreitung. In den USA wird die Zahl der HAV Infektionen mit 20'000 bis 30'000 pro Jahr angegeben. Die effektive Inzidenz könnte aber um den Faktor 2 bis 10 höher liegen (AWWA, 1999b). Die durch HAV ausgelösten Schäden der Leberzellen sind wahrscheinlich auf die Reaktion des Immunsystems zurückzuführen und nicht primär durch das Virus verursacht. Die Symptome wie Übelkeit, Erbrechen, Schüttelfrost, Fieber und Braunfärbung des Urins dauern meist vier Wochen. Die Mortalität ist kleiner als 0,3% (Sobsey, 1999). Bei Kindern verläuft die Krankheit in der Regel gutartig, bei Erwachsenen ab 50 Jahren sind zunehmend schwerere Krankheitsverläufe zu erwarten (Walter, 2000). Die Leberschäden sind reversibel und nach verschwinden des Viruses ist die Heilung vollständig (Sobsey, 1999).

Als Reservoir des HAV dient der Mensch. Die fäkal-orale Übertragung der Viren über das Wasser ist wichtig, die meisten Fälle werden aber über kontaminierte Nahrungsmittel und von Person zu Person verursacht.

Infizierte Personen scheiden im Stuhl bereits vor dem Ausbrechen der Krankheit 10^6 bis 10^9 Viruspartikel/g aus. Die 27 bis 30 nm grossen Viren haben eine hohe Persistenz in der Umwelt und können über vier Monate im Boden oder Wasser überleben (AWWA, 1999b). Sie weisen zudem eine hohe Resistenz gegenüber Desinfektionsmitteln auf (Walter, 2000).

In den USA wurden zwischen 1946 und 1980 mehr als 20% aller wasserbedingten Erkrankungen durch HAV verursacht. Zwischen 1989 und 1992 wurden vier Fälle registriert, die auf Fehler bei der Trinkwasseraufbereitung und verunreinigtes Grundwasser zurückzuführen waren (AWWA, 1999b; Sobsey, 1999).

Hepatitis-E-Virus

Hepatitis-E-Virus (HEV) gehört zu der Familie der Caliciviridae. Diese 22 bis 34 nm grossen Viren wurden erst 1980 entdeckt und haben keine Verwandtschaft zu einem bisher bekannten Hepatitisvirus (Walter, 2000). Das Krankheitsbild von HEV ist demjenigen von HAV sehr ähnlich. Auffallend ist allerdings die hohe Sterblichkeit von schwangeren Frauen insbesondere im dritten Drittel der Schwangerschaft von 15 bis > 20% (Hurst, 1999).

Die Übertragung erfolgt in der Regel fäkal-oral über kontaminiertes Trinkwasser. Mensch-zu-Mensch Kontakt spielt normalerweise keine Rolle (Sobsey, 1999). Neben mehreren Primatenarten kann HEV auch Hausschweine infizieren, so dass diese Tiere als Virusreservoir für Epidemien eine wichtige Rolle spielen können (Lemon, 1995). In der Umwelt sind HEV weniger resistent als HAV.

HEV kommen vor allem auf dem Indischen Subkontinent, Asien, Nordafrika, Mittelamerika und Osteuropa vor (Sobsey, 1999). In den tropischen Ländern ist HEV unter jugendlichen Erwachsenen für über 50% der akuten Hepatitiden verantwortlich. In Europa hingegen wird mit einer Erkrankungsrate von 1,5 bis 3% gerechnet (Walter, 2000).

Protozoen

Etwa ein Viertel bis ein Drittel aller Protozoen sind Parasiten. Diese leben auf oder in einem Wirt und schädigen ihn mehr oder weniger. Die meisten parasitären Protozoen besitzen wie die Bakterien bei optimalen Verhältnissen eine hohe Teilungsrate, was zu einem Massenbefall des Wirtes führt. Im Gegensatz zu Bakterien besitzen Protozoen einen durch eine Kernmembran vom Zytoplasma der Zelle abgegrenzten Zellkern, der die genetische Information enthält. Der Lebenszyklus der Protozoen kann über mehrere Entwicklungsstadien mit sexuellen und asexuellen Stufen führen. In die Umwelt gelangen oft resistente Zysten, Oozysten oder Sporen, die relativ lange überleben können (Hiepe, 1983). Diese Dauerstadien sind sehr resistent gegenüber der Desinfektion von Wasser und meist nur durch Filtration zu eliminieren (siehe dazu Kap. Trinkwasseraufbereitung).

Bei der Auswahl der untenstehend eingehender beschriebenen Protozoen handelt es sich um Pathogene, die in letzter Zeit an Bedeutung gewonnen haben, oder die als „emerging Pathogenes" eingestuft wurden und in Zukunft Probleme verursachen könnten. Erkannt wurden einige Protozoen, wie zum Beispiel Cryptosporidium, erst im Zusammenhang mit AIDS Patienten. Aufgrund der Immuninsuffizienz der Patienten kann es zu schweren Infektionen kommen oder sogar zum Tod führen. Medikamente zur Behandlung einer Cryptosporidiose sind bis heute nicht vorhanden (Pontius, 1995).

Bekannte pathogene Protozoen

Cryptosporidium parvum, *Giardia lamblia* und *Entamöbea histolytica* verursachen jährlich weltweit mehr als eine Milliarde Infektionen, von welchen eine erhebliche Anzahl wasserbedingt sind (Smith und Lloyd, 1996; Eckert, 1993) Bis vor kurzem war Giardiasis die Hauptursache wasserbedingter Ausbrüche (Marshall et al., 1997). In den USA wird deren Zahl auf 60% geschätzt (Bennet et al., 1987). Mittlerweile wurde Giardia von Cryptosporidium an der Spitze der Anzahl Ausbrüche abgelöst (Morris und Levin, 1995) und die Häufigkeit der Cryptosporidiosen nimmt sowohl in entwickelten als auch in weniger entwickelten Ländern zu (zit. Ford, 1999). *E. histolytica* verursacht vor allem bei niedriger Sanitärhygiene Probleme. In wenig entwickelten Ländern können bis zu 72% der Bevölkerung von der Erkrankung betroffen sein (Rose und Yates, 1995).

In der Schweiz und in Deutschland wurden vor allem die Cryptosporidien als mögliche Pathogene erkannt. In Oberflächengewässern, Grund- und Trinkwasser konnten die Oozysten nachgewiesen werden (Metzler et al., 1996; Svoboda et al., 1999, Exner et al., 2001). Giardia konnte ebenfalls in Oberflächengewässern bestimmt werden, die Verbreitung und damit die Bedeutung in der Schweiz und Deutschland sind allerdings noch unklar.

Tab. 2.6: Überblick ausgewählter Protozoenarten, die als Pathogene erkannt wurden und bereits grössere Ausbrüche verursacht haben.

Protozoen	**Infektionsdosis [Anzahl]**	**Vorkommen im Grund- und Trinkwaser**		**Quelle**
		weltweit	**Schweiz / Deutschland**	
Cryptosporidium parvum	1-30 Oozysten	weltweit verbreitet, grössere Ausbrüche	Oberflächengewässer, Grund- und Trinkwasser, keine Ausbrüche bekannt	*Metzler et al., 1996; MacKenzie et al., 1994; Ford, 1999; Auckenthaler et al., 2002*
Entamöbea histolytica	10-100 bis 2000 – 4000 Zysten	weltweit, vor allem wenig entwickelte Länder, Ausbrüche	k.A.	*Rose und Yates, 1995; Eckert, 1993*
Giardia lamblia	1-10 Zysten	weltweit verbreitet, Ausbrüche	Trinkwasser bedingter Ausbruch in Deutschland, Oberflächengewässer	*Gornik et al., 2001; Ford, 1999; Metzler et al., 1996; Regli et al., 1996*
Cyclospora cayetanensis	unbekannt	weltweit verbreitet, wasserassoziierte Ausbrüche	k.A.	*Ortega, 1999; AWWA, 1999b*
Microsporidien (z.B. *Enterozytozoon bieneusi*)	unbekannt	weltweit, ubiquitär, wasserassoziiert, Oberflächengewässer, Trinkwasser	k.A.	*AWWA, 1999b*
Toxoplasma gondii	unbekannt, ca. 250	wasserbedingter Ausbruch in Kanada	k.A.	*Isaac-Renton et al., 1998*

(k.A.: keine Angaben)

Cryptosporidium spp.

Cryptosporidium ist ein coccidischer Parasit mit einer weltweiten Ausbreitung (Girdwood, 1995). Humane Infektionen von Cryptosporidium wurden erstmals 1976 dokumentiert (Nime et al., 1976) und wasserbedingte Cryptosporidiosen sind erst seit 1984 bekannt (D'Antonio et al., 1985). Seither sind mehrere Epidemien vor allem in den USA und England beschrieben worden (Lisle und Rose, 1995), von denen die bekannteste diejenige von 1993 in Milwaukee (Wisconsin, USA) mit über 400'000 Erkrankten ist (MacKenzie et al., 1994).

Wichtige Erregerreservoire für *C. parvum* sind vor allem Kälber, doch kommen auch viele andere Haus- und Wildtiere in Betracht (Schleupen, 1996). Aufgrund unterscheidbarer Merkmale, wie Virulenz und genetische Marker, geht man heute davon aus, dass die Art *C. parvum* nicht einheitlich ist. Es wurden Stämme beschrieben, die möglicherweise nur beim Menschen vorkommen, während andere Stämme tatsächlich zwischen Menschen und Tieren zirkulieren können (Morgan et al., 1997; Pozio et al., 1992). In Australien werden die meisten Erkrankungen durch den humanen Genotyp von *C. parvum* verursacht (Morgan et al., 1997). In England und der Schweiz hingegen scheint eher der bovine Genotyp von *C. parvum* für die Erkrankungen verantwortlich zu sein (McLauchlin et al., 1999; Fretz et al., 2002).

Während der Krankheit werden bis zu 10^{10} Oozysten mit einer Grösse von ca. 5 µm mit dem Kot ausgeschieden, wie bei Kälbern festgestellt werden konnte (Metzler und Tabisch, 1997). Die Übertragung erfolgt auf fäkal-oralem Weg via Mensch-Mensch, Tier-Mensch oder über die Aufnahme von kontaminiertem Wasser (Pontius, 1995). Die Infektionsdosis bei gesunden erwachsenen Menschen beträgt rund 30 bis 130 Oozysten (DuPont et al., 1995; Pontius, 1995) wobei bereits bei einer Oozystenkonzentration von 0,1 bis 0,3/L und einem angenommenem Konsum von 2 L/Tag mit dem Auftreten epidemischer Krankheitsfälle gerechnet werden muss (LeChevallier und Norton, 1995). Die Krankheit äussert sich in wässriger Diarrhoe, die 2 bis 14 Tage nach der Exposition auftritt und 1 bis 26 Tage dauern kann. Bei immuninsuffizienten Personen kann eine Infektion zu einem chronischen Krankheitsverlauf mit schweren choleraähnlichen Diarrhöen und weiteren Symptomen führen (Girdwood, 1995). Die Prävalenzrate in der europäischen Bevölkerung liegt bei 2 bis 3% (Eckert, 1993). Eine Studie aus den USA zeigt Prävalenzraten von bis zu 58% bei jungen Erwachsenen (Kuhls et al., 1994).

Die Oozysten von *C. parvum* sind sehr umweltresistent und bleiben in kühlem Wasser monatelang lebensfähig. In der Schweiz konnten Oozysten in Oberflächengewässern (Regli, 1994), in Karstquellwasser (Schweizer und Metzler, 1997) und im Grundwasser (KLBL, 1999) nachgewiesen werden. Detaillierte Studien über das Vorkommen von *C. parvum* in einem Karstgebiet haben gezeigt, dass die Konzentration der Oozysten im Rohwasser von Niederschlagsereignissen abhängig sein kann (Auckenthaler et al., 2002). Die Oozysten konnten auch im Trinkwasser aus demselben Studiengebiet nach einer Aufbereitung mit Flockung, Sedimentation, Schnellsandfiltration und Chlorung nachgewiesen werden (Svoboda et al., 1999).

Entamöbea histolytica

E. histolytica ist weltweit verbreitet und tritt vor allem in Gebieten mit schlechten sanitären Einrichtungen auf. Bevor *E. histolytica* und *E. dispar* nicht unterschieden werden konnten, nahm man an, dass rund 10% der Weltbevölkerung mit Amöbiasis infiziert seien. In den USA ist *E. dispar* möglicherweise viel häufiger verbreitet als *E. histolytica* (Keene, 1999). Die Infektion mit Entamöbea verläuft denn auch meist asymptomatisch (Girdwood, 1995), was in vielen Fällen auf die harmlose Form *E. dispar* zurückzuführen sein könnte (Keene, 1999). Bei niedrigem Hygienestandard können jedoch bis zu 72% der Bevölkerung von *E. histolytica* betroffen sein (Rose und Yates, 1995). Das Krankheitsbild von *E. histolytica* zeigt sich in ulzerativen Dickdarmläsionen, Abszessen in der Leber und selten auch in Veränderungen anderer Organe (Eckert, 1993). Die Inkubationszeit beträgt zwischen 7 und 10 Tagen und einigen Jahren bis die ersten Symptome der Krankheit auftreten (Girdwood, 1995). Die Übertragung der 10 bis 20 µm grossen sphärischen Zysten erfolgt über den fäkal-oralen Weg durch Nahrungsmittel und Trinkwasser. Die Infektionsdosis ist tief. In einer Studie konnten mit 2000 bis 4000 Zysten 100% von 42 Freiwilligen infiziert werden (Keene, 1999). Die Zysten bleiben in Fäzies, Wasser und Erde bei 10°C etwa einen Monat infektiös. Das einzige relevante Erregerreservoir ist der Mensch (Eckert, 1993). Wasserbedingte Ausbrüche sind selten. In den USA wurde zwischen 1971 und 1994 nur ein trinkwasserbedingter Ausbruch beschrieben (Rose und Yates, 1995).

Giardia lamblia

G. lamblia wurde 1960 als humanpathogen erkannt (Boch und Supperer, 1983) und gehört zu den fakultativ pathogenen Parasiten, das heisst eine Infektion muss nicht zu einer Erkrankung führen und es gibt stille Ausscheider der Giardia Zysten. Die Infektionsdosis ist sehr niedrig, bereits 1 bis 10 Zysten können eine Infektion auslösen (Rose und Botzenhardt, 1990). Die Aufnahme der Zysten erfolgt fäkal-oral durch die Aufnahme von kontaminierten Nahrungsmitteln oder kontaminiertem Wasser (Eckert, 1993), wobei der Übertragung durch das Wasser eine besonders grosse Bedeutung zukommt (Bennet et al., 1987). In den USA, Australien und England gilt *Giardia lamblia* als häufigste Ursache für Durchfallerkrankungen und weltweit als häufigster Darmparasit beim Menschen (zit. Schleupen, 1996). Gemäss den Berechnungen der EPA (Environmental Protection Agency in den USA) darf das Trinkwasser im Jahresmittel nicht mehr als 0,7 Zysten pro 100'000 L enthalten, um das jährliche Erkrankungsrisiko auf maximal 1:10'000 bei einer täglichen Aufnahme von 2 L pro Person beschränken zu können (Botzenhardt, 1994).

Die Prävalenzrate von *Giardia lamblia* liegt in der europäischen Normalbevölkerung bei 3 bis 4% und in Entwicklungsländern bei mehr als 50% (Eckert, 1998). Mögliche Reservoirwirte können Wiederkäuer, Hunde, Katzen, Biber, Bisam und Nager sein. In einer Veterinärstudie in der Schweiz konnten bei Giardia-Infektionen bei Kälbern in 26%, bei Schaflämmern in 30%, bei Hunden in 6,5% und bei Katzen in 5,3% Zysten im Kot nachgewiesen werden (Eckert et al., 1990).

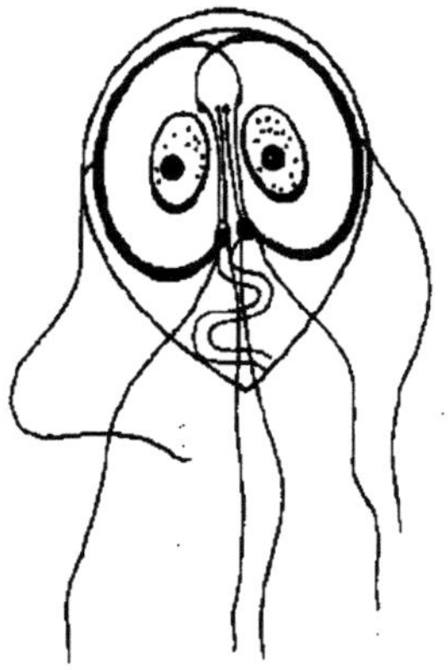

Abb. 2.7: Giardia Trophozoit, die Länge beträgt rund 12 bis 15 µm und der Durchmesser 6 bis 8 µm (www.taurangawater.govt.nz)

Die Persistenz der Zysten mit einer Länge von 10 bis 15 µm und einer Breite von 7 bis 10 µm in der Umwelt, die grosse Anzahl an ausgeschiedenen Zysten und das breite Wirtsspektrum führen dazu, dass *Giardia lamblia* lange in der Umwelt überleben kann (Rose und Botzenhardt, 1990). In der Schweiz konnten Zysten in Oberflächengewässern (Metzler et al., 1996) und bei einer privaten Trinkwasserversorgung nachgewiesen werden (Regli et al., 1996). In Deutschland ist im Jahr 2000 ein erster dokumentierter Giardiaausbruch in einer Wasserversorgung einer Ortsgemeinde des Landes Rheinland-Pfalz aufgetreten. Das zur Trinkwassergewinnung genutzte Quellwasser war durch Oberflächenwasser verunreinigt worden (Gornik et al., 2001).

Mögliche zukünftige pathogene Protozoen

Cyclospora cayetanensis, Microsporidien und *Toxoplasma gondii* sind in den USA als mögliche zukünftige Pathogene erkannt worden (AWWA, 1999b). Deren Übertragung über das Wasser und die Bedeutung für die Trinkwasserversorgung wird momentan abgeklärt. Ein erster wasserbedingter Ausbruch von *C. cayetanensis* wurde in Nepal unter britischen Soldaten festgestellt. *Toxoplasma gondii* hat kürzlich in Kanada zu einer kleineren Epidemie geführt (Isaac-Renton et al., 1998). Die grössere Gruppe der Microsporidien wurden erst in Oberflächenwässern nachgewiesen. Die hier vorgestellten „emerging pathogens" stellen nur eine Auswahl weiterer möglicher, in naher oder ferner Zukunft über das Trinkwasser verbreiteter, pathogener Protozoen dar.

Cyclospora cayetanensis

Cyclospora cayetanensis wurde erst 1979 von Ashford, 1979 zum ersten Mal beschrieben. Bis heute sind tausende von Humaninfektionen vor allem aus wärmeren Ländern bekannt und *C. cayetanensis* wird heute als möglicher wasserbedingter „emerging pathogen" gesehen. Der Lebenszyklus des Parasiten und die Epidemiologie sind noch weitgehend unbekannt (Girdwood, 1995). Die Krankheitssymptome von *C. cayetanensis* sind ähnlich wie diejenigen von *Cryptosporidium parvum,* ausser dass sie häufig unterbrochen aber länger andauernd sind. Die Grösse der Oocysten von *C. cayetanensis* mit 8 bis 10 µm unterscheiden sich wenig von denjenigen von *C. parvum* mit 4 bis 6 µm, was zu Verwechslungen der beiden Protozoen bei der Abklärung von Erkrankungen führen kann (Sterling, 1995).

Die Oocysten von *Cyclospora* sind umweltresistent. Nur sporulierte Oocysten sind infektiös. Die Sporulation ist temperaturabhängig und dauert bei 25°C rund zwei Wochen, bei 4°C bis zu 4 bis 6 Monaten. Während dieser Zeit müssen die Oocysten in einer feuchten Umgebung sein, was auf eine wasserbedingte Übertragung und die fäkal-orale Route hindeutet (Sterling, 1995).

Ausbrüche wurden ausgelöst durch den Verzehr von rohen Früchten (MMWR, 1996), Basilikum (MMWR, 1997) und Wasser (AWWA, 1999b). In den Jahren 1996 und 1997 fand in Nordamerika ein Cyclosporiasis Ausbruch statt, der auf Himbeeren aus Guatemala zurückzuführen ist. Als Ursache gilt höchst wahrscheinlich fäkal verunreinigtes Wasser, das zur Applikation von Fungiziden benutzt wurde (Bern et al., 1999). Eine wasserbedingte Übertragung von Cyclospora ist von britischen Soldaten in Nepal und aus einem Krankenhaus in Chicago, USA bekannt (AWWA, 1999b). Flusswasseruntersuchungen in Guatemala zeigten bei einer Belastung mit *E. coli* von 5 x 10^5 bis 2 x 10^7 pro 100 mL eine Anzahl von rund 15'000 Cyclospora Oozysten (Bern et al., 1999)

Microsporidia

Die Ordnung Microsporidia umfasst rund 20 Familien, 100 Gattungen und Hunderte von Spezies. Die Microsporidien sind ubiquitär vorhanden. Etwa 12 Spezies sind als humanpathogen bekannt (Cali, 1999). Mehreren dieser Protozoenarten wurde in letzter Zeit, wegen hohen Infektionsraten bei AIDS PatientInnen, vermehrt Aufmerksamkeit geschenkt. Sie sind obligate Zellparasiten mit typischem Sporenstadium und bei einer Grösse von 1 bis 5 µm relativ klein. (Eckert, 1993). Es wurden Koinfektionen mit Cryptosporidium festgestellt. Das Krankheitsbild zeigt sich in chronischem Durchfall. Microsporidien können aber auch Augen, Leber, Herz, Nieren und zentrales Nervensystem befallen. Die Sporen werden von Mensch-zu-Mensch oder auf fäkal-oralem Weg übertragen (Weidner, 1991). Trinkwasser wird als potentieller Übertragungsweg angesehen, da einzelne Spezies in der Umwelt, insbesondere in kaltem Wasser bis zu mehreren Wochen und Monaten überleben können (AWWA, 1999b). Potentiell pathogene Microsporidien wurden aus Oberflächengewässern isoliert (Sparfel et al., 1997).

Es bestehen bisher noch sehr wenige Informationen über die Verbreitung der Micosporidien im Wasser, da geeignete Analysenmethoden fehlen, um die Prävalenz in Rohwasser und aufbereitetem Wasser zu bestimmen. Die American Water Works Association sieht die Microsporidien als „emerging pathogens“ mit einer mittleren Priorität für die Wasserversorgungen (AWWA, 1999b).

Toxoplasma gondii

T. gondii kommt weltweit vor und infiziert praktisch alle warmblütigen Wirte. Toxoplasmose ist eine der am meisten verbreiteten Humaninfektionen. Die Krankheit verläuft bei Erwachsenen oft asymptomatisch. Eine Infektion von Geburt an kann bei Kindern zu mentaler Retardation, Verlust des Sehvermögens, Höhrschwäche und zum Tod führen (Dubey, 1999). Die Infektionsdosis beim Menschen ist nicht genau bekannt. In Mäusen konnte eine letale Dosis von 250 Oozysten ermittelt werden (Isaac-Renton et al., 1998). Das Reservoir der 10 bis 12 µm grossen Oozysten sind Wild- und Hauskatzen (Dubey, 1999). Die Oozysten sind sehr umweltresistent und widerstehen hohe Chlorkonzentrationen.

Ein wasserbedingter Ausbruch von *T. gondii* ist aus British Columbia, Kanada bekannt. Die epidemiologische Auswertung des Falles zeigte zwei Cluster von Personen, die wahrscheinlich während zwei Niederschlagsereignissen erkrankten. Die Aufbereitung des trüben Wassers erfolgte lediglich mit einer Chlorung. In den Wasserproben konnten keine *T. gondii* nachgewiesen werden. Die Proben wurden allerdings erst 12 Wochen nach dem Ausbruch erhoben (Isaac-Renton et al., 1998).

Schlussfolgerungen

Wasserbedingte Infektionskrankheiten, ausgelöst durch verschiedene pathogene Keime kommen weltweit vor und verursachen jährlich Millionen von Erkrankungen mit grossen sozialen und ökonomischen Folgen. Auch in der Schweiz und in Deutschland können im Rohwasser und teilweise ebenfalls im Trinkwasser mikrobielle Krankheitserreger festgestellt werden. Diese als „emerging pathogens“ bezeichneten Mikroorganismen wie *Cryptosporidium* spp., verotoxinbildende *E. coli* und Norwalk-like Viren haben auch in Industriestaaten bereits grosse Ausbrüche verursacht. Anstelle der Pathogene werden bei den Routineuntersuchungen des Trinkwassers in der Schweiz die fäkalen Indikatorkeime *E. coli* und Enterokokken und in Deutschland zusätzlich *C. perfringens* analysiert und teilweise auch festgestellt. Diese Befunde sind stets ein Indiz für eine humane oder animale fäkale Verunreinigung des Trinkwassers und es ist zu erwarten, dass Erkrankungen durch hoch infektiöse Mikroorganismen stattfinden. Solche wasserbedingten Erkrankungen festzustellen ist allerdings schwierig, da die meisten betroffenen Personen keinen Arzt aufsuchen und die Meldesysteme in der Schweiz und in Deutschland noch nicht alle wasserübertragbaren infektiösen Mikroorganismen erfassen. Dass aber in beiden Ländern wasserbedingte Ausbrüche stattfinden können, zeigen die Fälle von La Neuveville, Küblis und in Rheinland-Pfalz. Ursachen für diese Fälle waren regelwidrige Standorte von Anlagen innnerhalb der Gewässerschutzzonen sowie das Versagen technischer Systeme für die Wasseraufbereitung und und als Folge davon das Eindringen von verunreinigtem Oberflächen- oder Abwasser ins Trinkwasser.

Um die heutige gute Trinkwasserqualität in der Schweiz und in Deutschland bewahren zu können, sind Investitionen in der nahen Zukunft unumgänglich. Die Kenntnisse über die Herkunft und die Verbreitung pathogener Keime muss verbessert, die Nutzungsbeschränkungen (z.B. Gülleverbot) in den engeren Schutzzonen konsequent durchgesetzt und die lokalen hydrogeologischen Erkenntnisse bei der Nutzung von Grundwasser müssen erweitert werden. Dadurch lässt sich die Gefährdung beim Konsum von Trinkwasser abschätzen. Wo nötig, müssen geeignete technische Systeme zur Wasseraufbereitung installiert werden. Diese müssen der mikrobiologischen und chemischen Beschaffenheit des Rohwassers angepasst werden und einfach zu bedienen sein, damit sie auch von kleineren Wasserversorgungen sicher betrieben werden können. In Zukunft ist mit dem Auftreten weiterer pathogener Mikroorganismen im Grund- und Trinkwasser zu rechnen. Die Konsequenzen für den Grund- und Trinkwasserschutz und die Wasseraufbereitung bleiben aber die selben, wie für die heute bekannten Mikroorganismen.

Nachweismethoden für pathogene Mikroorganismen 3

Antonis Chatzinotas und Hauke Harms

Einleitung

Grundsätzlich sind zwei Vorgehensweisen für die Bewertung der mikrobiologischen Qualität von Trinkwasser denkbar. Die Praxis der Qualitätssicherung des Trinkwassers in der Schweiz und in Deutschland, wie auch in vielen anderen Ländern, beruht seit etwa einem Jahrhundert auf dem Nachweis ausgewählter Indikatorkeime, welche auf Verunreinigung des Wassers durch Fäkalien des Menschen oder anderer Warmblüter hindeuten. Dieses Vorgehen beruht auf der Annahme, dass mit dem Trinkwasser aufgenommene Krankheitserreger aus Darminhalt stammen, während aus anderen Quellen stammende oder indigene Mikroorganismen natürlichen Wassers nicht pathogen sind. Es trägt weiterhin der Erfahrung Rechnung, dass durch Trinkwasser erworbenen Pathogene typischerweise den Verdauungstrakt ihrer Träger befallen und diesen, oft nach drastischer Vermehrung, in infektiösem Zustand wieder verlassen. Als Kriterium für die mikrobielle Qualität von Trinkwasser haben sich die Keimzahlen der aeroben, mesophilen Bakterien, von Bakterien der Art *Escherichia coli* sowie der Enterokokken durchgesetzt, da das Auftreten pathogener Bakterien häufig mit dem Nachweis dieser Gruppen korreliert.

Die Verwendung von Indikatorbakterien stösst dort an ihre Grenzen, wo eine solche Korrelation nicht gegeben ist. Mögliche Gründe sind nichtfäkale Quellen der Krankheitserreger, deren Auftreten in höheren Zahlen oder längeres Überleben als die Indikatororganismen und Infektiösität in sehr kleinen Dosen. In diesem Zusammenhang seien beispielsweise besonders infektiöse Viren genannt sowie pathogene Protozoen, welche besonders überlebensfähige Dauerformen bilden.

Angesicht dieser bekannten Schwierigkeiten ist es naheliegend, dass momentan Verfahren entwickelt werden, welche den direkten spezifischen Nachweis pathogener Organismen im Trinkwasser ermöglichen sollen. Wie wir in diesem Kapitel zeigen werden, gibt es in dieser Hinsicht eine Vielzahl von Forschungsansätzen, welche entweder konventionelle

Kultivierungsabhängige Methoden, molekulare Methoden oder Kombinationen aus beiden verwenden. Insbesondere für den Nachweis von Viren oder bisher nur selten als Krankheitserreger in Erscheinung getretene Mikroorganismen werden vermehrt molekulare Nachweismethoden entwickelt. Es ist noch offen, wann diese Bemühungen zu zielgerichteten und dennoch praxistauglichen Tests für mikrobielle Trinkwasserqualität führen werden. Es ist jedoch offensichtlich, dass die beschriebenen Verfahren bei speziellen Fragestellungen, wie der Untersuchung von Trinkwasserseuchen oder zur Unterstützung von Indikatorkeim-Verfahren bei Gefährdungssituationen durch bestimmte Erreger, bereits Anwendung finden können.

Nachweis von pathogenen Bakterien und Indikatorkeimen

Grundsätzlich gibt es zwei Vorgehensweisen für die Bewertung der mikrobiologischen Qualität von Trinkwasser:

- Den Nachweis ausgewählter Indikatorkeime, die auf eine fäkale Kontamination hinweisen
- Den direkten Nachweis spezifischer pathogener Bakterien

Indikatorkeime

Die Auswahl von Indikatorkeimen unterscheidet sich von Land zu Land sowie zwischen internationalen Organisationen. Indikatorkeime sollten typisch für die Darmflora von Warmblütern sein und ihr Überleben ausserhalb des Darmtraktes sollte dem der pathogenen Zielbakterien ähnlich sein. Am häufigsten werden "total coliforme Keime", thermotolerante coliforme Bakterien, *Escherichia coli* und Fäkalstreptokokken bzw. Enterokokken als Indikatorkeime für Trinkwasser angeführt (Toranzos und McFeters, 1997).

Unter dem Begriff „total coliforme Keime“ fasst man verschiedene Genera der *Enterobacteriaceae* zusammen. Auf speziellen Medien isolierte Kolonien werden durch enzymatische Tests (Cytochromoxidase negativ, β-Galactosidase positiv) als coliforme Keime bestätigt (Dutka et al., 1974). Einige Mitglieder dieser Gruppe, z.B. *Klebsiella terrigena* oder *Serratia fonticola* kommen auch ausserhalb des Darmtraktes von Warm-blütern vor und sind somit für die Trinkwasserkontrolle nur eingeschränkt von Nutzen (Mathewson, 1998). Thermotolerante coliforme Bakterien (früher als Fäkalcoliforme be-zeichnet) gelten als zuverlässigere Indikatoren für Fäkalkontaminationen als die „total coliformen Keime“, da sie zum Teil diejenigen coliformen Bakterien ausschliessen, die ausserhalb des Darmtraktes überleben können (WHO, 1989). Der Nachweis dieser Indikatororganismen erfolgt hauptsächlich durch Kultivierung auf speziellen kommerziell erhältlichen Medien, oft in Verbindung mit spezifischen enzymatischen Tests, die auf dem Abbau von Substanzen in farbige oder fluoreszierende Bestandteile basieren (Mathewson, 1998; Toranzos und McFeters, 1997).

Tab. 3.1: Eine Auswahl molekularbiologischer Nachweismethoden für bakterielle Indikatororganismen und pathogene Bakterien

Bakterien	Nachweis- und Identifikationsmethoden	Detektionsgrenze
E. coli	PCR mit 16S rRNA Primern nach Anreicherung (Tsen et al., 1998)	In Leitungswasser und Umweltproben: 1 Zelle/100 ml
	Koloniehybridisierung mit 16S rRNA Oligonukleotidsonden nach kurzer Anreicherung (Lin und Tsen, 1999)	In Leitungswasser: 1CFU/100 ml
Enterotoxische (ETEC) und enterohämorrhagische (EHEC) *E. coli*	Multiplex-PCR zum Nachweis der Gene, die für (a) hitzelabile bzw. -stabile Toxine in ETEC, oder (b) Shiga-ähnliche Toxine in EHEC kodieren (Tsen und Jian, 1998)	In Leitungswasser und Umweltproben: 1 Zelle/100 ml
Enterococcus spp.	Hybridisierung fixierter Zellen mit rRNA Oligonukleotid-Sonden nach Anreicherung über Nacht (Meier et al., 1997)	$10\text{-}10^4$/100 ml Enterokokken konnten in Wasserproben innerhalb von 20 h nachgewiesen werden
	PCR mit rRNA-Primern nach kurzer Anreicherung und anschliessende Hybridisierung in Mikrotiterplatten mit 23S rRNA Oligonukleotidsonden unterschiedlicher Spezifität (Frahm et al., 1998)	In inokulierten Wasserproben: 1 CFU/ml; Dauer 26 h
Campylobacter jejuni, C. coli	Nichtselektive Anreicherung mit anschliessender seminested PCR-Amplifikation der intergenischen Sequenzen zwischen zwei Flagellin-Genen (flaA, flaB) (Waage et al., 1999a)	In inokulierten Wasserproben: 3-15 CFU/100 ml
Legionella spp.	Seminested PCR-Nachweis mit 16S rRNA-Primern (Miyamoto et al., 1997)	Detektion von 1fg extrahierter Legionella DNA möglich; sensitiver als kommerziell erhältlicher Legionella PCR-Kit
Salmonella spp.	Nested PCR-Nachweis nach Filtrierung und nichtselektiver Anreicherung (Waage et al., 1999b)	In inokulierten Wasserproben: 10 CFU/100 ml
Leptospira spp.	Seminested PCR-Nachweis mit 16S rRNA-Primern zur Unterscheidung pathogener und saprophytischer Leptospiren (Murgia et al., 1997)	In inokulierten Wasserproben: 5 Zellen/100ml

Tab. 3.1: Fortsetzung

Bakterien	Nachweis- und Identifikationsmethoden	Detektionsgrenze
Mycobacterium ulcerans	PCR-Nachweis einer hochrepetitiven Insertionssequenz mit anschliessender Hybridisierung (Ross et al., 1997)	1 Zelle/ml in Umweltproben
Helicobacter pylori	Immunomagnetische Anreicherung der Zellen mit anschliessender nested PCR spezifisch für das ureA Gen von *Helicobacter pylori* (Sasaki et al., 1999)	
	Übersichtsartikel über mögliche Detektionsmethoden von Velàsquez und Feirtag, 1999.	
Yersinia enterolitica	Nested PCR-Nachweis des yadA-Genes entweder nach immunologischer Anreicherung oder Zentrifugation (Kapperud et al., 1993)	In inokulierten Wasserproben: 8-17 CFU/100 ml, Dauer 2-3 Tage
	Nested PCR-Nachweis nach Filtrierung und nichtselektiver Anreicherung (Waage et al., 1999c)	
Vibrio cholerae	Anreicherung mit anschliessender seminested PCR spezifisch für das Cholera-Toxin (Theron et al., 2000)	In inokulierten Wasserproben: 1 CFU/ml

Obwohl sich die meisten kultivierungsabhängigen Tests für Indikatororganismen bewährt haben, sind nicht alle lebendigen coliformen Bakterien mit nur einer bestimmten Kultivierungsmethode zu erfassen. So können zum Beispiel beschädigte Bakterien auf selektiven Medien nicht nachgewiesen werden, da sie sich dort nicht vermehren (Bucklin et al., 1991; McFeters et al., 1986). Da sie sich aber möglicherweise wieder "erholen" können, besteht das Risiko, dass die tatsächliche Bakterienzahl unterschätzt wird. Die Anwendung von speziellen kommerziell erhältlichen Medien zum Nachweis von beschädigten Bakterien im Trinkwasser wurde daher empfohlen (Bucklin et al., 1991; APHA, 1992).

Es wurde wiederholt festgestellt und kritisiert, dass *E. coli*-spezifische Medien (Alonso et al., 1998; Brenner et al., 1996; Manafi and Kneifel, 1989) keine bzw. kaum Informationen über die qualitative Zusammensetzung der gesamten *E. coli* Population aus humanen und nicht humanen *E. coli*-Populationen in den untersuchten Wasserproben liefern (APHA, 1992). Molekularbiologische Methoden, wie das Ribotyping (Paarven et al., 1999) erlauben hingegen diese Diskriminierung. Molekulare Fingerprintanalysen, wie z.B. Denaturating Gradient Gel Electrophoresis (DGGE) oder Randomly Amplified Polymorphic DNA (RAPD) produzieren spezifische Bandenmuster, die die Identifizierung der jeweiligen *E. coli* Stämme und möglicherweise sogar die genaue Bestimmung einer Kontaminationsquelle erlauben (Farnleitner et al., 2000; Rothmeier et al., 1997).

Fäkalstreptokokken bzw. Enterokokken werden ebenfalls als Indikatorkeime benutzt, da sie generell persistenter und resistenter gegenüber Desinfektionsmassnahmen als *E. coli* sind, und sich im Gegensatz zu *E. coli* im Wasser nicht vermehren können (Sinton et al., 1993). Die üblichen Kultivierungsmethoden und biochemischen Tests für den Nachweis dieser Organismen sind jedoch sehr zeitaufwendig und oft unzuverlässig, da physiologische Unterschiede zwischen Stämmen einer Art zu beobachten sind. Diese Variationen können epidemiologische Studien erheblich erschweren (Devriese et al., 1993). Daher werden vermehrt molekularbiologische Methoden zum Nachweis dieser Indikatororganismen entwickelt und auf ihre routinemässige Anwendbarkeit hin getestet (Tabelle 3.1). Es konnte beispielsweise gezeigt werden, dass Enterokokken durch eine Hybridisierung mit gruppenspezifischen Sonden innerhalb von 20 h korrekt nachgewiesen werden können. Unter Hybridisierung versteht man die Bindung einer markierten Nukleinsäure, die sogenannte Sonde, an komplementäre Nukleinsäuresequenzen. Die biochemische Charakterisierung von Enterokokken mit einem kommerziell erhältlichen System kann in bis zu 21% der Fälle zu falschen Resultaten führen (Frahm et al., 1998).

Sulfit-reduzierende Clostridien *(Clostridium perfringens und Clostridium welchii)* können als mögliche Indikatorkeime für eine länger zurückliegende bzw. entfernte tierische Fäkalkontamination dienen, da sie als Sporen längere Zeit in der Umwelt überleben und somit auch verfrachtet werden können (WHO, 1989). Ihr Nachweis mit Kultivierungs-methoden ist jedoch aufwändig, da anaerobe Techniken anzuwenden sind. *Pseudomonas aeruginosa* kann zwar in fäkal kontaminierten Gewässern nachgewiesen werden, ist jedoch als Indikatororganismus ungeeignet, da sich diese Art auch ohne Fäkalkontamination in nährstoffreicher Umgebung vermehren kann. Das Vorkommen von *P. aeruginosa* in Wasserleitungssystemen von Krankenhäusern stellt jedoch ein Risiko für Patienten dar (Nguyen et al., 1998).

Direkter Nachweis pathogener Bakterien

Der Nachweis von Indikatorkeimen, wie z.B. der Fäkalcoliformen ist nur hinsichtlich des möglichen Vorkommens pathogener Mikroorganismen im Grund- oder Trinkwasser aussagekräftig, nicht aber hinsichtlich deren tatsächlichen Auftretens. Die für einen direkten Nachweis erforderliche Kultivierung pathogener Bakterien ist in vielen Fällen sehr zeitaufwändig. Dennoch gibt es eine Reihe von kultivierungsabhängigen Standardprozeduren (z.B. APHA, 1992). Der spezifische Nachweis einiger pathogener Organismen wird zudem oft dadurch erschwert, dass diese, wie z.B. *Helicobacter pylori*, je nach Umweltbedingungen ihre Zellmorphologie und ihren Stoffwechsel verändern (Velasquez und Feirtag, 1999), oder dass sie nur in geringen Zellzahlen vorkommen und nur sehr langsam wachsen (Waage et al., 1999). Serotyping von Isolaten bietet präzise Möglichkeiten zur Identifikation pathogener Organismen. Jedoch ist auch dieser Ansatz immer noch sehr zeit- und arbeitsaufwändig, sind doch beispielsweise für die genaue Differenzierung von *Salmonella* Stämmen über 150 Serumproben notwendig (Kilger und Grimont, 1993).

VBNC-Zellen (viable but not culturable; lebend, jedoch nicht kultivierbar) können sich auf Nährböden nicht teilen und sind somit nicht kultivierbar (Owen, 1995). Ihre Lebensfähigkeit und Virulenz (wie z.B. bei *Helicobacter pylori* oder *Campylobacter* spp.) bleibt potenziell weiterhin erhalten (Gribbon and Barer, 1995; Waage et al., 1999). Somit ist der Nachweis auch der VBNC-Formen zur Qualitätskontrolle von Wasser notwendig.

Alternativen zum kultivierungsabhängigen Nachweis vieler pathogener Organismen (und zum Nachweis von VBNC-Zellen) bieten auch hier molekularbiologische Methoden (Tabelle 3.1), wie z.B. die PCR und Hybridisierung mit spezifischen Gensonden, sowie immunologische Trenn- und Anreicherungsverfahren (z.B. Velasquez and Feirtag, 1999; Sasaki et al., 1999). Die PCR, d. h. die enzymatische Vervielfältigung ausgewählter Nukleinsäuresequenzen mit hoher Sensitivität und Spezifizität, wird dabei entweder für die schnelle Identifizierung isolierter pathogener Mikroorganismen oder für den direkten Nachweis in Umweltproben ohne vorherige Kultivierung angewandt. Verschiedene molekularbiologische Protokolle wurden zum Nachweis von pathogenen Organismen in Lebensmitteln oder in menschlichen Fäkalproben entwickelt, z.B. für *Salmonella* (Way et al., 1993), *Legionella* (Hay et al., 1995), *Yersinia* (Trebesius et al., 1998) und enterotoxische *E. coli* (Tamanai-Shacoori et al., 1996). Es kann jedoch davon ausgegangen werden, dass diese Protokolle (oft erhältlich zusammen mit kommerziellen PCR-Kits) für einen direkten Nachweis pathogener Organismen im Wasser angepasst (z.B. durch Entfernung im Wasser enthaltener PCR-inhibierender Substanzen) und verwendet werden können (z.B. Frahm und Obst, 1995; Fricker und Fricker, 1995). Einige Autoren kombinieren molekularbiologische Methoden mit nichtselektiver Anreicherung auf unspezifischen Medien (Waage et al., 1999), da gezeigt werden konnte, dass eine nichtselektive Anreicherung für gestresste und womöglich auch beschädigte Zellen zu bevorzugen ist (Humphre, 1986). Zudem verdünnt eine solche Anreicherung mögliche PCR-inhibierende Kontaminationen sowie tote Zellen, während die direkte PCR kaum eine Unterscheidung zwischen toten und lebenden Zellen erlaubt. Dennoch sind Methoden, die keine Anreicherung der Bakterien beinhalten, in vielen Fällen vorzuziehen, da VBNC-Stadien und nicht-kultivierbare pathogene Organismen sonst nicht erfasst werden. Enterohämorrhagische und enterotoxische *E. coli* (EHEC bzw. ETEC) haben in den letzten Jahren als Krankheitserreger an Bedeutung gewonnen. Der Nachweis dieser pathogenen Organismen im Wasser erfordert im allgemeinen deren Isolierung auf *E coli*-spezifischen Medien. Anschliessend müssen die pathogenen von den nichtpathogenen *E. coli* unterschieden werden. ETEC werden auf Grund der Gene nachgewiesen, die für Enterotoxine codieren. EHEC können mit Hilfe einer immunologischen Bestimmung des Serotypus oder auf Grund der Gene, die für die sogenannten ‚Shiga-ähnlichen' Toxine codieren, detektiert werden (Gray, 1995; Mathewson, 1998). Methoden, die auf Kultivierung dieser Bakterien verzichten, sind in ihrer Mehrzahl immunologische Verfahren, wie z.B. IMS (immunomagnetic separation; immunomagnetische Trennung) (Pyle et al., 1999; Porter et al., 1997; Tortorello und Stewart, 1994). Dabei werden mit spezifischem Antiserum beschichtete Kügelchen benutzt, um die Zielorganismen anzureichern. Eine anschliessende Inkubation mit spezifischen Antikörpern und Cyanotydol Tetrazoliumchloride (CTC) erlaubt einerseits die Identifikation, andererseits den direkten Nachweis von lebenden Zellen. Lebendige aerobe Bakterien veratmen CTC, was einen sichtbaren Farbniederschlag in diesen Zellen zur Folge hat. Mittels eines Laser Scanning Cytometers benötigten die Autoren einer Studie ca. 8 h für den Nachweis von lebenden *E. coli* O157:H7, einem enterohämorrhagischen *E. coli*-Serotypus (Pyle et al., 1999). Dies ist wesentlich schneller als konventionelle Methoden (Hawkins and Orme, 1995), welche eine 24-stündige Präinkubation benötigen mit anschliessender Bestätigung als EHEC. Zu beachten ist, dass einige Antikörper mit verschiedenen Antigenen kreuzreagieren können, da verschiedene *E. coli*-Serotypen dieselben Verotoxine bilden können, und somit mit diesen Antikörpern nicht unterscheidbar sind (O'Brien and Holmes, 1987). Molekularbiologische Methoden basierend auf PCR können *E. coli* O157:H7 von anderen *E. coli* Stämmen differenzieren, z.B. durch Restriktionsite-spezifische PCR (RSS-PCR) (Kimura et al., 2000).

Nachweis von Viren im Wasser

Die üblichen mikrobiellen Standardindikatoren für fäkale Wasserverunreinigung sind zur Beurteilung viraler Kontaminationen ungeeignet. Mehr als 140 Viren (Hurst, 1991; Schwartzbrod, 1991) befinden sich im menschlichen Darmtrakt und können im Wasser nachgewiesen werden. Der wichtigste Übertragungsweg für Viren ist fäkal-oral. Die Übertragung durch Einnahme von Trinkwasser aus öffentlicher Versorgung konnte bisher jedoch nur in wenigen Fällen schlüssig nachgewiesen werden. Viele Viren bleiben im Boden und in Gewässern länger infektiös als bakterielle Indikatororganismen. Zudem sind einige der angewandten Desinfizierungsmethoden für Trinkwasser nicht ausreichend, um virale Kontaminationen zu eliminieren (Blackmer et al., 2000). Da Viren im allgemeinen in geringeren Dosen infektiös sind als Bakterien, sind äusserst sensitive Methoden notwendig, um auch kleinste Mengen an Viren in grossen Volumina nachzuweisen. Die Übertragung durch Einnahme von Trinkwasser aus öffentlicher Versorgung konnte bisher Die Detektion von Viren erfolgt im allgemeinen nach einer Konzentrierung viraler Partikel aus grossen Probevolumina, z. B. mittels Filtration, physikalisch-chemischer Flockung oder Affinitätschromatographie. Diesen Methoden zur Virenanreicherung ist gemein, dass ein gewisser Verlust an Viren auftreten kann (Hurst, 1997a). Ausserdem kann es während der Konzentrierungsschritte auf Grund von pH-Veränderungen oder der Zugabe von potentiell toxischen Chelatoren zur Virusinaktivierung kommen. Der Nachweis der Viren im Konzentrat erfolgt in Zellkulturen, elektronenmikroskopisch und (oder) mittels molekularer und immunologischer Methoden. Da die klassischen Verfahren oft sehr zeitaufwändig oder auch unspezifisch sind, wird momentan insbesondere die Entwicklung molekularbiologischer Nachweismethoden weiter vorangetrieben (Tabelle 3.2).

Zellkulturen

Zellkulturen werden mit einer Virusanreicherung inkubiert (z.B. Abbaszadegan und Dowd, 1998; Hurst, 1997b) und auf das Auftreten sogenannter cytopathologischer Effekte (CPE) untersucht. Jedoch ist die alleinige Anwendbarkeit von Zellkulturen mit einigen Nachteilen behaftet:

- Ohne zusätzliche Analysen, z. B. molekularbiologische Charakterisierung, ist keine Aussage über einen bestimmten Stamm möglich
- Zellkulturen sind zeitaufwändig und dementsprechend kostenintensiv Standardzellkulturen für Enteroviren können 3-14 Tage dauern, oft ist sogar eine zweite Inkubationsrunde notwendig, die jedoch nur selten routinemässig angewandt wird (Blackmer et al., 2000)
- Die BGM-Zellkulturlinie, die am häufigsten verwendete Zellkulturlinie für Enteroviren (Buffalo Green Monkey Nierenzellen), kann nicht zwischen Enteroviren und Reoviren unterscheiden. Jedoch sind Reoviren häufiger in Umweltproben anzutreffen als Enteroviren (Puig et al, 1994). Somit können mit Hilfe dieser Zellkulturen beobachtete CPE nicht ohne weiteres den verursachenden Viren zugeordnet werden. Einige Enteroviren, wie z.B. Adenovirus 40 oder 41 lösen keine CPE aus, andere Viren, z.B. der humanpathogene Astrovirus können sich in diesen Zelllinien nicht vervielfältigen (Subsey, 1987; Chung et al., 1996; Chapron et al., 2000)
- Für viele Viren, z.B. Norwalk Viren sind Zellkulturlinien noch nicht entwickelt worden
- Ergebnisse basierend auf Zellkulturen können durch organische oder anorganische Kontaminationen, welche toxisch sind, verfälscht werden
- Die meisten Protokolle schreiben Inkubationszeiten vor, die möglicherweise für eine eindeutige Aussage nicht ausreichend sind. Es konnte z.B. gezeigt werden, dass Proben, die mit Standardmethoden als negativ eingestuft wurden, sich nach bedeutend längerer Inkubation letztlich doch als positiv erwiesen (Abbaszadegan et al., 1999)

Molekulare Nachweismethoden

Molekulare Nachweismethoden beruhen hauptsächlich auf der Amplifizierung diskreter Bereiche des viralen Genomes mittels PCR (Tabelle 3,2). Dadurch ist es möglich, Nukleinsäuren (RNA oder DNA) von Viren, die nur in sehr geringer Anzahl vorkommen, zu vervielfältigen. Sowohl die Sensitivität als auch die Spezifität wird durch PCR erhöht, vor allem wenn noch ein zusätzlicher zweiter PCR-Schritt (eine sogenannte nested-PCR) erfolgt. Wie schon erwähnt, schreiben viele Protokolle zum Nachweis von Viren eine Aufkonzentrierung grosser Anfangsvolumina vor. Da die für die PCR verwendeten Enzyme durch in den Proben enthaltene und aufkonzentrierte Substanzen inhibiert werden können, sind verschiedene Reinigungsschritte z.B. mit Sephadex G-100 (Abbaszadegan et al., 1999) notwendig. Selbst bei sehr hohen Viruskonzentrationen können beispielsweise 100 µg/ml Huminsäuren zu falschen Resultaten führen (Soule et al., 2000). Neuere Methoden, wie z.B. „nucleic acid capture", extrahieren selektiv die Nukleinsäuren aus einer aufkonzentrierten Lösung. Bei dieser Methode hybridisieren Biotin-markierte Sonden spezifisch mit viraler RNA. Die Biotin-Moleküle können durch biotinbindende Proteine, wie z.B. Streptavidin gebunden werden. Dies erfolgt in einem zweiten Schritt mit Streptavidin-beschichteten Magnetkügelchen, die dann eine Anreicherung der gewünschten Zielmoleküle erlauben (Regan und Margolin, 1997).

Für die Evaluierung der PCR-Protokolle sind Positiv- und Negativkontrollen unabdingbar, da eine nicht erfolgte Amplifizierung mittels PCR nicht zwangsläufig bedeutet, dass keine Viren vorhanden sind. Es wurde vorgeschlagen, dass die Sensitivität des angewandten PCR-Protokolls in einem zusätzlichen Versuch, z.B. durch Southern-Hybridisierung, spezifisch getestet werden sollte (Abbaszadegan et al., 1999). Hierbei werden die PCR-Produkte auf einen Filter geladen und anschliessend mit markierten Nukleotidsonden, die eine Sequenz innerhalb des PCR-Produktes erkennen können, spezifisch hybridisiert.

Einige Nachteile der molekularen Detektion müssen beachtet werden:

- Da nur die Anwesenheit von Viren-DNA nachgewiesen wird, können Aussagen über die Infektivität (infectivity) der erfassten Viren nicht getroffen werden. Für einen entsprechenden Nachweis ist ein verlässlicher Zellkulturassay notwendig (Gratacap-Cavalier et al., 2000)
- Durch die Konzentrierung der Proben werden auch Inhibitoren der molekularbiologischen Reaktionen angereichert (Abbaszadegan et al., 1999). Die notwendigen Reinigungsschritte können zum Abbau oder zum Verlust von RNA bzw. DNA führen (Reynolds, 1997). Auf Grund der Empfindlichkeit von RNA deutet jedoch der erfolgreiche Nachweis viraler RNA auf intakte Viren hin
- Vielfach ist ein zweiter PCR-Schritt notwendig, eine sogenannte nested-PCR, um falsch negative Resultate zu vermeiden und um eine sehr geringe Anzahl von Viren nachweisen zu können (Chapron et al., 2000)
- Viele Wasserqualitätslabors besitzen weder die Infrastruktur noch das Know-how, um "ausgefeilte" Nachweismethoden durchführen zu können

Viele Autoren bevorzugen aus den obengenannten Gründen eine Kombination von Zellkulturen (falls vorhanden) mit molekularbiologischen Methoden. Ein Vergleich zwischen dem von der EPA empfohlenem System, dem Nachweis aller kultivierbaren Viren in einem MPN-Ansatz (most probable number) mit der BMCK-Zellinie, und derselben Zellkultur mit anschliessenden PCR-Schritten (Chapron et al., 2000), zeigte, dass das konventionelle Zellkulturverfahren die Anzahl Viren, vor allem die der Adenoviren und Enteroviren unterschätzt. Der kombinierte Ansatz ist sensitiver, schneller (Nachweis vor dem Auftreten der CPE), und billiger (auf eine zweite Passage durch Zellkulturen kann verzichtet werden) (Reynolds et al., 1997).

Tab. 3.2: Eine Auswahl molekularbiologischer Nachweismethoden für Viren

Viren	Nachweis- und Identifikationsmethoden	Detektionsgrenze
Enteroviren: Poliovirus	Filtrierung und Anreicherung viraler RNA mit magnetischen Kügelchen, kombiniert mit RT-PCR einer nichtcodierenden Sequenz in der 5'-Region (Regan und Margolin, 1997)	1 PFU/250 µl konzentrierter Wasserprobe
	Ultrafiltration und RT-PCR einer nichtcodierenden Sequenz in der 5'-Region (Soule et al., 2000)	In inokulierten Wasserproben: 10^{-3} $TCID_{50}L^{-1}$ im Vergleich zu 1 $TCID_{50}L^{-1}$ für Zellkulturen
Enteroviren	Zellkulturen kombiniert mit RT-PCR (integrierter Ansatz), bzw. direkte RT-PCR (Reynolds et al., 1997)	In Trinkwasser: 10^{-3} MPN/l im integrierten Ansatz
	Filtration mit anschliessender RT-PCR (Abbaszadegan et al., 1999) Zellkulturen kombiniert mit RT- und nested PCR (Chapron et al., 2000)	
Hepatitis A	Zellkulturen kombiniert mit RT-PCR (integrierter Ansatz) (Reynolds et al., 1997)	In 300-400l aufkonzentrierten Wasserproben: 0.04 PFU
	Filtration mit anschliessender RT-PCR (Abbaszadegan et al., 1999)	In inokulierten Wasserproben: <1 $TCID_{50}L^{-1}$
	Ultrafiltration und RT-PCR (Soule et al., 2000)	
Rotavirus	Filtration mit anschliessender RT-PCR (Abbaszadegan et al., 1999)	1 $TCID_{50}L^{-1}$
	Ultrafiltration und RT-PCR (Gratacap-Cavallier et al., 2000)	In inokulierten Wasserproben: 1 $TCID_{50}L^{-1}$
	Ultrafiltration und RT-PCR (Soule et al., 2000)	
Astroviren, Adenoviren Typus 40 und 41	Zellkulturen kombiniert mit RT- und nested PCR (Chapron et al., 2000)	
Norwalk-like Virus, Calicivirus	Filtration und RT-PCR (Lodder et al., 1999)	10-100 RNA-enthaltende Partikel in aufkonzentrierten Abwässern

Nachweis pathogener Protozoen

Giardia intestinalis (früher *Giardia lamblia*) und *Cryptosporidium parvum* sind die wichtigsten pathogenen Protozoen im Wasser. Sie können als Dauerformen, sogenannte Cysten (*Giardia*) und Oozysten (*Cryptosporodium*), in der Umwelt überleben. Deren Nachweis ist jedoch schwierig, da sie in geringerer Zellzahl als beispielweise Bakterien vorkommen. Aufgrund häufiger Epidemien ist für beide Organismen schon eine Vielzahl von Nachweismethoden entwickelt worden.

Cryptosporidium spp.

Da die Infektionsdosis von *Cryptosporidium parvum* sehr gering ist, sind für Wasseranalysen äusserst sensitive Nachweismethoden erforderlich. Trinkwasserbedingte Infektionen mit Cryptosporidium werden durch die hohe Resistenz der Oozysten gegenüber Desinfektionsmitteln in der Trinkwasseraufbereitung begünstigt. Der Nachweis erfolgt üblicherweise immunologisch mittels Fluoreszenzmikroskopie oder Durchflusszytometrie (Wallis et al., 1996; Schaefer, 1997). Der einfache immunologische Nachweis der Oozysten ohne zusätzliche Lebendfärbungen ist nicht aussagekräftig bezüglich Lebendigkeit (viability) und Infektivität. Kreuzreaktionen artspezifischer Antikörper und falsch positive Resultate auf Grund von Autofluoreszenz z.B. durch in den Proben vorhandene Algenzellen sind möglich.

Eine Reihe von PCR-Methoden sind für den Cryptosporidien-Nachweis entwickelt worden (Wiedenmann et al., 1998; Tabelle 3.3). Als kritisch für die PCR erwies sich die Beseitigung möglicher PCR-inhibierender Substanzen und die Anreicherung der Organismen. Sluter et al., 1997 konnten zeigen, dass beispielsweise in Seewasser lösliche PCR-Inhibitoren von weniger als 27 kDa Molekulargewicht enthalten sind. Die notwendige Aufreinigung der Oozysten erfolgt entweder durch Sucroseflotation oder Percoll-Dichtegradienten-Zentrifugation. Weitere Möglichkeiten bieten ebenfalls die fluoreszenzaktivierte Zelltrennung (fluorescence activated cell sorting, FACS) und die immunomagnetische Trennung (immunomagnetic separation, IMS) (Vesey et al., 1993; Deng et al., 1997). Da mit der PCR auch Nukleinsäuren von toten Organismen und freie DNA amplifiziert werden können, wird zum Nachweis lebender Organismen die PCR mit der *in vitro*- Excystierung kombiniert. Die *in vitro*-Excystierung, ein biologischer Prozess, bei dem infektiöse Sporozoiten von Oozysten freigesetzt werden, dient als Beweis für das Vorkommen lebender Oozysten. Die DNA lebendiger, excystierter Sporozoiten kann selektiv extrahiert und nachgewiesen werden, während die DNA toter Organismen in den Oozysten eingeschlossen und somit nicht-detektierbar bleibt (Filkorn et al., 1994; Wagner-Wiening und Kimmig, 1995).

Alternativ wird die RT-PCR zum Nachweis von mRNA als Kriterium für lebende Organismen angewandt. Dieses Verfahren geht davon aus, dass die Halbwertszeit der mRNA sehr kurz ist, und dass in toten Organismen keine mRNA nachgewiesen werden kann. Jedoch erfüllen nicht alle mRNA-Typen diese Voraussetzungen (Mahbubani et al., 1991; Knippers, 1995). Ausserdem kann der mRNA-Gehalt einer Zelle auf Grund des metabolischen Zustandes variieren. Geeignet erscheint die mRNA der Hitzeschock-Proteine (hsp), weil angenommen wird, dass diese mRNA-Moleküle sehr instabil und somit in toten Oozysten nicht vorhanden sind (Stinear et al., 1996). Jedoch sind erst wenige Arbeiten darüber erschienen. Zudem wird darauf hingewiesen, dass lebendige Oozysten nicht mit infektiösen Organismen gleichzusetzen sind (Rochelle et al., 1997a, b). Um dies zu klären, entwickelten Rochelle et al. (1997a, b) als zeitsparenden Ersatz zu Tierversuchen einen Zellkulturtest kombiniert mit einem Nachweis von mRNA der hsp70-Gene von *Cryptosporidium parvum*.

Sowohl die bereits angesprochene *in vitro*-Excystierung als auch Zellkulturen erfordern entsprechend ausgebildetes Personal und vorhandene Infrastrukturen in den Wasserämtern. Ungeklärt ist auch, wie auf einen negativen Infektivitäts-Zellkulturtest reagiert werden soll, wenn gleichzeitig lebende Organismen nachgewiesen werden können. Die Verlässlichkeit aller PCR-Protokolle hängt von der Qualität der Primersequenzen ab. Voraussetzung dafür ist unter anderem eine umfassende Datenbank, um neuentworfene Primersequenzen auf ihre Spezifität hin testen zu können. Im Vergleich zu den vorhandenen Bakteriendatenbanken ist der Datensatz für Protozoen noch relativ klein. Eine sorgfältige Überprüfung der Spezifität neuer Primer ist somit nicht möglich, es sei denn man bezieht Kulturen anderer Spezies mit ein. Da jedoch die wenigsten Autoren ihre PCR-Protokolle mit möglichst vielen Cryptosporidium-Arten testen, müssen unspezifische Kreuzreaktion bisher durch zusätzliche Tests ausgeschlossen werden, z.B. durch eine anschliessende Hybridisierung mit internen Sonden (Chrisp und LeGendre, 1994), oder eine zusätzliche zweite PCR mit den Produkten der ersten PCR.

Generell muss beanstandet werden, dass keine der PCR-Methoden Tests in verschiedenen Labors unterzogen wurden, um sie auf ihre Reproduzierbarkeit hin zu überprüfen.

Giardia spp.

Die gegenwärtig am häufigsten angewandten Detektionsmethoden basieren auf der mikroskopischen Analyse von aufkonzentrierten Wasserproben in Kombination mit Immunofluoreszenz-Techniken (APHA, 1992; Riggs et al., 1983) oder speziellen Färbungen (Jakubowski und Ericksen, 1979). Vergleichende Studien über die Effizienz der Anreicherungsschritte offenbarten jedoch einen hohen Verlust sowohl an *Giardia* Cysten wie auch *Cryptosporidium* Oozysten in Abhängigkeit von der gewählten Methode (LeChevallier et al., 1995; Nieminski et al., 1995). PCR-Detektionsmethoden (Tabelle 3) werden als eine kostengünstigere, sensitivere und zeitsparende Methode angesehen. Voraussetzung für eine routinemässige Anwendung der PCR anstelle immunologischer Methoden ist auch hier die Beseitigung der oben beschriebenen Probleme, wie z.B. die Entfernung von Substanzen, die die PCR beeinträchtigen oder die Verbesserung der Anreicherungsschritte. Alternativen bieten z.B. die Anreicherung der *Giardia*-Cysten mit Hilfe von antikörperbeschichteten Kügelchen und anschliessender PCR (Mahbubani et al., 1998). Protokolle zum Nachweis lebender *Giardia*-Cysten mittels RT-PCR werden ebenfalls vermehrt entwickelt und auf ihre Anwendbarkeit zur Kontrolle von Wasserproben untersucht (Abbaszadegan et al., 1997; Kaucner und Stinear, 1998).

Tab. 3.3: Eine Auswahl molekularbiologischer Nachweismethoden für Protozoen

Protozoen	Nachweis- und Identifikationsmethoden	Detektionsgrenze
Cryptosporidium parvum	Übersichtsartikel von Wiedenmann et al. (1998)	
	PCR- bzw. nested PCR-Nachweis eines Gens, das für ein Zellwandprotein kodiert mit anschliessender Southern-Hybridisierung (Mayer und Palmer, 1996)	Nested PCR mit inokulierten Wasserproben: 370 Oozysten/l
	Aufkonzentrierung mit einer Durchlaufzentrifuge mit anschliessendem seminested PCR-Nachweis (Leisinger 1997)	In inokulierten Wasserproben: 117 Oozysten/100 l
	RT-PCR-Nachweis von HSP70 mRNA mit anschliessender Southern-Hybridisierung (Stinear et al., 1996)	In inokulierten Konzentraten von Wasserproben: 1 Oocyste
	PCR- bzw. RT-PCR-Nachweis (nach Infektion von Zellkulturen) von HSP70 mRNA mit anschliessender Southern-Hybridisierung (Rochelle et al., 1997b)	1 Oocyste
	Immunomagnetische Anreicherung von Oozysten mit anschliessender Infektion von Zellkulturen und RT-PCR von HSP70 mRNA (di Giovanni et al., 1999)	
Giardia muris	PCR-Nachweis des Giardin-Gens nach direkter DNA-Extraktion oder nach immunomagnetischer Anreicherung der Cysten (Mahbubani et al., 1998)	3-30 Cysten/ml in inokulierten Konzentraten von Wasserproben; abhängig von Wassertrübung.
Giardia intestinalis	PCR-Nachweis des Giardin-Gens (Kaucner und Stinear, 1998)	2 Cysten/100 µl in inokulierten Konzentraten von Wasserproben
Microsporidien: *Entero-cytozoon bieneusi, Ence-phalitozoon intestinalis, Vittaforma cornae*	PCR-Nachweis und Sequenzierung von Teilen der 18S rDNA (Dowd et al., 1998)	Erstmaliger direkter Nachweis humanpathogener Microsporidien in Wasserproben

Weitere pathogene Protozoen

Neben *Giardia intestinalis* und *Cryptosporidium parvum* sind weitere Protozoen aus gesundheitlichen Gründen von Interesse, so z. B. Microsporidien, *Naegleria fowleri*, *Acanthamoeba* spp., oder *Entamoeba histolytica* (Marshall et al., 1997). Sieben Microsporidien-Arten, insbesondere *Encephalitozoon intestinalis* und *Enterocytozoon bieneusi*, sind human-pathogen und treten vor allem bei immunsuppressiven Personen auf (Dowd et al., 1998; Marshall et al., 1997). Eine Schwierigkeit für den Nachweis pathogener Microsporidien stellt deren geringe Grösse dar (Durchmesser < 2µm). Microsporidien-Sporen sind aus Protein- und Chitinschichten aufgebaut und könnten so möglicherweise den Transport durch Kläranlagen überleben (Weber und Bryan, 1994). Da sie sehr klein sind, können sie ähnlich wie Bakterien durch die Bodenmatrix ins Grundwasser gelangen. Die Verwendung eines Transmissions-Elektronenmikroskops ist für Wasseranalysen nicht praktikabel. Die immunologische Unterscheidung zwischen humanpathogenen und fisch- oder insektenpathogenen Microsporidien hat sich auf Grund von Kreuzreaktion der Antikörper ebenfalls als nicht möglich erwiesen (Enriquez et al., 1997). Da bisher keine in vitro-Kultivierung von *E. bieneusi* möglich ist, müssen molekulare Detektionsmethoden angewandt werden (Dowd et al., 1998). Monoklonale Antikörper für andere Microsporidien-Arten sind jedoch entwickelt worden, müssen aber noch auf ihre Anwendbarkeit zur Qualitätskontrolle von Wasser getestet werden (Aldras et al., 1994; Beckers et al., 1996).

Naegleria spp. werden vor allem durch Filtration oder Zentrifugation für eine anschliessende Kultivierung angereichert (Sykora et al., 1983; Kyle und Noblet, 1985). Beide Methoden variieren hinsichtlich ihrer Ausbeute sehr und erschweren somit eine verlässliche Quantifizierung insbesondere niedriger Konzentrationen von *N. fowleri* in Umweltproben (Pernin et al., 1998). Die Konzentrierung hängt offensichtlich auch vom Zellstadium der Protozoen ab (vegetative Zelle, Cysten und flagellate Formen) (Pernin et al., 1998).

Schlussfolgerungen und Ausblick

Es ist gezeigt worden, dass ein ganzes Arsenal von spezifischen Methoden zur Unterstützung und Absicherung des üblichen Nachweis von Indikatorkeimen zur Verfügung steht oder sich in der Entwicklung befindet. Zum einen erhöhen diese Methoden die Spezifität des Nachweises pathogener Organismen, zum anderen bieten sie auch zeitliche und ökonomische Vorteile im Vergleich zu vielen konventionellen Methoden. Ein nicht zu unterschätzender Nutzen ist, dass empfindliche molekularbiologische Untersuchungsverfahren sehr frühzeitig Kontaminationen im Trinkwasser aufzeigen könnten. Es ist allerdings ebenso offensichtlich, dass sich bisher kein alleinstehend anwendbares Verfahren als Alternative zur gängigen Praxis der Trinkwasseruntersuchung aufdrängt. Die beschriebenen Verfahren werden vielmehr vorerst nur punktuell zur Anwendung kommen, z.B. in der epidemiologischen Forschung und zur Absicherung von Testergebnissen. Die Implementierung molekularbiologischer Standardmethoden in der Trinkwasserqualitätskontrolle erfordert eine gründliche Erprobung in der Praxis.

Dennoch soll an dieser Stelle versucht werden, Idealanforderungen an eine Methode zum Nachweis aller pathogenen Trinkwasserorganismen zu formulieren, und über eine mögliche Entwicklung zu spekulieren. Eine geeignete Methode sollte billig und schnell sein und pathogene Organismen innerhalb grosser Konzentrationsbereiche einschliesslich kleinster Dosen sicher detektieren. Sie sollte alle Trinkwasserpathogene gleichermassen identifizieren, quantitativ sein, und sie sollte infektiöse von nichtinfektiösen Organismen oder Organismenbestandteilen unterscheiden. Die meisten dieser Anforderungen könnten in Zukunft theoretisch von sogenannten Biochips, Arrays genetischer Information, erfüllt werden, welche entweder in der Entwicklung oder für bestimmte Anwendungen bereits kommerziell erhältlich sind. Grundsätzlich bestehen solche Biochips aus fixierter Einzelstrang DNA, welche der Ziel-DNA komplementär ist. Über den Chip wird eine DNA-Lösung geleitet. Ist die Ziel-DNA darin enthalten, bindet diese spezifisch an ihr Gegenstück und ein an der fixierten DNA befindlicher Fluoreszenzlabel wird aktiviert und registriert, beispielsweise über einen optischen Chip. Die Belegung eines einzelnen Chips mit Sequenzen aller bekannten Trinkwasserpathogenen ist theoretisch möglich, und dürfte letztlich durch Pixelgrösse und Signalzuordnung limitiert sein. Quantitative Bestimmungen könnten durch Redundanz der vorgelegten DNA erreicht werden. Es ist ebenfalls denkbar, dass Kombinationen aus molekulargenetischen und immunologischen Tests auf Biochips plaziert werden. Somit könnte zum einen der Nachweis von Problemorganismen sicherer gemacht werden und zum anderen auch Pyrogene (fieberauslösende Biomoleküle) erfasst werden. Grosse Bedeutung wird Verfahren zur gleichermassen quantitativen Extraktion von DNA aus Bakterien, Viren, Protozoen, Wurmeiern etc. beikommen. Die Entwicklung solcher Verfahren in Form von fertig verwendbaren Kits wird sicher durch das homogene Medium Trinkwasser begünstigt werden.

Die Entwicklung von Biochips oder ähnlichen Detektionssystemen für die Trinkwasseranalyse erscheint aus mehreren Gründen möglich und vielversprechend. Zum einen ist die Genomforschung als Lieferant notwendiger Sequenzdaten wesentlich durch biomedizinische Fragestellungen und Geschäftsinteressen getrieben und es ist zu erwarten, dass die Total- und Teilsequenzierung medizinisch relevanter Organismen weiterhin vorrangig betrieben wird. Trinkwasserchips sind fernerhin potentielle Massenartikel deren Handhabung sich praktisch beliebig vereinfachen lässt. Die Autoren möchten noch einmal auf den spekulativen Charakter dieser Vision hinweisen, gleichzeitig aber ihre Überzeugung ausdrücken, dass der herkömmliche Nachweis von Indikatorkeimen nicht ein weiteres Jahrhundert konkurrenzlos bleiben wird.

Transport von Mikroorganismen 4

Peter Huggenberger

Einleitung

Das Auftreten von Mikroorganismen im Grund- und Trinkwasser wird von den Wasserversorgern vermehrt als Problem erkannt. Die Herkunft und der Transport der Mikroorganismen unter den lokalen Gegebenheiten ist in vielen Fällen unbekannt. Weltweit gibt es jedoch mehrere in Forschungsprojekten durchgeführte Feldversuche zum Transport von Mikroorganismen im Grundwasser. Ein zur Zeit in der Fachliteratur häufig zitierter Feldversuch wurde in Virginia (USA) durchgeführt (Oyster-site, Zhang et al., 2001). Aber auch in der Schweiz sind ähnliche Feldversuche durchgeführt worden: Wilerwald im Emmental (Niehren et al., 1998), Röschenz im Laufental (Auckenthaler et al., 2002; Raso, 2001), Wiese bei Basel (Huggenberger, 2001). Ziel dieser Versuche war das Studium des Transportverhaltens von Mikroorganismen in Abhängigkeit der hydrogeologischen und hydrogeochemischen Inhomogenitäten des Untergrundes. Dabei zeigte sich, dass beim Transport von Mikroorganismen im Grundwasser die Grenzen der Beschreibbarkeit komplexer Systeme und Prozesse erreicht werden. Die Praxis muss sich deshalb darauf konzentrieren, Grössenordnungen möglicher Gefährdungen zu ermitteln. Die Resultate vieler Versuche zeigen, dass kleine Partikel im Feldversuch im allgemeinen weiter transportiert werden, als aufgrund von Laborexperimenten bisher angenommen wurde (Tabelle 4.1)

Mikrobiologische Verunreinigungen im Trinkwasser werden mit den gängigen Überwachungsprogrammen, die auf einem bestimmten Beprobungsrhythmus basieren, meist nur durch Zufall entdeckt. Das bedeutet, dass das Risiko von Verunreinigungen durch Mikroorganismen in Grundwasserfassungen, besonders in der Nähe von Flussläufen, überprüft werden sollte, wenn der Untergrund aus gut durchlässigen Flusskiesen aufgebaut ist, oder wenn die Standortverhältnisse es nicht erlauben, die engere Grundwasserschutzzone S2, welche verhindern soll, dass Keime und Viren in die Fassung gelangen, vorschriftsgemäss auszuscheiden (z.B. Karst).

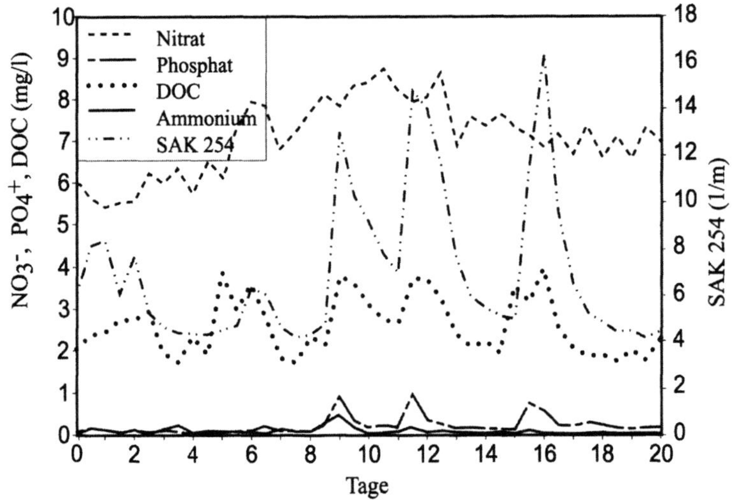

Abb. 4.1: Variable Qualität eines Oberflächengewässers (SAK 254 = UV Extinktion bei 254 nm; 1/m). Systematische Beprobungen von Oberflächengewässern bei Hochwasserereignissen zeigen, dass die Belastung nicht kontinuierlich verläuft, sondern kurzfristig, z.B. bei bestimmten Ereignissen auftritt (Wüthrich et al. 2001).

Im Abschnitt „Prozesse des Transports von Mikroorganismen" wird ein Überblick gegeben über die wesentlichen Vorgänge, vom Eintrag von Mikroorganismen via Oberfläche oder Oberflächengewässer ins Grundwasser bis zu ihrem potentiellen Auftreten in einer Grundwasserfassung.

Im Unterschied zu Karst- und heterogenen Kluftsystemen, wo sich die Filtrationsprozesse bezüglich Mikroorganismen meist auf Bodenpassage und Verwitterungsschichten beschränken, spielen in Lockergesteinsgrundwasserleitern die Interaktion mit den Kornoberflächen und die Geometrie der Poren (Lawrence und Hendry, 1996) eine Rolle. Die Interaktion zwischen Mikroorganismen und Mineraloberflächen wird im allgemeinen durch die Kolloid-Filtrationstheorie (Hunter, 1981; Everett, 1992; Eimelech et al., 1995) und durch die DLVO-Theorie (Derjaguin und Landau, 1941; Verwey und Overbeck, 1948) beschrieben. Diese Herleitungen betreffen Prozesse im kleinen Massstabsbereich. Aussagen im makroskopischen oder grossmassstäblichen Bereich sind sehr beschränkt aufgrund von Daten oder Prozesskenntnissen im mikroskopischen Bereich herzuleiten (Abb. 4.2). Obschon theoretische Ansätze zum „Upscaling" untersucht wurden (Renard und de Marsily, 1997; Gelhar 1993), bestehen noch keine brauchbaren Modelle zur Erfassung von Prozessen beim Transport von Mikroorganismen im grossen Massstabsbereich (Rehmann et al., 1999).

Untersuchungen zum Schutz von Trinkwasserfassungen im näheren Einflussbereich von Fliessgewässern („Riparian Zone" nach Stanford, 1998; Pusch et al., 1998; Brunke, 1998) machten deutlich, dass zwischen Oberflächengewässern und Grundwasser zeitlich und räumlich komplexe biogeochemische Prozesse ablaufen. Solche Prozesse wurden an naturbelassenen Flusssystemen studiert (Pusch et al., 1998; Gonser, 1994)

Tab. 4.1: In Feldexperimenten beobachtete Transportdistanzen von Mikroorganismen (modifiziert nach Rehmann et al., 1999)

Kolloid	Max. Transport-distanz [m]	Ort	Aquifer-Material, Mächtig-keit	K-Wert, [m d^{-1}]	v, [m d^{-1}]	v_{col} [m d^{-1}]	Referenz
Phage T4	920	Neuseeland	Kies				*Noonan und McNabb, 1979*
Bacillus sterothermophilus	900	Neuseeland	Kies	* 10^4	164	200m/d	*Martin und Noonan, 1977*
E. coli	350-830	Kasachstan	Sand mit Kies, Gerölle, 4-8 m	* 10^5	160		*Anan'ev und Demin, 1971*
Coxsackie B3 und unidentifizierte Viren	408	Babylon, N.Y.	Grobsand mit Feinkies				*Vaughn und Landry, 1977*
Phage	400	Lake George, N.Y.	Feinsand mit etwas Kies, Grobsand	4.6-19.5	3-12		*Aulenbach, 1979*
Poliovirus, Coxsackie B3 und Echovirus	250	Vineland, N.J.	Sand mit Grobkies				*Koerner und Haws, 1979*
Coliphage f2, Enteroviren, fäkale Streptokokken	183	Fort Devens, Mass.	Siltiger Sand und Kies	8,6			*Schaub und Sorber, 1977*
Echovirus 6, 21, 24 und 25 und unidentifizierte Viren	45,7	Holbrook, N.Y.	Grobsand mit Feinkies, 1-2% Silt				*Vaughn und Landry, 1977*

(*Geschätzte Grössenordnung der Grundwassergeschwindigkeit und den Werten des hydraulischen Gradienten. v: mittlere Grundwassergeschwindigkeit, v_{col}: Kolloidgeschwindigkeit, K-Wert: Angabe der Grössenordnung)

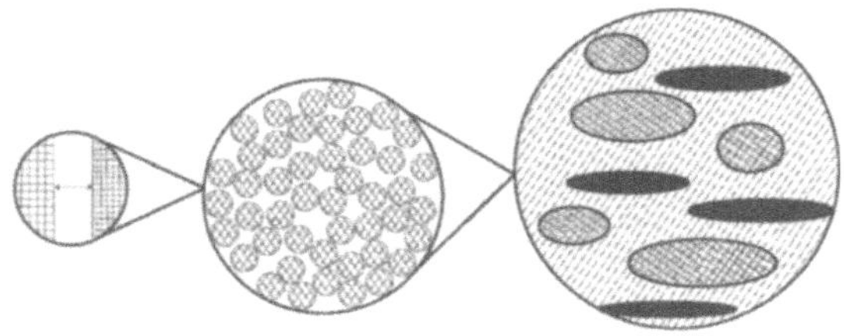

Abb. 4.2: Für unterschiedliche Transportprozesse massgebende Skalenbereiche (Niehren, 1998)

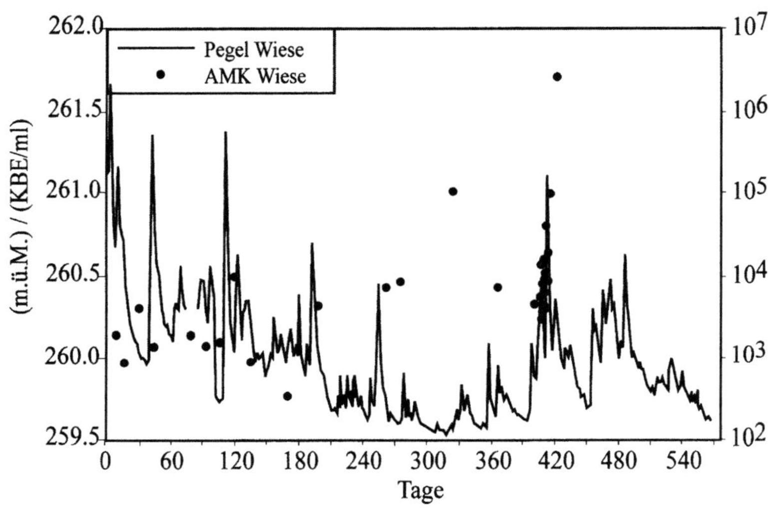

Abb. 4.3: Belastung des Flusswassers der Wiese bei Basel mit aeroben mesophilen Keimen und *E. coli* (Huggenberger, 2001)

Im Zusammenhang mit der Einleitung von Kläranlagenabwässern, der Abschwemmung von Hofdüngern aus der Landwirtschaft und den Regenwasserentlastungen bei Starkniederschlägen gelangen zeitweilig grosse Mengen an Mikroorganismen in die Gewässer (Abb. 4.3). Die Konzentration von Mikroorganismen ist im allgemeinen in intensiv genutzten Gebieten wesentlich höher als in Gebieten mit geringer landwirtschaftlicher Aktivität oder geringer industriell-gewerblicher Nutzung. Bei infiltrierenden Flusssystemen kann ein Teil der Mikroorganismen in den darunterliegenden Grundwasserleiter gelangen. In revitalisierten Flüssen bzw. Flussabschnitten, charakterisiert durch einen intensiveren Austausch von Wasser und Partikeln zwischen Fluss- und Grundwasser, können Mikroorganismen unter gewissen Bedingungen noch einfacher in den Untergrund und damit ins Grundwasser gelangen (Huggenberger, 2001). Durch die Erhöhung der Anzahl von Mikroorganismen im Grundwasser erhöht sich, in Abhängigkeit der Durchlässigkeit des

Grundwasserleiters, die Wahrscheinlichkeit, dass Mikroorganismen in eine nahe gelegene Wasserfassung gelangen.

Prozesse des Transports von Mikroorganismen

Der Schwerpunkt des Abschnittes „Prozesse des Transports von Mikroorganismen" liegt nicht in der Zusammenfassung bzw. Herleitung der wichtigsten in der Literatur beschriebenen Ansätze des Partikel- oder Mikroorganismen-Transports im mikroskopischen Bereich, sondern in einer semiquantitativen Beschreibung von Phänomenen und Prozessen im Feldmassstab. Die wichtigsten Gleichungen werden in einem separaten Kasten angegeben. Der Feldmassstabbereich ist für viele Wasserversorger von zentraler Bedeutung. In den folgenden Abschnitten wird meist von Mikroorganismen gesprochen, solange die Prozesse gleichermassen für Viren, Bakterien und Protozoen gelten.

Wegen den Unterschieden des Transports von Mikroorganismen in Karst- und heterogenen Kluftgrundwasserleitern sowie Lockergesteins-Grundwasserleitern werden die Transportprozesse in zwei separaten Abschnitten behandelt.

Transport von Mikroorganismen in Karst- und heterogenen Kluftsystemen

Der Transport von Mikroorganismen in Karstgebieten umfasst eine breite Palette von unterschiedlichen Teilprozessen mit unterschiedlichen zeitlichen Dimensionen:

- Transport durch den Boden (im pedologischen Sinn)
- Transport in der oberflächennahen Verwitterungszone
- Eintritt und Transport in Schluckstellen
- Transport im eigentlichen Karströhrennetz
- Austauschprozesse zwischen den obigen Elementen und dem Porenraum der durchflossenen geologischen Formationen

Beim Eintritt von Wasser in Schluckstellen und bei der Passage durch Karströhrennetze oder grosse Klüfte ist der Transport von Mikroorganismen besonders schnell.

Im regionalen Massstab beeinflussen die Topographie, die Lithologie und die Tektonik (Lage und Ausbildung von Störungszonen) das Fliessverhalten des Sickerwassers und des Grundwassers. Der Einfluss der Topographie auf die regionale Grundwasserzirkulation wurde von J. Tóth 1963 beschrieben (Abb. 4.4a-c). Er unterscheidet lokale, intermediäre und regionale Fliesssysteme. Ein lokales System hat sein Anreicherungsgebiet auf dem nächstgelegenen topographischen Hoch und den Ausfluss in der nächstgelegenen Depression. In intermediären und regionalen Systemen liegen Anreicherungs- und Ausflussgebiete in grösserer Entfernung voneinander. Bisher konnten in allen drei Systemen Mikroorganismen nachgewiesen werden (Chappelle, 1992).

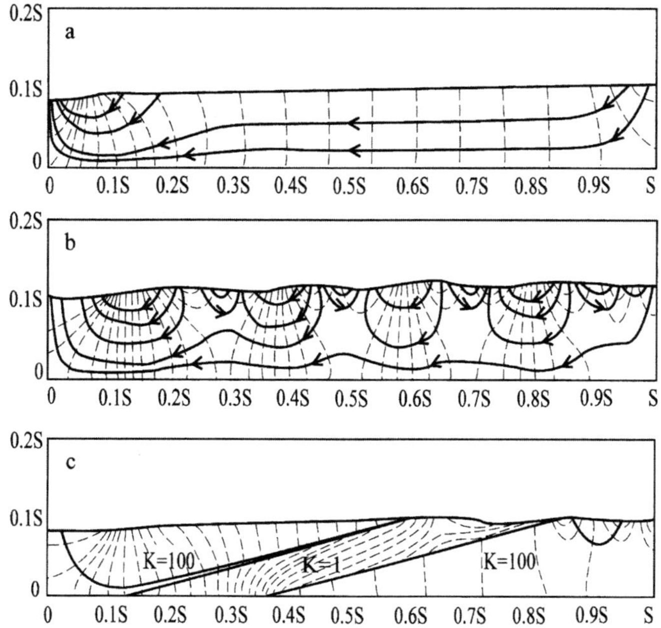

Abb. 4.4: Einfluss der Topographie auf die regionale Grundwasserzirkulation. a) Tafelland mit randlichem Tal (links) führt zu einem ausgedehnten Grundwasseranreicherungsgebiet auf dem Plateau. Das Exfiltrationsgebiet liegt im randlichen Tal; b) Eine hügelige Topographie führt zu einer Serie von lokalen, oberflächennahen bis regionalen, tieferliegenden Grundwasserzirkulations-systemen; c) Einfluss unterschiedlicher Durchlässigkeiten von geologischen Formationen auf die regionale Grundwasserzirkulation (nach Freeze und Witherspoon, Water Res. Res., V.3, 623-634, 1967. Reproduziert mit Genehmigung der American Geophysical Union)

Die räumliche Verteilung von durchlässigen und weniger durchlässigen Formationen beeinflusst zusätzlich die Verteilung der hydraulischen Potentiale (Freeze und Witherspoon, 1967). Diese beiden Phänomene zusammen ergeben komplizierte Grundwasser-Zirkulations-systeme auf engstem Raum (Abb. 4.4c).

Die typischen Hydrogramme von Karstquellen liefern wichtige Hinweise über den Charakter des Einzugsgebietes (Herold, 1997). Charakteristisch ist die rasche Zunahme der Quell-schüttung bei Niederschlagsereignissen, respektive die anfänglich rasche und danach verlangsamte Abnahme der Schüttung beim Übergang zu Trockenzeiten. Die Spitzen-abflüsse sind Indikatoren für die schnellen Wasserwege (Jeannin, 1998).

Experimente bei einer Karstquelle in der Nordwestschweiz (Auckenthaler et al., 2002) zeigen, dass das erste Auftreten von Mikroorganismen häufig schneller erfolgt, als der Anstieg der Trübung. Noch vor dem Auftreten der Mikroorganismen nimmt jedoch die Quellschüttung zu. Entsprechend ist die Zunahme der Quellschüttung für mikrobiologisch besonders empfindliche Quellen ein Schlüsselindikator für ein auf die Mikrobiologie ausgerichtetes Quellwassermanagement.

Transportprozesse in Karst- und heterogenen Kluftsystemen verlaufen nicht wesentlich anders als diejenigen in porösen Medien. Hingegen ist der Grad an Inhomogenität in Karstsystemen, ausgedrückt durch die Varianz der hydraulischen Eigenschaften, im Vergleich zu porösen Medien im allgemeinen derart hoch, dass die Beschreibung eines mittleren Verhaltens oder ein „Upscaling" nur beschränkt möglich ist. Erfahrungen von Markierversuchen mit Partikeln (Auckenthaler et al., 2002) und Bakteriophagen (Rossi et al., 1994, 1998) zeigen, dass das Verhalten von natürlichen Mikroorganismen mit diesen künstlichen Partikeln gut simuliert werden kann.

Im Karst des Schweizer Juras infiltrieren zwischen 50 und 75% des wirksamen Anteils der Niederschläge konzentriert in das Karstsystem und die restlichen 25-50% versickern in das mehr oder weniger klüftige oder poröse Karbonatgestein (Doerfliger und Zwahlen, 1998; Doerfliger, 1996).

Transport von Mikroorganismen in Lockergesteins-Grundwasserleitern

Die DLVO- oder Zweischicht-Theorie wurde entwickelt, um die Anziehungs- und Abstossungskräfte zwischen Partikeln und Oberflächen zu beschreiben. Die Summe dieser Kräfte bestimmt, ob Partikel an Oberflächen haften (Van Loosdrecht et al., 1987, 1989). Daneben spielt der Radius eines sich der Oberfläche nähernden Partikels eine Rolle. Die Sorption von Partikeln auf Oberflächen wird mit der Theorie der Freien Energie erklärt. In ihrer einfachsten Form besagt diese, dass, falls die freie Energie eines Systems beim Kontakt einer Zelle mit einer Oberfläche reduziert wird, das Partikel an der Oberfläche adsorbiert (Fletcher und Loeb, 1979). Die Effizienz der Filtration von Partikeln ist zudem von der Zusammensetzung des geologischen Mediums abhängig (Atteia und Kozel, 1997). Im Allgemeinen nimmt die Filtrations-Effizienz zu, wenn die mittlere Zellgrösse der Mikroorganismen grösser ist als 5% der Komponenten des geologischen Mediums (Jang et al., 1983). Die Porendurchmesser in Sand, Silt und Tongemischen liegen etwa in der Grössenordnung von Bakterien, entsprechend ist der Transport von Bakterien in solchen Gemischen stark limitiert. In den Schottervorkommen hingegen sind Makroporen und stark durchlässige Zonen weit verbreitet und Bakterientransport in Abhängigkeit der Heterogenität deshalb viel einfacher möglich (Tab. 4.1, Rehmann et al., 1999).

Oberflächenladungen und der Grad an Hydrophobizität (Scholl et al., 1990) stellen die wichtigsten Eigenschaften von Sedimentoberflächen dar, die das Sorptionsverhalten von Mikroorganismen beeinflussen. Generell kann beobachtet werden, dass der Sorptionsgrad bei positiv geladenen Oberflächen wie Kalzit oder mit Fe-Hydroxiden überzogenen Quarzoberflächen höher ist, als bei negativ geladenen Quarz- oder Glimmer-Oberflächen. Zudem kann das Vorhandensein von organischem Material die Sorption von Bakterien dadurch beeinflussen, dass es die Oberfläche vergrössert und damit mehr Plätze für das Anhaften von Mikroorganismen schafft (Harvey et al., 1987).

Das hydrodynamische Fliessregime hat ebenfalls einen Einfluss auf die Sorption von Mikroorganismen. In Kolonnenexperimenten konnte gezeigt werden, dass mit zunehmender Strömungsgeschwindigkeit ein stärkerer Durchbruch von Mikroorganismen erfolgte (Tan et al., 1994). Im allgemeinen reduzieren grosse Fliessgeschwindigkeiten die Kontaktzeit und reduzieren dadurch die Wahrscheinlichkeit der Adsorption von Mikroorganismen (Yates und Yates, 1988).

Eine Erhöhung der Retention von Mikroorganismen mit zunehmender Ionenstärke konnte bei Transportexperimenten von Fontes et al., 1991 und Sharma et al., 1985 beobachtet werden. Diese Effekte sind konsistent mit der Doppelschicht-Theorie (Van Loosdrecht et al., 1987), die eine stärkere Sorption bei zunehmender Ionenstärke voraussagt.

Der Einfluss der Ionenstärke des Wassers auf die Loslösungseffizienz von Mikroorganismen ist im Feld noch weitgehend unerforscht. Nach Niehren, 1999 können Kolloide in hohen Konzentrationen auftreten, beispielsweise infolge von Wassernutzungen (z.B. Pumpbetrieb), Druck- und Temperatur-Schwankungen oder Einleitung von saurem Wasser (z.B. Infiltration von saurem Oberflächenwasser). Die Kolloidkonzentrationen korrelieren mit dem Kalzium- und Natriumgehalt des Wassers, wobei die Kolloidkonzentration sensitiver auf die Kalziumkonzentration reagiert. Vereinfachend gilt: je geringer die Ionenstärke, desto höher ist der Kolloidgehalt des Grundwassers (Niehren, 1999).

Eine hydrogeologische Besonderheit im Zusammenhang mit Mikroorganismen stellt der ungesättigte Untergrund dar. Dieser enthält in den Porenräumen Gas- und Wasseranteile. Der Porenwasserdruck ist kleiner als der Atmosphärendruck. In dieser Zone herrscht ein freier Austausch zwischen Gasen (O_2, CO_2) und dem vorhandenen Porenwasser. Es besteht auch ein freier Austausch mit der Geländeoberfläche. Im ungesättigten Untergrund findet eine Vielfalt von für die Mikrobiologie relevanten Aktivitäten statt: Sauerstoffzehrung, CO_2-Produktion, Nitrifikation, Denitrifikation etc. Diese Prozesse beeinflussen die chemische Zusammensetzung der infiltrierenden Porenwässer. Neben der Filtration stellt die partielle Wassersättigung im ungesättigten Untergrund ein Hindernis für den Transport von Mikroorganismen dar. Entsprechend kann vielerorts ein markanter Konzentrationsabfall der Mikroorganismen bei der Passage im ungesättigten Untergrund beobachtet werden (Chapelle, 1992). Die meisten Ansätze zur Modellierung des Transports von Mikroorganismen wurzeln in der bekannten Advektions-Dispersions-Gleichung (Corapcioglu et al., 1984; Tan et al., 1992). Diese wurde durch zusätzliche Terme bzw. Prozesse für Wachstum, Absterben, Frass, Zell-Dichte und Sorptionsverhalten erweitert (Hornberger et al., 1992).

Korngrössenverteilungen und Ionenstärken wurden als wichtige Grössen betrachtet, die den Transport von Mikroorganismen in porösen Medien wesentlich beeinflussen. Sukzessiv wurde auch die Filtrationstheorie mit der Advektions-Dispersions-Gleichung verknüpft (z.B. Rajagopalan und Tien, 1976). Erst in den letzten Jahren wurde jedoch erkannt, dass die Inhomogenität des Untergrundes, ausgedrückt durch die Varianz des natürlichen Logarithmus der Durchlässigkeitswerte und die Korrelationslänge (σ_{lnK}, λ), beim Transport von Mikroorganismen ebenfalls eine wesentliche Rolle spielt (Rehmann et al., 1999).

Die relevanten Phänomene, die den Transport von Mikroorganismen in porösen Medien bestimmen, sind die Geschwindigkeit des Porenwassers sowie die Haftung und Lösung von Mikroorganismen an Mineraloberflächen. Haftung- und Lösungsraten von Mikroorganismen an Oberflächen werden gewöhnlich als kinetische Prozesse erster Ordnung modelliert, d.h. die Haftungsrate kann als Produkt von zwei Wahrscheinlichkeiten betrachtet werden: die Wahrscheinlichkeit, dass Mikroorganismen mit Sedimentkörnern kollidieren (collector efficiency), wenn sie in die Nähe von Kornoberflächen geraten und die Wahrscheinlichkeit, dass sie bei dieser Kollision haften bleiben (collision efficiency). Sowohl Porenwassergeschwindigkeit und Kollektoreffizienz sind sehr stark von der Porosität und der Korngrösse abhängig. Für nicht-konsolidierte Lockergesteine, wie fluviale Schotter, die einen grossen Teil der Grundwasserleiter in den Flusstälern der Schweiz bilden, korrelieren diese Parameter mit der hydraulischen Leitfähigkeit (Rehmann et al., 1999). Die nicht-physikalischen Parameter, die den Mikroorganismen-Transport beeinflussen, werden in den Kollisionseffizienz-Term gepackt. Dort spielen die Zellen- und Korneigenschaften eine Rolle, z.B. die Summe der elektrostatischen Kräfte und die Van der Waal's-Kräfte zwischen einem Mikroorganismus, der sich einer Mineraloberfläche nähert und dem Sedimentkorn (Simoni, 1999).

Mikroorganismen zeigen typischerweise negative Oberflächenladungen. D.h. in einem Quarz-Dolomit dominierten System ergibt sich dadurch eine elektrostatische Abstossung zwischen Mikroorganismus und Sedimentoberfläche. Es existieren jedoch mineralische Phasen wie z.B. die Eisenoxide, welche positiv geladene Oberflächen besitzen, die Bakterien elektrostatisch anziehen können und deshalb höhere Kollisionseffizienzen aufweisen. Gerade solche eisen- und manganreichen oxidischen Zonen sind charakteristisch für das hyporheische Interstitial (beispielsweise zeigen die Flüsse Töss und Wiese und Grundwasseranreicherungsgebiete in der Region Basel Fe- und Mn-Hydroxidausfällungen unterhalb der Sohle bzw. im Überflutungsbereich). Die Bedeutung dieser Zonen, deren räumliche Ausdehnung und die dort wirkenden Prozesse sind im Feld noch weitgehend unerforscht. Man weiss aber, dass in diesen Zonen intensive biogeochemische Reaktionen stattfinden. Bekannt sind die Umwandlung von Ammonium zu Nitrit und Nitrat (Brunke, 1998). In anaeroben Zonen kann Nitrat zu Stickstoff denitrifiziert oder zu Ammonium reduziert werden (Pusch et al., 1998). Zudem ist weiss man, dass dort auch ein Abbau der für Mikroorganismen leichtverfügbaren DOC-Fraktionen stattfindet. Weitere beschriebene Phänomene sind die Lösung und Fällung von Mn und Fe sowie die Adsorption von schwerlöslichen Uranhydroxiden und Arsenverbindungen an Hydroxidoberflächen. Zudem kann Nitrat das Algenwachstum fördern und damit für eine gewisse Zeit zur Bedeckung von positiv geladenen Mineraloberflächen führen. Reaktionsraten von Haftungs- und Lösungsprozessen von Mikroorganismen an Oberflächen reagieren empfindlich auf hydrogeologische und geochemische Änderungen im System (Barber et al., 1992).

Generell werden Lockergesteinsgrundwasserleiter durch versickernde Niederschläge und Infiltration von Oberflächengewässern gespiesen. Insbesondere die Flusswasserinfiltration spielt bei einer Reihe von Flüssen eine dominante Rolle (Aare, Rhein, Töss, Wiese, etc.). Das Eindringen von Mikroorganismen in einen Grundwasserleiter ist abhängig von der Konzentration der Mikroorganismen im Oberflächengewässer, der Durchlässigkeit der Flusssohle sowie der Beschaffenheit und den hydraulischen Eigenschaften des Grundwasserleiters. Generell weisen verbaute Flüsse über weite Strecken eine kolmatierte Flusssohle auf. Nur bei ausserordentlichen Hochwasserereignissen kann die kritische Sohlenschubspannung ausreichen, um für eine gewisse Zeit die Flusssohle aufzureissen und entsprechend ihre Durchlässigkeit zu erhöhen. Diese Prozesse verlaufen in natürlichen Flusssystemen grundsätzlich anders. Die Dynamik dieser Flusssysteme bewirkt, dass der eigentliche Flusslauf zusammen mit der biologisch aktiven, hyporheischen Zone räumlich-zeitlich variiert. Um herauszufinden, unter welchen Bedingungen wie viele der im Oberflächenwasser vorhandenen Mikroorganismen die hyporheische Zone passieren können, müssen neben der Inputfunktion auch die Durchlässigkeit und der Grad an Heterogenität des Grundwasserleiters bekannt sein. Das in Abschnitt „Modell des Transports von Mikroorganismen zwischen Oberflächengewässern und Grundwasserfassungen“ vorgeschlagene makroskopische Modell wird in der Folge am Beispiel von Feldversuchen am Fluss Wiese (Zufluss des Rheins aus dem Schwarzwald) veranschaulicht. Die an diesem System beobachteten Vorgänge haben gewissermassen Modellcharakter. Vergleichbare Phänomene des Austausches Fluss-Grundwasser werden von verschiedenen Autoren an der Töss, am Brenno, im Rosegtal und an der Aare beschrieben.

Transportgleichungen für Mikroorganismen in Porengrundwasserleitern (Rehmann et al., 1999. WRR., 35/7 1987-2006)

1. Transportgleichung

$$\frac{\partial C}{\partial t} = \frac{\partial}{\partial x_i}(v_i C) + \frac{\partial}{\partial x_i}(D_{ij}\frac{\partial C}{\partial x_i}) - k_{dc}C - k_c C - \left[\frac{3(1-n)}{2d}\alpha_c \eta\right] v_i C + S\frac{1}{\rho}\frac{\rho_n}{n}k_y$$

Advektion, Dispersion, Inaktivierung, Filtration Loslösung
und Adsorption

C: Fraktion frei bewegl. Mikroorganismen

v_i: *Porenwassergeschwindigkeit, m/d*

D_{ij}: *lokaler Dispersionstensor,* $m^2 d^{-1}$

k_{dc}: *Inaktivierungsrate der freien Mikroorganismen,* d^{-1}

k_c: *Mikroorganismen – Adsorptionskoeffizient,* d^{-1}

n: effektive Porosität

α_c: *Kollisionseffizienzfaktor*

η: *Kollektoreffizienz,* d^{-1}

d: effektiver Korndurchmesser (d_{10}), *m*

S: Anteil an haftenden Mikroorganismen

ρ_n: *mittlere Dichte des porösen Mediums,* kgm^{-3}

ρ: *Dichte der Lösung (Wasser und Inhaltsstoffe),* kgm^{-3}

k_y: *Mikroorganismenloslösungsrate,* d^{-1}

2. Gesetz der Massenerhaltung für an Oberflächen adsorbierte Mikroorganismen

$$\frac{\rho_b}{n}\frac{\partial S}{\partial t} = -k_{ds}\frac{\rho_b}{n}S + k_c \rho C + \left[\frac{3}{2}\frac{(1-n)}{d}\alpha_C \eta\right] v_i \rho C - S\frac{\rho_b}{n}k_y$$

k_y: *Mikroorganismenlösungsrate,* d^{-1}

α_C: *Kollisionseffizienzfaktor*

k_{ds}: *Inaktivierungsrate für haftende Mikroorganismen,* d^{-1}

Vereinfachend wird angenommen, dass die Mikroorganismen sich wie gelöste Stoffe verhalten (Rehmann et al., 1999; Bear, 1972). Damit kann der Koeffizient der hydrodynamischen Dispersion D_{ij} verwendet werden.

Die Filtration ist von mehreren Grössen abhängig. Die wichtigsten Parameter sind:

der Kollisionseffizienzfaktor α_c, der angibt, welcher Anteil an Mikroorganismen an Sedimentoberflächen haften, nachdem sie mit diesen Oberflächen in Kontakt gekommen sind und die einfache Kollektoreffizienz η, die angibt welche Fraktion von Partikeln in der Lösung mit Oberflächen in Kontakt gekommen ist. In der eigentlichen Herleitung spielen Diffusionsmechanismen, elektrostatische Anziehungskräfte zwischen Partikeln, van der Waal's- und Gravitationskräfte eine Rolle (Rajagopalan und Tien, 1976). Nach Rehmannn et al., 1999 ist die Variabilität der hydraulischen Leitfähigkeit die unabhängige Eingabegrösse in einem solchen komplexen stochastischen System. Aufgrund der Form der wichtigsten Gleichungen wird üblicherweise der natürliche Logarithmus der hydraulischen Leitfähigkeit (ln *K*) verwendet. Durch die Anwendung und Kombination verschiedener publizierter Ansätze (z.B. Gorelick et al., 1993; Kauffmann, 1996) werden als Vereinfachung des Modells von Rajagopalan und Tien, 1976, die Faktoren Porosität und effektive Korngrösse mit ln *K* korreliert. Durch diese Substitutionen wird die einfache Kollektoreffizienz eine einfache Funktion der hydraulischen Leitfähigkeit und der mittleren Geschwindigkeit des Porenwassers.

In ähnlicher Weise wird angenommen, dass Mikroorganismenhaftungs- bzw. Loslösungsparameter mit den verschiedenen Aquifermaterialien über den ln *K* korrelieren. Darin nicht berücksichtigt sind die Einflüsse der jeweiligen Lösung (pH, Ionenstärke und Eigenschaften der Mikroorganismen, z.B. Oberflächeneigenschaften). Laborexperimente (Rehmann et al., 1997) zeigen, dass eine solche Korrelation nicht vollkommen ist und mindestens eine Komponente der Variabilität nicht korreliert.

Nach Garabedian et al., 1988 kann jedoch angenommen werden, dass

$$k_c = a_1 + b_1 \ln K + \delta_1$$

wobei a_1 und b_1 Konstanten sind und δ_1 eine Zufallskomponente des Adsorptionsterms darstellt, die im ln *K* Feld nicht berücksichtigt werden konnte. Mit diesem Ansatz ist es möglich, dass der Adsorptionskoeffizient k_c für $b_1>0$ positiv mit ln *K* korreliert. Negative Korrelation würde bedeuten $b_1<0$ und $b1=0$ bedeutet keine Korrelation.

Ähnliche Ausdrücke können auch für die Loslösungs- und Kollisionseffizienz Parameter hergeleitet werden (Rehmann et al., 1999).

$$k_y = a_2 + b_2 \ln K + \delta_2$$

$$\alpha_c = a_3 + b_3 \ln K + \delta_3$$

Modell des Transports von Mikroorganismen zwischen Oberflächengewässern und Grundwasser

Das vorliegende Modell des Transports von Mikroorganismen beruht auf Feldmessungen und Modellrechnungen, die im Grundwasseranreicherungsgebiet der Wasserversorgung Basel durchgeführt wurden. Ziel dieser Felduntersuchungen war es, Erfahrungen über den Transport von Mikroorganismen im Zusammenhang mit Revitalisierungsmassnahmen im Zustrombereich von Trinkwasserfassungen zu sammeln. Die Wiese ist ein kanalisierter Fluss, wie er in der Schweiz sehr häufig vorkommt. Im Einflussbereich einer nicht mehr benötigten flussnahen Grundwasserfassung wurde in einem 600 m langen Abschnitt der Flusslauf neu gestaltet. Damit kann der Transport von Mikroorganismen zwischen Fluss-Grundwasserleiter bei unterschiedlichen Abflüssen und Belastungsmustern studiert werden. Das Grundwasserüberwachungsprogramm umfasst eine Kombination von Feldexperimenten und nummerischen Modellierungen der Grundwasserströmung inkl. Eichung der Modelle an Messdatensätzen. Vor der Experimentierphase wurde im Zustrombereich eines Grundwasserbrunnens ein erster Markierversuch sowie ultrahochauflösende Georadar- und Geoelektrik-Messungen durchgeführt. Basierend auf diesen Resultaten konnten die notwendigen Grundwasserbeobachtungsrohre lokalisiert und installiert werden. Die Anordnung der Piezometer erfolgte in mehreren Klustern, die eine gezielte Beprobung (inkl. Grundwasserfauna) unterschiedlicher Grundwasserstockwerke erlaubten. Speziell interessierte der Transport von Mikroorganismen bei mittleren und grösseren Hochwässern. Einzelne dieser Experimente wurden gezielt von Markierversuchen und Radonmessungen begleitet. Die Inhomogenitäten des Grundwasserleiters wurden durch die Kombination von Bohrgutanalysen und Georadarmessungen ermittelt (Regli et al., 2002). Die Prozesse des Modells sind im Detail ziemlich kompliziert und werden in diesem Abschnitt vereinfachend diskutiert.

Der Transport von Mikroorganismen von einem Fliessgewässer zu einer Wasserfassung wird im wesentlichen von folgenden Prozessen beeinflusst.

- Konzentration der Mikroorganismen im Fliessgewässer
- Filterwirkung der Sohle des Fliessgewässers (K-Wert)
- Hydraulische Randbedingungen (Pegel des Fliessgewässers, Grundwasserentnahmen, etc.)
- Filterwirkung der Grundwasserleiter
- Einfluss der Grundwassereigenschaften

Konzentration der Mikroorganismen im Flusswasser

Die Daten der Grundwasserüberwachung des Flusses Wiese zeigen deutlich, dass Spitzenkonzentrationen von Mikroorganismen vorwiegend bei grossen Hochwässern auftreten. Die Spitzenkonzentrationen betragen dann bis zu 265‘000 (KBE/ml) für aerobe mesophile Keime, 10'600 (KBE/100ml) für *E. coli* und 2’600 (KBE/100ml) für Enterokokken (Abb. 4.3). Diese hohen Konzentrationen werden im Normalfall bei den wöchentlichen bis zweiwöchentlichen Beprobungsrhythmen nicht erfasst. Wegen der riesigen Unterschiede der Konzentration an Mikroorganismen bei Normal-, Mittel- und Hochwasser ist der Einfluss der Filterwirkung von Flussohle und Grundwasserleiter von grosser Bedeutung.

Filterwirkung von Flusssohle und Vorland

Die Filterwirkung der Flusssohle und des Vorlandes ist charakterisiert durch ein komplexes Zusammenwirken von:

- biologischen Prozessen, z.B. Filtration von Partikeln durch Algen und Bewuchs auf den Gesteinskomponenten der Flusssohle, Biofilme
- der Kolmation der Flusssohle
- der Infiltration durch das Wiesevorland (Flussvorland des verbauten Flusses)

Die biologischen Prozesse in der hyporheischen Zone sind so komplex, dass erst wenige gesicherte Erkenntnisse darüber vorliegen (Pusch et al., 1998). Hingegen ist der Prozess der Kolmation (Verringerung der Durchlässigkeit der Gewässersohle durch Ablagerung von Schwebstoffen in oder auf der Sohle) etwas besser bekannt (Schälchli, 1993; Mitteilungen VAW 124). Die Kolmation bewirkt einerseits eine Verringerung des Porenraums bei gleichzeitiger Verfestigung des Sohlensubstrates und andererseits eine Reduktion der Sohlendurchlässigkeit. Kolmatierte Fliessgewässersohlen verringern die Grundwasserneubildung durch die Flusssohle und verändern den Lebensraum der Gewässerfauna.

Abb. 4.6 zeigt, wie eine kolmatierte Flusssohle bei zunehmendem Abfluss sukzessive aufgerissen werden kann. Das konzeptionelle Modell von Schälchli, 1993 gliedert sich in drei Phasen. In einer Anfangsphase wird mit zunehmendem Abfluss vorerst Feinmaterial in die Poren eingespült. Dies bewirkt eine Verringerung der Sohlendurchlässigkeit. Bei weiterem Anstieg des Flusspegels erreicht die kritische Sohlenschubspannung (Mass für die Scherkraft an der Flusssohle) einen Wert, bei dem eine begrenzte Dekolmation stattfindet. Dabei erhöht sich die Sohlendurchlässigkeit um ein Mass, das vom vorgängigen Kolmationsgrad abhängig ist. Bei extrem hohen Abflüssen wird im Modell die Deckschicht vollständig aufgerissen und die Durchlässigkeit der Flusssohle erreicht einen maximalen Wert. Mit der Abnahme des Abflusses setzt dann erneut der Kolmationsprozess ein. Die Aufnahme während des Baus der Revitalisierungsstrecke zeigt ein Baggerloch von 2 m Tiefe, direkt neben der Wiese. Die Schotter, die im Baggerloch angetroffen wurden, sind durchlässig (K-Werte in der Grössenordnung von 3.0 10^{-3} m/s), einzelne Zonen mit extrem durchlässigen Rollkieslagen (Openframework Zonen, K-Wert bis zu 10^{-1} m/s) können eindeutig identifiziert werden. Erstaunlich ist, dass dennoch kein Wasser vom Flussbett in das unmittelbar neben der Wiese gelegene Baggerloch eindringen konnte. Bei Hochwasserereignissen wurde demzufolge die Flusssohle entweder nicht aufgerissen oder sie kolmatierte sogleich wieder bei abnehmendem Abfluss. Im Gegensatz dazu ist die Durchlässigkeit des Vorlandes sehr gross. Es dominiert sauberer Kies, praktisch ohne Silt und Tonanteil. Durch die geringe Bodenmächtigkeit und die relativ hohe Bodendurchlässigkeit (oberste Lage besteht vorwiegend aus Sand), wird verständlich, dass die Infiltration bei einem Abfluss von 60 m^3/s vorwiegend auf dem Vorland erfolgt.

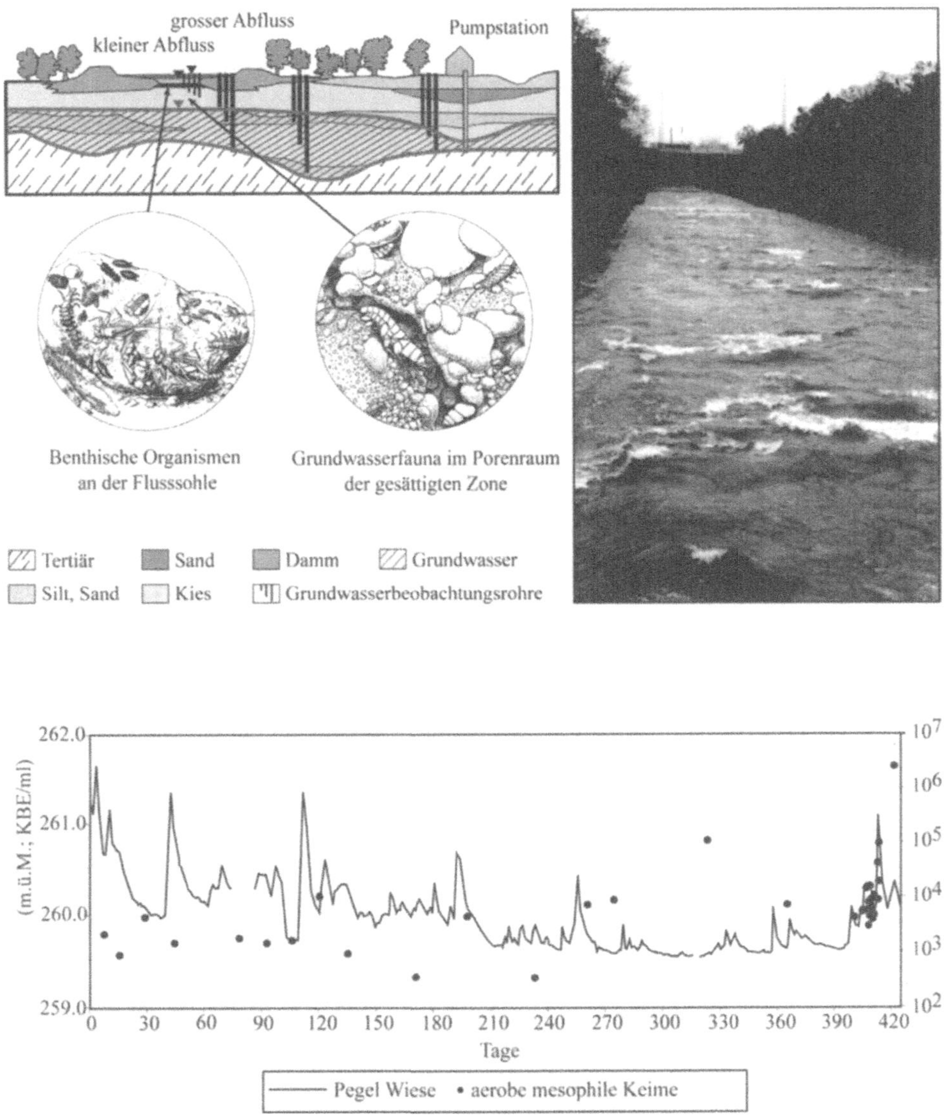

Abb. 4.5: Transport von Mikroorganismen bei grossen Abflüssen (Beispiel Wiese, Basel, Foto: P. Huggenberger)

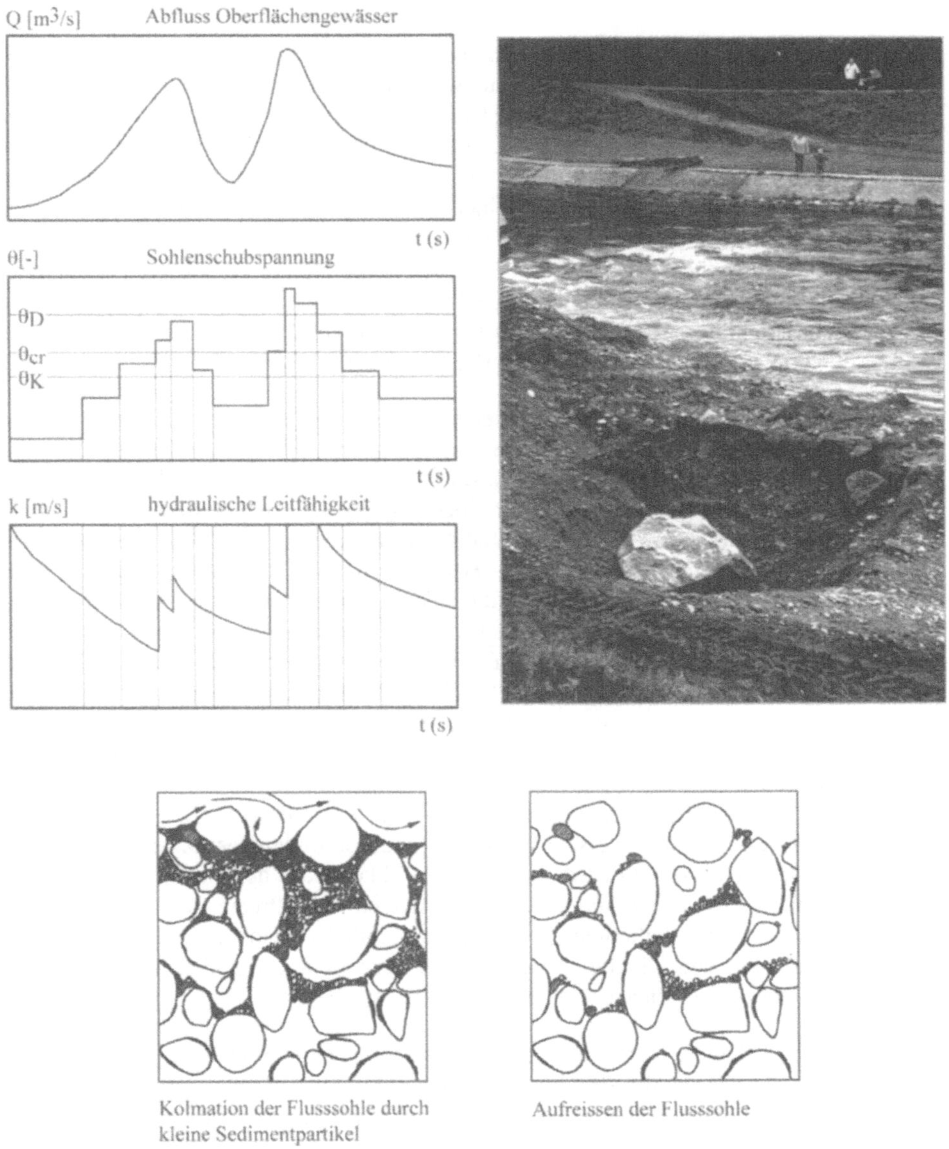

Abb. 4.6: Veränderung der Flusssohle bei einem Hochwasser. Oben links: Abfluss der Oberflächengewässer. θ_K: Beginn Dekolmation, θ_{cr}: Beginn des Aufreissens der Flusssohle, θ_D: Geschiebetransport. Bild oben rechts: Wieseufer illustriert den fehlenden Austausch zwischen Fluss und Grundwasser, trotz hoher Durchlässigkeit in den Schottern. ($>10^{-3}$m/s). Unten: Mechanismus der Kolmation der Flusssohle (umgezeichnet nach Schälchi, 1993, Foto: P. Huggenberger)

Hydraulische Randbedingungen

Der Einfluss unterschiedlicher Randbedingungen wie unterschiedliche Wasserstände, mehrere Grundwasserentnahme-Brunnen, etc. wurde mit Hilfe eines 3D Grundwasser- und Transportmodells abgeschätzt (Abb. 4.7). Das Modellgebiet umfasst einen Bereich von 1800 m x 1200 m (südwestlichste Ecke: 612'200/ 268'900). Das Modell besteht aus einem regelmässigen Gitternetz mit Zellen von 20 m x 20 m x 5 m (vertikal) und einer Verfeinerung des Gitternetzes im Bereich einer Revitalisierungsstrecke bis zu Zellenweiten von 5 m x 5 m x 5 m. Die totale Anzahl von Zellen des Modells beträgt 19'516. Die östlichen und westlichen Modellränder sind als Fixpotentiale (fixed head boundary) definiert, die nördlichen und südlichen Modellbegrenzungen sind undurchlässig (no flow boundary). Der mittlere K-Wert wurde aus einer Anzahl von Simulationen mit den Daten vom 9.12.98 ermittelt (homogener K-Wert). Der Wert liegt bei 5.75 x 10^{-3} m/s.

Abb. 4.7 zeigt die Resultate von Simulationen bei grossen und kleinen Abflüssen sowie vor und nach der Revitalisierung. Die Resultate illustrieren, wie sich der Anströmbereich eines Brunnens in Abhängigkeit des Wieseabflusses und ändernder Sohlendurchlässigkeit (die Sohlendurchlässigkeit der Wiese wurde für Hochwässer jeweils um den Faktor 10 erhöht) verändern kann. Insbesondere entsteht im Modell bei Hochwasser praktisch ein Kurzschluss zwischen Wiese und betrachtetem Brunnen (8013). Deshalb ist zu erwarten, dass sich nach der Revitalisierung der Anteil an Uferinfiltrat mindestens anfänglich erhöht. Erfahrungen bei anderen Flüssen zeigen, dass sich mit der Zeit wieder eine gewisse Kolmation der Flusssohle einstellt, welche bei Niedrigwasser den Eintrag von pathogenen Keimen limitiert. Im Zustand vor der Revitalisierung kann eine Erhöhung des Eintrages an pathogenen Keimen nur durch Versickerung auf dem Vorland erklärt werden. Die Grundwasserströmungsmodelle liefern Trends des möglichen Verhaltens eines solchen Systems bei unterschiedlichen Randbedingungen. Wie rasch die Flusssohle bei Niedrigwasser wieder kolmatiert, kann mit dem bestehenden Messprogramm nicht endgültig beantwortet werden. Immerhin deuten die Daten des Markierversuchs (Uranin) und der Radonmessungen an, dass die erwarteten Phänomene, wie z.B. Erhöhung der Durchlässigkeit der Flusssohle bei Hochwasser, entlang der Revitalisierungsstrecke auch tatsächlich auftreten. Die mit den Resultaten der Grundwassermodellierung berechneten Aufenthaltszeiten bei Hochwasser von 5 Tagen, stimmen mit den Resultaten des Markierversuchs und mit den, auf Grund der Radonkonzentrationen berechneten Aufenthaltszeiten, gut überein.

Der Wiesewasseranteil des beim Brunnen 8013 geförderten Grundwassers liegt während des auslösenden ersten Dezember-Hochwassers bei 10%. Unabhängig davon zeigen die Radondaten (Abb. 4.8) deutlich, dass beim unmittelbar nachfolgenden grösseren Hochwasser der Anteil an Uferinfiltrat massiv ansteigt und praktisch zu einer vollständigen vertikalen Durchmischung des Grundwassers führt.

Generell wird erwartet, dass auf Grund der neuen Strukturierung der Flusssohle die Dekolmation nicht überall gleichzeitig und gleich intensiv einsetzt. Die K-Wert Verteilung der Flusssohle wird insbesondere bei Hochwasser in Zukunft stärker variieren als bisher.

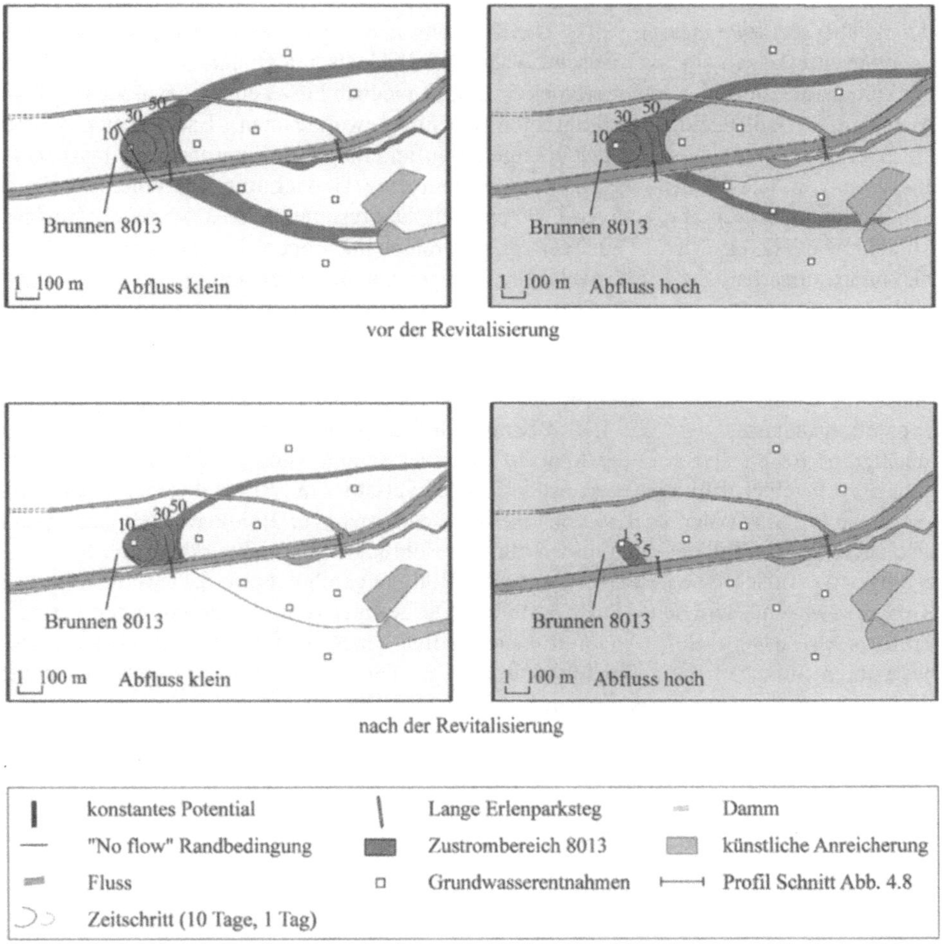

Abb. 4.7: Anstrombereich von Wasserfassungen in Abhängigkeit veränderter hydraulischer Verhältnisse an der Flusssohle vor und nach der Revitalisierung und Berücksichtigung der Infiltration von Oberflächenwasser auf dem Vorland bei Hochwasser. Vor der Revitalisierung und bei kleinem Abfluss praktisch keine Inflitration von Oberflächenwasser. Nach der Revitalisierung und Hochwasser vorwiegend Infiltration von Oberflächenwasser.

Filterwirkung des heterogenen Grundwasserleiters

Geologische Aufschlüsse in Kiesgruben oder in Baustellen (Abb. 4.9) geben einen guten Einblick in den strukturellen Aufbau von Grundwasserleitern. Deutlich erkennbar sind Inhomogenitäten in verschiedenen Massstabsbereichen (Schrägschichtungen und ausgedehnte Zonen ohne Anteil an Feinsediment, deren Durchlässigkeiten in der Grössenordnung von bis zu 10^{-1} m/s liegen können). Ausführliche Analysen von Korngrössenverteilungen zeigen, dass nur eine beschränkte Anzahl von Schottertypen auftreten (Huggenberger, 1993; Jussel, 1992). Die hydraulischen Eigenschaften der unterscheidbaren Schottertypen zeigen charakteristische Wahrscheinlichkeitsdichtefunktionen (Abb. 4.9). Die Funktionen der einzelnen Schottertypen überlappen jedoch bis zu einem gewissen Grad, mit Ausnahme der sogenannten Rollkieslagen (Openframework Zonen). Entsprechend werden die überlappenden Teile zu einer einzigen Häufigkeitsfunktion zusammengefasst. Bildanalysen von Kiesgruben ermöglichten zudem eine Gewichtung der unterschiedenen Schottertypen nach ihrem relativen Flächenanteil. Die resultierende Kurve zeigt bimodalen Charakter. Die Zonen ohne Feinanteil (Openframework Zonen) bilden dabei ein kleines Maximum ausserhalb der Kurve, welche die übrigen Schottertypen umfasst.

Die Häufigkeit und die Ausdehnung der starkdurchlässigen „Rollkieslagen" bestimmt die Varianz der hydraulischen Leitfähigkeit, $\sigma_{\ln K}$ und ihre Korrelationslänge λ. Die Korrelationslänge ist ein Mass für die Korrelation von zwei Messpunkten in Abhängigkeit ihrer Separationsdistanz. Die Abb. 4.10 (oberes Bild) zeigt deutlich, dass die Varianz davon abhängt, ob Rollkieslagen in der Analyse mitberücksichtigt werden. Die Variabilität innerhalb der einzelnen Lithofazies (konstanter oder variabler Wert) spielt eine etwas untergeordnete Rolle. Aus der Analyse von einer grossen Anzahl von Bohrkernbeschreibungen ist bekannt, dass die Rollkieslagen in den meisten Bohrungen gar nicht erkannt werden. Viele geologische Aufschlüsse (Abb. 4.10 mittleres Bild) zeigen hingegen, dass Rollkieslagen in Kiesgrundwasserleitern sehr häufig auftreten. Die Rollkieslagen haben nur einen geringen Einfluss auf das mittlere Ausbreitungsverhalten eines Stoffes im Grundwasser. Sie bestimmen hingegen die schnellen Wasserwege. Die Sedimentanalysen bei den neuen Bohrungen und die geophysikalischen Messungen (Radar) im Anstrombereich von Brunnen 8013 zeigen eine Varianz der hydraulischen Leitfähigkeiten ($\sigma_{\ln K}$) in der Grössenordnung von 1.8 bis 2.3. Die Schwankungsbreite der K-Werte beträgt mehrere Grössenordnungen. Diese grosse Streuung ist eine mögliche Erklärung für das rasche Auftreten von Mikroorganismen im Brunnen 8013 bei grossen Hochwässern.

Neue Resultate von Feldexperimenten über den Transport von Mikroorganismen in inhomogenen Grundwasserleitern wurden in einer Arbeit von Rehmann et al., 1999 veröffentlicht. Abb. 4.10 zeigt Durchbruchskurven (Step-Funktion) von Transportexperimenten mit Mikroorganismen. Geprüft wurde die Abhängigkeit des Mikroorganismen-Transports von der Varianz der hydraulischen Leitfähigkeit des Schottergrundwasserleiters, ausgedrückt als $\sigma_{\ln K}$. Nach Rehmann et al., 1999 (Abb. 4.10. unten) erfolgt in inhomogenen Systemen der Durchbruch von Mikroorganismen nach einer Transportdistanz von 30 m rascher, als der Durchbruch von konservativen Markierstoffen. Dies kann eigentlich nur damit erklärt werden, dass die wenigen Mikroorganismen, die nicht schon an der Oberfläche oder vom porösen Medium herausgefiltert werden (z.B. an einer Mineraloberfläche zurückgehalten werden), bis zum Ziel (Beobachtungsrohr), im wesentlichen die schnellen Wasserwege, d.h. die grösseren und durchgängigen Poren benutzt haben. Inhomogene geologische Systeme (z.B. fluviale Schotter) weisen dabei merklich grössere Häufigkeiten an starkdurchlässigen Zonen auf (siehe Abb. 4.9), als vergleichsweise homogene Systeme (Sandablagerungen von distalen Flüssen). Diese Erkenntnisse stehen bezüglich Filtrationsvorgänge und dem raschen Auftreten von Mikroorganismen in flussnahen Brunnen im Grundwasseranreicherungsgebiet von Basel in guter Übereinstimmung.

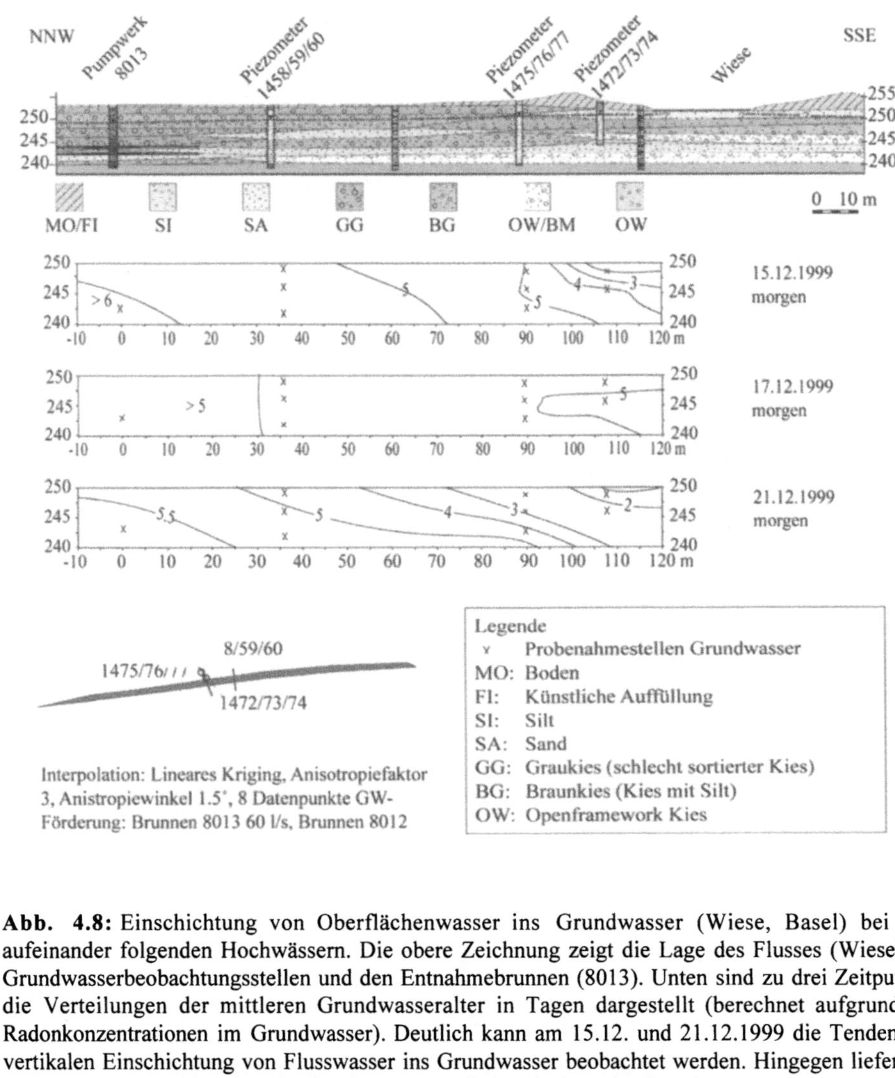

Abb. 4.8: Einschichtung von Oberflächenwasser ins Grundwasser (Wiese, Basel) bei zwei aufeinander folgenden Hochwässern. Die obere Zeichnung zeigt die Lage des Flusses (Wiese), die Grundwasserbeobachtungsstellen und den Entnahmebrunnen (8013). Unten sind zu drei Zeitpunkten die Verteilungen der mittleren Grundwasseralter in Tagen dargestellt (berechnet aufgrund von Radonkonzentrationen im Grundwasser). Deutlich kann am 15.12. und 21.12.1999 die Tendenz zur vertikalen Einschichtung von Flusswasser ins Grundwasser beobachtet werden. Hingegen liefern die Daten vom 17.12.1999 Hinweise für eine gute vertikale Mischung im Grundwasser.

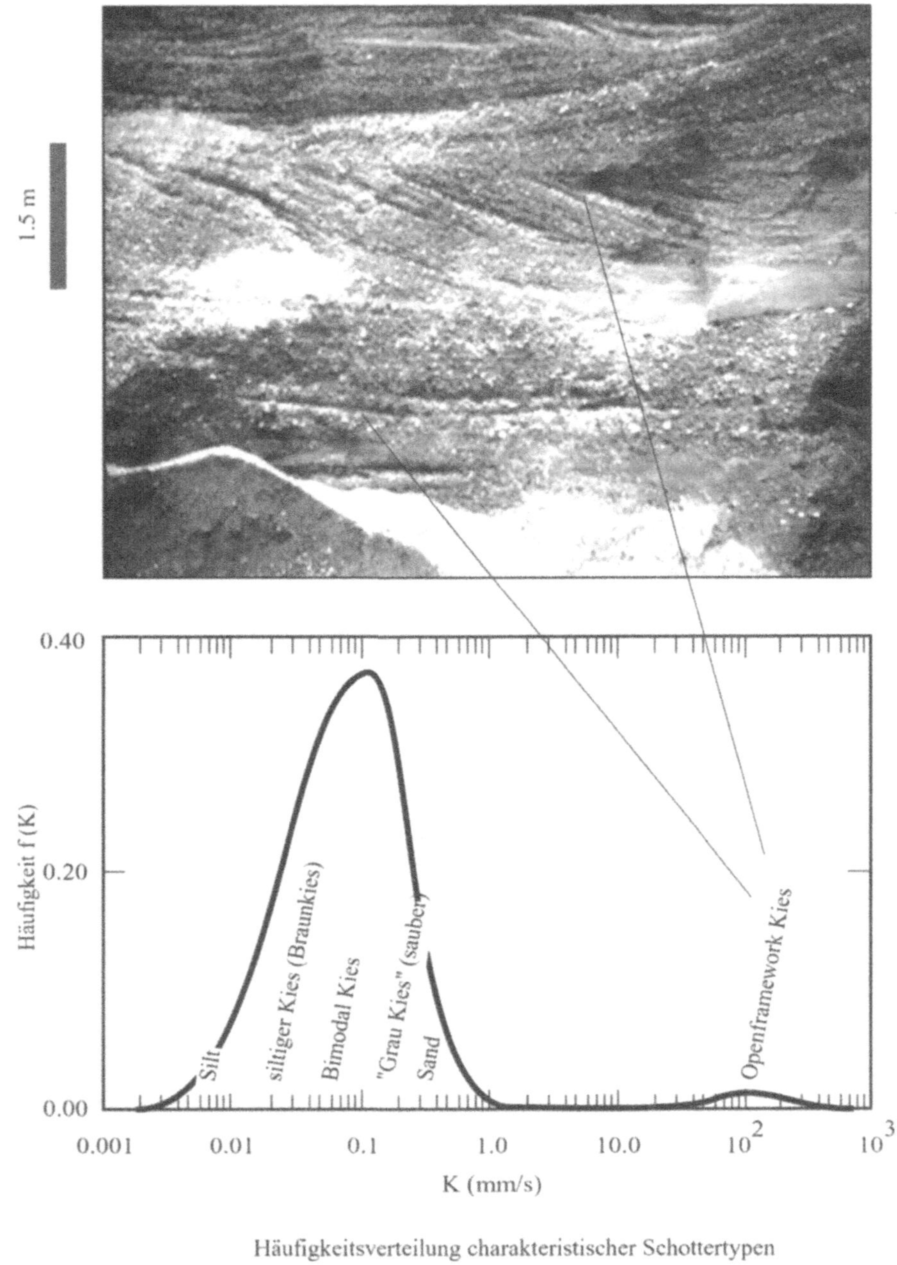

Abb. 4.9: Hydraulische Eigenschaften von Schottern in Abhängigkeit des Kiestyps. Kies ohne Feinanteil (Openframework-Zonen) weisen eine markant höhere hydraulische Leitfähikgeit auf. (K-Wert Verteilung nach Jussel, 1994; und Huggenberger et al., 1988; Foto: P. Huggenberger)

Charakter der Inhomogenitäten von fluvialen Schottern. Baustelle COOP Basel, Foto: P. Huggenberger

Geostatistische Analyse von Kiesgrubenaufschlüssen	Varianz des natürlichen Logarithmus der K-Wert-Verteilung (σ_{lnK})	Korrelationslänge (Angabe über die Distanz, über welche Daten Messpunkte miteinander in Beziehung stehen (λ))
Alle Lithologien, ausser "Open-framework-Zonen" (K-Werte von Kiestypen ohne Streuung)	0.4	4.6 m
Alle Lithologien, ausser "Open-framework-Zonen" (K-Werte von Kiestypen mit Streuung)	0.9	2.4 m
Alle Lithologien, inklusive"Open-framework-Zonen" (K-Werte von Kiestypen ohne Streuung)	1.7	1.3 m
Alle Lithologien, inklusive"Open-framework-Zonen" (K-Werte von Kiestypen mit Streuung)	2.3	1.3 m

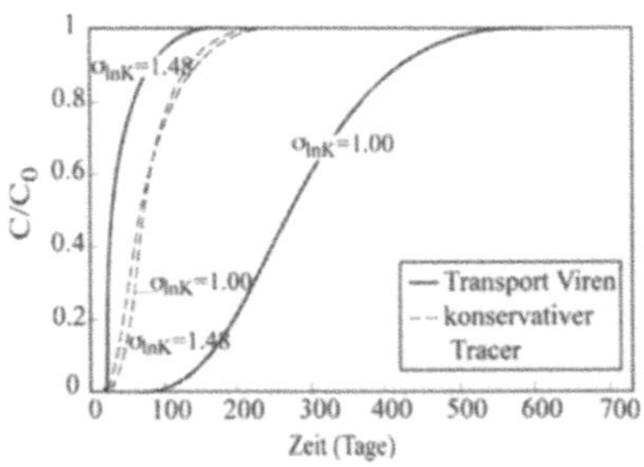

Vergleich Durchbruchskurven von Markierstoffen und Mikroorganismen (Distanz 30 m) in Abhängigkeit der Inhomogenitäten der Schotter, ausgedrückt durch σ_{lnK} (Varianz des natürl. Logarithmus der hydraulischen Leitfähigkeiten, modifiziert nach Rehmann et al., WRR. 35(7), pp.1987-2006, 1999)

Abb. 4.10: Oben: Häufiges Auftreten von Rollkieslagen in Kiesaufschlüssen. Mitte: Zusammenhang Korrelationslänge und Varianz der hydraulischen Leitfähigkeiten in Abhängigkeit der Berücksichtigung von Rollkieslagen (Openframework Zonen) und der K-Wert-Variabilität der unterscheidbaren Lithologien. Unten: Transport von Mikroorganismen in Abhänigkeit der Varianz der hydraulischen Leitfähigkeit im Vergleich mit Durchbruckskurven von konservativen Tracern (Experimente von Rehmann et al., 1999)

Schlussfolgerung und Empfehlungen

Die Interaktion zwischen Mikroorganismen und Mineraloberflächen wird im Allgemeinen durch die Kolloid-Filtrations-Theorie (Hunter, 1981; Everett, 1992; Eimelech et al., 1995) und durch die DLVO-Theorie (Derjaguin und Landau, 1941; Verwey und Overbeck, 1948) beschrieben. Diese Theorien erlauben keine einfachen Korrelationen zwischen Laborergebnissen und den tatsächlichen Verhältnissen im Feld. Mit den Arbeiten von Rehmann et al., 1999 werden Lücken in diesem Zusammenhang zwar nicht geschlossen, jedoch Wege aufgezeigt, in welcher Richtung Feldexperimente gezielt auf den Grundwasserschutz ausgerichtet werden können. Feldexperimente zeigen, dass in inhomogenen, Lockergesteins-Grundwasserleitern der Transport von Mikroorganismen über beträchtliche Distanzen in relativ kurzer Zeit erfolgen kann (Abb. 4.11). In stark inhomogenen Systemen ist der Transport vom Mikroorganismen sogar rascher als jener von konservativen Markierstoffen wie beispielsweise Uranin. Damit wird klar, dass die Gefährdung vieler flussnaher Wasserfassungen wesentlich grösser ist als bisher angenommen wurde. Diese Gefährdung ist vielerorts am grössten bei Hochwässern, wenn sich der Eintrag an Mikroorganismen um mehrere Zehnerpotenzen erhöhen kann.

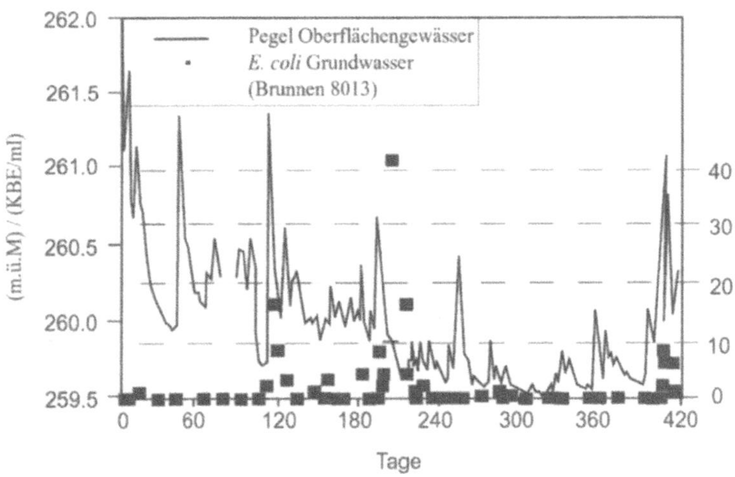

Abb. 4.11: Auftreten von *E. coli* im Grundwasser (Brunnen 8013), Grundwasseranreicherung Basel-Stadt. Pegelverlauf des Oberflächengewässers (Fluss Wiese). Die Distanz Fluss-Grundwasserbrunnen beträgt 120 m. Der Brunnen 8013 ist nicht mehr in Betrieb.

In solchen Fällen wird der Transport von Mikroorganismen auch durch erhöhte hydraulische Gradienten zwischen Fluss- und Grundwasser beschleunigt. In der Übergangszone Fluss-Grundwasser werden die meisten Mikroorganismen zurückgehalten. Diejenigen, die jedoch diese Passage durchbrechen, können in stark inhomogenen Systemen über weite Strecken transportiert werden. Neben den Inhomogenitäten spielt natürlich auch die Anzahl der ins Grundwassersystem eindringenden Mikroorganismen eine wichtige Rolle. Mit zunehmender Aufenthaltszeit des Grundwassers steigt aber auch die Chance, dass die Mikroorganismen irgendwo in den Poren zurückgehalten werden. Das bedeutet, dass der Strategie „Maximierung der Aufenthaltszeit des Grundwassers" eine bedeutende Rolle zukommt und unterstreicht gleichzeitig die Notwendigkeit ausreichender Aufentaltszeiten in der engeren Grundwasserschutzzone S2.

Der Eintrag von Mikroorganismen in Grundwasserfassungen könnte weiter verringert werden durch:

- Verbesserung der Kenntnisse der Gefährdungsmuster (Ereignisse mit hohen Konzentrationen an Mikroorganismen) und entsprechender Anpassung der Grundwasserentnahme
- Verbesserung der Gewässergüte durch emissionsseitige Verminderung des mikrobiologischen Eintrages
- Das Erkennen der wichtigsten biogeophysikalischen Vorgänge in der Übergangszone Fluss- Grundwasser würde es erlauben, gezielt natürliche Prozesse zum Schutz von Grundwasserfassungen auszunützen
- Kenntnisse über die Heterogenität der Grundwasserleiter ($\sigma_{\ln K}$, λ) spielen die Schlüsselrolle bei der Beurteilung des Transports von Mikroorganismen im Grundwasserleiter
- Die Kenntnisse der Verteilung der Aufenthaltszeiten in Flussebenen bilden die Basis für eine Optimierung der Sicherheit flussnaher Wasserfassungen

Es gibt somit verschiedene Massnahmen, die ergriffen werden können, wenn Mikroorganismen in flussnahen Grundwasserfassungen und Karstgebieten auftreten.

Partikeltracer 5

Adrian Auckenthaler

Einleitung

In der Hydrogeologie werden Tracer, auch Markier- oder Färbstoff genannt, zur Erkundung von ober- und unterirdischen Wasserwegen zwischen einer Eingabestelle und einer oder mehreren hydrologisch tiefer liegenden Austrittsstellen eingesetzt. Tracer bestehen aus Partikeln oder Substanzen (biologisch oder chemisch) die mit dem Wasser mittransportiert werden können. Tracerversuche dienen der Beantwortung verschiedener hydrogeologischer Fragestellungen:

- Abgrenzung des hydrogeologischen Einzugsgebietes einer Wasserfassung zur Bestimmung des Zuströmbereichs
- Bestimmung von Fliessgeschwindigkeiten im Grundwasserleiter und daraus die Ableitung der Fliesszeit für die Ausscheidung von Grundwasserschutzzonen
- Simulation von Verunreinigungswolken, zum Beispiel aus Altlasten
- Abflussmessung in Bächen (über die Verdünnung des Tracers)

Die Ausscheidung von Schutzzonen für Trinkwasserfassungen ist wohl die wichtigste Anwendung von Tracern. Die engere Schutzzone S2 in der Schweiz oder die Schutzzone II in Deutschland sollen verhindern, dass pathogene Mikroorganismen in das geförderte Grundwasser gelangen. Die Abklärungen für die Bestimmung dieser Zonen werden heute meist mit fluoreszierenden Tracern durchgeführt, deren Transportverhalten im Grundwasser sich jedoch in mancher Hinsicht von dem der Mikroorganismen unterscheidet. Die Verwendung von Partikeltracern kann deshalb zur zuverlässigeren Abschätzung der Gefährdung von Trinkwasserfassungen beitragen. Das Transportverhalten von Partikeln, bestimmt durch ihre Grösse, Oberflächenstruktur und Ladung, widerspiegelt das Verhalten von Mikroorganismen viel genauer als die Moleküle der fluoreszierenden Tracer. In der Schweiz wurden deshalb bereits in den 70er Jahren Versuche mit Bakterien in porösen Grundwasserleitern durchgeführt (Marti, 1975). Neuere Studien Ende der 90er Jahre mit unterschiedlichen Partikeltracern fanden sowohl in Karst- als auch in Lockergesteinsgrundwasserleitern statt.

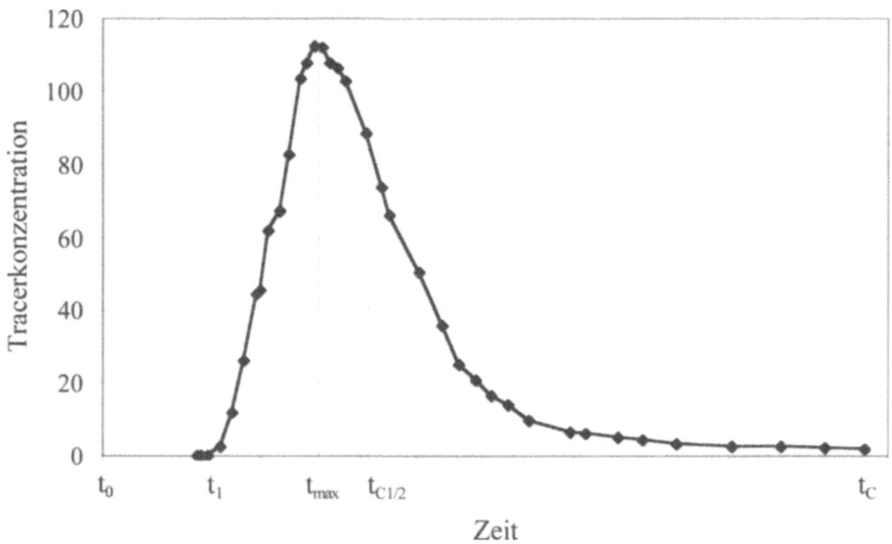

Abb. 5.1: Idealisierte Durchbruchskurve (verändert nach SGH, 2002). (t_0: Einspeisung des Tracers; t_I: Zeitpunkt des ersten messbaren Auftretens des Tracers; t_{max}: Zeitdauer bis zum Konzentrationsmaximum; $t_{C1/2}$: Zeitdauer bis 50% des insgesamt wiedergefundenen Tracers in der Beobachtungsstelle angekommen ist (daraus errechnet sich die mittlere Abstandsgeschwindigkeit); t_C: Zeitdauer bis der Tracer wieder verschwunden ist, resp. bis zu einer geringen, quasi stabilen Konzentration)

Tab. 5.1: Künstliche und natürliche Tracer

Art des Tracers	Parameter
künstlich wasserlöslich	Farbstoffe (Uranin, Eosin, Sulforhodamin, Naphthionat, Lyssamin etc.)
	Salze
	Tenside, Aromen, andere Chemikalien
künstlich partikulär	fluoreszierende Mikrosphären
	Bakteriophagen
	Bakterien
	Lycopodium Sporen
	Geobombs
	andere partikuläre Tracer
natürliche Wasserinhaltsstoffe	Isotope aus der Umwelt
	Umweltchemikalien
	physikalische Umwelteinflüsse

Mit der Wahl des eingesetzten Tracers, der Art der Eingabe und dem Ausgangssignal können Rückschlüsse auf den Grundwasserleiter gezogen werden. Aus der Durchbruchskurve errechnet sich die maximale und die mittlere Fliessgeschwindigkeit. Breite und Höhe der Kurve geben Auskunft über die Filterwirkung und die Porosität des Grundwasserleiters. Die Integration der gemessenen Konzentration des Tracers über die Versuchsdauer multipliziert mit der Wassermenge (z.B. Schüttung bei einer Quelle) ergibt die Wiederfindung des Tracers beim Output, zeigt also, wieviel des eingesetzten Tracers an einer bestimmten Austrittsstelle angekommen ist. Je grösser die Wiederfindung, desto besser ist die Verbindung von Input zu Output und desto grösser auch die Möglichkeit einer Gefährdung durch einen unerwünschten Input in den Untergrund zum Beispiel durch das Ausbringen von Jauche. Abbildung 5.1 zeigt die Durchbruchskurve eines Tracers in einem Karstgebiet.

Es gibt eine grosse Anzahl verschiedener natürlicher und künstlicher, wasserlöslicher Tracer (siehe Tab. 5.1). Davon wird meist nur eine kleine Auswahl eingesetzt. Es handelt sich dabei um künstliche Substanzen in Form von fluoreszierenden Farbstoffen wie Uranin (Fluorescein), Eosin und Sulforhodamin. Das Uranin gilt als bester Farbtracer, da es sehr gut wasserlöslich, sehr mobil und zudem in tiefen Konzentrationen kostengünstig nachzuweisen ist. Künstliche Partikel werden nur vereinzelt und bisher meist für wissenschaftliche Fragestellungen gebraucht. Aufgrund ihrer Eigenschaften als Partikel oder Kolloide repräsentieren sie die natürlichen Mikroorganismen viel besser als fluoreszierende Farbstoffe. Für die Minimierung mikrobiologischer Risiken in Grundwasserschutzzonen können sie in Zukunft einen wichtigen Beitrag zur Sicherung der Trinkwasserfassungen leisten. Natürliche Wasserinhaltsstoffe wie zum Beispiel Isotope können ebenfalls für die Beschreibung von Wasserwegen dienen. Die Inputstellen lassen sich aber nicht so genau lokalisieren wie mit künstlichen Tracern. Es geht dabei mehr um eine Abschätzung beispielsweise des Alters von Grundwässern oder der Austauschprozesse zwischen Oberflächengewässern und Grundwasservorkommen (Talamba et al., 2000; Uhlenbrook et al., 2000).

An einen guten Tracer werden mehrere Anforderungen gestellt. In der untenstehenden Auflistung sind acht Eigenschaften erwähnt, die für biologische, chemische oder fluoreszierende Tracer und Isotope gelten.

Gewünschte Eigenschaften für einen Tracer (nach Keswick et al., 1982a):

- nicht toxisch oder pathogen (d.h. keine negativen Effekte auf Mensch, Tier und Umwelt)
- natürlicherweise nicht im Grundwasser vorhanden, gut unterscheidbar von ähnlichen Substanzen oder Partikeln (Mikroorganismen) in der Umwelt
- keine Beeinflussung des Grundwasserflusses
- stabil in der Umwelt für eine genügend lange Zeit
- Fluss mit dem Grundwasserstrom, keine Filtration, Adsorption oder andere Verminderung während des Transports
- einfach und kostengünstig zu analysieren, bereits in sehr geringen Konzentrationen nachweisbar
- keine Interaktion mit der Matrix, Wasserinhaltsstoffen oder Mikroorganismen
- bei gleichzeitiger Verwendung mehrerer Tracer, gute Unterscheidbarkeit im Labor

Tab. 5.2: Unterschiede partikulärer zu molekularen Tracern

Partikel	**Molekül / Ion**
suspendiert	gelöst
verschiedene Grössen von 10 nm – 10 μm	Ionen (z.B. Cl^-) bis grössere Moleküle (z.B. Uranin)
Filtrationseffekte, Sorption	Sorption
Ausschluss von Feinporen, dadurch schnellerer Transport	Matrixdiffusion
unsichtbar in grossen Konzentrationen	Fluoreszenz von Auge sichtbar ab ca. 4 ppm (Uranin)
Nachweis ab 0,5 – 1 Partikel / mL (teilweise tiefer bei Filtration)	Nachweis ab ca. 3 x 10^9 Molekülen / mL (für Uranin, andere Tracer erst ab höherer Anzahl)
Analyse trübungsunabhängig bei Bakterien und Bakteriophagen	Nachweisgrenze von Trübung des Wassers abhängig
Partikelladung positiv, negativ oder neutral	reaktive oder nicht reaktive Stoffe (v.a. nicht reaktive Stoffe werden eingesetzt)
repräsentativer für Transportverhalten von Mikroorganismen	eher zur Simulation des Transportes chemischer Stoffe geeignet
toxikologisch unbedenklich bei marinen Bakteriophagen und Mikrosphären, bei Bakterien abhängig von der Art und der Markierung	Uranin und Eosin toxikologisch unbedenklich, andere Stoffe (z.B. Sulforhodamin B, Rhodamin WT) toxikologisch bedenklich od. nicht zu empfehlen (Umweltbundesamt, 1997)

Es gibt sowohl molekulare als auch partikuläre Tracer, die diese Eigenschaften erfüllen. Je nach Fragestellung kann somit ein möglichst idealer Tracer gewählt werden. Es gibt jedoch einige wesentliche Unterschiede zwischen partikulären und molekularen Tracern (Tab. 5.2). Die Partikeltracer sind nicht im Wasser gelöst, sondern bilden eine Suspension von mikroskopischen Partikeln. Durch ihre Grösse können sie Feinporen nicht passieren und werden deshalb bevorzugt in den schnellen Wasserwegen transportiert. Selbst in hohen Konzentrationen sind Partikel von blossem Auge nicht sichtbar, im Gegensatz zu fluoreszierenden Markierstoffen. Die Analytik von Partikeln ist zwar meist aufwändig, ihre Nachweisgrenze jedoch sehr tief und trübungsunabhängig, was insbesondere bei Karstwasseruntersuchungen ein Vorteil gegenüber fluoreszierenden Tracern sein kann. Partikel repräsentieren Transporteigenschaften von pathogenen Mikroorganismen besser und eignen sich deshalb zu deren Simulation. Im folgenden wird nur auf Partikeltracer eingegangen.

Partikeltransport

Eine Reihe von Umweltfaktoren und Partikeleigenschaften beeinflusst den Transport von natürlichen und künstlichen Partikeln durch den Boden, die schützende Überdeckung und den Grundwasserleiter. Der folgende Abschnitt gibt einen kurzen theoretischen Überblick des Transportverhaltens von Partikeln.

Der Transport von Partikeln wird von der Filtrationstheorie beschrieben. Yao et al., 1971 haben ein Modell für die Wasserfiltration aufgestellt und in Laborexperimenten verifiziert. Sie führten die Unterschiede im Adsorptionsverhalten von Partikeln auf ihre Grösse und Dichte und einen „collision efficiency factor", der die Gesamtheit der chemischen Eigenschaften des Systems berücksichtigt sowie auf die Viskosität und die Fliessgeschwindigkeit des Transportmediums zurück. Die Adsorption von Mikroorganismen an einen Feststoff ist ein physikalisch-chemischer Prozess und kann durch die erweiterte DLVO Theorie der Kolloidstabilität erklärt werden (van Oss et al., 1986; Grasso et al., 1996). Diese Theorie beschreibt die Interaktionsenergie zwischen zwei Oberflächen als die Summe von drei abstandsabhängigen Kräften: den auf grössere Distanzen wirkenden Lifshitz-van der Waals Kräften, den elektrostatischen Interaktionen und den auf kürzere Distanzen wirkenden Lewis Säure-Base Interaktionen, auch hydrophobe/-phile Kräfte genannt. Van der Waals Kräfte sind meistens anziehend, während elektrostatische Interaktionskräfte wegen der negativen Ladung der meisten Bakterienoberflächen und natürlichen Oberflächen in der Regel abstossend sind. Dagegen können Lewis Säure-Base Interaktionen sowohl abstossend als auch anziehend sein. Einzig die elektrostatische Interaktion kann durch die Änderung der Ionenstärke und des pH-Wertes der Lösung beeinflusst werden. Bei hohen Ionenstärken ist die diffuse Ionenschicht um das Bakterium und die Matrix komprimiert, was zu verminderter Abstossung und damit zu erhöhter Adsorption führt (Gerba, 1984). Die Verringerung der Ionenstärke der Lösung kann zu Desorption führen und den Partikeltransport erhöhen (Fontes et al., 1991; Martin, 1992).

Der pH-Wert verändert die Oberflächenladung von Mikroorganismen und der Bodenmatrix und hat damit einen Einfluss auf das Adsorptionsverhalten. Bei tiefem pH-Wert ist die Adsorption wegen der reduzierten Oberflächenladungen stärker als bei hohem pH-Wert (Bales et al., 1995; Scholl und Harvey, 1992; Goyal und Gerba, 1979). Hydrophobe Eigenschaften der Bakterien und des Aquifermaterials führen zu verstärkter Adsorption. Je höher der Anteil an organischem Material ist, desto grösser kann die Adsorption sein und die elektrostatischen Kräfte um Grössenordnungen übertreffen (Bales et al., 1993).

Die Adsorption von Bakterien an die Luft-Wasser Grenzfläche ist generell sehr bedeutend und nimmt mit steigender Hydrophobizität der Bakterien noch zu (Schäfer, 1997), was stärkere Retention von hydrophoben Bakterien in ungesättigtem Boden bewirken dürfte. Die Verteilung der Korngrössen in Sandsäulen ist ein dominierenderer Faktor im Transport von Mikroorganismen als deren Partikelgrösse und die Ionenstärke der Lösung (Fontes et al., 1991). In gepackten Säulen aus gesiebtem Sand kann der Transport um Grössenordnungen kleiner sein als in ungestörten, natürlichen Proben (Smith et al., 1985). Selbst in identisch gepackten Säulen können markante Unterschiede im Transport auftreten (Harvey et al., 1993). Feldversuche zeigen denn auch kein homogenes Fliessmuster, sondern präferenzielle Fliesswege, entlang derer hohe Fliessgeschwindigkeiten auftreten. Daneben gibt es Zonen, die kaum am Transport beteiligt sind (Natsch et al., 1996). Die gesättigte hydraulische Leitfähigkeit ist der Parameter mit der grössten Variabilität im Boden (Jury et al., 1991). Die heterogene Verteilung der Porengrösse, hauptsächlich in geklüfteten Materialien bewirkt, dass Mikroorganismen, da sie von einem Teil des Porenvolumens ausgeschlossen sind, schneller transportiert werden als konservative Tracer (Harvey, 1997, Bales et al., 1995).

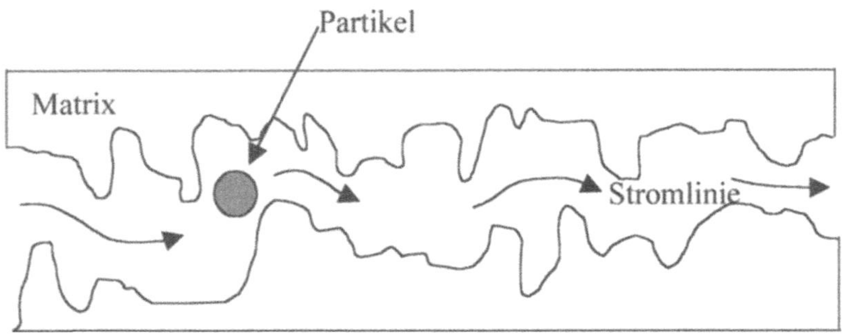

Abb. 5.2: Partikeltransport entlang einer Stromlinie. Es findet keine Matrixdiffusion statt

In Feldversuchen mit Partikeltracern sind die meisten der oben erwähnten chemisch-physikalischen Grössen kaum auf der gesamten Fliessstrecke messbar. Für die Interpretation und das Verständnis der Ergebnisse von Tracerversuchen ist aber die Berücksichtigung der verschiedenen Faktoren unter Umständen hilfreich.

Arten von Partikeltracern

In den letzten 100 Jahren wurden in der Hydrogeologie sehr unterschiedliche Partikeltracer verwendet. Die heute am häufigsten eingesetzten sind Bakterien und Bakteriophagen. Sie sind einfach zu kultivieren und können deshalb in grossen Mengen ausgebracht werden. Ihr Nachweis ist in geringen Konzentrationen kostengünstig möglich. Fluoreszierende Mikrosphären wären ideale Tracer, da sie als künstliche Partikel während dem Transport keinem Abbau unterliegen und in unterschiedlichen Grössen und Oberflächenladungen erhältlich sind. Die Kosten für eine ausreichende Partikelzahl sind allerdings heute noch recht hoch und der Nachweis im Wasser teilweise sehr aufwändig. Andere Partikeltracer haben kaum Potential routinemässig für hydrogeologische Versuche eingesetzt zu werden.

Bakterien

Gegen Ende des 19. Jahrhunderts wurden Mikroorganismen zum ersten Mal zur Bestimmung von präferenziellen Fliesswegen in Grundwasserleitern, wie z.B. Karst- oder Kluftgestein eingesetzt. Die Resultate der Studien, durchgeführt mit rot-gelb pigmentierten Bakterien wurden in den 1890er Jahren publiziert (Abba, et al. 1899; Pfuhl, 1897). Bakterien sind nach wie vor die am häufigsten verwendeten biologischen Tracer in hydrogeologischen Studien (Harvey, 1997). In den letzten Jahrzehnten kamen weltweit denn auch in etlichen weiteren Tracertests Bakterien zum Einsatz. Die Tests fanden meist in Lockergesteinsgrundwasserleitern statt. So führte Marti, 1975 und Marti et al., 1979 in den 70er Jahren in der Schweiz mehrere Versuche mit verschiedenen Bakterien und fluoreszierenden Tracern durch. Die Keime konnten dabei in einer Fliessdistanz von bis zu 90 m nachgewiesen werden. Weitere Experimente sind in Tabelle 5.3 zusammengestellt.

Aus Karstgrundwasserleitern sind nur sehr wenige Arbeiten publiziert worden (Käss, 1998). Ein Versuch mit *Serratia marcescens* und Bärlapp Sporen (Lycopodium) zwischen der oberen Donau und der Aachquelle ist von Batsche et al., 1970 dokumentiert.

Serratia marcescens waren und sind die am häufigsten eingesetzten Bakterien für Tracertests (Käss, 1998). Die Absterberate dieser Keime in Batchexperimenten betrug 99,9% innerhalb 32 bis 137 Tagen (Merkli, 1974). Käss et al., 1983 fanden eine Überlebensfähigkeit der Keime im Grundwasser von 3 Monaten. Für Tracerversuche werden Suspensionen von 10^{12} bis 10^{13} Keimen benötigt. Versuche in teilweise ungesättigter Zone erfordern eine um eine bis zwei Grössenordnungen höhere Anzahl.

Serratia sp. wurde in den 70er Jahren als potentiell humanpathogen erkannt und daher seine Anwendung für Tracertests nicht mehr empfohlen (Gerba, 1984). Käss, 1998 hingegen ist der Meinung, dass bisher noch kein Fall von Enteritits erwähnt wurde und *Serratia marcescens* ohne Bedenken für hydrogeologische Tracertests eingesetzt werden kann.

Der klassische Indikatorkeim für fäkale Verunreinigungen, *E. coli*, wurde ebenfalls für Tracerversuche eingesetzt. In Lockergesteinsgrundwasserleitern konnten Fliessdistanzen von bis zu 60 m festgestellt werden (Marti et al., 1979). Versuche in quartären Kiesen Oberbayerns mit *E. coli* zeigten starke Schwankungen in der Anzahl Keime im Tracerdurchgang nach 10 m Fliessstrecke. Diese Schwankungen sind vermutlich auf den Strömungsdruck des Grundwassers zurückzuführen.

In Karst- und Kluftgrundwasserleitern ist *E. coli* häufig „natürlicherweise" festzustellen, ebenso in Lockergesteinsgrundwasser mit Interaktion mit Oberflächenwasser. Verschiedene Autoren haben das Problem der Unterscheidbarkeit natürlicher und als Tracer injektierter *E. coli* dadurch gelöst, dass sie Antibiotika resistente *E. coli* verwendet haben (Champ und Schroeter, 1988; Rahe et al., 1979). Dabei stellt sich das Problem der Resistenzübertragung auf andere Bakterien. Freisetzungsversuche mit Antibiotika resistenten Keimen sind grundsätzlich abzulehnen.

Vorteile für die Anwendung von Bakterien in Tracertests sind die einfache Kultivierung grosser Mengen und die Detektion mit mikrobiologischen Standardmethoden. Die Detektionslimite ist abhängig von der filtrierten Wassermenge und liegt bei rund einer koloniebildenden Einheit pro 100 mL.

Die Überlebensfähigkeit von Bakterien ist von verschiedenen Umwelteinflüssen abhängig. So führt eine tiefere Temperatur, ein höherer Wassergehalt des Bodens und ein grösserer Anteil organischer Substanz zu einem längeren Überleben von Bakterien (Yates und Yates, 1988). Die effektive Lebensdauer beträgt je nach Bakterium und externen Einflüssen einige Tage bis Wochen. Batch- und Säulenversuche von Merkli, 1974 ergaben für eine 99,9% Reduktion Eliminationszeiten von 70 bis 205 Tagen für Salmonella, 30 bis 32 Tagen für *E. coli* und 17 bis 44 Tagen für Streptococcus. Eine 99,9% Reduktion scheint beachtlich zu sein, bedenkt man aber einen möglichen Eintrag von Keimen in der Grössenordnung 10^7 bis 10^9 (z.B. Anzahl E.coli/m^3 Jauche) bedeutet dies, dass nach den angegebenen Zeiten noch 10^4 bis 10^6 Keime vorhanden sind. *E. coli* konnte in einem Lockergesteinsgrundwasserleiter denn auch bis über 180 Tage nach der Injektion nachgewiesen werden (Seiler, 1988). In einem feuchten Boden fanden Young und Greenfield, 1923 *E. coli* sogar noch 5 Jahre nach der Eingabe.

In den letzten Jahrzehnten wurden verschiedene Bakterien wie z. B. *Escherichia coli*, *Streptococcus faecialis*, *Bacillus subtilis* und *Chromobacterium violaceum* für Tracerversuche im Grundwasser verwendet. Keines dieser Bakterien konnte aber mit Zufriedenheit eingesetzt werden, da sie nicht immer eindeutig vom natürlichen Hintergrund unterschieden werden können. Selbst mit markierten Bakterien (z.B. Antibiotikaresistenz) ist eine Unterscheidung oft schwierig, da Marker verloren gehen können. Zudem stellen Marker ein potentielles Umwelt- und Gesundheitsrisiko dar und dürfen deshalb heute höchstens beschränkt eingesetzt werden (Keswick et al., 1982a). In der Schweiz sind solche Versuche im Grundwasser aus gewässerschutzrechtlichen Gründen nicht zulässig.

Bakteriophagen

Bakteriophagen, auch einfach Phagen genannt, sind Viren, die nur ein spezifisches Bakterium befallen und weder auf den Menschen noch auf Tiere und Pflanzen eine Wirkung haben. Wie alle Viren können sich die Bakteriophagen nicht allein vermehren. Sie brauchen dazu ein spezifisches Wirtsbakterium, in welches sie eindringen, dessen Kontrolle sie übernehmen und mit der Produktion von neuen Phagen beginnen, die entweder permanent über die Zelloberfläche ausgeschieden oder am Ende des Infektionszyklus durch die Zerstörung des Bakteriums (durch Lysis der Zellmembran) freigesetzt werden (Rossi et al., 1998, Rossi und Käss, 1998). Bakteriophagen haben eine Grösse von 10 nm bis mehreren 100 nm.

Erste erfolgreiche Tracerversuche mit Bakteriophagen in der Schweiz wurden in einem Karstgebiet im Jura durchgeführt (Aragno und Müller, 1982). Weitere Versuche mit marinen Phagen in diversen Grundwasserleitern von Rossi, 1994 und Doerfliger, 1996 zeigten die breite Anwendbarkeit dieser Tracermethode.

Bei den Phagen handelt es sich um kolloidale Partikel im Gegensatz zu den gelösten (fluoreszierenden) Tracern und damit um eine eigene spezielle Form von hydrogeologischen Tracern. Zudem sind sie „natürliche" biologische Tracer und repräsentieren mit ihrer Struktur und Form am ehesten pathogene Viren. Sie eignen sich damit ideal zur Simulation des Transportes von Viren und in eingeschränktem Mass auch von Bakterien.

Aufgrund ihrer spezifischen Affinität zu einem bestimmten Wirtsbakterium können Multitracerversuche mit mehreren Phagen ohne Beeinträchtigung des Nachweises gleichzeitig durchgeführt werden. Die Wirtsspezifität führt dazu, dass die Phagen weder pathogen noch toxisch sind und sich somit auch zur Untersuchung von Trinkwasserfassungen eignen. Mit der Verwendung von marinen Phagen, wie sie von der Universität Neuenburg angeboten werden, kann ein Hintergrundsignal in kontinentalen Grundwasserleitern ausgeschlossen werden.

Die Nachweisgrenze liegt bei ca. 1 Phagen pro 2 mL. Diese Menge Wasser wird auch zur Analyse gebraucht (Rossi, 1994). Für ein Tracerexperiment müssen meist 10 bis 20 Liter einer Phagensuspension mit 10^{14} bis 10^{15} Phagen eingesetzt werden. Der Tracercocktail enthält ca. 1 gr. Protein, einige Gramm Mineralstoffe und verschiedene organische Substanzen (Aminosäuren, Wachstumsmedium). Der Einfluss auf den Grundwasserleiter ist auch bei geringen Wassermengen vernachlässigbar (Rossi, 1994; Käss, 1998).

Verschiedene Umweltfaktoren beeinflussen die Inaktivierung der Phagen. Als Inaktivierung wird der Verlust der Virulenz bezeichnet, also die Möglichkeit des Phagen, sich im Wirtsbakterium zu vermehren. Die Inaktivierung ist nicht gleichbedeutend mit dem Absterben des Phagen. Yates et al., 1985 sehen eine Erhöhung der Temperatur als Hauptgrund für die Inaktivierung an. Bereits eine Temperaturerhöhung von 15°C auf 20°C führt zu einer Reduktion der Virulenz (Rossi, 1994; Käss, 1998).

Die UV-Strahlung der Sonne kann ebenfalls zur Inaktivierung führen. Bei hydrogeologischen Abklärungen im Untergrund spielt dies weniger eine Rolle, wohl aber bei der Lagerung der Wasserproben. Diese müssen kühl (bei 4°C) und dunkel gelagert werden. pH-Werte im Neutralbereich, wie sie im Grundwasser in der Schweiz vorkommen, haben keinen Einfluss auf die Inaktivierung. Bei $pH < 5$ wurden die meisten Viren allerdings gut absorbiert (Goial und Gerba, 1979) und werden somit durch Filtration zurückgehalten. Dieser Prozess ist jedoch meist reversibel (Babich und Stotzky, 1980).

Bei ungesättigten Verhältnissen reduziert sich nicht nur die Transportgeschwindigkeit, sondern die Oberflächenspannung an der Luft-Wasser-Grenzfläche ist stark genug, um das Capsid eines Phagen zu zerstören (Rossi, 1994).

Die Inaktivierung der Bakteriophagen scheint in den zahlreichen Experimenten in Karst- und Lockergesteinsgrundwasserleitern der Schweiz keine oder nur eine geringe Rolle zu spielen. Die Inaktivierungsraten aus Laborversuchen liegen zwischen $K = 6{,}92 \times 10^{-4}$ und $1{,}59 \times 10^{-2}$ [min^{-1}]. Die Überlebensfähigkeit in der Umwelt liegt bei einigen Tagen (Rossi, 1994). Meist kann jedoch nicht zwischen Filtration und Inaktivierung unterschieden werden. Die Wiederfindungsraten sind vergleichbar mit denen fluoreszierender molekularer Tracer sowohl in Grund-, als auch in Oberflächengewässern (Rossi et al., 1998). Die Fliessdistanz in Lockergesteinsgrundwasserleitern betrug weniger als 100 m, im Karst wurden nach einer Fliessdistanz von über 6000 m noch Phagen nachgewiesen (Tab. 5.4).

Die Universität Neuenburg, Laboratoire de Microbiologie bietet Tracerversuche mit Bakteriophagen an. Informationen sind via Internet unter www.unine.ch zu erhalten. Das Institut für Medizinische Mikrobiologie an der Universität Neuenburg führt ebenfalls Forschung auf dem Gebiet der Bakteriophagen durch (Rossi und Käss, 1998)

Fluoreszierende Mikrosphären

Mikrosphären sind künstliche Partikel aus reinem Polystyrol. Sie sind in Grössen von 0,02 µm bis 15 µm erhältlich und haben eine Dichte von 1,06 g/mL. Die Oberfläche der Mikrosphären wird mit verschiedenen funktionellen Gruppen modifiziert, so dass negativ oder positiv geladene Partikel entstehen. Die negative Oberfläche wird mit Aldehyd-Sulfat, Sulfat oder Carboxyl, die positive Oberfläche mit Aminogruppen erzeugt. Von beiden

Ladungsarten gibt es hydrophile Partikel. Die durch die Modifikation entstandene externe Schicht ist nur einige Å dick und verändert die Grösse der Mikrosphären nicht.

Den Mikrosphären kann beim Polymerisationsprozess ein fluoreszierender Farbstoff beigegeben werden. Die Wirkung des Farbstoffs auf die Oberflächeneigenschaften ist vernachlässigbar. Der Farbstoff ist im Partikel gebunden und die Mikrosphären können dadurch bis zu einem Jahr bei 4°C aufbewahrt werden (Molecular Probes, 2000).

Mikrosphären haben gegenüber biologischen Tracern einige Vorteile. Die Grösse, die Fluoreszenz und die Oberflächenladung sind frei wählbar. Da es sich um künstliche Partikel handelt und ihr Farbstoff fest eingebunden ist, erfahren sie weder einen Abbau oder eine Inaktivierung noch eine Vermehrung während des Experiments und bei der Analyse. Ihre Anzahl wird nur durch Filtration und Verdünnung reduziert. Mikrosphären sind weder toxisch und noch kanzerogen. Tracerversuche in Einzugsgebieten von Trinkwasserfassungen sind somit möglich. Aufgrund der wählbaren Grösse und Fluoreszenz sind Multitracerexperimente denkbar und wurden in kleinerem Massstab bereits durchgeführt.

In den 80er Jahren wurden in Deutschland und den USA einige Tracerversuche mit Mikrosphären in Lockergesteins- und Karstgrundwasserleitern durchgeführt (Käss, 1998). In einem sandigen Grundwasserleiter bei Cape Cod (Massachusetts, USA) konnten auf einer Distanz von 6,9 m verschieden grosse carboxylierte (negativ geladene) und ungeladene Mikrosphären nachgewiesen werden (Harvey et al., 1989). Neuere Versuche stammen aus der Schweiz. Niehern, 1999 konnte in einem Porengrundwasserleiter im Mittelland in einem Abstand von 64 m von der Eingabestelle carboxylierte (negativ geladene, hydrophile) Mikrosphären nachweisen. Erfolgreiche Tracerversuche ebenfalls mit carboxylierten Mikrosphären konnten in einem Karstgebiet in der Nordwestschweiz von Auckenthaler et al., 2002 durchgeführt werden (siehe Tabelle 5.3 und 5.4).

Die Nachweisgrenze für Mikrosphären ist abhängig von der verwendeten Nachweismethode. Mit einem online Mikrosphärenzähler (nach dem Prinzip der Flusszytometrie) konnte Niehren, 1999 eine Mikrosphäre/mL detektieren. Die Effizienz des Zählers hängt jedoch stark von der Justierung des Lasers ab und beträgt bei Feldversuchen ca. 30 bis 40%. Als zweite Methode bietet sich die Filtration von Wasserproben an. Die Nachweisgrenze ist hier abhängig von der filtrierten Wassermenge und beträgt einige Mikrosphären pro 100 mL. Je kleiner der Durchmesser der Mikrosphären gewählt wird, desto kleiner müssen auch die Filterporen sein und desto weniger Wasser kann filtriert werden, bevor die Filterporen durch Wasserinhaltsstoffe verstopft werden.

Für Tracerversuche eignen sich Mikrosphären mit einem Durchmesser von 1 μm am besten. Diese Grösse ergibt sich aus dem Filtrationsminimum der drei Prozesse Diffusion, Interzeption und Sedimentation (siehe Kap. Trinkwasseraufbereitung, Filtration). Zudem ist dies der Grössenbereich vieler Bakterien. Bei der Wahl der Fluoreszenz ist auf eine mögliche Hintergrundfluoreszenz durch Kieselalgen, Sporen (Chlorophyll fluoresziert im Rotbereich) oder Kalzitkristalle (Fluoreszenz im Blaubereich) zu achten (Käss, 1998, Käss et al., 1994) (siehe Abb. 5.3). In Lockergesteinsgrundwasserleitern sind diese Störeffekte teilweise anders als in Karst- oder Kluftgrundwasserleitern.

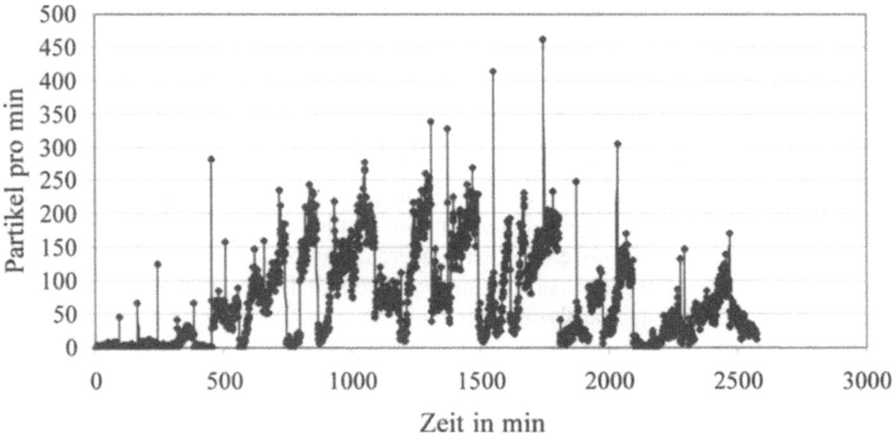

Abb. 5.3: Durchgang von Algen und Sporen (<40 µm) bei einem Regenereignis in einer Karstquelle. Daten aufgenommen mit dem Mikrosphärenzähler von Niehren, 1999 (Anregung Laser 625 nm, Emissionsfilter 645 nm). Die starken Schwankungen sind bedingt durch die Art der Probenahme (Auckenthaler, 2000 unpubliziert).

Die Oberflächenladung hat einen Einfluss auf den Transport. Es gibt Autoren, die vom Gebrauch geladener Partikel abraten und nur mit neutralen Partikeln befriedigende Resultate erzielen konnten (Käss, 1998). Andere Arbeiten ergaben jedoch gute Wiederfindungsraten auch mit negativ geladenen, hydrophilen Partikeln (Niehren, 1999), welche repräsentativer für gleichfalls negativ geladene Bakterien sind. Mit positiv geladenen Partikeln wurden noch keine Tracerversuche durchgeführt.

Die Mikrosphären sind in wässriger Lösung (2,5% Feststoffanteil) erhältlich. Bei einem Durchmesser von 1 µm sind in einem mL Suspension 3,6 x 10^{10} Mikrosphären enthalten. Bei kleineren Durchmessern sind die Konzentrationen entsprechend höher.

Fluoreszierende Mikrosphären werden von den Firmen Molecular Probes und Polysciences Ltd. in verschiedenen Grössen und Fluoreszenzen angeboten. In der Schweiz gibt es keine Stelle, die eine Analytik für Mikrosphären kommerziell anbietet.

Weitere partikuläre Tracer

In den letzten Jahrzehnten wurde für hydrogeologische Studien eine grosse Anzahl weiterer partikulärer Tracer eingesetzt, unter anderem jedoch sehr seltsam anmutende wie z.B. Sägemehl, Korkbällchen, Spreu und Konfetti darunter sind (Käss, 1998). Zu den seriöseren gehören die Sporen des Bärlapp (Lycopodium) und die Hefen. Lycopodium-Sporen werden meist von Speleologen eingesetzt. Die Sporen haben eine Grösse von 30 µm. Sie werden eingefärbt, in einen Grundwasserleiter eingegeben und im Quellwasser mittels Filtern wieder aus dem Wasser geholt. Die Analyse erfolgt mikroskopisch (Keswick, 1982b). Wegen ihrer

Grösse sind Lycopodium-Sporen nur für Karstgrundwasserleiter geeignet. Ein weiterer Nachteil ist, dass die getrockneten Sporen explosiv sind. Bei der Anwendung ist deshalb Vorsicht geboten.

Zur Eignung von Hefen als Grundwassertracer wurde nur wenig Forschung betrieben. Hefezellen haben eine Grösse von 2 bis 3 µm. In Karstgrundwasserleitern können Hefen als sekundäre Tracer eingesetzt werden (Käss, 1998). Wood und Ehrlich, 1978 injektierten Hefezellen in ein Bohrloch in einem Lockergesteinsgrundwasserleiter. Die Zellen konnten in einer Fliessdistanz von 7 m gefunden werden. Sie wurden schneller transportiert als Iodid und Bromid. In jedem Fall müssen vor der Tracereingabe Nullproben entnommen werden, um einen möglichen natürlichen Hintergrund zu bestimmen. Die Proben müssen zudem so schnell wie möglich im Labor analysiert werden (Keswick, 1982a).

Tracerversuche mit Partikeln

Sowohl in Lockergesteins- als auch in Karstgrundwasserleitern konnten Partikel bei Tracerversuchen erfolgreich eingesetzt werden. Die Fliessgeschwindigkeit der Partikel bezogen auf das erste Auftreten und das Peakmaximum ist meistens etwas grösser als bei den gleichzeitig eingesetzten fluoreszierenden Farbstoffen (Tab. 5.3 und 5.4). Für den Vergleich der Fliessgeschwindigkeiten zwischen Molekülen und Partikeln ist das erste Auftreten des Tracers, der Anstieg der Kurven und das Peakmaximum wichtig. Beim ersten Auftreten kann es aufgrund der unterschiedlichen Nachweisgrenzen zu beträchtlichen Fliessgeschwindigkeitsunterschieden kommen, da Partikel bereits ab einem Partikel/mL, fluoreszierenden Farbstoffe aber erst ab 10^9 Molekülen/mL nachweisbar sind. Der effektive Unterschied im Transportverhalten wird in der Steigung der Kurven und im zeitlichen Auftreten des Peakmaximums sichtbar.

Aus den Literaturangaben (Tab. 5.3 und 5.4) ist nicht immer klar, ob die Transportgeschwindigkeiten aus der Zeit bis zum ersten Auftreten, bis zum Konzentrationsmaximum oder bis zum Durchgang von 50% der gesamten wiedergefundenen Tracermenge ermittelt wurden. Die Angaben der einzelnen Autoren für Partikel und Moleküle sollten aber in sich konsistent sein.

Die Wiederfindungsrate der Partikel ist durchweg kleiner als die von Molekülen. Die Partikel werden durch die Filtrationseffekte Sedimentation, Interzeption und Diffusion verbunden mit einer Sorption im Boden und im geologischen Untergrund relativ stark zurückgehalten. Biologische Partikel können zudem während der Versuchsdauer absterben. Die in der Hydrogeologie eingesetzten fluoreszierenden Tracer hingegen sind persistent über die Versuchsdauer und sorbieren kaum an Oberflächen, was wesentlich höhere Wiederfindungsraten als bei Partikeln zur Folge hat.

Tab. 5.3: Tracerversuche mit Partikeln und Markierstoffen in Lockergesteinsgrundwasserleitern

Tracer	Eingabe [Menge]	Art der Eingabe	Distanz [m]	V_{MITTEL} [m/h]	Wiederfindung	Quelle
Fluorescein	k.A.	Bohrloch	70	1,8	ja	*Marti, 1975*
Coliforme Keime	$5,4 \times 10^{12}$			2,1	ja	
Streptokokken	$10,4 \times 10^{6}$			3,0	ja	
Sulforhodamin	5 kg	Bodenoberfläche 24 m²	15	k.A.	nein	*Marti, 1975*
E. coli	5×10^{13}			k.A.	nein	
Streptococcus faecialis	$1,5 \times 10^{13}$			k.A.	nein	
Fluorescein	5 kg	k.A.	60	2	ja	*Marti et al., 1979*
E. coli	$3,2 \times 10^{11}$	vermutlich Bohrloch		1,9	ja	
Staphylococcus aureus	$1,2 \times 10^{11}$			k.A.	nein	
Streptococcus faecialis	$8,0 \times 10^{10}$			2	ja	
Serratia marcescens	$8,4 \times 10^{11}$			k.A.	nein	
Fluorescein	5 kg	Bohrloch	90	9	ja	*Marti et al., 1979*
Bacillus pumilus	$7,5 \times 10^{8}$			k.A.	nein	
Bacillus subtilis	$2,25 \times 10^{13}$			7	ja	
Streptococcus faecialis	$4,5 \times 10^{9}$			10	ja	
Serratia marcescens	$2,33 \times 10^{13}$			7,8	ja	
apathogene Polioviren	$2,0 \times 10^{9}$			k.A.	nein	
Uranin	20 g	k.A.	12,5	k.A.	15 ppm	*Käss et al., 1983*
E. coli	$8,0 \times 10^{11}$			0,17	86/100 ml	
Uranin	$1,0 \times 10^{10}$	Bohrloch	10	2,5	k.A.	*Seiler, 1988*
E .coli	k.A.			3	$1,3 \times 10^{4}$	
Mikrosphären	$1,92 \times 10^{11}$	Bohrloch	1,5	k.A.	$<2 \times 10^{7}$	*Higgo et al., 1993*
Uranin	20 g	Bohrloch	100	0,16	k.A.	*Schaber, 1993*
Mikrosphären	$4,6 \times 10^{10}$			0,67	102 / L	
Uranin#	300 g	Bohrloch	12	6,7	1300 ppb	*Rossi et al., 1998*
			64	4,9	88 ppb	
Bakteriophagen H40/1	1×10^{13}		12	8,0	71'000 / mL	
			64	7,0	980 / mL	
Uranin#	18 g	Bohrloch	64	3,4	k.A.	*Niehren, 1999*
Mikrosphären	$3,6 \times 10^{10}$			5,8	40/mL	
Uranin	k.A.	Bohrloch	60	0,47	k.A.	*Kennedy, 2000*
Bakteriophagen H40	k.A.			0,96	0.005%	

(#gleiche Teststelle Wilerwald BE; k.A.: keine Angaben; Wiederfindungsrate: „ja" Partikel bei der Beobachtungsstelle nachgewiesen; Wiederfindungsrate: „nein" kein Nachweis von Partikeln)

Tab. 5.4: Tracerversuche mit Partikeln und Markierstoffen in Karst

Tracer	Eingabe [Anzahl]	Art der Eingabe	Distanz [m]	V_{MITTEL} [m/h]	Wiederfindung	Quelle
Bakteriophage H4	6,1 x 10^{14}	Grube	1250	35	2,5%	*Auckenthaler et al. 2002*
Backteriophage H40	2,9 x 10^{14}			35	2,5%	
Mikrosphären	2,88 x 10^{11}			255	1,0%	
Uranin	50 g			32	23%	
Sulforhodamin[§]	8550 g	Doline	6250	18,1	80%	*Rossi et al., 1998*
Bakteriophage H6/1	1,49 x 10^{15}				1%	
Bakteriophage H40/1	9,18 x 10^{15}				21%	
Uranin[§]	5000 g	Doline	6550	168		*Rossi et al., 1998*
Bakteriophage H6/1	5,71 x 10^{13}			168		
Bakteriophage H40/1	1,68 x 10^{14}			177		
Uranin	5 kg	Schacht	2823	48	38,7%	*Käss, 1994*
Mikrosphären	4,55 x 10^{11}			43	0,061%	
NaCl	250 kg	Bohrloch	450	k.A.	65%	*Doerfliger, 1996*
Bakteriophage H6/1	5,43 x 10^{12}				3%	
Bakteriophage H4/4	3,63 x 10^{11}				6%	
Bakteriophage H40/1	1,57 x 10^{13}				40%	
Uranin	5 kg	Schluckloch	6150	k.A.	56%	*Doerfliger, 1996*
Bakteriophage H6/1	5,71 x 10^{13}				25%	
Bakteriophage H40/1	1,68 x 10^{14}				51%	
Bakteriophage Psf2	1,17 x 10^{13}				10%	

([§]Verschiedene hydrologische Bedingungen im selben Gebiet k.A.: keine Angaben)

Dass die Partikel trotz der erheblichen Filtration über weite Distanzen im Grundwasser transportiert werden können, kann den Tabellen 5.3 und 5.4 entnommen werden. Da insbesondere in Kluft- und Karstwassergrundleitern nach dem Eintrag aus dem Boden oder der ungesättigten Zone in die wasserführenden Schichten (gesättigte Zone) kaum mehr eine Filtration stattfindet, können Partikeln grosse Fliessdistanzen von über 6000 m zurücklegen. Im Gegensatz dazu stehen Lockergesteinsgrundwasserleiter, wo die gemessenen Fliessdistanzen in der Regel wesentlich kürzer sind, aber immerhin 100 m oder einige 100 m betragen können.

Ein weiterer Unterschied im Transportverhalten zeigt sich in der höheren Fliessgeschwindigkeit der Partikel gegenüber den Molekülen. Durch den Ausschluss der Partikel aus Feinporen findet keine Matrixdiffusion statt, das heisst, die Partikel werden bevorzugt in den präferenziellen Fliesswegen transportiert. Der Ausschluss von der Matrixdiffusion bewirkt zudem eine geringere Dispersion und somit ein schnelleres Aufsteigen und Abfallen der Durchgangskurven von Partikeln. Das Transportverhalten von Partikeln ist entscheidend bei der Festlegung von Grundwasserschutzzonen. In der Schweiz wurden deshalb bereits in den 70er Jahren Versuche mit partikulären Tracern in Grundwasser führenden Schottern durchgeführt (Marti, 1975; Marti et al., 1979).

Das Ziel der Arbeiten war, die Schutzzonen optimal der Gefährdung durch Mikroorganismen anzupassen. Die heutigen gesetzlichen Regelungen zum Trinkwasserschutz mit der Ausscheidung von Grundwasserschutzzonen in der Schweiz und in Deutschland basieren auf der Fliesszeit des Grundwassers, der Filterwirkung des Bodens und der Beschränkung des Eintrags von Mikroorganismen in den Schutzzonen (siehe dazu Kap. Schutz von Wasserfassungen). Da Partikel in Karstgebieten über weite Distanzen transportiert werden können, ist ein Schutz dieser Grundwasservorkommen daher hauptsächlich über die Beschränkung respektive die Verhinderung des Eintrages von Mikroorganismen zu erreichen. Die bisher durchgeführten Experimente mit Partikeltracern zeigen, dass in Zukunft bei der Ausscheidung von Grundwasserschutzzonen vermehrt auf das spezifische Verhalten von Partikeln geachtet und neben fluoreszierenden Tracern ebenfalls Partikel eingesetzt werden sollten.

Schlussfolgerungen

Die Abgrenzung hydrologischer Einzugsgebiete von Trinkwasserfassungen und die Ausscheidung von Grundwasserschutzzonen können meist nur aufgrund von Tracerversuchen vorgenommen werden. Dabei ist der Einsatz von Partikeltracer neben fluoreszierenden Stoffen zur Abschätzung der Gefährdung des Trinkwassers durch Mikroorganismen z.B. aus der Landwirtschaft oder lecken Kanalisationen empfehlenswert. Partikel widerspiegeln den Transport von Mikroorganismen realistischer als chemische Tracer. Sowohl in Lockergesteins- als auch in Karst- und Kluftgrundwasserleitern konnten Partikeltracer erfolgreich eingesetzt werden. Bei der Anwendung von Partikeltracern sind einige Punkte zu beachten. Die Dauer der Experimente sollte aufgrund der Überlebensfähigkeit der Keime einige Tage jedoch nicht überschreiten, obwohl teilweise bis zu einem halben Jahr nach der Tracereingabe biologische Partikel nachgewiesen werden konnten. Da die Wiederfindungsrate von Partikeln in der Regel kleiner und der analytische Aufwand grösser ist als bei fluoreszierenden Tracern, sollte vor dem Einsatz partikulärer Tracer das hydrogeologische Einzugsgebiet bekannt sein. Vorgängige Tracerversuche mit fluoreszierenden Tracern sind deshalb zweckmässig. Partikeltracer sollten immer zusammen mit chemischen Tracern verwendet werden. Aufgrund des unterschiedlichen Transportverhaltens können Aquifereigenschaften wie präferenzielle Fliesswege oder Matrixdiffusion dadurch besser sichtbar gemacht werden. Von den derzeit nutzbaren Partikeltracern eignen sich die Bakteriophagen am besten. Marine Phagen sind weder toxisch noch pathogen, kommen im Grundwasser in der Schweiz und in Deutschland natürlicherweise nicht vor (keine

Hintergrundbelastung) und sind relativ einfach zu analysieren. Sie erfüllen die von Keswick, 1982a genannten Kriterien für Tracer am besten. Mikrosphären haben ein Potential zur Verwendung als hydrogeologische Tracer. Wie die marinen Bakteriophagen sind sie weder toxisch noch pathogen. Durch die Wahl verschiedener Grössen und Fluoreszenzen sind Multitracerexperimente möglich. Die Nachweismethoden für Mikrosphären sind aber heute für einen Einsatz ausserhalb wissenschaftlicher Zwecke noch nicht anwendbar. Sie müssten dahingehend verbessert werden, dass tiefe Konzentrationen rasch und eindeutig von der Hintergrundbelastung unterschieden werden können.

6 Trinkwasseraufbereitung

Adrian Auckenthaler

Einleitung

Die Trinkwasserversorgungen in der Schweiz decken ihren Bedarf zu 39% aus Quellwasser, zu 43% aus gepumptem Grundwasser und zu 18% aus Seewasser (Michel und Schweizer, 1998). Etwa 38% des gefassten Quell- und Grundwassers erfüllen die gesetzlichen Anforderungen an Trinkwasser und bedürfen keiner Aufbereitung. Die übrigen 62% müssen in mehr oder weniger aufwändigen Aufbereitungsschritten vor dem Konsum behandelt werden. Im Vordergrund steht dabei die Inaktivierung und die Eliminierung von Mikroorganismen. In Deutschland stammt das Trinkwasser der öffentlichen Wasserversorgungen zu 65% aus Grundwasser, zu 26% aus Oberflächenwasser, einschliesslich angereichertes und uferfiltriertes Grundwasser und zu 9% aus Quellwasser (Anonym, 2002). Insbesondere bei den genutzten Oberflächenwässern und dem Quellwasser können mikrobiologische Belastungen des Rohwassers ein Problem darstellen.

Die Aufbereitung erfolgt meist mit einem einstufigen Verfahren, wie die Daten aus einigen Kantonen der Nordwestschweiz beispielhaft zeigen (Abb. 6.1). Den grössten Anteil bilden heute die UV-Anlagen mit rund 70%, gefolgt von den Aufbereitungen mit Chlor (Javelle, Chlorgas und Chlordioxid) Filteranlagen und der Ozonung. Während des Aufbaus der Wasserversorgungen in der Schweiz und in Deutschland im letzten Jahrhundert, wurde zur Desinfektion hauptsächlich Chlor eingesetzt. Die UV-Bestrahlung für die Trinkwasseraufbereitung war noch unbekannt. Mitte der 70er Jahre wurde das Problem der Bildung von unerwünschten chlororganischen Verbindungen bei der Chlorung des Trinkwassers erkannt. Dies ist mit ein Grund, weshalb nach Alternativen zum Chlor gesucht wurde. Der Vorteil der UV-Anlagen ist der geringe Kontroll- und Wartungsaufwand und eine Aufbereitung ohne Beimischung von Chemikalien ins Trinkwasser. Der Nachteil gegenüber dem Chlor ist, dass im Trinkwassernetz kein Desinfektionsmittel mehr wirkt und es deshalb zu Wiederverkeimungen kommen kann. Das genutzte Grund- und Quellwasser in der Schweiz weist aber im allgemeinen einen geringen DOC- und AOC-Gehalt (Dissolved Organic Carbon, Assimilable Organic Carbon) auf und hat somit ein geringes Wiederverkeimungspotential.

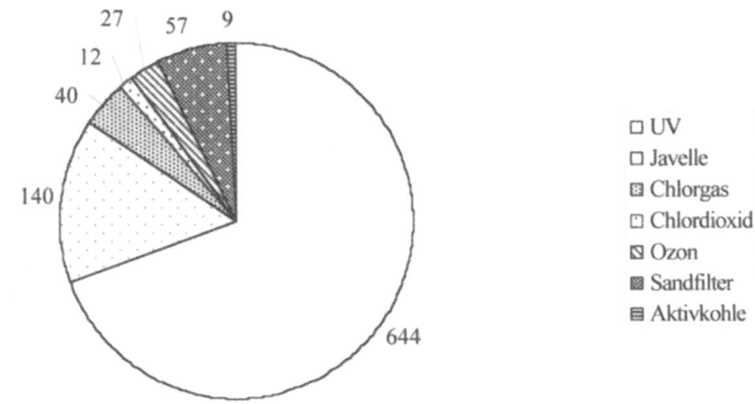

Abb. 6.1: Anzahl der Trinkwasseraufbereitungsanlagen in den Schweizer Kantonen Aargau, Solothurn, Basel-Land und Bern (Stand 2000). „Filter" und „Aktivkohle" bedeuten mehrstufige Verfahren, die Übrigen sind einstufige Verfahren. (Daten freundlicherweise von den kantonalen Trinkwasserinspektoren zur Verfügung gestellt)

Unter der Bedingung einer optimalen mehrstufigen Aufbereitung kann selbst in grossen Wasserversorgungen, die Oberflächenwasser nutzen, auf eine Chlorung als Netzschutz verzichtet werden (Forster, 1998).

Bei der Ozonung von DOC-haltigen Wässern bilden sich hingegen mikrobiologisch verfügbare Substanzen (AOC), die zu einer Wiederverkeimung im Trinkwassernetz führen können. Ozon sollte deshalb nur in Kombination mit einem Aktivkohlefilter eingesetzt werden, um die organischen Anteile zu entfernen. Dies entspricht jedoch nicht immer der heutigen Praxis.

Bei klarem, wenig turbidem Wasser können die einstufigen Verfahren im Hinblick auf die Indikatorkeime gute Dienste leisten. Neben diesen Keimen können aber weitere pathogene Mikroorganismen vorhanden sein, die nur mit kombinierten Verfahren wirksam eliminiert oder inaktiviert werden können. Weist das aufzubereitende Wasser starke Trübungsschwankungen auf, wie dies in Karstwässern oft der Fall ist, müssen vor der Desinfektion zuerst die Trübstoffe mittels Filtration aus dem Wasser entfernt werden. Je stärker die Belastung des Rohwassers mit Mikroorganismen und anderen Partikeln ist, desto aufwändiger wird das Verfahren, um jederzeit die gesetzlichen Anforderungen erfüllen zu können und desto professioneller muss auch die Betreuung der Anlage sein. Für eine Wasserversorgung gilt es abzuschätzen, ob sich der finanzielle Aufwand lohnt oder ob nicht günstiger Fremdwasser bezogen werden kann.

Als häufigstes Kriterium für den Einbau einer Aufbereitungsanlage gilt der Nachweis der fäkalen Indikatorkeime *E. coli* und Enterokokken im Trink- resp. Rohwasser bei Routinekontrollen. Mit diesen Kontrollen werden die Belastungsspitzen bei Niederschlagsereignissen aber meistens nicht oder nur unvollständig erfasst. Zudem wird nur in Ausnahmefällen auf pathogene Mikroorganismen, wie Protozoen (z.B. Cryptosporidium und Giardia) oder Viren (z.B. SRSV und Enteroviren) untersucht. Die Entscheidung für den Einbau einer bestimmten Aufbereitungsanlage wird somit oft aufgrund von praktischen Überlegungen gefällt und weniger nach genauen Kenntnissen des Einzugsgebietes und der mikrobiologischen Belastung des Wassers.

Die Wahl des adäquaten Verfahrens für eine Wasserversorgung muss im Einzelfall abgeklärt werden und ist von verschiedenen Faktoren abhängig:

- Risikoabschätzung (Einfluss aus Landwirtschaft, Kanalisation, Oberflächengewässern, etc.)
- Rohwasserqualität (Grundbelastung mit Fäkalkeimen, Trübung)
- Variabilität der Wasserinhaltsstoffe (z.B. in Karstgebieten)
- Wasserbezug von einer anderen Wasserversorgung (Nutzung des eigenen Wassers nur bei geringer Belastung)

Die Aufbereitung des Trinkwassers sollte nur die letzte Massnahme zur Sicherstellung einer einwandfreien Trinkwasserqualität darstellen. Vorrang hat die Reduzierung der Risiken durch die Verbesserung der Rohwasserqualität, welche durch die Ausscheidung von Schutzgebieten und die Verringerung der Emissionen erreicht werden muss.

Wirkung von Desinfektionsmitteln

Die Wirkung eines Desinfektionsmittels auf einen Mikroorganismus hängt ab vom Desinfektionsmittel selbst, vom physiologischen Zustand des Mikroorganismus, vom eingesetzten Reaktortyp und von der Wasserqualität. In diesem Abschnitt werden diese Aspekte für die am häufigsten verwendeten Desinfektionsmittel diskutiert.

Desinfektionsmittelreaktion

Als Mass für die Inaktivierungsleistung eines Desinfektionsmittels für einen bestimmten Mikroorganismus wird oft der C x t-Wert benutzt. Der Wert beschreibt den Zusammenhang zwischen der Desinfektionsmittelkonzentration C [mg/L] und der Einwirkungszeit (oder Kontaktzeit) t [min], die zur Inaktivierung eines bestimmten Mikroorganismus um eine bestimmte Reduktionsrate, beispielsweise 99%, notwendig ist.

$$k = C^n \times t$$

k [mg min/L] ist eine Konstante für ein gegebenes Desinfektionsmittel und eine bestimmte Reduktion des zu eliminierenden Organismus. n ist eine Konstante, auch als Verdünnungskoeffizient bezeichnet, mit Werten zwischen etwa 0,5 und 2,0 (Hoff, 1987). Bei der Erhöhung der Konzentration C verkürzt sich die Einwirkungszeit t bei gleichbleibender

Reduktionsrate, bei Verringerung von C verlängert sie sich. Soll die Reduktionsrate für einen Mikroorganismus erhöht werden und der Wert von n = 1, spielt es keine Rolle, ob C erhöht oder t verlängert wird. Ist der Wert von n > 1 heisst das, dass der Mikroorganismus mit Erhöhung von C schneller abstirbt als mit einer längeren Kontaktzeit t. Für n < 1 lohnt es sich hingegen, die Kontaktzeit zu verlängern an Stelle der Erhöhung der Desinfektionsmittelkonzentration (Langelais, 1991).

Bei der direkten Übertragung von C x t-Werten auf Trinkwasserversorgungen ist Vorsicht geboten, da die Wasserqualität einen starken Einfluss auf die Stabilität des Desinfektionsmittels haben kann. Das Desinfektionsmittel reagiert nicht nur mit den Mikroorganismen sondern auch mit anderen Wasserinhaltsstoffen. Die Desinfektionsmittelzehrung kann in „künstlichen (Labor-)Wässern" nicht richtig simuliert werden, weshalb auch die so ermittelte Absterbekinetik der Mikroorganismen Unsicherheiten unterworfen ist. Versuche zur Inaktivierungsleistung eines Desinfektionsmittels sollten deshalb mit genutztem Trinkwasser durchgeführt werden.

Physiologischer Status

Der physiologische Status der untersuchten Mikroorganismen, wie z.B. die Ausbildung der Zellwand und das Verhältnis von Oberfläche zu Volumen ist für die Bestimmung der Desinfektionsleistung ebenfalls sehr wichtig. Schnell und unter optimalen Bedingungen mit reichlichem Nahrungsangebot gezüchtete Stämme sterben rascher ab, als unter erschwerten, „natürlichen" Bedingungen gezüchtete (Berg, 1987). Ebenso scheinen Wildtypen von Mikroorganismen resistenter gegen Desinfektionsmittel zu sein, als Laborstämme.

Reaktortyp

Der Reaktortyp spielt bei der Desinfektion in einem Wasserwerk eine entscheidende Rolle und kann bei konzentrationsmässig gleicher Desinfektionsmittelzugabe zu erheblichen Differenzen in der Inaktivierungsleistung führen. Pfropfströmreaktoren haben einen viel höheren Wirkungsgrad als ideal gerührte Reaktoren (completely stirred tank reactor, CSTR). Für die CSTR sollte anstatt der C x t-Wert besser der C x ϑ-Wert benutzt werden, wobei ϑ die Aufenthaltszeit im Reaktor ist (ϑ = Volumen / Durchfluss). Die C x t- und C x ϑ-Werte sind nicht austauschbar. Severin, 1987 hat die Inaktivierung von *E. coli*, *C. parasilosis* und f2-Virus mittels UV-Strahlung in einem CSTR (Iϑ; I: Intensität der Bestrahlung) und einem Batchreaktor (It) untersucht. Die I x ϑ-Werte lagen für alle drei Mikroorganismen um den Faktor 1000 höher als die It-Werte.

Die gesetzlichen Vorschriften für die Wasseraufbereitung in den USA (Surface Water Treatment Rule) definieren die Kontaktzeit für die Berechnung der C x t-Werte als die Zeit, innerhalb der 10% des Wassers durch den Reaktor strömt (Von Huben, 1991). Diese Definition ist allerdings eher zur Beschreibung der Desinfektion mit Chlor oder Chlordioxid geeignet, bei Ozon kann es Schwierigkeiten geben, falls nicht ein ideal gerührter Reaktor

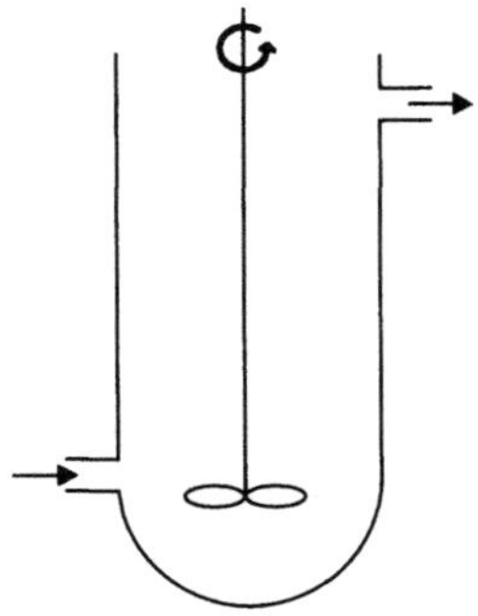

Mischreaktor (CSTR)

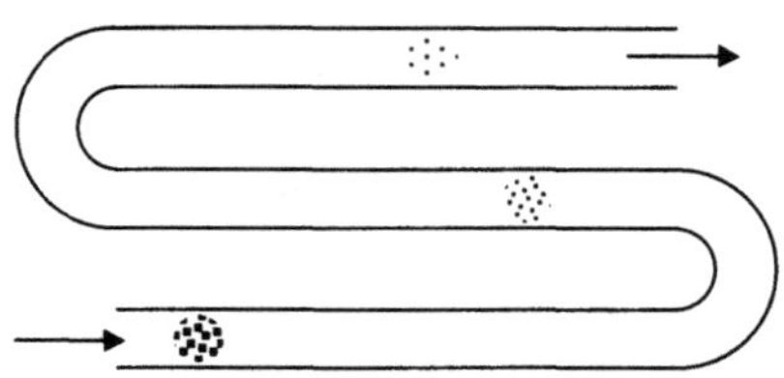

Pfropfströmreaktor oder Batchreaktor

Abb. 6.2: Beim Mischreaktor können Kurzschlüsse zwischen Ein- und Austritt bestehen, was dazu führt, dass die Reaktion innerhalb einzelner Wasserpakete noch nicht abgeschlossen ist, wenn sie den Reaktor verlassen. Im Pfropfströmreaktor findet die Reaktion innerhalb eines Wasserpaketes statt. Die Reaktionszeit beim Austritt aus dem Reaktor ist für alle Wasserpakete die gleiche.

betrachtet wird. Die Ozonkonzentration kann im Fall von Pfropfströmreaktoren aufgrund der Ozonzehrung auf Null absinken (von Gunten, 1998b).

Die Abb. 6.3 zeigt, dass die beiden Reaktoren bei kleinen Umsätzen vergleichbar sind. Bei gewünschten hohen Abbau- und Inaktivierungsraten von mehreren Grössenordnungen von Mikroorganismen in Desinfektionsprozessen, benötigen gerührte Reaktoren eine deutlich höhere Konzentration an Desinfektionsmittel als Pfropfströmreaktoren.

In der Literatur sind eine Vielzahl von C x t-Werten für verschiedene Desinfektionsmittel und Mikroorganismen zu finden (z.B. Sobsey, 1989). Die Tabelle 6.1 zeigt deutliche Unterschiede der Effizienz einzelner Desinfektionsmittel. Auffallend sind insbesondere die grossen Differenzen in den C x t-Werten für die 99% Reduktion der Mikroorganismen.

Der zugelassene Höchstwert für Chlor im Trinkwassernetz liegt gemäss eidgenössischer Lebensmittelgesetzgebung in der Schweiz bei 0,1 mg/L (Anh. 4, FIV). Der Parameter Restchlor im Netzwasser ist in der neuen Trinkwasserverordnung in Deutschland, welche am 1. Januar 2003 in Kraft tritt, nicht mehr enthalten (Oehmichen et al., 2001). Wie der Tabelle 6.1 entnommen werden kann, ist die Inaktivierung der Protozoen Cryptosporidium und Giardia mit einer einfachen Chlorung nicht möglich. Hier sind weitergehende Aufbereitungsschritte erforderlich. Es ist jedoch zu beachten, dass die Unterscheidung von lebenden und abgestorbenen Oozysten von Cryptosporidien analytisch heikel ist. Die in der Literatur enthaltenen Reduktionsraten unterscheiden sich je nach Nachweisverfahren stark. In-vivo-Tests mit neonatalen Mäusen zeigen viel höhere Inaktivierungsraten als in vitro Tests mit der DAPI-Färbung oder dem Excystationsverfahren (Korich et al., 1990, Clancy et al., 1998).

Tab. 6.1: C x t –Werte für 99% Inaktivierung von Mikroorganismen für verschiedene Desinfektionsmittel bei 5°C

Mikroorganismus	Chlor [min x mg/L]	Chloramin [min x mg/L]	Chlordioxid [min x mg/L]	Ozon [min x mg/L]	Quelle
Giardia lamblia	47-150			0,5-0,6	*Hoff, 1987*
				0,53	*Wickramanayake et al., 1984b*
	104-122$^{\#}$		23$^{\#}$	1,4$^{\#}$	*Botzenhardt, 1994*
Giardia muris			11,2	0,27	*Langelais et al., 1991*
	30-630	1400	7,2-18,5	1,8-2,0	*Hoff, 1987*
				1,94	*Wickramanayake et al., 1984a*
Giardia sp.	240				*Langelais et al., 1991*
Cryptosporidium sp.	7200	7200§	78§	5-10	*Korich et al., 1990*
C. *parvum*	1440		>120	>5	*Botzenhardt, 1994*
E. coli	0,034-0,05	95-180	0,4-0,75	0,02	*Hoff, 1987*
	3-4‡		1,2‡	0,012-0,04‡	*Botzenhardt, 1994*
				0,009‡	*Wuhrmann et al., 1955*
	0,032-0,05$^{\nabla}$				*Kaneko, 1998*
Streptokokken	0,17-0,84‡				*Weidenmann et al., 1997*
Polio 1 Virus	1,1-2,5	700-3740	0,2-6,7	0,1-0,2	*Hoff, 1987*
		567			*Shin et al., 1998*
Rotavirus				0,006-0,06	*Vaughn et al., 1987*
	0,01-0,05	3810-6480	0,2-2,1	0,006-0,06	*Hoff, 1987*
Norwalk Virus		360			*Shin et al., 1998*
Phage f2	0,08-0,18				*Hoff, 1987*
Phage MS2		1167			*Shin et al., 199*

(§ für 90% Inaktivierung in vivo, $^{\#}$ für 99,9% Inaktivierung, ‡ für 99,99% Inaktivierung$^{\nabla}$ bei 30°C, Trübung mit 1-5 mg/L Kaolin)

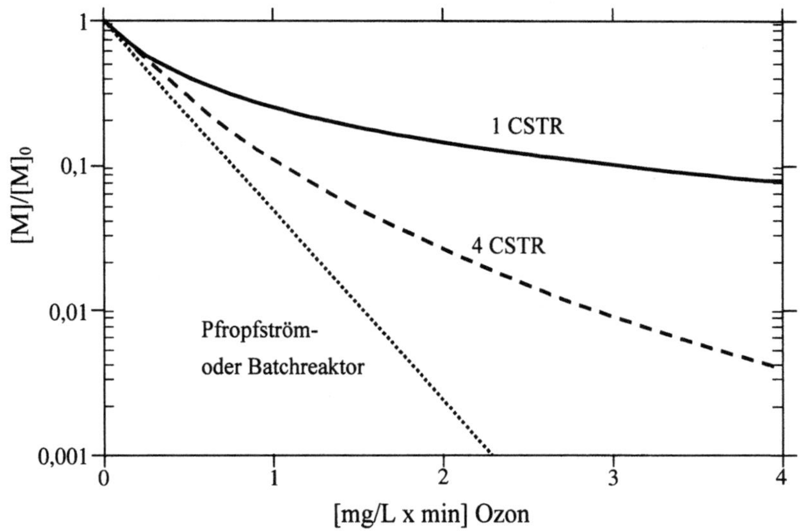

Abb. 6.3: Inaktivierung von Mikroorganismen als Funktion des C x t –Wertes bei der Ozonung mit Einfluss des Reaktortyps. CSTR: completely stirred tank reactor (gemischter Reaktor), bei mehreren CSTR sind diese in Serie geschaltet (nach Stettler et al., 1998, gwa, 78(11):876-890)

In der Schweiz und in Deutschland werden zur hygienischen Kontrolle des Trinkwassers die klassischen Indikatoren für fäkale Verunreinigungen, *E. coli* und Enterokokken untersucht. *E. coli* stirbt bereits bei einer geringen Desinfektionsmittelkonzentration und nach kurzer Kontaktzeit ab. Die C x t-Werte liegen bei 0,034 bis 0,05 mg min/L für Chlor, die meisten anderen Mikroorganismen weisen höhere C x t –Werte für die 99% Inaktivierung auf. *E. coli* eignet sich somit nicht für eine Endproduktekontrolle von aufbereitetem Trinkwasser. Um sich ein Bild über den mikrobiologischen Zustand eines Trinkwassers machen zu können, ist deshalb die Untersuchung des Rohwassers (unaufbereitetes Trinkwasser) notwendig. Eine erhöhte Anzahl Indikatorkeime im Rohwasser deutet darauf hin, dass weitere pathogene Keime vorhanden sein können. In Zukunft sollte das Trinkwasser deshalb auch auf pathogene Keime untersucht werden, denn nur so lässt sich ein klares Bild der möglichen Gefährdung zeichnen und die Trinkwasseraufbereitung der vorherrschenden Situation anpassen.

Ozonungsverfahren

Ozon (O_3) ist ein aus dreiatomigen Molekülen zusammengesetzte Form des Sauerstoffes. In höheren Konzentrationen ist Ozon ein Gas mit stechendem, durchdringendem Geruch. Es bildet sich unter Einwirkung von atomarem Sauerstoff (O) auf molekularen Sauerstoff (O_2), das aber leicht wieder zerfällt nach:

$$2\,O_3 \rightarrow 2\,O_2 + 2\,O \quad \text{und} \quad 2\,O \rightarrow O_2$$

Durch das Auftreten von molekularem Sauerstoff ist Ozon eines der stärksten Oxidationsmittel und in höheren Konzentrationen stark giftig. Ozon wird als Oxidations- und Bleichmittel und in der Wasseraufbereitung benutzt.

Erste Experimente in der Wasseraufbereitung mit Ozon fanden 1886 in einer Wasserversorgung in Frankreich statt. In der Schweiz wurde 1950 die erste Desinfektion von Grundwasser mit Ozon in Füllinsdorf BL errichtet. Das Ozon wurde damals noch direkt in den Brunnenschacht eingegeben (Courbat et al., 1999). Die Ozonkonzentration betrug 0,5 mg/L bei der Abgabe ins Trinkwassernetz und 0,2 mg/L nach 300 m.

Bei neuen Anlagen wird heute normalerweise nach der Ozonung ein Aktivkohlefilter eingebaut. Dadurch kann für die Desinfektion eine genügende Menge Ozon ins Wasser zugegeben, und die Konzentration des Ozons im Trinkwassernetz trotzdem tief gehalten werden. Bei älteren Ozonanlagen ohne Aktivkohle ist eine wirkungsvolle Desinfektion ohne erhöhte Ozonkonzentration im Netz kaum möglich. In der Schweiz gilt im Trinkwassernetz eine Höchstkonzentration von 0,05 mg Ozon/L (Anh. 4, FIV). In Deutschland dürfen aufgrund der neuen Trinkwasserverordnung, welche am 1. Januar 2003 in Kraft tritt, zukünftig für die Aufbereitung von Trinkwasser nur noch Stoffe verwendet werden, die vom Bundesministerium für Gesundheit (BMG) in einer Liste bekannt gemacht wurden (Castell-Exner et al., 2001). Die Liste wird auch die Restkonzentrationen enthalten.

Die Herstellung von Ozon erfolgt aus molekularem Sauerstoff oder aus Luft durch die Korona-Entladung in Röhren oder Plattenkondensatoren. Eine klassische Produktionslinie umfasst:

Gasquelle → Staubfilter → Gastrockner → Ozon Generator
→ Kontaktkolonne mit Wasser → Aktivkohle für überschüssiges Ozon

Bei Luft als Gasquelle und bei guten Betriebsbedingungen entsteht rund 1 mol N_2O_5 pro 100 mL Ozon, das weiter zu Salpetersäure reagieren kann. Bei normalem Betrieb entsteht ca. 3 bis 5 g Salpetersäure (HNO_3) pro kg Ozon. Es ist ebenfalls wichtig, dass die Luft trocken ist, da sonst OH-Radikale entstehen können, die die Ozonproduktion hemmen (von Gunten, 1998b).

Geschichtlicher Hintergrund

Christian F. Schönbein, Professor in Basel, entdeckte 1839 durch Zufall das Ozon. Er ging dem bei der Elektrolyse des Wassers und bei Funkenentladung auftretenden „schwefligen Geruch“ nach und identifizierte ihn ausschliesslich anhand physikalischer Parameter als eigenständigen Stoff. Schönbein erbrachte den bis heute noch nutzbaren Nachweis durch sein Ozonometer, mit Kaliumjodid und Stärke präpariertes Papier, das durch Blaufärbung Ozon anzeigt. Er stellte das Vorkommen des Ozons in der Atmosphäre fest und initiierte 1853 in Österreich die erste Langzeitmessung des Ozongehaltes in der Atmosphäre. Über Jahre bemühte sich Schönbein auch die Zusammensetzung des Ozons aufzuklären. Dies gelang ihm jedoch nicht und er musste widerstrebend die Erklärung anderer anerkennen, das Ozon sei eine Modifikation des Sauerstoffs (Nolte, 1999).

Reaktion von Ozon mit Molekülen und Mikroorganismen

Ozon ist im Wasser unstabil. Die Zerfallsgeschwindigkeit hängt von der Zusammensetzung des Wassers ab (Hoigné und Bader, 1994). Ein hoher pH-Wert, ein hoher DOC und eine tiefe Alkalität führen zu einer geringen Stabilität. Umgekehrt führt ein tiefer pH-Wert, ein tiefer DOC und eine hohe Alkalität zu einer hohen Stabilität.

Dieser Umstand kann in Wasserversorgungen, die bezogen auf diese Parameter sehr unterschiedliche Wässer nutzen, Probleme verursachen (Abb. 6.4). Bei mehreren Wasserbezugsquellen wird je nach Wasserverbrauch oder Wasserqualität beispielsweise nur die eine oder andere Quelle genutzt. Dadurch ändert sich jeweils die Wasserzusammensetzung bezüglich der Ozonung was eine zu geringe Desinfektion des Trinkwassers zur Folge haben kann.

Neben der Reaktion mit Molekülen oder Mikroorganismen wird Ozon auch in einer Kettenreaktion, ausgehend von dem Initiator Hydroxid (OH^-, erhöhter pH-Wert) abgebaut (siehe Abb. 6.5). Das Hydroxid reagiert mit Ozon (O_3) zu Superoxidion ($^\bullet O_2^-$ / HO_2). $^\bullet O_2^-$ transferiert sein Elektron auf weiteres Ozon worauf sich das Ozonidion $^\bullet O_3^-$ bildet, das zum Teil zu HO_3 protoniert wird und sofort zu Sauerstoff (O_2) und einem OH-Radikal zerfällt. $^\bullet OH$ reagiert sehr schnell mit organischen Wasserinhaltsstoffen. Das so gebildete organische Radikal überträgt, nach Anlagerung von Sauerstoff (O_2), häufig noch ein Wasserstoffatom (H) auf Sauerstoff. Somit wird wieder ein Superoxidion ($^\bullet O_2^-$ / HO_2) gebildet und die Kettenreaktion setzt sich fort. Bicarbonat und Carbonat fangen OH-Radikale ohne Bildung von Superoxidionen und wirken so als Inhibitoren der Kettenreaktion (von Gunten, 1998a). Die Reaktionen von OH-Radikalen mit Molekülen oder Mikroorganismen bezeichnet man als weitergehende Oxidation (advanced oxidation).

Verschiedene Parameter beeinflussen die Desinfektionseffizienz von Ozon. Die Geschwindigkeit von chemischen Reaktionen nimmt bei einem Temperaturanstieg von 10°C um das Zwei- bis Dreifache zu. Bei höherer Temperatur nimmt aber die Löslichkeit von Ozon im Wasser ab. In einem Temperaturbereich von 0 bis 30 °C wird jedoch bei höherer Temperatur eine bessere Desinfektionswirkung erzielt und die Ozonreaktion überwiegt alle anderen Effekte der Ozoninstabilität. Der pH-Wert scheint nur einen geringen Einfluss auf die bakterizide oder virizide Wirkung zu haben (Diaper, 1972, Langelais et al., 1991). In der Literatur zeigen die Werte jedoch eine grosse Streuung. Die Ionenstärke hat im Bereich des Mineralisationsgehaltes von Trinkwasser (um 1000 mg/L in vielen Wässern) keinen signifikanten Einfluss (von Gunten, 1998b).

Die Mikroorganismen sind meist nicht frei im Wasser, sondern an Partikel, wie beispielsweise Aggregate oder mineralisches oder organisches Material, gebunden. Für die Desinfektion ist es entscheidend, um welche Art von Partikel es sich handelt. Unoxidierbares mineralisches Material hat keinen negativen Effekt, ebenso zur Flockung von Partikeln zugegebenes Aluminiumsulfat. Organisches Material aber führt zu einer starken Ozonzehrung und hat somit eine negative Wirkung auf die Desinfektion. Allgemein sollte das Wasser möglichst frei von Trübung und organischer Substanz sein, um eine optimale Desinfektion zu erreichen (Langelais et al., 1991).

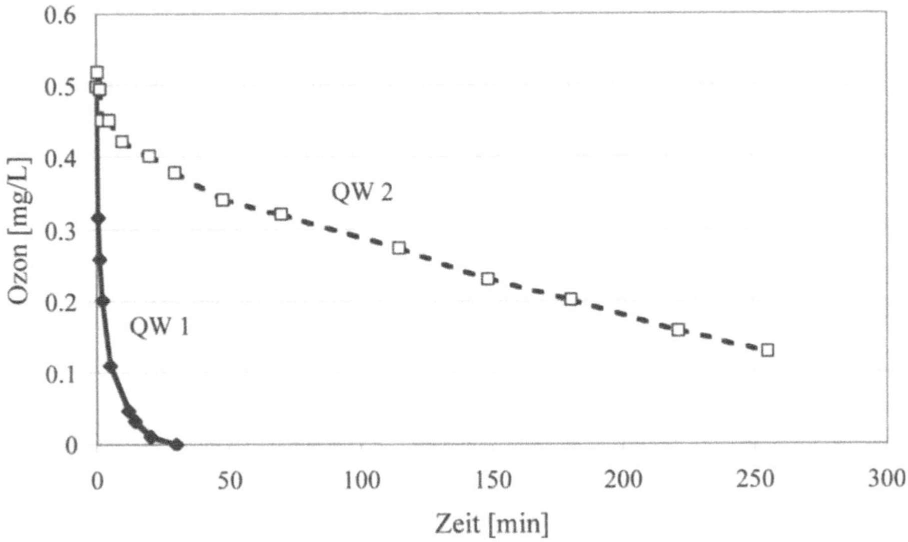

Abb. 6.4: Ozonabnahme in zwei Quellwässern in der Nordwestschweiz bei einer Dosierung von 0,5 mg/L Ozon. Beide Quellen (QW 1 und QW 2) werden von derselben Wasserversorgung genutzt. Die Wasserqualitätsparameter der Quellen sind: QW 1: pH 7,58; Ges. Härte 24,5 frH°; Ca^{2+} 90 mg/L; UV-Absorption 4,5/m; DOC 1,5 mg/L; QW 2: pH 7,33; Ges. Härte 49,4 frH°; Ca^{2+} 152 mg/L; UV-Absorption 1,0/m; DOC 0,5 mg/L

Das Ozon verfügt über verschiedene Mechanismen für die Inaktivierung von Mikroorganismen. Der erste Angriffsort bei Bakterien ist die Membran. Im weiteren Verlauf der Reaktion werden auch Enzyme inaktiviert. Es kommt schliesslich zur Lysis der Zellmembran und auch die DNA wird zerstört. Bei den Viren werden zuerst hauptsächlich die Proteine des Capsids zerstört (Cronholm et al., 1976). In einem zweiten Schritt wird ebenfalls die DNA oder RNA zerstört.

Koagulationseffekte von Ozon

Die Koagulationswirkung von Ozon für die Partikelentfernung wird kontrovers diskutiert (Langelais et al., 1991) und kann je nach Rohwasserbeschaffenheit bezüglich des DOC und des Calciumgehaltes sehr unterschiedliche Ergebnisse zeigen. Das Koagulieren von Partikeln bei der Aufbereitung von Ozon in Kombination mit Sandfiltern ist erwünscht, um die Wirkung des Filtrationsprozesses zu verbessern.

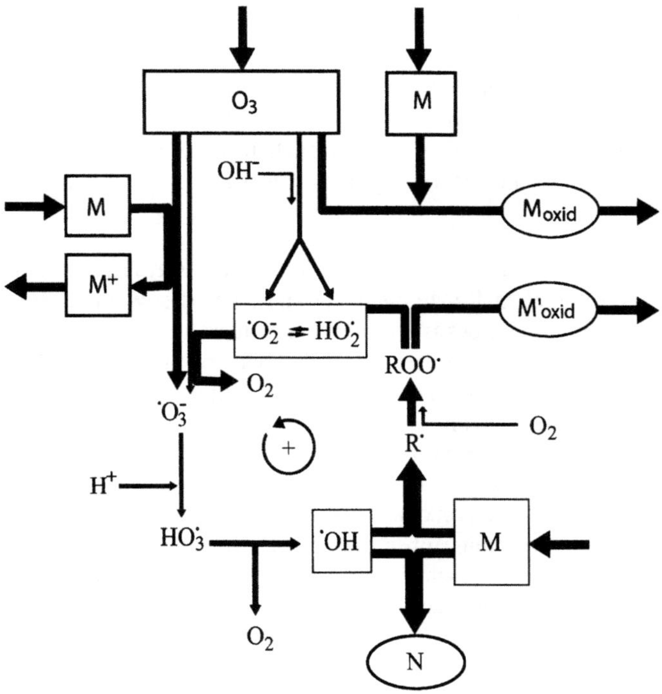

Abb. 6.5: Reaktionen von Ozon mit Molekülen und Mikroorganismen und Ozonabbau über die Kettenreaktion beginnend mit OH^- (vereinfacht nach von Gunten, 1998a, Mitt. Gebiete Lebensm. Hyg., 89:669-683)

Ein möglicher Mechanismus für die Koagulationswirkung des Ozons erfolgt über die organische Substanz (DOC). Die Kolloide erhalten vor allem durch höher molekulare Stoffe wie Huminsäuren eine Stabilisierung im Wasser. Je höher die Beladung mit DOC ist, desto mehr geht der relative Kollisionswirksamkeitsfaktor zurück. Als verantwortlicher Stabilisierungsmechanismus ist hier weniger die Erhöhung der negativen Partikelladungsdichte, als vielmehr die sterische Stabilisierung durch eine teilweise oder vollständige Bedeckung der Oberfläche mit Huminstoffmolekülen anzunehmen. Durch geringe Ozondosierung (0,2 bis 0,5 mg O_3/mg DOC) kann die Wirkung des DOC vollständig gebrochen werden, da die adsorbierbaren DOC-Anteile zu niedermolekularen, stärker negativen Produkten oxidiert werden, die kaum noch an Teilchenoberflächen adsorbieren. Damit könnte der Mikroflockungseffekt des Ozons teilweise erklärt werden. Bei zu hoher Dosierung von Ozon (0,5 bis 2,0 mg O_3/mg DOC) können aber die organischen Substanzen zu niedermolekularen Produkten mit mehr Säurefunktion oxidiert werden. Dadurch können zu viele stabilisierende Carboxylatgruppen gebildet werden. Werden in einer Trinkwasseraufbereitung neben dem Ozon auch noch Flockung und Filtration eingesetzt, kann sich in diesem Fall der benötigte Flockungsmittelbedarf ohne weiteres verdoppeln (Jeckel, 1995). Die Koagulationseffekte werden ebenfalls über den Gehalt der Calciumionen beeinflusst. Der genaue Mechanismus dieser Reaktion ist aber noch unklar (Winzenbacher, 1999).

Nebenprodukte bei der Ozonung

Als Nebenprodukte bilden sich häufig Aldehyde, Ketone und organische Säuren (Langelais et al., 1991). Die ozonolytisch oder durch OH-Radikale entstandenen Oxidationsprodukte sind meist polarer und deshalb besser wasserlöslich als die Ausgangsprodukte. Viele dieser Produkte sind mikrobiell besser verfügbar und werden in Aktivkohlefiltern schlechter zurückgehalten.

Bei bromhaltigen Wässern kann durch den Oxidationsprozess aus Bromid Bromat entstehen. Bromat wird als komplettes Kanzerogen klassiert, da es sowohl initiierende als auch promovierende Eigenschaften besitzt. In den Guidelines for Drinking Water Quality von 1996 errechnet die WHO auf der Basis der Risikohöhe von 10^{-5}, das heisst ein zusätzlicher Krebsfall pro 100‘000 Personen bei lebenslanger Exposition, eine zulässige Bromatkonzentration im Trinkwasser von 3 µg/L. Da aufgrund von analytischen Schwierigkeiten dieser Wert nicht zuverlässig überwacht werden kann, wurde pragmatisch ein vorläufiger Richtwert (Guideline level) von 25 µg/L festgelegt. Bei der Bewertung wurde eine lebenslange Aufnahme von 2 Litern Trinkwasser zugrunde gelegt (WHO, 1996; Rosskamp und Dieter, 1999). In der Verordnung über Fremd- und Inhaltsstoffe in der Schweiz ist ein Toleranzwert von 10 µg/L festgeschrieben (Anh. 4, FIV). Bei Bromidkonzentrationen von Trinkwasser unter 100 µg/L sollte der Richt- resp. Toleranzwert nicht überschritten werden (von Gunten und Hoigné, 1994).

Untersuchungen zur Ozonung von Rohwasser ergaben, dass die Bromidkonzentration in Schweizer Trinkwässern meist unter 25 µg/L liegt und die Bromatbildung in Bezug auf den Toleranzwert von 10 µg/L kein Problem darstellt (von Gunten und Sahli, 2000). In Deutschland wird der Bromatwert von 10 µg/L in einigen Fällen überschritten. Bromatkonzentrationen von über 25 µg/L sind allerdings in den untersuchten Trinkwässern nicht aufgetreten (Sacher et al., 1997). In der novellierten Trinkwasserverordnung in Deutschland gilt bis Ende 2007 ein Bromatgrenzwert von 25 µg/L, ab 2008 ein Wert von 10 µg/L (Oehmichen et al., 2001).

Wird neben Ozon gleichzeitig auch Chlor verwendet und die beiden Desinfektionsmittel kommen miteinander in Kontakt, kann Chlorat (ClO_3^-) entstehen. Chlorat wird ebenfalls als kanzerogen eingestuft. In der Schweiz gilt ein Toleranzwert von 0.2 mg Chlorat/L (Anh. 4, FIV). Das Ozon (O_3) geht mit Chlor (Hypochlorit, OCl^-) bei pH-Wert > 8 folgende Reaktion ein:

$$2O_3 + OCl^- \rightarrow 2O_2 + ClO_3^- \quad (1)$$

$$O_3 + OCl^- \rightarrow 2O_2 + Cl^- \quad (2)$$

wobei die Reaktion (2) mit 77% überwiegt und somit hauptsächlich Sauerstoff (O_2) und Chlorid (Cl^-) entstehen. Die Reaktion von Chloramin mit unterchloriger Säure (HOCl) führt zur Bildung von Nitrat (NO_3^-) und Chlorid (Cl^-), nicht aber zu Chlorat (ClO_3^-) (Hoigné und Bader, 1983).

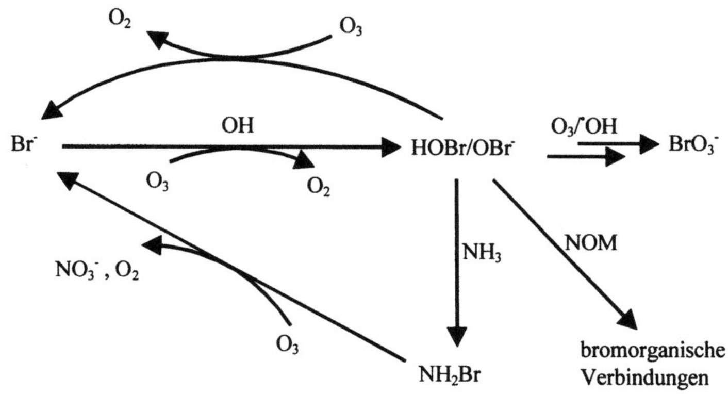

Abb. 6.6: Schematische Darstellung der Bromatbildung bei der Ozonung von bromhaltigen Wässern. Ozon und OH-Radikale oxidieren gelöstes Bromid (Br^-) langsam zu unterbromiger Säure, die weiterreagiert: 1) durch Oxidation zu Bromat oder Rückbildung von Bromid; 2) mit natürlichem organischen Material (NOM) zu bromorganischen Verbindungen und 3) mit Ammoniak unter Bildung von Bromamin, das wiederum mit Ozon unter Freisetzung von Bromid und Nitrat reagiert (nach von Gunten, 1998a, Mitt. Gebiete Lebensm. Hyg., 89:669-683)

Chlorungsverfahren

Chlor ist das weltweit am häufigsten zur Trinkwasseraufbereitung eingesetzte Desinfektions- und Oxidationsmittel. Primäre Ziele des Chlorungsprozesses sind: Desinfektion, Entfernung von Geruchs- und Geschmacksproblemen, Verhinderung des Wachstums von Algen und Mikroorganismen, die bei der Koagulation und der Flockung interferieren, Verhinderung des Keimwachstums in und auf Filtern, Abbau von Hydrogen-Sulfid, Entfernung von Eisen und Mangan und die Entfärbung des Wassers (White, 1972).

Die Chlorung ist im Vergleich zu anderen Verfahren relativ billig und bei genügender Dosierung recht zuverlässig bei der Bekämpfung von pathogenen Keimen abgesehen von den Protozoen Cryptosporidium und Giardia (siehe Tab. 6.1). Der Vorteil von Chlor gegenüber Ozon liegt bei der hohen Persistenz, was eine lange Desinfektionswirkung bedeutet, die für den Trinkwassernetzschutz ausgenutzt wird.

Weidenmann et al., 1997 haben die Desinfektionswirkung von geringen Chlorkonzentrationen, wie sie im Netzschutz eingesetzt werden auf fäkale Streptokokken untersucht. Die Konzentration des freien Chlors (unterchlorige Säure und Hypochlorit) bei 10°C und pH 7,7 betrugen 0,05 mg/L, 0,1 mg/L, 0,2 mg/L und 0,4 mg/L. Die Zeitdauer für eine 99,99% Reduktion betrug bei aufsteigendem Chlorgehalt 3,4 bis 5,2 min beziehungsweise 2,8 bis 4,1 min, 1,7 bis 3,1 min und 0,8 bis 2,1 min. Diese Resultate zeigen, dass selbst geringe

Chlorkonzentrationen bei aufbereitetem Trinkwasser eine beträchtliche Reduktion von Bakterien haben können.

Reaktionen von Chlor

Das Chlor wird in der Trinkwasseraufbereitung hauptsächlich in Form von Javelle, Chlorkalk und Chlorgas zugesetzt. Die entsprechenden Reaktionen lauten:

$$NaOCl \rightarrow OCl^- + Na^+ \quad (1)$$

$$CaCl(ClO) \rightarrow Cl^- + OCl^- \, Ca^{2+} \quad (2)$$

$$Cl_2 + H_2O \rightarrow HOCl + H^+ + Cl^- \quad (3)$$

Die Reaktion (3) kann zu einer pH-Senkung, die Reaktionen (1) und (2) können zu einer pH-Erhöhung führen, wobei die Pufferkapazität der Wässer in der Schweiz meist ausreichend ist um die Protonen und Hydroxidionen abzupuffern (von Gunten, 1998b).

Die Speziierung der unterchlorigen Säure (HOCl) ist pH abhängig. Bis zu einem pH-Wert von 7 überwiegt HOCl, darüber überwiegt das Hypochlorit (OCl^-). Dies ist entscheidend, da die unterchlorige Säure wesentlich effizienter ist in der Abtötung von Mikroorganismen als das Hypochlorit. Die unterchlorige Säure und das Hypochlorit werden zusammen als freies oder aktives Chlor bezeichnet.

Das Chlor geht eine Reihe von Reaktionen mit Wasserinhaltsstoffen ein, eine davon ist die Bildung von Chloramin (NH_2Cl, $NHCl_2$, NCl_3). Die unterchlorige Säure (HOCl) bildet in Gegenwart von Ammoniak (NH_3) Chloramine, die auch als gebundenes Chlor bezeichnet werden. Die Chloramine haben eine viel geringere Desinfektionsleistung für Mikroorganismen, die sich frei im Wasser bewegen, das heisst nicht an Oberflächen anhaften als das freie Chlor (Tab. 6.1 und Abb. 6.7).

Ein Vorteil der Chloramine liegt darin, dass kaum Desinfektionsnebenprodukte entstehen. Bei der Bekämpfung von Biofilmen scheint Monochloramin zudem besser geeignet zu sein als das freie Chlor. LeChevallier, 1987 konnte zeigen, dass Monochloramin bis zu 600 Mal besser in Biofilme eindringt. Dort reagiert es relativ spezifisch mit Nukleinsäure, Tryptophane und schwefelhaltigen Aminosäuren (Jacangelo und Olivieri, 1985). Das freie Chlor hingegen reagiert bereits an der Oberfläche und dringt deshalb schlechter in den Biofilm ein.

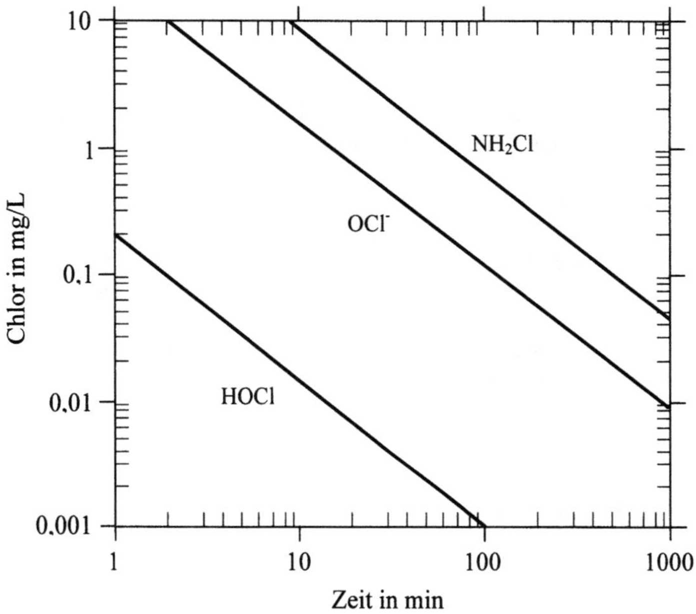

Abb. 6.7 : Reduktion von *E. coli* um 99% mit unterchloriger Säure (HOCl), Hypochlorit (OCl^-) und Monochloramin (NH_2Cl) bei 2 bis 6°C (vereinfacht nach White, 1992, Handbook of Chlorination and alternative Desinfection, 3ed. ,van Nostran Reinhold, S.239)

Nebenprodukte der Chlorung

Das Chlor kann im Wasser mit gelöstem natürlichem organischem Material (DNOM / NOM) Chloroform und andere Trihalomethane (THM) bilden, die potentiell gesundheitsgefährdend sind oder die Geruchs- und Geschmacksprobleme verursachen können (z.B. Chlorphenol). Mitte der 70er Jahre wurde diese Problematik erkannt und bis heute sind über 500 haloorganische Desinfektionsnebenprodukte identifiziert worden (Boorman et al., 1999).

Die Art und Konzentration des NOM, die Chlorkonzentration und die Kontaktzeit beeinflussen die Bildung der THM. Je höher der Chlorgehalt und je länger die Kontaktzeit, desto mehr THM werden gebildet.

Reaktionen von Chlordioxid

Chlordioxid ist ein relativ unstabiles Gas, das nicht verflüssigt oder komprimiert werden kann. Es muss deshalb direkt auf der Anlage produziert werden. Die Herstellung erfolgt beispielsweise aus Natriumchlorit mit Chlor (1) oder Natriumchlorit mit Salzsäure (2) (Sauter, 1994).

$$2\ ClO_2^- + HOCl \rightarrow 2\ ClO_2 + OH^- + Cl^- \quad (1)$$

$$5\ ClO_2^- + 4\ HCl \rightarrow 4\ ClO_2 + 5\ Cl^- + 2\ H_2O \quad (2)$$

Chlordioxid (ClO_2) ist ein sehr giftiges explosives Gas und sämtliche für die Herstellung benötigten Chemikalien gelten als gefährliche Arbeitsstoffe.

Die Speziierung des Chlordioxids ist im Unterschied zu Chlor pH unabhängig. Die oxidierende Wirkung des Chlordioxids beruht auf einem Elektronentransfer und die Reaktion läuft meist in zwei Schritten ab. Im ersten Schritt wird ein Radikal gebildet, das im zweiten Schritt wiederum mit Chlordioxid oder mit anderen Wasserinhaltsstoffen reagiert.

Die Lebensdauer des Chlordioxids im Wasser hängt stark von dessen chemischen Zustands ab. In reinem Wasser beträgt sie für konzentrierte Lösungen mehrere Wochen. Eine Nachdesinfektion mit Chlordioxid eignet sich deshalb als Netzschutz für aufbereitete Wässer.

Nebenprodukte von Chlordioxid

Der Vorteil von Chlordioxid gegenüber Chlor besteht darin, dass es Phenolverbindungen oxidiert und dadurch den Chlorphenolgeruch entfernt, kein Bromat und Trihalomethan bildet, Ammonium nicht oxidiert, Geruchs- und Geschmacksstoffe oxidiert, Färbungen durch Huminstoffe entfernt und kaum chlororganische Nebenprodukte bildet.

Das wichtigste Nebenprodukt ist das Chlorit. Chlorat kann sich ebenfalls bereits bei der Herstellung von Chlordioxid bilden. Chlorat kann aber auch wie beim Chlor bei gleichzeitiger Anwendung von Ozon entstehen (Haag und Hoigné, 1983).

UV- Strahlung

Zur Desinfektion des Wassers mit UV-Strahlen werden Niederdruck und Mitteldruckstrahler eingesetzt. Die Niederdruckstrahler emittieren Strahlung mit einer Wellenlänge von 254 nm. Die Mitteldruckstrahler verfügen über verschiedene Emissionsmaxima zwischen ca. 200 und 380 nm. Der Wirkungsmechanismus der UV-Strahlung beruht auf der Veränderung der DNA. Mitteldruckstrahler verändern ebenfalls die Proteine. Die DNA hat bei 254 nm ein Absorptionsmaximum, d.h. sie nimmt bei dieser Wellenlänge am meisten Energie auf. Dadurch werden die Basensequenzen verändert. Das hat einerseits einen Einfluss auf die Bildung von Aminosäuren und dadurch auf die Proteine. Die Zelle kann ihren Stoffwechsel nicht mehr aufrechterhalten und stirbt ab. Andererseits führt die Veränderung der Basensequenz auch zu einem Replikationsstopp der DNA. Die Zelle kann sich nicht mehr teilen (Kolch, 1998).

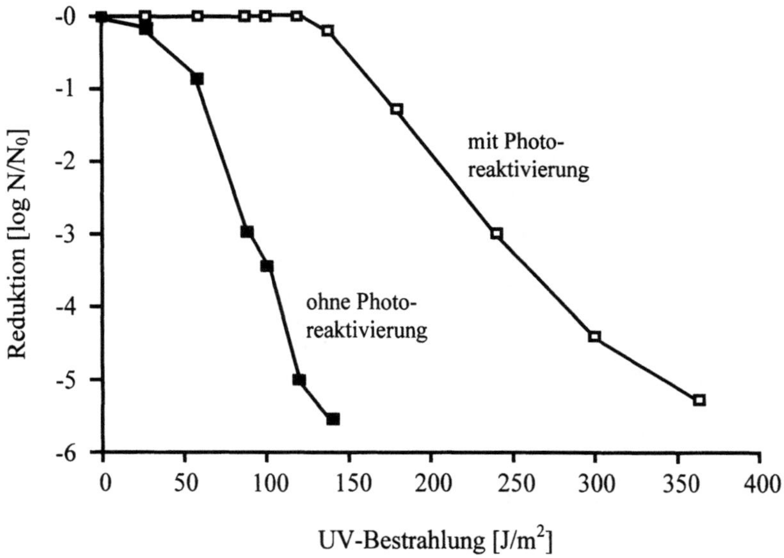

Abb. 6.8: Inaktivierung von *E. coli* ATTC 11229 nach UV-Bestrahlung ohne und mit Photoreaktivierung (nach Bernhardt, H. et al., 1994, gwf, 135(12):677-689)

Bakterien haben ein Reparatursystem, um Fehler in der DNA zu beheben. Die Reaktivierung der DNA kann bei Licht oder Dunkelheit stattfinden, wobei die Photoreaktivierung bei *E. coli* um mehr als 2 log-Einheiten grösser ist als in Abwesenheit von Licht (Shaban et al., 1997).

Bernhardt et al., 1994 haben die UV-Bestrahlungswerte für eine Reduktion um 4 log-Einheiten ohne und mit Photoreaktivierung für verschiedene Bakterienspezies untersucht. Die notwendige Dosis in J/m^2 liegt mit Photoreaktivierung um einen Faktor 1,2 bis 4,2 höher als ohne Photoreaktivierung.

Die Effizienz einer UV-Desinfektion für einen bestimmten Mikroorganismus ist abhängig vom Wasserfluss, resp. der Expositionszeit, der Intensität des Strahlers und der Transmission des Wassers. Während in der Praxis der Wasserfluss gemessen wird, werden die anderen Parameter nur über eine relative Transmissionsmessung bestimmt.

Sommer et al., 1997 haben festgestellt, dass in Systemen mit einem Niederdruckstrahler (12 W) die Herabsetzung der UV-Strahlung zu einer grösseren Abnahme der Reduktionsrate führte, als die Zunahme der Transmission des Wassers. In UV-Systemen mit drei Niederdruckstrahlern (80 W) konnte dieser Effekt nicht beobachtet werden. Hingegen sank die Reduktionsrate deutlich beim Ausfall eines von drei Strahlern. Die Transmission wurde bei diesen Experimenten konstant gehalten.

Das Produkt aus Leistung mal Expositionszeit ist nicht zwingend konstant für eine bestimmte log-Reduktion. Bei konstanter Dosis haben Sommer et al., 1998 bei höherer

Strahlungsleistung bei *E. coli* eine stärkere Inaktivierung gemessen als bei längerer Expositionszeit und niedrigerer Leistung. Für *B. subtilis* und Bakteriophagen war dies nicht der Fall.

Die rein rechnerische Ermittlung der Strahlendosis, die bei unterschiedlicher Wasserzusammensetzung auf Mikroorganismen wirkt, ist schwierig und mit Fehlern behaftet (Snozzi et al., 1999). In der österreichischen Prüfvorschrift Ö-Norm M 5873 und dem deutschen Arbeitsblatt des Deutschen Vereins des Gas- und Wasserfaches, DVGW W 294 wird deshalb die Prüfung der Desinfektionsleistung der Anlagen mit Biodosimetrie verlangt. Der Schweizerische Verein des Gas- und Wasserfaches, SVGW, hat diese Anforderungen in der Prüfanordnung W/TPW 152 „Anforderungen an Anlagen zur Desinfektion von Wasser mittels Ultraviolettstrahlern" auch für die Schweiz als verbindlich erklärt. Die geforderte UV-Dosis wird von 250 J/m^2 auf 400 J/m^2 erhöht. Bei dieser Dosis ist insbesondere die Photoreaktivierung der Mikroorganismen berücksichtigt und es kann eine log-Reduktion um 4 Einheiten von humanpathogenen Bakterien und Viren erreicht werden (Sommer et al., 1998; Bernhardt et al. 1994).

Neuere Untersuchungen von Bukhari et al., 1999 zeigen auch, dass mit Mitteldruckstrahlern (1 kW) bei einer Dosis von 410 J/m2 eine Cryptosporidieninaktivierung um mehr als 4 log-Einheiten möglich ist. Die Autoren verglichen eine in vivo Methode (neonatale Mäuse) mit zwei in vitro Methoden (DAPI-PI und Excystation). Die in vivo Methode zeigte eine deutlich grössere Inaktivierung (vgl. Tab. 6.2).

Da Mikroorganismen oft an Partikel haften, sollte die Trübung bei der Desinfektion mit UV-Strahlung wie auch bei den anderen einfachen Aufbereitungsverfahren den Wert von 1,5 TE/F nicht überschreiten (Bernhardt et al. 1994). Die Überwachung der Transmission gibt ebenfalls einen Hinweis auf die Trübung. Praktische Erfahrungen deuten aber darauf hin, dass UV-Anlagen mit einer garantierten Dosis von $\geq$ 400 J/m^2 teilweise relativ trübes Wasser nicht verwerfen, da der Intensitätssensor noch einen genügend hohen Wert misst. Als Folge davon können mit den Partikeln auch Mikroorganismen ins Leitungsnetz gelangen (Burren, 2000).

Clancy et al., 1998 haben ein erweitertes Verfahren (advanced UV) zur Inaktivierung von Cryptosporidien-Oozysten in klarem Wasser getestet. Die Verfahrenseinheit besteht aus zwei Reaktionskammern, die beide in der Mitte je einen 2 µm Filter enthalten. Beide Seiten des Filters werden bestrahlt. Die Oozysten werden im ersten Filter zurückgehalten und über eine bestimmte Zeit bestrahlt. Anschliessend wird der Wasserfluss in der Reaktionskammer umgekehrt und die Oocysten gelangen in den zweiten Filter, wo sie wiederum bestrahlt werden, bis die gewünschte Dosis erreicht ist. Die Filtergeschwindigkeit ist unabhängig von der Bestrahlungsdosis. Die erreichte hohe Bestrahlungsdosis von über 87'000 J/m^2 (14,6 mW/cm^2 x 10 min) führte zu einer Reduktion von über 4 log-Stufen.

Tab 6.2: Inaktivierung von Mikroorganismen mit UV-Strahlung

Mikroorganismus	UV Dosis [J/m2]	Inaktivierung [%]	Quelle
Cryptosporidium	190	>99,99[§]	*Bukhari et al.,1999*
	250	>99,996[§]	*Bukhari et al.,1999*
	87'000	99,9999	*Clancy et al., 1998*
Giardia	630	70	*Sobsey, 1998*
	630	<90	*Rice et al.,1981*
B. suptilis Sporen	100	30,8	*Sommer et al.,1998*
	300	94,87	*Sommer et al.,1998*
	500	99,9	*Sommer et al.,1998*
E. coli ATCC 11229	40	98,95	*Sommer et al.,1998*
	70	99,98	*Sommer et al.,1998*
	90	99,994	*Bernhardt et al., 1994*
	100	99,99	*Sommer et al.,1998*
	630	99,9	*Rice et al., 1981*
E. coli Wildtyp	40	75,5	*Sommer et al.,1998*
	70	99,96	*Sommer et al.,1998*
	100	>99,999	*Sommer et al.,1998*
L. pneumophila	18	99,9	*Buttler et al.,1987*
Y. enterolytica	27	99,9	*Buttler et al.,1987*
C. jejuni	50	99,9	*Buttler et al.,1987*
Phage MS2	100	66,8	*Sommer et al.,1998*
	300	97,7	*Sommer et al.,1998*
	500	99,7	*Sommer et al.,1998*
Phage ϕ174	15	79,1	*Sommer et al.,1998*
	60	99,7	*Sommer et al.,1998*
	100	99,99	*Sommer et al.,1998*
Phage B40-8	90	83,0	*Sommer et al.,1998*
	150	96,3	*Sommer et al.,1998*
	210	99,5	*Sommer et al.,1998*

([§] Inaktivierung bestimmt mit neonatalen Mäusen, Mitteldruckstrahler (1kW))

Bildung von Nebenprodukten

Auch bei der UV-Behandlung des Wassers stellt sich die Frage nach unerwünschten Nebenprodukten. So kann beispielsweise die Bildung von Nitrit aus Nitrat in Mitteldruckstrahlern von Bedeutung sein. Unterhalb von 254 nm nimmt der Absorptionskoeffizient des Nitrat-Ions stark zu. Bei einer Bestrahlung von 50 mg/L Nitrat mit 500 J/m^2 bilden sich 147 µg/L Nitrit, der Toleranzwert von 100 µg/L wird dadurch überschritten (Anl. 2, Teil 2, Nr. 9, TrinkwV; Anh. 4, FIV). Um dies zu verhindern, muss die Strahlung unterhalb 235 nm abgeschnitten werden (Bernhardt et al., 1994).

Bei der Bestrahlung von Herbiziden, ausgehend von einer Konzentration von 0,1 µg/L, fanden von Sonntag und Schuchmann, 1992 nur eine sehr geringe Bildung von Nebenprodukten. Die Bestrahlungswerte bei der Trinkwasseraufbereitung haben ebenfalls nur einen unbedeutenden Einfluss auf den Abbau von Huminstoffen und führen damit nicht zu einer Erhöhung des Wiederverkeimungspotentials des Trinkwassers (Bernhardt et al. 1994).

Filtration

Der Prozess der Filtration dient dazu, Partikel vom Wasser abzutrennen. Die Filtration wird der Desinfektion vorgeschaltet, wenn eine zu starke Trübung des Wassers auftreten kann und dadurch die Desinfektion mit Ozon, Chlor oder UV nicht effektiv wirkt.

Langsam-Sandfiltration

Langsam-Sandfilter bestehen aus feinem Sand und werden bei kleinen Filtrationsraten betrieben (siehe Tab. 6.4). Meistens wird keine Vordesinfektion mit Chlor oder Ozon vorgenommen. Die geringe Filtrationsrate bewirkt eine lange Verweilzeit des Wassers über der Filterschicht und im Filter, was zur Ausbildung einer biologischen Aktivität im Filtrationsprozess führt, die entscheidend zur Reduktion der Verschmutzung beiträgt.

Die Langsam-Sandfilter (LSF) können in drei biologische Zonen eingeteilt werden, vergleichbar mit einem flachen See mit einem sandigen Untergrund. Im Wasser über dem Sand bildet sich eine Planktonschicht aus. An der Grenzschicht Sand/Wasser siedeln sich fadenförmige Organismen (filamentous attached organisms) und Detritus an. Die Gemeinschaft an der Grenzschicht wird auch als Schmutzdecke bezeichnet und weist die höchste Algenkonzentration im Filter auf. Im Interstitial der Sandschicht besteht ebenfalls eine beträchtliche biologische Aktivität. Die Mächtigkeit der aktiven Schicht unter der Schmutzdecke hängt unter anderem von der Sandkorngrösse und der Filtrationsgeschwindigkeit ab, beträgt aber meistens nur einige Zentimeter (meistens 10 bis max. 40 cm).

Die benthische Gemeinschaft im Filter operiert in einer aeroben Umgebung, da von oben kontinuierlich Sauerstoff dazukommt, wohingegen natürliche Sedimente meistens anaerob sind. Anaerobe Filterbedingungen führen zu erheblichen Qualitätsproblemen und sollten auf alle Fälle vermieden werden (Haarhoff und Cleasby, 1991).

Da die biolgische Aktivität im Langsam-Sandfilter eine entscheidende Rolle spielt, brauchen die Filter zum Erreichen ihrer vollen Leistung eine Vorlaufzeit, bis sich die Schmutzdecke ausgebildet hat. Die Entfernung der Schmutzdecke (filter scraping) führt zu einer Reduktion der Leistung.

Rumm et al., 1997 haben für einen Langsam-Sandfilter im Wasserwerk Echthausen, Deutschland die Organismendichten der Meiofauna (tierische Organismen im Grössenbereich von 42 µm bis 1 mm) untersucht. Ihre Hochrechnungen ergaben im Mittel 2670 Individuen pro Kubikmeter Filtersand (Min. 150 Individuen, Max. 15'388 Individuen). Der Austrag aus dem Filter lag bei 9 Individuen pro Kubikmeter Wasser. Die ausgetragenen Organismen sind sehr widerstandsfähig gegen die in der Trinkwasseraufbereitung gängigen Desinfektionsmittelkonzentrationen. Die Autoren messen dem Vorkommen von Invertebraten im Trinkwasser aber keine hygienische Relevanz bei.

Die physikalischen Mechanismen, die zur Partikelentfernung beitragen, sind die Sedimentation, die Interzeption und die Diffusion (siehe dazu Abschnitt Schnell-Sandfiltration und Abb. 6.8).

Die Länge der Filtrationszyklen, das heisst die Filterlaufzeit zwischen der Entfernung der Schmutzdecke (Scraping), ist stark von der Rohwasserqualität abhängig. Die Filterlaufzeit sollte mindestens 1,5 bis 2 Monate dauern. Kürzere Laufzeiten führen zu erhöhten Kosten und Aufwand. Das ideale Rohwasser sollte folgende Werte aufweisen (Cleasby, 1991):

- Trübung < 5 FNU
- Algen: keine starken Algenblüten, Chlorophyll-a < 5 mg/m^3
- Eisen: < 0,3 mg/L
- Mangan: < 0,3 mg/L

Um diese Anforderungen zu erfüllen, können unterschiedliche Aufbereitungsstufen vorgeschaltet werden. Ein Ziel bei der Verwendung der Langsam-Sandfiltration ist es, ein einfaches Aufbereitungsverfahren auch für kleinere Wasserversorgungen zu betreiben. Es ist deshalb aus technischer und ökonomischer Sicht nicht wünschenswert, einen Schnell-Sandfilter mit Flockungsmittel einzusetzen, um die Trübung vor dem Langsam-Sandfilter zu reduzieren. Eine Alternative ist die Kiesfiltration. Wegelein et al., 1987 haben einen horizontalen Kiesfilter (Horizontal-flow Roughing Filtration, HRF) mit drei Kammern betrieben. Die Kiesgrössen der einzelnen Kammern lagen bei 20 mm, 10 mm und 5 mm und die Filtergeschwindigkeiten bei 0,5 bis 2,0 m/h. Die Autoren kamen zum Schluss, dass mit diesem Verfahren die suspendierten Partikel befriedigend zurückgehalten werden und dadurch die Filterlaufzeit von Sandfiltern auf mehrere Monate verlängert werden kann. Selbst Oberflächenwässer mit zeitweiligen Trübungen von 1000 NTU können soweit aufbereitet werden, dass sie sich für die Langsam-Sandfiltration eignen (Cleasby, 1991).

Tab. 6.3: Reduktionsraten für Mikroorganismen bei der Langsam-Sandfiltration

Mikro-organismus	Biolog. Zustand des Filters	Filtrations-rate [m/h]	Sandkorn-grösse [mm]	Reduktions-rate [%]	Quelle
Giardia lamblia	ausgebildet	0,12	0,615	>99,94	*Hendricks et al., 1991*
	neu	0,12	0,615	>99,99	*Hendricks et al., 1991*
	neu	0.47	0,615	>99,06	*Hendricks et al., 1991*
Giardia sp.	neu	0,04 – 0,40	0,28 – 0,41	>98	*Bellamy et al., 1985a*
	ausgebildet	0,04 – 0,40	0,28 – 0,41	100	*Bellamy et al., 1985a*
Cryptospori-dium sp.	k.A.	k.A.	k.A.	99,997	*Timms, et al., 1995*
	k.A.	k.A.	k.A.	99,99	*Schuler, et al., 1991*
Coliforme Keime	Neu	0,04 – 0,40	0,28 – 0,41	85	*Bellamy et al., 1985a*
	gechlort	0,12	0,29	60*	*Bellamy et al., 1985b*
E. coli	neu	0,4	k.A.	95,1	*Pointer et al., 1977*
	ausgebildet	0,4	k.A.	99,2	*Pointer et al., 1977*
Gesamt-keimzahl (Standard plate count bacteria)	neu	0,04 – 0,40	0,28 – 0,41	88-91	*Bellamy et al., 1985a*
	ausgebildet	0,04 – 0,40	0,28 – 0,41	96-98	*Bellamy et al., 1985a*
Polio 1	ausgebildet	0,2	k.A.	92,46-99,98	*Pointer et al., 1977*
	ausgebildet	0,1	k.A.	99,997	*Pointer et al., 1977*
Bakteriophage T7	neu	0,4	k.A.	99,87	*Pointer et al., 1977*
	ausgebildet	0,4	k.A.	99,97	*Pointer et al., 1977*

(Zwischen den Testläufen wurde der Filter mit 5 mg/L NaOCl gespült um das Ausbilden der biologischen Aktivität im Filter zu verhindern; k.A. keine Angaben)

Tab. 6.4: Unterschiede zwischen Langsam- und Schnell-Sandfiltration (Haarhoff und Cleasby, 1991)

	Langsam-Filter	**Schnell-Filter**
Filtrationsrate	0,1 m/h	10 m/h
Wasser über Sandschicht	1,5 m	1,5 m
Sandtiefe	80 cm	80 cm
Retentionszeit über dem Sand	15 h	9 min
Retentionszeit im Sand (bei einer Porosität von 0,40)	3,2 h	2 min
Laufzeit	1-6 Monate	1-4 d

Schnell-Sandfiltration mit Flockung/Sedimentation

Bei der Schnell-Sandfiltration (auch Tiefenfiltration) ist die Grösse der Trübstoffe im Mittel wesentlich kleiner als die Porenweite der Filterschicht, so dass im Gegensatz zur Langsam-Sandfiltration Siebeffekte mit der Bildung einer Schmutzdecke auf dem Filter meist von untergeordneter Bedeutung sind. Die Wirkung der Tiefenfiltration beruht auf der Anlagerung der Trübstoffe an die innere grosse Filteroberfläche. Zur Abscheidung müssen die Teilchen an die Oberfläche der Filterkörner transportiert werden, um durch einen anschliessenden Haftvorgang zu einer stabilen Ablagerung kommen zu können (Gimbel, 1985).

Ausgehend von einer laminar durchströmten Filterschicht können drei Transportmechanismen zur Partikelabscheidung aus Flüssigkeiten unterschieden werden. Ein Partikel, das sich auf einer Stromlinie bewegt, kann durch seine eigne Grösse mit dem Filterkorn in Kontakt treten (Abb. 6.9, Fall A), dieser Prozess wird als Interzeption bezeichnet. Falls das Partikel eine grössere Dichte als die Flüssigkeit besitzt, wird es aufgrund der Gravitationskraft die Stromlinie verlassen und auf dem Filterkorn sedimentiert (Abb. 6.9, Fall B). Dieser als Sedimentation bezeichnete Prozess ist für Partikel mit relativ grossem Dichteunterschied zur Flüssigkeit und Durchmessern über 1 µm bedeutend. Für abnehmende Partikelgrössen unter 1 µm spielt zunehmend die Brown'sche Bewegung eine Rolle. Durch Zusammenstösse von Partikeln und Molekülen verursacht durch die thermische Bewegung können die Partikel die Stromlinien ebenfalls verlassen und mit der Filterkornoberfläche in Kontakt kommen. Dieser Prozess wird als Diffusion bezeichnet (Abb. 6.9, Fall C). Aufgrund der drei Prozesse ergibt sich für Partikelgrössen um 1 µm ein Minimum in der Kollektor-Effizienz (single collector efficiency) eines Filters. Das Minimum kann durch die Filtergeschwindigkeit, die Temperatur und die Dichte der Partikel leicht verschoben werden. Die meisten in der Trinkwasseraufbereitung unerwünschten Partikel liegen genau im Grössenbereich des Minimums der Kollektor-Effizienz (Yao et al., 1971).

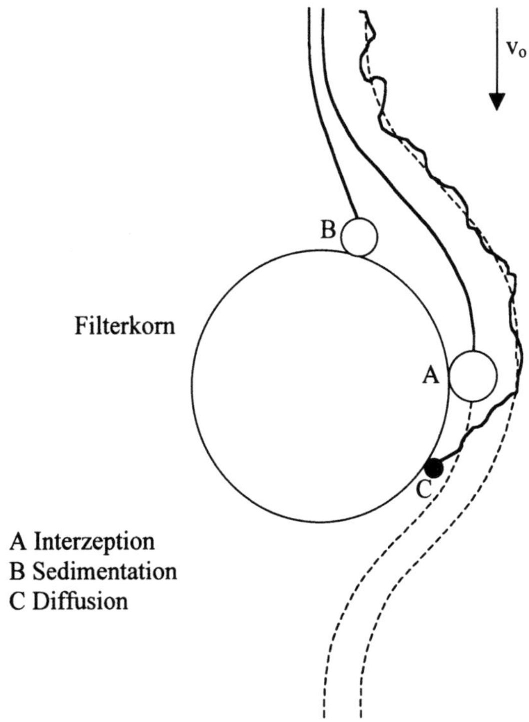

Abb. 6.9: Filtrationsmechanismen von Partikeln an Filterkörnern (verändert nach Yao et al., 1971, Environ. Sci. Technol., 5(11):1105-1112)

Die Oberflächen der im Wasser vorhandenen Partikel und der Filterkörner sind meist negativ geladen. Um eine Haftung der Partikel am Filterkorn zu erreichen, werden dem Wasser Flockungsmittel - meist Aluminium- und Eisensalze - zugesetzt, die zu zwei Destabilisierungsvorgängen führen. Bei der Adsorptions-Ladungsneutralisation wird durch die Adsorption positiv geladener Metallhydroxokomplexe die negative Oberfläche der Partikel verändert, so dass die elektrostatischen Abstossungskräfte verringert werden und eine Anlagerung der Partikel aneinander und an die Filteroberfläche stattfindet. Bei der Mitfällung werden voluminöse Hydroxid-Niederschläge gebildet, in die Partikel miteingeschlossen werden.

Je nach dem Ort der Flockungsmittelzugabe wird bei der Schnell-Sandfiltration zwischen der Flockungsfiltration (inline filtration) und der Flockenfiltration (direct filtration) unterschieden. Bei der Flockungsfiltration erfolgt die Beimischung des Flockungsmittels unmittelbar vor dem Sandfilter. Die Aggregation und die Flockenbildung finden in starkem Mass im Überstauraum des Filters sowie im Porenraum des Filters statt.

Tabelle 6.5: Reduktionsraten für Mikroorganismen bei der Schnell-Sandfiltration

Mikro-organismus	Flockungs-mittel	Menge [mg/L]	Filtrations-art	Reduktionsrate [log]	[%]	Quelle
Giardia sp.	Aluminium	6	Flo./Filt.	3,3	>99,999	*Nieminski, 1995*
	Poly-Al-Cl	5,9-8	Flo./Filt.	3,87	>99,999	*Nieminski, 1995*
	Aluminium	12	Sed./Filt.	3,40	>99,998	*Nieminski, 1995*
	Poly-Al-Cl	5,9-8	Sed./Filt.	3,26	>99,998	*Nieminski, 1995*
			Diverse		99,99–99,96	*LeChevallier, 1991*
Cryptosporidium sp.	Aluminium	6	Flo./Filt.	2,97	>99,999	*Nieminski, 1995*
	Poly-Al-Cl	5,9-8	Flo./Filt.	2,79	>99,999	*Nieminski, 1995*
	Aluminium	12	Sed./Filt.	2,98	>99,998	*Nieminski, 1995*
	Poly-Al-Cl	5,9-8	Sed./Filt.	2,25	99,994	*Nieminski, 1995*
Fäkal Coliforme	k.A.	k.A.	Sed./Filt.	>4,3	>99,999	*Payment, 1985*
Total Coliforme	k.A.	k.A.	Sed./Filt.	>5,0	>99,999	*Payment, 1985*
P. aeruginosa	k.A.	k.A.	Sed./Filt.	>3,3	>99,999	*Payment, 1985*
Gesamtkeimzahl (Standard plate count bacteria)	k.A.	k.A.	Sed./Filt.	>4,6	>99,999	*Payment, 1985*
Enteroviren	k.A.	k.A.	Sed./Filt.	1,05	99,98	*Payment, 1985*

(Poly-Al-Cl: Polyaluminium-Chlorid; Flo./Filt: Flockungsfiltration; Sed./Filt.: Flockung mit Sedimentation und anschliessender Filtration; Diverse: Durchschnitt von 66 Filteranlagen)

Von einer Flockenfiltration spricht man, wenn die Ausbildung der Flocken in einem getrennten Schritt erfolgt und schon weitgehend abgeschlossen ist, bevor es zur Partikelabscheidung kommt (Winzenbacher, 2000). Die Dosierung für Fe (III) und AL (III) liegt üblicherweise zwischen 0,6 und 12 mg/L (Nieminski et al., 1995; Winzenbacher, 1999).

Die Effizienz des Flockungsvorganges wird durch eine Vielzahl von chemisch-physikalischen Parametern, wie Wasserbeschaffenheit, Hydrodynamik der Einmischung der Metallsalzlösungen oder pH-Wert und DOC beeinflusst. Bei der Unterschreitung eines bestimmten Verhältnisses von Flockungsmittelzugabe zum DOC-Gehalt können feinstkolloidale Fällungsprodukte entstehen, die im Filter nur unzureichend zurückgehalten werden und zusammen mit Metallspezies hohe Flockungsmittelrestgehalte im Filter verursachen können (Winzenbacher et al., 1999).

Zur Überwachung des Filtrationsprozesses eignen sich online Partikelzähler oder Trübungsmesser. Bei optimalem Betrieb des Filters bei Trübungen von 0,1 bis 0,2 NTU im

Reinwasser kann eine Reduktion um 3 log Einheiten, beispielsweise bei Giardia, erreicht werden (Nieminski et al., 1995). Für die Cryptosporidien ist die Reduktion etwas geringer, wie Tabelle 6.5 zeigt.

Bei der Aufbereitung mit Schnell-Sandfiltern erfolgt vor dem Filter oft eine Desinfektion mit Chlor oder Ozon. Nach dem Filter findet immer eine zweite Desinfektion, ebenfalls mit Chlor, Ozon oder UV statt, um die nicht filtrierten Mikroorganismen und diejenigen, die im Filter entstanden sind (wie z.B. *Pseudomonas aeruginosa*), zu inaktivieren.

Die Reduktionsrate für Mikroorganismen im gesamten Aufbereitungsprozess ist durch die Desinfektion deshalb noch etwas höher als in Tabelle 6.5 angegeben. Durch die Kombination der verschiedenen Aufbereitungsschritte kann eine beträchtliche Reduktion von Mikroorganismen erreicht werden. Es ist jedoch unrealistisch, eine Reduktion von 12 log-Einheiten und damit eine Nulltoleranz für Mikroorganismen in aufbereitetem Trinkwasser zu fordern. In diesem Fall sollten eher höhere Anforderungen an das Rohwasser gestellt und definiert werden, welches Wasser noch aufbereitungswürdig ist (Payment et al., 1985).

Membranfiltration

Mit Membranfiltern können sehr kleine Partikel wie Viren und je nach Membran sogar gelöste inorganische Salze vom Wasser abgetrennt werden. Membranfilter werden in drei bis vier Kategorien aufgeteilt: Mikrofiltration 0,1 – 10 µm, Ultrafiltration 0,001 – 0,1 – (10) µm, Nanofiltration 0,0005 – 0,007 µm und Umkehrosmose 0,0001 – 0,001 µm (Mallevialle et al., 1993). Die gewählte Membranporengrösse bestimmt die im aufbereiteten Wasser vorhandenen Partikelgrössen. Eine Entfernung gelöster Wasserinhaltsstoffe erfolgt nur, wenn die Stoffe partikelgebunden sind. Damit unterscheiden sich Membranverfahren deutlich von den klassischen Flockungs- und Filtrationsverfahren, mit denen neben einer Partikelentfernung gleichzeitig auch eine Elimination gelöster höhermolekularer Wasserinhaltsstoffe erreicht werden kann (Lipp, 2000). Die Membranfiltration ist im Gegensatz zur Flockungs- resp. Flockensedimentation eine rein physikalische Methode zur Partikelabtrennung.

Als Materialien werden Hohlfasern aus Zellulose oder Acryl oder keramische Membranen mit Titan verwendet. Die Membranen aus Hohlfasern haben sich gegenüber Platten oder Rohrmodulen durchgesetzt (Birou, 2000). Die Hohlfasern haben beispielsweise einen Durchmesser von 0,5 mm. Die Filtermodule bestehen aus bis zu 20'000 dieser Hohlfasern. Der benötigte Prozessdruck, um das Wasser durch die Membran zu pressen, liegt bei der Mikro- und Ultrafiltration bei 0,7 bis 8 bar, bei der Nanofiltration bei 5-10 bar (Boller, 1994).

Die von der Membran zurückgehaltenen Partikel werden mit Druckluft oder Reinwasser (aufbereitetes Wasser) wieder entfernt und als Schlammwasser weggespült. Eine typische Spülfrequenz beträgt bei einem Karstwasser zwei bis vier Spülungen pro Stunde und Membran (Birou, 2000). Bei der Rückspülung mit Druckluft werden rund 10-15% des Rohwassers, bei der Rückspülung mit Wasser rund 5-10% des Reinwassers gebraucht (Jacangelo et al., 1991). Für den Betrieb einer Membranfilteranlage muss eine Wasserversorgung also auch bei tiefen Wasserständen oder geringer Quellschüttung eine zusätzliche Reserve von 5-15% an Rohwasser haben.

Falls die Druckdifferenz nach der Spülung der Membran nicht auf einen Basiswert zurückkommt, muss die Membran chemisch gereinigt werden. Die verwendeten Chemikalien sind je nach Rohwasserbeschaffenheit Säuren, Laugen, Wasserstoffperoxid oder Detergentien. Zwischen zwei chemischen Reinigungen liegt ein Zeitraum von zwei Wochen bis 6 Monaten je nach Rohwasserbeschaffenheit (Lipp, 2000).

Für die Aufbereitung mit Membranverfahren spielt die Rohwasserqualität weniger eine Rolle, sie hat einzig einen Einfluss auf die Durchflussmenge durch die Membran. Die Mikro- und Ultrafiltration zeigen denn auch sehr hohe Reduktionsraten von Mikroorganismen, teilweise von über 7 log-Einheiten. Der Tabelle 6.6 ist aber auch zu entnehmen, dass Oozysten von Cryptosporidien selbst nach einem Mikrofilter mit 0,2 µm Porendurchmesser nachgewiesen werden konnten. Als mögliche Ursache kommen Fehler in der Membran oder die Deformationsfähigkeit der Oozysten in Frage (Drozd et al., 1997).

Bei der Wahl von Membranfiltrationsverfahren zur Trinkwasseraufbereitung müssen zuerst Pilotversuche im halbtechnischen Massstab durchgeführt werden, um die verfahrens technischen Eckdaten zu ermitteln. Allein aufgrund der Kenntnis der Wasserzusammensetzung kann im Einzelfall keine Aussage über die prinzipielle Eignung eines Membransystems gemacht werden (Lipp, 2000).

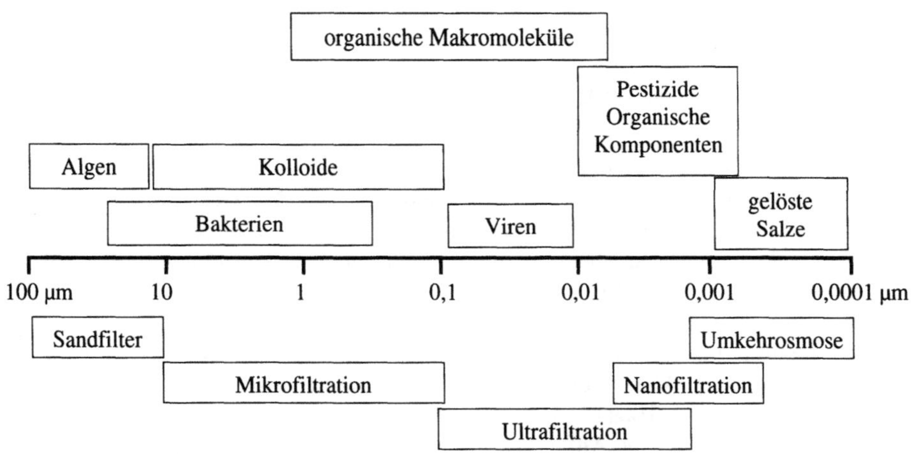

Abb. 6.10: Partikelgrössen und Filtrationsverfahren (verändert nach Mallevialle, 1993, Membrane filtration in drinking water treatment. European Water Filtration Congress, Proceedings, 15-17 March 1993, 17:4.1-4.16.)

Tab.6.6: Reduktionsraten von Mikroorganismen mit Membranverfahren

Mikroorganismus (Partikelgrösse)	Porengrösse der Membran	Partikel im Rohwasser [Anzahl/L]	Reduktions-rate [log]	Quelle
Virus MS2	80 kDa	10^5	≥5	*Jacangelo et al., 1991*
(0,02 µm)	80 kDa	10^7	≥7	*Jacangelo et al., 1991*
E. coli	80 kDa	10^5	≥5	*Jacangelo et al., 1991*
(1-3 µm)	80 kDa	10^7	≥7	*Jacangelo et al., 1991*
Giardia	80 kDa	10^5	≥5	*Jacangelo et al., 1991*
(7-14 µm)	80 kDa	10^7	≥7	*Jacangelo et al., 1991*
Cryptosporidium	0,25 µm	10^7	≥7	*Hirata et al., 1998*
(2-6 µm)	0,25 µm	10^9	≥7	*Hirata et al., 1998*
	13 kDa	10^7	7	*Hirata et al., 1998*
	13 kDa	10^9	7,6	*Hirata et al., 1998*
	0,2 µm		>4,3	*Drozd et al., 1997*

Überwachung von Aufbereitungsverfahren

Das eidgenössische Lebensmittelgesetz von 1992 verpflichtet die Betreiber von Wasserversorgungsanlagen in der Schweiz zur Selbstkontrolle (Art. 23, LMG). Früher wurden der Umgang mit Lebensmitteln und Prozesse, wie die Trinkwasseraufbereitung nur punktuell kontrolliert und mit Analysen (chemisch, mikrobiologisch, physikalisch) verifiziert. Ein Ziel des aktuellen Gesetzes ist es, Prozesse mit Hilfe der Selbstkontrolle dauernd zu überwachen, um bei Schwierigkeiten sofort eingreifen zu können. Als Mittel dazu dient die Risikoanalyse und die Bestimmung von kritischen Kontrollpunkten (CCP) bei der Herstellung, Behandlung, Lagerung, Abgabe und beim Transport von Lebensmitteln. In Tabelle 6.7 ist dieses HACCP-System (Hazard Analysis of Critical Control Points) gemäss Codex Alimentarius für verschiedene Aufbereitungsverfahren für Trinkwasser zusammengestellt. Werden die festgelegten CCP beachtet, und die Massnahmen bei Überschreitung der Grenzwerte eingehalten, sollten die gesetzlich vorgeschriebenen Toleranzwerte der Hygieneverordnung (HyV, 1995) eingehalten werden können.

Der Aufbau des HACCP-Systems ist schematisch in Tab. 6.8 für die Chlorung des Trinkwassers aufgeführt.

Tab. 6.7: Kurzfassung HACCP-Qualitätssicherung nach Raster Codex Alimentarius für die Trinkwasseraufbereitung mittels Ozon (nach KLBL, 1999)

Verfahrens-schritt	Gefahr	Para-meter	CCPs	Toleranzwerte	Überwachungs-system	Massnahmen	Dokumentation
Trübungs-messung ohne Filtration	Unzureichende Desinfektion wegen Wassertrübung	Trübung	CCP1	1,5 FNU bei Quell- und Grundwasser	Trübungsmessung, kontinuierlich mit Alarm	Verwurf, Kontrolle durch Brunnenmeister	Kontinuierlicher Schreiber, Verwurf und Massnahmen dokumentieren
Trübungs-messung nach Filtration	Unzureichende Desinfektion wegen zu geringer Partikel-entfernung	Trübung	CCP2	Trübung max. 0,3 FNU[#]	Trübungsmessung, kontinuierlich mit Alarm	Verwurf, Kontrolle durch Brunnenmeister	Kontinuierlicher Schreiber, Verwurf und Massnahmen dokumentieren
Ozonung	Unzureichende Desinfektion wegen zu geringer Ozon-konzentration	Konzen-tration in Kontakt-kolonne	CCP3	mind. 0,1 mg/L Bei Cryptosporidien mind. 1 mg/L	Online Sonde, mit Alarm	Verwurf, Kontrolle durch Brunnenmeister	Datenblatt, 1 Mal täglich, falls nicht automatisch aufgezeichnet
Ozonung	Gesundheitsgefährdung wegen zu hoher Ozon-konzentration im Netz-wasser	Ozon-messung im Reinwas-serbecken	CCP4	≤ 0,05 mg/L	Online Sonde, mit Alarm	Verwurf, Kontrolle durch Brunnenmeister	Datenblatt, 1 Mal täglich, falls nicht automatisch aufgezeichnet

Überprüfen des Systems (Text für Ozon, Chlor, UV und Filtration): Mikrobiologie Trinkwasser mind. 4 Mal pro Jahr; Ozon online Sonden wöchentliche Vergleichsmessung manuell, Kalibration, falls Abweichung; Chlor online Sonden wöchentliche Vergleichsmessung manuell, Kalibration, falls Abweichung; Kalibration Intensitätsmessung 1 Mal pro Jahr mit Dokumentation; Kalibration Trübungsmessung 1 Mal pro Monat mit Dokumentation, wenn das Gerät keine automatische Kalibration durchführt, oder gemäss Angaben des Herstellers.

[#]Anmerkung: In der Fremd- und Inhaltsstoffverordnung (Anh. 4, FIV) in der Schweiz ist ein Toleranzwert von 1 TE/F für Trinkwasser angegeben. In der neune Trinkwasserverordnung in Deutschland (Anl. 3) gilt ein Grenzwert von 1 NTU für Trinkwasser am Ausgang des Wasserwerks. Die angegebenen Werte stammen aus neuerer Literatur.

Tab. 6.7 (Fortsetzung): Kurzfassung HACCP-Qualitätssicherung nach Raster Codex Alimentarius für die Trinkwasseraufbereitung mittels Chlor

Verfahrensschritt	Gefahr	Parameter	CCPs	Toleranzwerte	Überwachungssystem	Massnahmen	Dokumentation
Trübungsmessung ohne Filtration	Unzureichende Desinfektion wegen Wassertrübung	Trübung	CCP1	1,5 FNU bei Quell- und Grundwasser[#]	Trübungsmessung kontinuierlich mit Alarm	Verwurf, Kontrolle durch Brunnenmeister	Kontinuierlicher Schreiber, Verwurf und Massnahmen dokumentieren
Trübungsmessung nach Filtration	Unzureichende Desinfektion wegen zu geringer Partikelentfernung	Trübung	CCP2	Trübung max. 0,3 FNU[#]	Trübungsmessung kontinuierlich mit Alarm	Verwurf, Kontrolle durch Brunnenmeister	Kontinuierlicher Schreiber, Verwurf und Massnahmen dokumentieren
Chlorung	Unzureichende Desinfektion wegen zu geringer Chlorkonzentration	Konzentration an einer Referenzstelle (konstante Chlorkonz.) bei der Aufbereitung oder im Netz	CCP3	Aufbereitung: 0,05 - 0,1 mg/L Netz: 0,02-0,08 mg/L	Online Sonde mit Alarm bei Aufbereitung oder im Netz	Verwurf, Kontrolle durch Brunnenmeister	Kontinuierlicher Schreiber, Verwurf und Massnahmen dokumentieren
Chlorung	Gesundheitsgefährdung wegen zu hoher Chlorkonzentration im Netzwasser	Chlormessung im Netzwasser	CCP4	max. 0,1 mg/L (LMB Kap. 27) Toxikologisch WHO max: 5 mg/L	Online Sonde mit Alarm bei Aufbereitung oder im Netz	Verwurf, Kontrolle durch Brunnenmeister	Kontinuierlicher Schreiber, Verwurf und Massnahmen dokumentieren
Chlorung	Unzureichende Desinfektion oder zu hohe Dosierung	Verbrauch Javelle pro aufbereitete Wassermenge	CCP5	14% Javellelauge mind. 0,2 dl/100 m^3 max. 0,8 dl / 100 m^3	Wassermengen- und Füllstandsmessung wöchentlich	Justierung der Dosierung	Datenblatt

Tab. 6.7: (Fortsetzung): Kurzfassung HACCP-Qualitätssicherung nach Raster Codex Alimentarius für die Trinkwasseraufbereitung mittels UV-Desinfektion

Verfahrens-schritt	Gefahr	Parameter	CCPs	Toleranzwerte	Überwachungssystem	Massnahmen	Dokumentation
Trübungs-messung	Unzureichende Desinfektion wegen Wasser-trübung	Trübung	CCP1	1,5 FNU bei Quell- und Grundwasser[#] 0,3 FNU bei auf-bereitetem Wasser nach Filter[#]	Trübungsmessung, kontinuierlich mit Alarm	Verwurf, Kon-trolle durch Brunnenmeister	Kontinuierlicher Schreiber, Verwurf und Massnahmen dokumentieren
UV-Desin-fektion	Unzureichende Desinfektion wegen zu altem Strahler	Strahlungs-dauer	CCP2	Max. Brenndauer (z.B. 10'000 Std.)	Betriebsstundenzähler	Strahler auswechseln	Datenblatt
UV-Desin-fektion	Unzureichende Desinfektion wegen zu altem Strahler	Strahlungs-dauer	CCP3	Anzahl Einschaltvorgänge	Einschaltzähler	Strahler auswechseln	Datenblatt
UV-Desin-fektion	Unzureichende Desinfektion wegen zu geringer Transmission	Intensität Strahlung	CCP4	Richtwert z.B. 80% Grenzwert z.B. 60% (die Dosis beim GW und max. Durchfluss muss mind. 400 J/m^2 betragen)	kontinuierliche Trans-missionsmessung, mit Alarm wenn Grenzwert unterschritten	Verwurf, Kon-trolle und Be-hebung der Störung durch Brunnenmeister	Datenblatt, 1 Mal wöchentlich und gegebenenfalls Massnahmen dokumentieren
UV-Desin-fektion	Partiell verunreinigte Strahler	Verschmutzte Strahler	CCP5	von Auge sichtbare Verunreinigung	regelmässige Reinigung gemäss Herstellerangaben	Strahler aus-wechseln falls Reinigung un-möglich	Datenblatt

Tab. 6.7 (Fortsetzung): Kurzfassung HACCP-Qualitätssicherung nach Raster Codex Alimentarius für die Trinkwasseraufbereitung mittels Filtration

Verfahrens-schritt	Gefahr	Para-meter	CCPs	Toleranzwerte	Überwachungs-system	Massnahmen	Dokumentation
Trübungs-messung	Zu starke Wassertrübung, dadurch keine garantierte Partikelentfernung	Trübung	CCP1	20 FNU bei Quell- und Grundwasser, wenn ein Filter nachgeschaltet	Trübungsmessung kontinuierlich mit Alarm	Verwurf, Kontrolle durch Brunnenmeister	Kontinuierlicher Schreiber, Massnahmen bei Alarm handschriftlich
Flockungs-mittel-dosierung	Zu geringe Partikelentfernung wegen zu geringer Dosierung	Ver-brauch Flockungs mittel	CCP2	Kalibration anlageabhängig	Wassermengen- und Füllstandsmessung wöchentlich	Justierung der Flockungs-mittelzugabe	Datenblatt, handschriftlich
Flockungs-mittel-dosierung	Ungenügende Desinfektion wegen zu geinger Partikel-entfernung	Trübung	CCP3	1,5 FNU vor Desinfektionsmittel-zugabe[#]	Trübungsmessung kontinuierlich mit Alarm	Erhöhung der Flockungsmittelzugabe oder Drosselung des Wasserflusses	Kontinuierlicher Schreiber, Verwurf und Massnahmen dokumentieren
Funktion Sandfilter / Aktivkohle-filter	Ungenügende Filtration	Mano-meter-druck	CCP4	Kalibration anlageabhängig	Manometer Druck-differenz wöchentlich	Filter rückspülen	Datenblatt
Funktion Sandfilter / Aktivkohle-filter	Zu geringe Filterschicht	Filter-schicht	CCP5	Kalibration anlageabhängig	Optische Kontrolle, monatlich	Filtermaterial ersetzen	Datenblatt
Funktion Sandfilter / Aktivkohle-filter	Ungenügende Filtration	Durch-fluss-geschwin-digkeit	CCP6	Kalibration anlageabhängig	Durchflussmesser	Drosselung des Durchflusses	Datenblatt, 1 Mal wöchentlich handschriftlich
Sandfilter	Ungenügende Filtration	Trübung	CCP7	0,3 FNU nach Filtration[#]	Trübungsmessung kontinuierlich mit Alarm	Verwurf, Kontrolle durch Brunnenmeister	Kontinuierlicher Schreiber, Verwurf/ Massnahmen dokumentieren
Aktivkohle	Ungenügende Adsorption da Aktivkohle beladen	spez. Stoffüber wachung	CCP8	Anforderung der FIV für bestimmten Stoff	periodische Messung im Labor	Auswechseln od. Regeneration der Aktivkohle	Datenblatt

Tab. 6.8: Vereinfachter Aufbau des HACCP-Systems

Abfolge für HACCP	Beispiel
Verfahrensschritt:	Zugeben von Javelle
Mögliche Gefahr:	Zu geringe Desinfektion wegen zu geringer Javellezugabe
Festlegen des CCP:	CCP 1
Grenz- und Toleranzwert festlegen:	Bei der Aufbereitung muss der Chlorgehalt 0,08 bis 0,10 mg/L betragen
Überwachungssystem:	Online Messung oder regelmässige Handmessung
Massnahmen, wenn Grenz- und Toleranzwerte unter- oder überschritten:	Chlorzugabe erhöhen
Dokumentation:	Ereignisse schriftlich festhalten, um die Rückverfolgbarkeit zu gewährleisten

Kosten der Trinkwasseraufbereitung

Wie bereits eingangs erwähnt, ist die Wahl des adäquaten Verfahrens für die Trinkwasseraufbereitung in einer Wasserversorgung von verschiedenen Faktoren abhängig und muss im Einzelfall abgeklärt werden. Entscheidend ist unter anderem, ob es sich um ein Karstgrundwasser mit kurzzeitig auftretenden hohen Trübungen und mikrobiologischen Belastungen oder um ein Grundwasser aus fluvialen Ablagerungen handelt. Die genutzten Schottergrundwässer liegen in der Schweiz und in Deutschland meist in dicht besiedelten Gebieten und im Zuströmbereich der Fassungen können neben mikrobiologischen auch chemische Risiken oder die Oxidation reduzierter Wässer eine Rolle spielen. Diese Aspekte sind bei der Wahl des Aufbereitungsverfahrens zu berücksichtigen.

Die Investitionen für die Trinkwasseraufbereitung können insbesondere für kleinere Wasserversorgungen sehr hoch sein und den Betrag von einer Million Franken leicht überschreiten. Langfristig betrachtet liegen die Betriebskosten mit Amortisation aber selbst für aufwändigere Verfahren durchschnittlich bei 10 bis 20 Rp./m^3, wobei die produzierte Wassermenge und die Auslastung der Anlage berücksichtigt werden muss (Widmer, 2000).

In der Tabelle 6.9 sind die Kosten für einfache Aufbereitungsverfahren aufgelistet, ausgehend von einer kleineren Wasserversorgung mit einem Jahresverbrauch von 100'000 m^3. Es handelt sich dabei nur um die Anschaffungs- und Betriebskosten mit Amortisation, nicht aber um Gebäudekosten. Einfache Aufbereitungsanlagen lassen sich meist in bestehende Gebäude einbauen.

Tab. 6.9: Kosten für einfache Aufbereitungsverfahren bei einem Jahresverbrauch von 100'000 m^3

Verfahren	Anlagekosten [SFr.]	Betriebskosten [SFr./Jahr]	Kosten pro m^3 [Rp./m^3]
Javelle (Dosierpumpe, Behälter, Auffangwanne, Injektor, ohne Wassermesser für mengenabhängige Dosierung)	4000 – 5000	300-500	2,0
Chlorgas (mit Gasdetektor, ohne separatem Raum, welcher jedoch gesetzliche Vorgabe ist)	13'000 – 14'000	10'000	15,0
Chlordioxid (mit Impfarmatur und Zubehör)	12'000 – 15'000	4'500	10,0
UV-Strahler	13'000 – 18'000	1200 – 2500	6,0

(Betriebskosten mit Amortisation auf 10 Jahre, Kapitalverzinsung 6%, Amortisation linear. 1,5 SFr. entsprechen ca. 1 Euro (Stand 2002), 100 Rp. sind 1 SFr.)

Um die optimale Leistung der Desinfektion bei den Chlor-, Ozon- und UV-Anlagen erreichen zu können, muss das Wasser klar sein. Die Trübung darf höchstens 1,5 FNU betragen. Wird dieser Trübungswert überschritten, sollte ein Trübungsmesser mit automatischer Steuerung der Verwurfklappe für das Rohwasser eingebaut werden. Die Kosten für eine Trübungsmesseinrichtung betragen zwischen SFr. 6000 und 8000.-. Falls der Richtwert für die Trübung oft überschritten wird, und in dieser Zeit kein Fremdwasser bezogen werden kann, muss das Rohwasser vor der Desinfektion filtriert werden.

In der Abbildung 6.11 sind die Kosten für mehrstufige Aufbereitungsverfahren aufgezeigt. Die Kurven zeigen die Kosten pro Kubikmeter bei unterschiedlicher Auslastung der jeweiligen Anlage. Die Wartungs-, Unterhalts- und Amortisationskosten sind unabhängig von der aufbereiteten Wassermenge. Da die Kosten für Strom und Chemikalien nicht so stark zu Buche schlagen, können sich die Betriebskosten pro Kubikmeter bei Halbierung der produzierten Wassermenge praktisch verdoppeln.

Für Membranverfahren zur physikalischen Partikelabtrennung liegen die Kosten noch deutlich höher als bei den klassischen mehrstufigen Verfahren. Je nachdem ob für das Rückspülwasser (2 bis 4 Rückspülungen pro Stunde für Karstwässer) Abwassergebühren erhoben werden, muss mit 25 bis 50 Rp./m^3 bei einer Anlage für 500'000 m^3 gerechnet werden (Lipp, 2000). Es ist zu erwarten, dass die Preise für diese relativ neue Technik in Zukunft sinken werden.

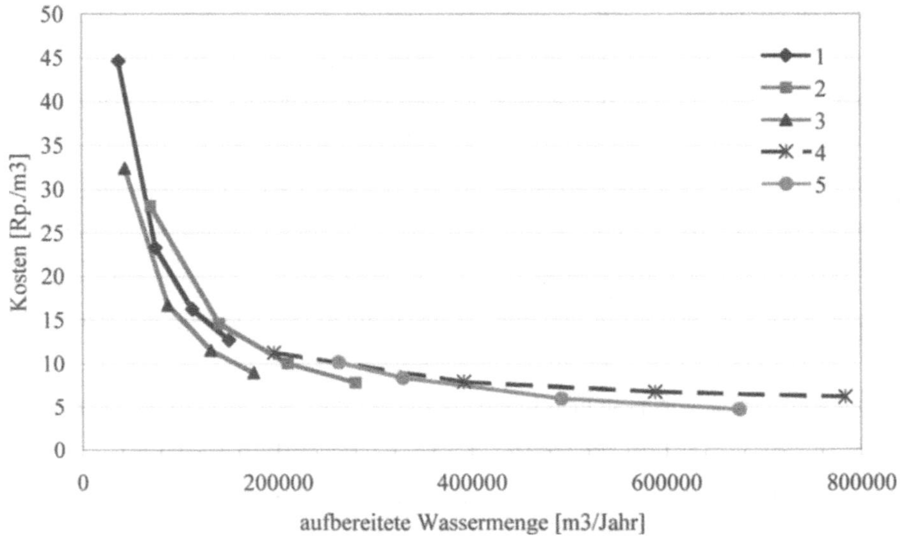

Abb. 6.11: Betriebskosten mit Amortisation auf 20 Jahre, Kapitalverzinsung 6%, Amortisationsrate 8,75%. Die Linien geben die Kosten bei 100% (grösste mögliche Wassermenge), 75%, 50% und 25% Auslastung für die Anlagen 1, 2, 4 und 5. Bei Anlage 3 ist die Auslastung 100%, 75%, 50% und 40%. Die Linien 1, 2 und 3 bedeuten Anlagen mit Flockung, Filtration und UV-Bestrahlung, die Linie 4 eine Anlage mit Flockung, Filtration und Chlordioxid und die Kurve 5 eine Anlage mit Flockung, Filtration, Ozonung, Aktivkohlefilter und UV-Bestrahlung (Widmer, 2000)

Schlussfolgerungen

Die Trinkwasseraufbereitung ist das Bindeglied zwischen dem Rohwasser (Grundwasser einschliesslich Quellen und Oberflächengewässer) und dem Trinkwasser. Sie steht auch als letzte Barriere, die Verunreinigungen vom Trinkwasser fernhalten soll. Damit die hohen Qualitätsanforderungen gemäss Schweizerischer und Deutscher Lebensmittel- und Trinkwassergesetzgebung an das Endprodukt Trinkwasser jederzeit eingehalten werden können, sind verschiedene Aspekte zu betrachten und die richtigen Massnahmen zu treffen. So muss als Erstes das hydrogeologische Einzugsgebiet der Trinkwasserfassung bekannt sein. Daraus kann abgeleitet werden, welche Faktoren wie zum Beispiel die Landwirtschaft, Siedlungs- oder Industriegebiete und Altlasten die Wasserqualität beeinflussen können. In Karst- und stark heterogenen Kluftgebieten sind neben den feinen Klüften im Grundwasserleiter meist auch schnelle präferenzielle Fliesswege vorhanden, die zu starken Schwankungen in der Konzentration einzelner Parameter, insbesondere der Mikroorganismen und der Trübung bei der Trinkwasserfassung führen können. Die Beherrschung dieser Belastungsspitzen ist für den Trinkwasseraufbereitungsprozess entscheidend. Lockergesteinsgrundwasserleiter weisen dagegen meistens eine relativ gute Filterwirkung für Partikel auf, so dass die Belastung mit Mikroorganismen in den Trinkwasserfassungen hier eher gering ist. Im Gegensatz zu

Partikeln können sich chemische Stoffe im Grundwasser akkumulieren und über weite Strecken transportiert werden.

Die Rohwasserqualität kann über die Analyse von mikrobiologischen, chemischen und physikalischen Parametern, die den Faktoren des Einzugsgebietes entsprechend ausgewählt sind, ermittelt werden. Für die Wahl des adäquaten Aufbereitungsverfahrens müssen in der Mikrobiologie neben den klassischen Indikatoren auch pathogene Keime, wie Viren oder Protozoen analysiert werden. Dadurch kann die notwendige Reduktion der relevanten Mikroorganismen bestimmt werden. Mit einem Screening der chemischen Stoffe kann die mögliche Bildung von Desinfektionsmittelnebenprodukten abgeleitet werden. Bei der Beprobung sollten Belastungsspitzen, verursacht durch Niederschläge und Hochwasser mit eingeschlossen werden, denn nur so kann die zeitliche Veränderung der Wasserqualität befriedigend beurteilt werden.

Aufgrund der zusammengestellten Reduktionsraten für verschiedene Mikroorganismen bei den unterschiedlichen Verfahren, kann mit einer Reduktion von 3-5 log-Einheiten gerechnet werden. Theoretisch könnten diese Werte höher sein, wenn eine lange Kette von Verfahren hintereinander geschaltet wird. Es ist aber in der Praxis unrealistisch und zu teuer, Reduktionsraten der Aufbereitung von 12 log-Einheiten zu fordern. Vielmehr sollte, auch aus Gründen der betrieblichen Sicherheit, die Rohwasserqualität verbessert werden.

Nur mit einer kontinuierlichen Überwachung des Aufbereitungsprozesses gemäss den gesetzlichen Vorgaben können die beschriebenen Reduktionsraten für Mikroorganismen erreicht und damit die gesetzlichen Anforderungen an das Trinkwasser eingehalten werden. Die Kosten für die Anschaffung und den Betrieb einer Aufbereitungsanlage sollten erst in Betracht gezogen werden, wenn zwei vergleichbare Verfahren zur Aufbereitung des Rohwassers vorliegen. Es ist wichtig, die Aufbereitung im Zusammenhang mit der Rohwasserqualität und deren Schwankung und den Risikofaktoren im Einzugsgebiet zu sehen. Dabei ist auch die Fremdwasserbeschaffung zu berücksichtigen.

Schutz von Wasserfassungen 7

Peter Huggenberger und Adrian Auckenthaler

Einleitung

Eine Vielzahl der Trinkwasserfassungen stammt noch aus einer Zeit, lange bevor die heute gültigen gesetzlichen Bestimmungen und technischen Normen in Kraft gesetzt worden sind. Der Ausbaustand vieler Wasserversorgungen entspricht deshalb, auch wenn seit ihrem Bau gewisse Erneuerungen vorgenommen wurden, dem Standard der 50er und 60er Jahre. Damals standen jedoch wesentlich weniger Hilfsmittel zur Beurteilung der Grundwasserqualität und der Dynamik der komplexen hydrogeologischen und hydraulischen Verhältnisse zur Verfügung. Zudem konzentrierten sich viele hydrogeologische Abklärungen auf die Brunnenergiebigkeiten. Die Entwicklung des Grundwasserschutzes erfolgte mit zeitlicher Verzögerung zu den neusten Erkenntnissen über Grundwasservulnerabilität und -gefährdung.

Die Trinkwassergewinnungsanlagen bei Lockergesteinsgrundwasserleitern werden heute durch zwiebelschalenartig ausgeschiedene Grundwasserschutzzonen geschützt, wobei von aussen nach innen immer strengere Schutzvorschriften gelten. Die grösste Bedeutung, auch im Hinblick auf die Mikrobiologie, kommt dabei der engeren Grundwasserschutzzone S2 zu. Laut schweizerischer Gewässerschutzverordnung soll die engere Schutzzone (S2) verhindern, dass Keime oder Viren in die Grundwasserfassungen oder –anreicherungsanlagen gelangen (Anh. 4 Ziff. 123 Abs. 1 Bst. a GSchV). Deshalb wurde bei der Revision der Stoffverordnung 1999 ein Verbot zum Ausbringen von Klärschlamm und flüssigem Hofdünger in der Zone S2 erlassen (Anh. 4.5 Ziff. 33 Abs. 2 StoV). Wasserteilchen dürfen vom Rand der Zone S2 erst nach einer Verweilzeit von 10 Tagen in die Trinkwasserfassung gelangen können. Als zweites Kriterium muss der Abstand von der Zone S1 (Fassungsbereich) bis zum äusseren Rand der Zone S2 in Zuströmrichtung mindestens 100 m betragen, falls nicht hydrologische Verhältnisse eine kleinere Dimensionierung zulassen (Anh. 4 Ziff. 123 Abs. 2 Bst. a und b GSchV). Wie der Tabelle 5.3 (Kap. Partikeltracer) entnommen werden kann, ist zu erwarten, dass Keime diese Fliesszeiten und Distanzen durchaus zurücklegen können. Eine Elimination der Keime muss somit hauptsächlich über die Filtration erfolgen. Die Resultate von neueren hydrogeologischen

Abklärungen von Anstrombereichen von flussnahen Wasserfassungen zeigen zudem, dass sich die Fliesszeiten zwischen infiltrierendem Flusswasser und Grundwasserfassung bei Hochwasser unter 10 Tage verkürzen können (Mattle et al. 2000, Huggenberger 1991).

Die deutsche Gesetzgebung zum Schutz von Grundwasserfassungen erfüllt die Richtlinien der EU. Die engere Schutzzone (Zone II) dient ebenfalls dem Schutz von Wasserfassungen vor pathogenen Mikroorganismen. Vom Rand der Zone II bis zur Trinkwasserfassung soll die Verweildauer des Grundwassers mindestens 50 Tage betragen. Die Schutzgebiets- und Ausgleichsverordnung verbietet das Ausbringen von flüssigen Wirtschaftsdüngern tierischer Herkunft und Sekundärrohstoffdüngern in der Zone II (§ 4 Abs. 2 Nr. 1 und 2 SchALVO). Das Weidenlassen von Tieren und das Ausbringen von Festmist ist allerdings, soweit zum Schutz des Grundwassers nicht erforderlich, erlaubt. Verschärfend können hingegen auch in der weiteren Schutzzone (Zone III) zusätzliche Beschränkungen oder gegebenenfalls in flurstückscharf abgegrenzten Bereichen ein ganzjähriges Verbot des Aufbringens von Klärschlamm, Fäkalien und ebenfalls auch von Jauche, Gülle und Gärsaft in Betracht gezogen werden (Pkt. 4.2.13 VwV-WSG).

Grössere Grundwasserschutzzonen mit Aufenthaltszeiten des Grundwassers von 50 Tagen, wie in Deutschland vorgeschrieben, bieten zwar einen besseren Schutz vor mikrobiellen Verunreinigungen, sind aber nicht in allen Ländern umsetzbar. Die vielfach schnellen Wasserwege in den gut durchlässigen Tal- und Karstgrundwasserleitern der Schweiz würden Grundwasserschutzzonen von viel grösserem Umfang bedingen, was sich in der Praxis kaum realisieren liesse. Nicht zuletzt aus diesem Grund und im Hinblick auf eine nachhaltige Sicherstellung einer guten Grund- und Trinkwasserqualität werden in Zukunft auch die unter- bzw. oberirdischen Anstrombereiche in die Schutzüberlegungen einbezogen werden müssen, wie dies in der Schweiz und in der EU bereits angestrebt wird.

Das Konzept des Trinkwasserschutzes bei Lockergesteinsgrundwasserleitern ist in Karst- und Kluftgrundwasserleitern nur bedingt anwendbar. Die Daten aus diversen Tracerversuchen mit Partikeln im Karstgrundwasser zeigen grosse Fliessdistanzen, hohe Fliessgeschwindigkeiten und damit kurze Aufenthaltszeiten (siehe Tab. 5.4). Ein Schutz dieser Grundwasservorkommen ist daher hauptsächlich über die Beschränkung respektive die Verhinderung des Eintrages von Mikroorganismen in die Karstsysteme zu erreichen, was in beiden Ländern mit den gesetzlichen Massnahmen angestrebt wird.

In Gebieten mit Karst- und Kluftgesteinsgrundwasser ist deshalb die Vulnerabilität des Einzugsgebietes für die Ausscheidung der Schutzzonen massgebend. Dabei wird in der Schweiz besonderes Gewicht auf die oberflächennahen Deckschichten (Boden), die Infiltrationsverhältnisse und die hydrologische Verbindung der ungesättigten und der gesättigten Zone gelegt (Anh. 4 Ziff. 123 Abs. 1 GSchV). Die Universität Neuenburg hat dazu im Auftrag des BUWAL die EPIK-Methode zur Kartierung der Vulnerabilität von Grundwasserleitern in Karstgebieten entwickelt. Die Methode erfasst die folgenden Kriterien: 1) Entwicklung des Epikarstes, 2) Schutzwirkung der Deckschicht, 3) Infiltrationsverhältnisse und 4) Entwicklung des Karstnetzes (BUWAL, 1998; Doerfliger et al., 1999).

In Deutschland bestehen für Karstgebiete keine weitergehenden Regelungen. Aufgrund der besonderen örtlichen hydrogeologischen Situation können die vorhandenen Vorschriften in den Zonen II und III aber verschärft werden (Pkt. 4.2.13 VwV-WSG).

Generell besteht von Seiten der Gemeinden, Wasserversorger und von Seiten der Kontrollorgane der Wunsch nach einfachen Handlungsanweisungen oder Beurteilungskriterien von Gefährdungsmustern, um daraus die erforderlichen konkreten Massnahmen ableiten zu können. Bereits in der Planungsphase mit der Auslegung des Betriebes neuer Wasserversorgungsanlagen müssen deshalb die wesentlichen Gefährdungsmuster erkannt und berücksichtigt werden können.

Für all diejenigen Fälle, wo trotz Einhaltung der gesetzlichen Anforderungen des Gewässerschutzes nicht zu jedem Zeitpunkt eine gute Rohwasserqualität gewährleistet werden kann, müssen Handlungsstrategien entwickelt werden. Solche Handlungsstrategien können sein:

- Aufgabe von Trinkwasserfassungen verbunden mit Ersatz aus anderen Wasserbezugsquellen
- Aufbereitung des Rohwassers
- Differenziertes Wasserentnahmemanagement gestützt auf Beurteilung des Gefährdungsmusters
- Kombinationen von Wasserentnahmemanagement und Trinkwasseraufbereitung
- Neupositionierung von Wasserfassungen auf Grund von eingehenden Abklärungen über Anstrombereiche und Wasserqualität. Optimierung der Lage neuer Brunnen auf eine möglichst grosse Aufenthaltszeit des Grundwassers (mindestens 10 Tage)

Die Aufgabe von lokalen Trinkwasserfassungen und der Wasserbezug aus regionalen Wasserwerken wird zur Zeit in einigen Karstgebieten praktiziert, wobei diese Strategie im Hinblick auf den Gewässerschutz nicht als nachhaltig bezeichnet werden kann. Aufgrund des analytischen Nachweises von fäkalen Verunreinigungen von Netzwässern nach stärkeren Niederschlägen wird von den Kontrollbehörden meistens die Aufbereitung des Rohwassers verlangt. Die Dimensionierung der Aufbereitungsanlagen wird dabei auf die festgestellten Verunreinigungen ausgelegt. Weitergehende Abklärungen über die Variabilität der Rohwasserqualität und das mögliche Auftreten pathogener Mikroorganismen werden oft nicht durchgeführt. Ein weiterer Schritt für eine sichere Trinkwassergewinnung ist die Etablierung eines Wasserentnahmemanagements. Dies setzt fundierte wissenschaftliche Erkenntnisse der lokalen Hydrogeologie und dem zeitlichen Auftreten der mikrobiologischen oder chemischen Gefährdung voraus. Dieses Wissen muss in Feldversuchen jedoch zuerst erarbeitet werden. Die Kombinationen von Wasserentnahmemanagement und Trinkwasseraufbereitung gewährleistet in vielen Fällen den weiteren sicheren Betrieb von Wasserversorgungen. Dadurch müssen die Probleme des Eintrags von Mikroorganismen mit zusätzlichen einschränkenden Auflagen für die landwirtschaftlich genutzten Gebiete nicht kurzfristig gelöst werden. Mittel- bis langfristig sind die Schutzzonen mit den damit verbundenen Nutzungsbeschränkungen allerdings durchzusetzten. Die Neupositionierung von Wasserfassungen könnte in denjenigen Fällen eine Option sein, wo grössere Wasserversorgungen ihre veralteten Wasserfassungen erneuern wollen. Zu berücksichtigen ist dabei insbesondere die erforderliche Aufenthaltszeit des Grundwassers für die ordnungsgemässe Ausscheidung der Schutzzonen. Der Rückgang des Wasserverbrauchs und die damit verbundenen geringeren Grundwasserentnahmen wirken sich zusätzlich positiv auf die Aufenthaltszeit aus. Eine Analyse von Wasserversorgungsgebieten zeigt, dass der Spielraum für eine Optimierung bezüglich der Grundwasserquantität und -qualität bei vielen Wasserversorgern keineswegs ausgeschöpft ist (Mattle et al. 2000, Huggenberger 2001). Dabei steht die Herkunft und die Aufenthaltszeit der Grundwässer bei unterschiedlichen hydrologischen Bedingungen verbunden mit der Anpassung der Pumpraten und der Pumpsysteme im Vordergrund. Teilweise sind Pumpsysteme im Einsatz, die keine Feinsteuerung der Pumpraten und damit auch kein ereignisorientiertes Rohwassermanagement zulassen. Damit stehen in Zukunft nicht nur technische Neuerungen in den eigentlichen Fassungen zur Diskussion, sondern auch eine Abklärung geeigneter Standorte für neue Wasserfassungen.

Im Folgenden werden Karstgrundwasserfassungen und Lockergesteinsgrundwasserfassungen mit Infiltration von Flusswasser näher beschrieben. Diese beiden Systeme sind, abgesehen von stark heterogenen Kluftsystemen, durch einer Verunreinigung durch Mikroorganismen am meisten gefährdet. Wasserfassungen in Porengrundwasserleitern ohne Oberflächenwasserinfiltration sind bei Einhaltung der üblichen Schutzmassnahmen und ordnungsgemäss ausgeschiedenen Grundwasserschutzzonen vor mikrobiologischen Verunreinigungen im allgemeinen besser geschützt. Chemische Verunreinigungen wie zum Beispiel durch Nitrat können allerdings bei diesen Fassungen sehr wohl eine grosse Bedeutung haben. Zur Bewältigung von chemischen Belastungen sind Grundwasserschutzzonen bzw. technische Massnahmen bei der Wassergewinnung in der Regel nicht geeignet oder aufwändig. Die notwendigen und geeigneten Massnahmen müssen daher im Anstrombereich derart belasteter Trinkwasserfassungen ergriffen werden.

Karstgrundwasserleiter

Viele Quellwässer in Karstgebieten zeichnen sich aus durch stark schwankende Qualitätseigenschaften mikrobiologischer, chemischer und physikalischer Art. Diese Schwankungen sind meist an Niederschlagsereignisse, Schneeschmelze oder Hochwassersituationen gebunden und eine Folge des heterogenen Aufbaus der Karstgrundwasserleiter. Diese Heterogenität der Transporteigenschaften erklärt sich durch die Variabilität der Poren der Gesteinsmatrix und die durch Lösungsprozesse oder tektonische Vorgänge entstandenen Spalten, Risse und Karströhrensysteme. Letztere wirken als schnelle Fliesswege. Zusammen mit dem Niederschlags- und Schmelzwasser können Verunreinigungen über diese schnellen Fliesswege mit Geschwindigkeiten von mehr als 75 m/h in die Quellen transportiert werden. In Trockenperioden werden die Quellen von Wasser aus der kompakten Gesteinsmatrix mit wesentlich langsameren Fliessgeschwindigkeiten gespiesen. Die Fliessgeschwindigkeit in Karstgrundwasserleitern ist deshalb stark von meteorologischen und hydrogeologischen Verhältnissen abhängig.

Das Kriterium der Grundwasserfliesszeit ist für die Ausscheidung von Schutzzonen in Karstgebieten grundsätzlich ungeeignet (Doerfliger und Zwahlen, 1998), weil als Folge der teilweise grossen Fliessgeschwindigkeiten und der oft geringen Filterwirkung des Bodens, die Mechanismen, die in Lockergesteinsgrundwasser für die Elimination sowohl von mikrobiellen als auch abbaubaren oder rückhaltbaren stofflichen Belastungen dienen, weitgehend wirkungslos bleiben. Dies macht Karstgrundwasser besonders vulnerabel, das heisst anthropogene Verunreinigungen können rasch und in hohen Konzentrationen in das Grundwasser und das daraus gewonnene Trinkwasser gelangen.

Abb.7.1: Aufbau eines Karstgrundwasserleiters nach Doerfliger, 1996. Die Buchstaben entsprechen den vier Kriterien von EPIK: Epikarst, Protection (Schutz), Infiltration und Karst. (Mit freundlicher Genehmigung, Doerfliger, N., 1996. Advances in karst groundwater protection strategy using artificial tracer tests analysis and multiattribute vulnerability mapping (EPIK method). Diss. Universität Neuenburg, pp. 308.)

Um Trinkwasserfassungen in Karstaquiferen wirksam zu schützen, sind verschiedene Massnahmen notwendig:

- Ausscheidung von Schutzzonen, die die Heterogenität und die teilweise geringe Filterwirkung des Bodens berücksichtigen
- Management von Quellwasserfassungen, und damit die Möglichkeit auf mögliche Verunreinigungen bei Niederschlagsereignissen rechtzeitig reagieren zu können
- Trinkwasseraufbereitung, die eine genügende Keimreduktion bewirkt
- HACCP-Konzept (Hazard Analysis of Critical Control Points) zur Überwachung der gesamten Trinkwasserentstehung

Bestehende Schutzzonen in Karstgebieten weisen mangels Kentnissen der hydrogeologischer Grundlagen oftmals nur beschränkte Schutzwirkung auf, so dass Trinkwasserverschmutzungen in Karstgebieten keine Seltenheit darstellen (Doerfliger und Zwahlen, 1998). Das Bundesamt für Umwelt, Wald und Landschaft (BUWAL) hat in Zusammenarbeit mit der Landeshydrologie und –geologie (LHG) deshalb eine neue Methode zur Ausscheidung von Schutzzonen in Karstgebieten entwickeln lassen (EPIK-Methode, „Centre d'hydrogéologie der Universität Neuenburg").

Die EPIK-Methode ist eine Multikriterienmethode, die auf der Beurteilung unterschiedlicher hydrogeologischer Gegebenheiten im Einzugsgebiet der Trinkwasserfassung beruht. Die nach dieser Methode ausgeschiedenen Schutzzonen berücksichtigen die Vulnerabilität der Karstgrundwasserleiter. Beurteilt werden vier Kriterien (Abb. 7.1): Ausbildung des Epikarstes („E"; oberflächennaher, stark verkarsteter Bereich des Kalkuntergrundes, mit stark erhöhter Wasserdurchlässigkeit), Schutz durch die Deckschicht („P", Protection), Infiltrationsverhältnisse („I" diffuse resp. konzentrierte Infiltration) und Entwicklung des Karstnetzes („K").

Die Ausdehnung der Schutzzonen gemäss EPIK wird in der Praxis hauptsächlich über bestehende Kartengrundlagen, Hydrogramme und Tracerversuche ermittelt. Die systematische Bestimmung der Vulnerabilität mittels Tracerversuchen oder geophysikalischen Methoden ist nicht immer notwendig. Bei Unsicherheiten müssen allerdings weitergehende Feldarbeiten durchgeführt werden. Eine genaue Beschreibung der EPIK-Methode gibt die „Praxishilfe Kartierung der Vulnerabilität in Karstgebieten (Methode EPIK)" herausgegeben vom BUWAL, 1998.

Ein weiterer Schritt für eine sichere Trinkwassergewinnung in Karstgebieten ist das Management von Wasserfassungen. Darunter wird die online Überwachung von Systemparametern wie Niederschlag, Schüttung, Trübung, UV-Absorption und allenfalls weiterer Parameter verstanden. Anhand dieser Aufzeichnungen kann der Betrieb der Wasserentnahme optimal gesteuert werden. Bei einer Veränderung der gemessenen Parameter gegenüber einem fassungsspezifischen Referenzzustand ist eine Verunreinigung des Karstquellwassers durch Mikroorganismen zu erwarten. Durch ein geeignetes Management soll verhindert werden, dass während Risiko-Phasen belastetes Wasser ins Versorgungsnetz eingespiesen wird. Sattdessen muss die Wasserversorgung temporär mit Wasser aus einer anderen Quelle, einem anderen Brunnen oder von einer benachbarten Wasserversorgung sichergestellt werden.

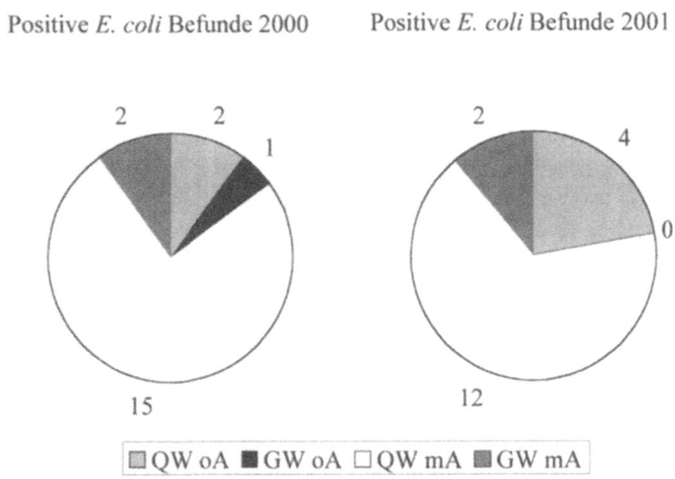

Abb. 7.2: Anzahl positiber *E. coli* Befunde in verschiedenen Wasserversorgungen im Kanton Basel-Landschaft (ohne Mehrfachnennungen). QW oA: Quellwassernutzung ohne Aufbereitung; GW oA: Grundwassernutzung ohne Aufbereitung; QW mA: Quellwassernutzung mit Aufbereitung; GW mA: Grundwassernutzung mit Aufbereitung (KLBL, 2001)

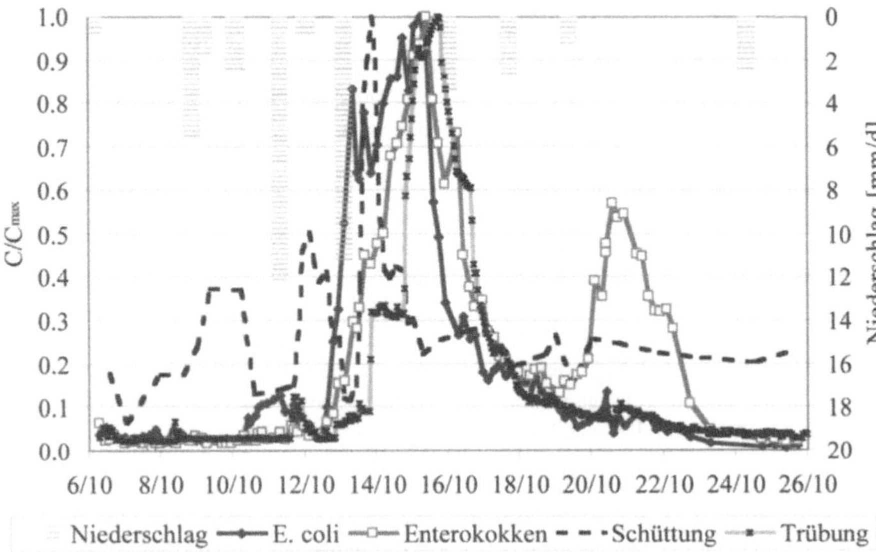

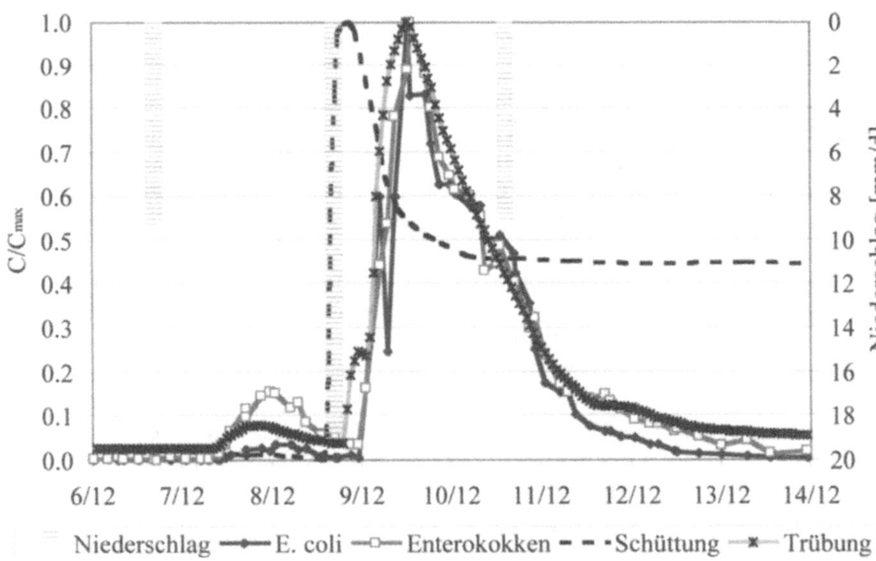

Abb.7.3: Veränderung von *E. coli*, Enterokokken, Schüttung und Trübung einer Karstquelle in der Nordwestschweiz im Oktober und Dezember 2000 in Abhängigkeit des Niederschlags. In der oberen Grafik ist ersichtlich, dass die Mikroorganismen vor der Trübung ansteigen, in der unteren ist der Anstieg gleichzeitig. Die Schüttung steigt in beiden Fällen vor dem Hauptpeak der Mikroorganismen an (Auckenthaler et al., 2002; Raso, 2001)

In der Schweiz wird oft die Trübung online gemessen. Deren Anstieg soll die Präsenz von Mikroorganismen anzeigen. Gleichzeitig ist sie ein wichtiger Parameter zur Überwachung des Aufbereitungsprozesses. Trübes Wasser lässt sich durch einstufige Aufbereitungsverfahren (Chlor, UV), wie sie heute in Karstgebieten vielerorts angewendet werden, nicht genügend aufbereiten. Ein Anstieg der Trübung über einen bestimmten Richtwert (i.d.R. > 1.5 FNU) führt deshalb zu einem automatischen Verwurf des Quellrohwassers.

Umfangreiche und detaillierte Untersuchungen einer Karstquelle in der Nordwestschweiz haben jedoch gezeigt, dass ein Anstieg der im Rohwasser vorhandenen Mikroorganismen vor dem Anstieg der Trübung erfolgen kann. Als besserer Systemparameter hat sich die Schüttung erwiesen, da sie stets vor dem vermehrten Auftreten der Mikroorganismen angestiegen ist (Auckenthaler et al., 2002). Dies lässt sich dadurch erklären, dass das „alte Wasser" von früheren Niederschlägen im Karströhrensystem zuerst durch das neue Niederschlagswasser verdrängt werden muss und erst anschliessend das „neue, verunreinigte Wasser" in die Quelle gelangen kann. Die Trübung kann dann benutzt werden, um den Rückgang der Belastung durch Mikroorganismen zu überwachen.

Diese einfache online Überwachung funktioniert relativ gut, ist teilweise jedoch etwas zu wenig sensitiv bei geringen Niederschlägen, die keine Veränderung der Systemparameter bewirken, jedoch bereits zum Transport von Mikroorganismen führen können. Die Maximalkonzentrationen der Mikroorganismen in diesen Fällen sind jedoch vergleichsweise gering und sollten mit einstufigen Aufbereitungsverfahren genügend reduziert werden können, da es sich dann um Wasser mit einer geringen Trübung handelt.

Bei genauer Kenntnis des hydrogeologischen Einzugsgebietes der Quellfassung ist die Voraussage des zeitlichen Auftretens von Mikroorganismen unter Umständen noch genauer möglich. Über die Systemparameter Niederschlag und Schüttung können die für die Form der Mikroorganismenkurve wichtigen Grössen (dominante Aufenthaltszeit und Dispersion) bestimmt werden. Dadurch lässt sich sagen, wann und wie lange Mikroorganismen im Quellwasser auftreten.

Die letzte Stufe zur Sicherung von einwandfreiem Trinkwasser ist die Aufbereitung des Rohwassers. Eine Trinkwasseraufbereitung kann nur richtig dimensioniert werden, und entsprechend funktionieren, wenn Art und Ausmass der mikrobiologischen, chemischen und physikalischen Verunreinigung bekannt sind. Für die Dimensionierung einer Aufbereitungsanlage und die optimale Wahl des geeigneten Verfahrens sind deshalb die Kenntnisse aus den hydrogeologischen Abklärungen zu den Schutzzonenausscheidungen sowie die online Überwachung des Rohwassers für das Wassermanagement nötig. Die Schutzzonenabklärungen zeigen auf, mit welchen Risiken gerechnet werden muss, d.h. woher das Wasser stammt und welche anthropogene Nutzung dort vorhanden ist. Das Management zeigt auf, mit welchen unvermeidbaren Mikroorganismen und Stoffkonzentrationen gerechnet werden muss.

Dass die Aufbereitungsanlagen nicht immer richtig dimensioniert sind, zeigen Routinekontrollen des Kantonalen Laboratoriums Baselland. Bei meist kleinen Wasserversorgungen mit Quellwassernutzung und einstufiger Trinkwasseraufbereitung treten immer wieder Fäkalindikatorkeime im Netzwasser auf (Abb. 7.2). Die Ursache ist hauptsächlich im fehlenden Management und einer ungenügenden Aufbereitung zu suchen. Die positiven

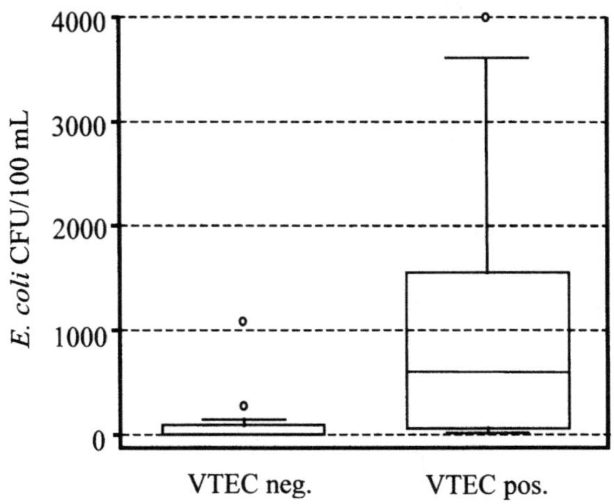

Abb. 7.4: Nachweis von verotoxinbildenden *E. coli* in Abhängigkeit der Anzahl *E. coli* im Quellwasser. Ab einer Anzahl von rund 300 *E. coli*/100 mL Probenvolumen konnten praktisch immer VTEC nachgewiesen (Auckenthaler et al., 2002)

Befunde von *E. coli* in Trinkwässern haben durchaus eine gesundheitliche Relevanz. Wie in einer Fallstudie gezeigt werden konnte, traten bei Konzentrationen ab ca. 300 *E. coli*/100 mL im Quellrohwasser meist auch verotoxinbildende *E. coli* auf (Auckenthaler et al., 2002). Andere Untersuchungen zeigen, dass auch Cryptosporidien, Campylobacter und Viren im Quellwasser gefunden werden (Svoboda et al., 1999; Metzler et al., 1996). Gerade bei Trübungen des Quellwassers, die eine Desinfektionsleistung stark vermindern, sind deshalb vorgeschaltete Filtersysteme unabdingbar.

Die drei beschriebenen Massnahmen, Schutzzonen, Management des Quellwassers und Trinkwasseraufbereitung bieten nur dann den genügenden Schutz vor Verunreinigungen, wenn sie auch richtig überwacht werden und bei Störungen rasch eingegriffen werden kann. Das schweizerische Lebensmittelgesetz verlangt von den Betreibern der Wasserversorgungen deshalb den Aufbau einer Qualitätssicherung (QS) und der Selbstkontrolle bei der Bereitstellung und Verteilung des Trinkwassers (Art. 11 Abs. 1-3 HyV). Dieses QS-Konzept richtet sich im Idealfall nach dem HACCP-Konzept gemäss Codex Alimentarius. Dabei werden vereinfacht zuerst die Risiken für eine mögliche Beeinträchtigung von Lebensmitteln abgeschätzt und anschliessend kritische Kontrollpunkte im Produktionsprozess mit dazugehörigen Richtwerten festgelegt. Beim Überschreiten der Richtwerte muss entsprechend der vordefinierten Massnahmen in den Prozess eingegriffen werden. Somit ist das Ziel, eine Optimierung des gesamten Herstellungsprozesses der Trinkwassergewinnung beginnend bei der Ausscheidung und Überwachung der Grundwasserschutzzonen über das Management der Wasserentnahmen und die Trinkwasseraufbereitung bis hin zur Verteilung und Abgabe des Wassers an die Konsumentinnen und Konsumenten.

Grundwassergefährdung infolge Fluss-Grundwasser Interaktionen

Die jüngsten Untersuchungen des Flusses Wiese in Basel zeigen, dass die erhöhte mikrobiologische Belastung dieses Gewässers deutlich an einzelne Ereignisse gebunden ist. Eingriffe und sich ändernde Randbedingungen im Anstrombereich eines Brunnens können das Rohwasser wesentlich beeinflussen. Dabei spielen Abflusstiefe im Fluss, Korngrössenzusammensetzung, Bewuchs und Grad der Kolmation der Flusssohle, hydraulischer Gradient im Grundwasserkörper, die Infiltration von Oberflächenwasser via Vorland ins Grundwasser bei Hochwässern und der Pumpbetrieb eine Rolle.

Die Untersuchungen des Flusssystems der Wiese in Basel zeigen insbesondere eine Reduktion der Aufenthaltszeit des Grundwassers zwischen Fluss und Wasserfassung von etwa 30 bis 40 Tagen bei Normalbedingungen auf knappe 10 Tage oder weniger bei Hochwasser. Durch eine Veränderung der Sohlenstruktur (z.B. durch Revitalisierung) kann sich diese Zeit nochmals um 4 bis 5 Tage verringern. Weil zu Beginn der Hochwässer im Fall der Wiese erhöhte mikrobiologische Belastungen des Flusswassers auftreten, gelangen auch wesentlich grössere Mengen an Mikroorganismen ins Grundwasser. Die meisten Mikroorganismen werden aber bereits in den ersten Metern zurückgehalten. Wenn sie jedoch in genügend hoher Anzahl auftreten, können einzelne auch weiter ins Grundwasser gelangen. In stark inhomogenen Grundwasserleitern können diese wenigen Mikroorganismen in kurzer Zeit Distanzen von über 100 Metern zurücklegen und entsprechend nach wenigen Tagen bei flussnahen Brunnen eintreffen. Die Aufenthaltszeit des Grundwassers zwischen Fluss und Brunnen ist deshalb eine für das optimale Management von flussnahen Trinkwasserfassungen wesentliche Grösse für den Schutz vor mikrobiologischer Verunreinigung. Viele Feldexperimente machen deutlich, dass Aufenthaltszeiten von mindestens 10 Tagen angestrebt werden sollten.

In der Praxis ist nur wenig über die Veränderung der Anstrombereiche in Abhängigkeit der Wasserführung der Oberflächengewässer bekannt. Das gleiche gilt für die ereignisbezogenen Gefährdungsmuster in Wasserfassungen. Die Anordnung von mehreren, beieinander liegenden, flussnahen Grundwasserentnahmestellen kann sich besonders in gering mächtigen Grundwasserleitern negativ auf den Eintrag von Mikroorganismen aus Oberflächengewässern auswirken. In solchen Fällen erfolgt eine laterale Verbreiterung der Anstromhyperbel, und somit überlappt der Anstrombereich von einzelnen Brunnen mit dem Einflussbereich der Infiltration eines Oberflächengewässers. Bei der Planung und bei der Festlegung der Betriebsreglemente sind diese Aspekte nur in den allerwenigsten Fällen berücksichtigt worden. Oft werden aber derartige, kritische Brunnen bei Hochwassereignissen ohnehin abgeschaltet, bzw. deren Rohwasser verworfen, was durchaus im Sinne eines einfachen, empirischen Rohwassermanagement ist. Es ist jedoch durchaus möglich, dass die Trinkwassergewinnung in solchen Situationen mit einer Neupositionierung von Entnahmebrunnen optimiert werden könnte.

Die neue Gewässerschutzverordnung in der Schweiz schützt erstmalig auch die ober- und unterirdischen Zuströmbereiche. Aus diesen Überlegungen wird deutlich, dass die Optimierung der Schutzzonen auch die Frage nach der optimalen Lage der Entnahmebrunnen nicht ausschliessen darf. Ein umfassender und nachhaltiger Gewässerschutz und die Sicherung der Wasserqualität verlangen die gesamtheitliche Betrachtung der regionalen Wasserkreisläufe.

Schlussfolgerungen

Die schnellen, sogenannt präferenziellen Wasserwege bei der Infiltration von Niederschlagswasser und Oberflächengewässern sowie die Inhomogenität der Grundwasserleiter bestimmen im wesentlichen die Filterwirkung für Mikroorganismen. Hydrogeologische Abklärungen, insbesondere sedimentologisch-strukturelle Überlegungen zum Aufbau der Grundwasserleiter, können wesentliche Angaben zur erwarteten Filterwirkung liefern.

In Karst- und Kluftgrundwasserleitern ist der Wasserentnahmeort in der Regel durch den Standort der Quelle vorgegeben. Damit ist auch das Einzugsgebiet zwangsläufig festgelegt, wobei es in Abhängigkeit vom Niederschlagsverlauf durchaus variieren kann. Die Ausscheidung von Schutzzonen sollte für Karstgebiete nach der EPIK-Methode (Doerfliger und Zwahlen, 1998) erfolgen, die verschiedene Eigenschaften der Überdeckung und des Grundwasserleiters und damit deren Schutzwirkung gegen Verunreinigungen berücksichtigt. Wegen der teilweise sehr hohen Fliessgeschwindigkeiten und der gleichzeitig oft nur sehr kurzen Aufenthaltszeiten können jedoch auch bei ordnungsgemäss ausgeschiedenen Schutzzonen in Karst- und Kluftgrundwasserleitern mikrobiologische Verunreinigungen des geförderten Rohwassers nicht gänzlich ausgeschlossen werden. Aus diesem Grund ist ein Rohwassermanagement notwendig. Damit dies einfach und optimal erfolgen kann, müssen zuerst gute Systemkenntnisse erarbeitet werden. Nur so sind präzise Aussagen über die Bedeutung beispielsweise einer Schüttungs- oder Trübungsmessung möglich. Die Aufbereitung von Karstwasser sollte dabei der letzte Schritt zur Gewährleistung von einwandfreiem Trinkwasser sein. Die Untersuchungsergebnisse zur Abklärung der Wasserherkunft und zur Ausscheidung der Schutzzonen sowie die Systemkenntnisse aus dem Monitoring der Quelle dienen der Wahl des optimalen Verfahrens. Die Selbstkontrolle der Wasserversorger auf Basis des HACCP-Konzeptes sollte dann nur noch die Bestätigung einer guten Trinkwasserqualität liefern.

Viele Anlagen für die Grundwassergewinnung in der Nähe von Flussläufen weisen konzeptionelle Mängel auf, die sich auf die mikrobielle Gefährdung oder gar Beeinträchtigung einzelner Brunnen, insbesondere während Hochwasserereignissen, auswirken. Gerade solche Mängel erfordern beim Auftreten von Nutzungskonflikten, z.B. in dicht besiedelten Gebieten, eine Erarbeitung von konsensfähigen Lösungen. Es empfiehlt sich deshalb eine differenzierte Betrachtung von Gesamtsystemen. Dabei ist zwischen kurz-, mittel- und langfristigem Handlungsbedarf zu unterscheiden. Die dauernde Gewährleistung der einwandfreien Trinkwasserqualität hat bei der Abwägung unterschiedlicher Interessen erste Priorität.

Schlussfolgerungen und Empfehlungen

8

Adrian Auckenthaler und Peter Huggenberger

Schlussfolgerungen

Im Grundwasser kommen natürlicherweise viele verschiedene Mikroorganismen vor, von denen die meisten für den Menschen harmlos sind. Daneben gibt es jedoch einige seit langem bekannte pathogene Keime sowie „emerging pathogens“ (neu entdeckte wasserübertragbare Pathogene), die in der Schweiz und in Deutschland verschiedentlich im Grundwasser und teilweise sogar im Trinkwasser nachgewiesen werden konnten. Die Bedeutung dieser Befunde ist zur Zeit noch unklar, da einerseits die Daten über das Ausmass der mikrobiologischen Belastung (Zusammensetzung der Mikroorganismenpopulation, Konzentrationsmaxima, zeitliches Auftreten) nicht vorhanden sind und andererseits Erkrankungen nicht systematisch erfasst werden können, unter anderem wegen fehlender oder erst im Aufbau begriffener Meldesysteme für viele der pathogenen wasserübertragbaren Keime. Zudem war bisher der Nachweis von pathogenen Krankheitserregern - allen voran Viren - analytisch sehr aufwändig oder nahezu unmöglich. In den kommenden Jahren sind auf dem Gebiet der mikrobiologischen Analytik jedoch erhebliche Fortschritte zu erwarten. Bereits heute wird die PCR-Methode (Polimerase Chain Reaction) teilweise routinemässig angewandt und es sind weitere analytische Vereinfachungen und Präzisierungen im Nachweis zu erwarten. Der Vorteil der neuen Methoden ist es, dass man sich nicht mehr nur auf das System der Indikatorkeime stützen muss, sondern in kurzer Zeit effektiv die gesuchten pathogenen Keime analysieren kann. Die Gefährdung durch wasserübertragbare Erkrankungen wird dadurch wesentlich besser beurteilbar.

Die verbesserten Nachweismethoden unterstützen eine optimale Anpassung der Trinkwasseraufbereitung an die jeweiligen Risiken. Die heutige Praxis für die Aufbereitung von Trinkwasser mittels einstufiger Verfahren wie UV oder Chlor ist in vielen Fällen, insbesondere in Karstgebieten, zu wenig effektiv. Von den Kantonalen Laboratorien können immer wieder Fäkalkeime in Netzwässern beobachtet werden. Aus diesem Grund aber stärker auf technische Lösungen zu setzten und Reduktionsraten von 12 log-Einheiten bei Aufbereitungsprozessen zu fordern (was theoretisch möglich wäre), ist unverhältnismässig und nicht praktikabel. Bei einem technischen Defekt der Anlagen wäre zudem mit einem höheren wasserbedingten Epidemierisiko zu rechnen. Viel wichtiger ist der präventive Schutz der Wasserfassungen vor Verunreinigungen mittels Nutzungsbeschränkungen innerhalb des Anströmbereichs und in den Grundwasserschutzzonen sowie das Management des Pumpenbetriebes mittels online Messung von Systemparametern (Niederschlag, Wasserstand, Schüttung Trübung, etc.).

Das Auftreten von Mikroorganismen in Trinkwasserfassungen und der Transport im Grundwasser hängen von der lokalen Hydrogeologie, den spezifischen Eigenschaften der Mikroorganismen selbst sowie von den anthropogenen Belastungsmustern im Einzugsgebiet der Fassung ab. Der Aufbau der Aquifere sowie deren Heterogenität bestimmen in einem hohen Mass das Transportverhalten von Mikroorganismen. Die Verteilung der hydraulischen Leitfähigkeit, ausgedrückt in ln K (Logarithmus naturalis von K) scheint bei der Beurteilung von Feldversuchen eine wichtigere Grösse zu sein, als die aus der DLVO-Theorie abgeleiteten Parameter zum Partikeltransport wie pH-Wert, Ionenstärke und Temperatur. In der DLVO-Theorie fehlt zudem die aktive Rolle der Mikroorganismen, wie zum Beispiel die Adhäsion an Oberflächen für die Bildung von Biofilmen.

Bei der Dimensionierung von Schutzzonen gegen mikrobiologische Verunreinigungen können partikuläre Tracer hilfreich sein. Als Partikel werden sie wie die Mikroorganismen über die schnellen Wasserwege transportiert und gelangen dadurch rascher in die Trinkwasserfassungen als chemische Tracer, die durch Matrixdiffusion im Transport retardiert werden. Die Wiederfindungsrate von Partikeltracern kann in Porengrundwasserleitern, bei der Bodenpassage oder bei grösseren Fliessdistanzen sehr gering sein. In Karstgrundwasserleitern sind sie aber wiederholt erfolgreich eingesetzt worden.

Der Schutz von Wasserfassungen vor Verunreinigung durch Mikroorganismen ist eine interdisziplinäre Aufgabe. Sie erfordert die Zusammenarbeit von Hydrogeologie, Mikrobiologie, Umwelt- und Ingenieurwissenschaften. Der umfassende Schutz beinhaltet die Ausscheidung und Überwachung der Grundwasserschutzzonen sowie eine optimierte Steuerung der Trinkwassergewinnung und -aufbereitung begleitet von einer wirkungsvollen Qualitätskontrolle.

Empfehlungen

Die untenstehenden Empfehlungen sind in kurz-, mittel- und langfristige Strategien eingeteilt. Das Ziel ist, die bestehenden Wasserentnahmen zu optimieren und bei neuen Wasserfassungen die notwendigen Abklärungsschritte aufzuzeigen. Für einige Wasserfassungen sind wohl keine weiteren Schritte zu einer Verbesserung des Zustandes notwendig, andere jedoch sind durch ihren ungünstigen Standort auch bei verstärkter Kontrolle und Schutzanstrengungen durch Verunreinigungen mikrobiologischer aber auch chemischer Art gefährdet. Die jeweiligen Schritte zur Lösung eines Problems können nur skizziert werden, da lokale Gegebenheiten immer eine wesentliche Rolle beim Partikeltransport spielen.

Kurzfristige Ziele:

- Präventiver Schutz des Grundwassers vor mikrobiologischen Verunreinigungen. Dies beinhaltet die Ausscheidung von Grundwasserschutzzonen nach den neusten Erkenntnissen (z.B. EPIK-Methode in Karstgebieten) und den Vollzug der notwendigen Nutzungsbeschränkungen in diesen Zonen. Dadurch kann das Risiko einer Verunreinigung der Wasserfassung durch Mikroorganismen erheblich reduziert werden. Diese meist kostengünstigen präventiven Massnahmen sind in der Regel anderen Massnahmen zur Sicherstellung einer guten Trinkwasserqualität wie das vorübergehende Verwerfen von Quellwasser oder die Aufbereitung des Rohwassers vorzuziehen.
- Ereignisorientierte Probennahmen, insbesondere in Karstgebieten, bei sehr heterogenen Kluftaquiferen und in Gebieten mit intensiver Oberflächengewässer-Grundwasserinteraktion. Durch geeignete Probennahmen können Gefährdungen rechtzeitig erkannt werden. Die gängige Praxis mit wenigen, nicht situationsgerechten, sondern nach einem starren Zeitraster geplanten Proben, sind für eine Beurteilung ungenügend. Primär sind die Proben dann zu nehmen, wenn das Risiko einer Belastung aufgrund früherer Untersuchungsresultate als am höchsten beurteilt wird.
- Aufbau und Etablierung von QS-Systemen. Werden aufgrund der Selbstkontrolle oder Laboranalysen Mängel in der Trinkwassergewinnung festgestellt, müssen diese behoben werden. Die meisten QS-Systeme der Wasserversorgungen wurden bis heute ohne umfassende Systemkenntnisse über die Wasserherkunft und die mikrobiologischen oder chemischen Belastungsmuster entwickelt. Diese Voraussetzungen sollten baldmöglichst erarbeitet werden.

Mittelfristige Ziele:

- Erkennen der Zusammenhänge zwischen Einzugsgebiet, Niederschlag, Schüttung und physikalischer und chemischer Belastung von Karstquellen sowie der Dynamik der Austauschprozesse Fluss-Grundwasser im Zustrombereich von Grundwasserbrunnen. Am geeignetsten dafür sind online Messungen, denn mit Hilfe von ihnen lassen sich Konzentrationsverläufe am zuverlässigsten ermitteln.
- Identifikation der Belastungsmuster von Karstquellen und von Oberflächengewässern in Gebieten mit Flussinfiltration. Es gilt zu eruieren, welche Mikroorganismen wann und in welchen Konzentrationen auftreten. Ein interdisziplinäres Vorgehen von Hydrogeologie und Mikrobiologie ist hier wichtig, um Eintragungsorte, Fliesszeiten im Untergrund und Art der mikrobiologischen Verunreinigung beurteilen zu können.
- Entwicklung des Managements von Quellwasserfassungen aus den Erkenntnissen der online Messungen und der Erfassung der mikrobiologischen Belastung des Wassers.
- Optimierung der bestehenden Grundwasserentnahmesysteme und Behebung von konzeptionellen Mängeln (z.B. Erkennen von defekten Abwasserleitungen als Ursache von Verunreinigungen).
- Im Bedarfsfall Aufbau oder Ausbau von Wasseraufbereitungssystemen. Zur Wahl der optimalen Verfahren und der Dimension der Anlagen dienen die Erkenntnisse

der online Messungen und der Erfassung der mikrobiologischen Belastung des Wassers.
- Anpassung der Analyseverfahren der Untersuchungslabors an das Gefährdungsmuster. Neben Indikatorkeimen sollen auch gezielte Untersuchungen auf pathogene Mikroorganismen durchgeführt werden. Durch die neuen PCR-Methoden ist eine bessere Erkennung der effektiven Gefährdungen möglich.

Langfristige Ziele:

- Optimierung von Grundwasserentnahmen auf die für eine Eliminierung der Mikroorganismen benötigte Aufenthaltszeit (minimal 10 Tage). Dies kann einerseits eine reine Optimierung des Pumpenregimes bedeuten oder andererseits die Versetzung des Entnahmebrunnens an einen geeigneteren Standort.
- Entwicklung von Biochips oder ähnlichen Detektionsystemen für die Trinkwasseranalyse. Die Belegung eines einzelnen Chips mit DNA-Sequenzen aller bekannten Trinkwasserpathogene ist theoretisch denkbar. Dadurch würden online Messungen von Mikroorganismen möglich, d.h. die Resultate von Wasseruntersuchungen würden nicht erst Stunden oder Tage nach der Probennahme vorliegen, sondern es wäre eine Echtzeit-Überwachung des Trinkwassers machbar.
- Beseitigung von Belastungsquellen für das Grundwasser (z.B. durch Regen- und Mischwasserentlastungen in Oberflächengewässer im Rahmen der regionalen Entwässerungsplanung (REP)).
- Etablieren möglichst intakter, regionaler hydrologischer Kreisläufe.

Zur Erreichung dieser Ziele sind neben den Wasserversorgern vor allem auch die Behörden (Bund, Länder, Kantone und Gemeinden), Geologie- und Ingenieurbüros sowie die Forschung angesprochen. Die lokale Problemerkennung muss von den Wasserversorgern und den Kontrollorganen in den Kantonen und Ländern ausgehen. Die Zurverfügungstellung geeigneter Mittel und Lösungsstrategien zur Minimierung der mikrobiologischen Belastung des Grundwassers wird hauptsächlich durch die Forschungsinstitutionen und die Behörden erfolgen. Geologie- und Ingenieurbüros sollten schliesslich für die Umsetzung der Erkenntnisse in den Wasserversorgungen besorgt sein.

Da die Trinkwassergewinnung ein dynamischer Prozess ist und sich immer wieder verändern kann, beispielsweise durch geänderte Landnutzung, Revitalisierung von Fliessgewässern oder neue Erkenntnisse zum Auftreten pathogener Mikroorganismen, müssen die Massnahmen zur Gewährleistung der einwandfreien Trinkwasserqualität mittels Grundwasserschutzzonen, Grundwassermanagement, Trinkwasseraufbereitung und Qualitätssicherung immer wieder überprüft, und wenn nötig den neuen Begebenheiten angepasst werden.

Referenzen

Abba, F., Orlandi, E. und Rondelli, A., 1899. Über die Filtrationskraft des Bodens und die Fortschwemmung von Bakterien durch das Grundwasser. Z. Hygiene und Infektionskrankheiten, 31:66-84. Leipzig: Veit & Comp.

Abbaszadegan, M., Huber, M. S., Gerba, C. P., und Pepper, I. L., 1997. Detection of viable *Giardia* cysts by amplification of heat shock-induced mRNA. Applied and Environmental Microbiology, 63:324-328.

Abbaszadegan, M. und Dowd, S. E., 1998. Detection of enterovirus and bacteriophages. In: Microbial Pathogens within Aquifers: Principles and Protocols. Pillai, S. D., ed. (Springer Verlag), 59-69.

Abbaszadegan, M., Stewart, P., und LeChevallier, M., 1999. A strategy for detection of viruses in groundwater by PCR. Applied and Environmental Microbiology, 65:444-449.

Abbaszadegan, M., 1999. Rotaviruses. Waterborne Pathogens. Manual of Water Supply Practices, America Water Works Association, 1st Edition, 253-256.

Adler, I. und Zickel, R., 1969. Winter vomiting disease. J. Infect. Dis., 119:668-673.

Aldras, A. M., Orenstein, J. M., Kotler, D. P., Shadduck, J. A., und Didier, E. S., 1994. Detection of microsporidia by indirect immunofluorescence antibody test using polyclonal and monoclonal antibodies. Journal of Clinical Microbiology, 32:608-612.

Aleksic, S. und Bockemühl, J., 1988. Serological and biochemical characteristics of 416 *Yersinia* strains from well water and drinking water plants in the Federal Republic of Germany: lack of evidence that these strains are of public health importance. Z. Bak. Mik. Hyg., 185:527-533.

Alonso, J. L., Soriano, K., Amoros, I., und Ferrus, M. A., 1998. Quantitative determination of *E. coli* and fecal coliforms in water using a chromogenic medium. Journal of Environmental Science and Health, Part A Toxic-Hazard. Subst. Environ. Eng., 33:1229-1248.

American Public Health Association (APAH), 1992. Standard Methods for the Examination of Water and Wastewater, 19th ed. (APHA, Washington D. C.)

Ammon, A. und Bräunig, J., 2002. Lebensmittelbedingte Erkrankungen in Deutschland. Gesundheitsberichterstattund des Bundes Heft 01/02, Robert Koch Institut, 20.

Anan'ev, N. I. und Demin, N. D., 1971. On the spread of pollutants in subsurface waters. Hyg Sarit, 36:292-294.

Andersen, J. K., Sorensen, R. und Glensbjerg, M., 1991. Aspects of the epidemiology of Yersinia enterolitica: a review. Int. J. Food Microbiol., 13:231-238.

Andersson, Y., de Jong, B. und Studahl, A., 1997. Waterborne Campylobacter in Sweden: The cost of an outbreak. Wat. Sci. Tech., 35(11-12):11-14.

Anonym, 2002. Wasserfakten im Überblick. www.bundesverband-gas-und-wasser.de

Aragno, M. und Müller, I., 1982. Premières expériments de tracage des eaux souterraines dans le karst du jura neuchâtelois (Suisse) à l'aide de bacteriophages. Bull. Centre d'Hydrogéologie, 4:59-81. Neuchâtel.

Atteia, O. und Kozel, R., 1997. Particle size distributions in waters from a karstic aquifer: from particles to colloids. J. of Hydrology 201, 102-119.

Ashford, R.W., 1979. Occurence of an undiscribed coccidian in man in Papua New Guinea. Ann. Trop. Med. Parasitol., 73:497-500.

Auckenthaler, A., Raso, G. und Huggenberger, P., 2002. Particle transport in a karst aquifer: natural and artificial tracer experiments with bacteria, bacteriophages and microspheres, Water Science and Technology, 46(3):131-138.

Aulenbach, D.B., 1979. Long-term recharge of trickling filter effluent into sand. Rep. EPA-600/2-79-068, Environ. Prot. Agency, Ada, Okla.

AWWA, 1999a. Comittee Report: Emerging Pathogens – bacteria. J. AWWA, 91(9):101-109.

AWWA, 1999b. Comittee Report: Emerging Pathogens – viruses, protozoa and algal toxins. J AWWA, 91(9):110-121.
AWWA, 1999c. Waterborne Pathogens. Manual of Water Supply Practices, America Water Works Association, 1st Edition, 285.
Babich, H. und Stotzky, G., 1980. Reductions in inactivation rates of bacteriophages by clay minerals in lake water. Water. Res., 14:185-187.
BAG, 2001. Meldungen Infektionskrankheiten. Bundesamt für Gesundheit, Bulletin, 49:954-955.
Bales, R. C., Li, S., Maguire, K. M., Yahya, M. T. und Gerba, C. P., 1993. MS-2 and poliovirus transport in porous media: hydrophobic effects and chemical perturbations. Water Res. Resear., 29(4):957-963.
Bales, R. C., Li, S., Maigure, K. M., Yahya, M. T., Gerba, C. P. und Harvey, R. W., 1995. Virus and bacteria transport in a sand aquiver, Cape Cod, MA. Ground Water, 33(4):653-661.
Barber II, L.B. et al., 1992. Geochemical heterogeneity in a sand and gravel aquifer: Effect of sediment mineralogy and particle size on the sorption of chlorobenzenes. J. of Contaminant Hydrology 9, 35-54.
Barwick, ; R. S., Levy, D. A., Craun, G. F., Beach, M. J. und Calderon, R. L., 2000. Surveillance for waterborne disease outbreaks United States, 1997-1998. MMWR, 49(SS04):1-35.
Batsche, H., Bauer, F., Behrens, H., Buchtela, K., Dombrowski, H. J., Geisler, R., Geyh, M. A., Hötzel, H., Hribar, F. , Käss, W., Mairhofer, J., Maurin, V., Moser, H., Neumaier, F., Schmitz, J., Schnitzer, W. A., Schreiner, A., Vogg, H. und Zötel, J., 1970. Kombinierte Karstwasser-untersuchungen im Gebiet der Donauversickerung (Baden-Württemberg) in den Jahren 1967-1969. Steir. Beitr. Hydrogeol., 4:4-165.
Baumgartner, A., 2001. „Norwalk-like"-Viren (NLV) und Lebensmittel – eine Situationsanalyse für die Schweiz. Lebensmittel-Technologie, 34(12):503-509.
Bear, J., 1972. Dynamics of Fluids in Porous Media. Dover, Mineola, N.Y.
Beck, H., 1997. Lebensmittelinfektionen und –intoxikationen: eine Übersicht. Amtstierärztlicher Dienst und Lebensmittelkontrolle, 4(1):11-14.
Beckers, P. J. A., Derks, G. J. M. M., van Gool, T., Rietveld, F. J. R., und Sauerwein, R. W., 1996. *Encephalitozoon intestinalis*-specific monoclonal antibodies for laboratory diagnosis of microsporidia. Journal of Clinical Microbiology, 34:282-285.
Bellamy, W. D., Hendricks, D. W. und Logdson, G. S., 1985b. Slow sand filtration: Influences of selected process variables. Journal AWWA, 12:62-66.
Bellamy, W. D., Silverman, G. P., Hendricks, D. W. und Logdson, G. S., 1985a. Removing Giardia cysts with slow sand filtration. Journal AWWA, 2:52-60.
Bennet, E. J., Holmberg, S. D., Rogers, M. F. und Solomon, S. L., 1987. Infectious and parasitic diseases. Am. J. Prev. Med., 3:102-114.
Berg, J. D., 1987. Disinfection: The physiological fate of the test organism. Proc. AWWA Seminar, Assurance of Adequate Disinfection, or C*t or not C*t. American Water Works Assn., Denver, CO, 85-99.
Bern, C., Hernandez, B., Lopez, M. B., Arrowood, M. J., de Mejia, M. A., de Merida, A. M., Hightower, A. W., Venczel, L., Herwaldt, B. L. und Klein, R. E., 1999. Epidemiologic studies of Cyclospora cayetaniensis in Guatemala. Emerging Infectious Diseases, 5(6):766-774.
Bernhardt, H. et al., 1994. Desinfektion aufbereiteter Oberflächenwässer mit UV-Strahlen. gwf, 135(12):677-689.
Beutin, L., 1996. Infektionen mit enterohämorrhagischen Escherichia coli (EHEC). Bundes-gesundheitsblatt, 11:426-429.
Birou, B., 2000. Membranfiltration und Trinkwasseraufbereitung. gwa, 80(1):29-36.
Blackmer, F., Reynolds, K. A., Gerba, C. P., und Pepper, I. L., 2000. Use of integrated cell culture-PCR to evaluate the effectiveness of Poliovirus inactivation by chlorine. Applied and Environmental Microbiology, 66:2267-2268.

Boch, J. und Supperer, R., 1983. Veterinärmedizinische Parasitologie. Paul Parey Berlin und Hamburg, 3. Auflage.
Bonde, G. J., 1962. Bacterial indicators of water pollution. 1.ed., Bonde, G. J. (ed), Teknisk Forlag Copenhagen.
Boller, M., 1994.Trends in water filtration technology. J. Water SRT-Aqua, 43(2):65-75.
Boorman, G. A., Dellaroc, V., Chapin, R. E., Hunter, S., Hauchman, F., Gardner, H., Cox, M. und Sills, R. C., 1999. Drinking water disinfection byproducts: Review and Approach to toxicity evaluation. Environmental Health Perspectives, 107(1):207-217.
Bosma, T. N. P. und Zehnder, 1994. Behavior of microbes in aquifers. Transport and Reactive Processes in Aquifers, Dracos & Staufer (eds), Balkema, Rotterdam, 37-41.
Botzenhardt, K., 1994. Die Beherrschung mikrobiologischer Belastungen bei der Oberflächenwasseraufbereitung nach der SWTR (USA). gwf-Wasser/Abwasser, 135:201-206.
Botzenhart, K., 2000a. Trinkwasser. In Umweltvirologie: Viren in Wasser und Boden, Walter, R. (ed.), Springer Wien New York, 57-84.
Botzenhart, K., 2000. Viren als Erreger wasserbedingter Infektionen. Mitt. Lebensm. Hyg., 91:26-43.
Brenner, K. P., Rankin, C. C., Sivagenasan, M. und Scarpino, P. V., 1996. Comparison of recoveries of *Escherichia coli* and total coliforms from drinking water by the MI agar method and the U. S. Environmental Protection Agency-approved membrane filter method. Applied and Environmental Microbiology, 62:203-208.
Brockmann, S. und Botzenhart, K., 2000. Clostridium perfringens als Indikator für fäkale Verunreinigung des Trinkwassers. Wasser-Abwasser gwf, 141(1):22-33.
Brunke, M., 1998. The Influence of Hydrological Exchange Patterns on Environmental Gradients and Community Ecology in Hyporheic Interstices of a Prealpine River. Dissertation ETH Zürich 12734.
Bucklin, K. E., McFeters, G. A., und Amirtharaja, A., 1991. Penetration of coliforms through municipal drinking water filters. Water Research, 25:1013-1017.
Bukhari, Z., Hargy, T. M., Bolton, J. R., Dussert, B. und Clancy, J. L., 1999. Medium-pressure UV for oocyst inactivation. Journal AWWA, 91(3):86-94.
Bundesgesetz über den Schutz der Gewässer (Gewässerschutzgesetz GSchG), vom 24. Januar 1991 (Stand 21. Dez. 1999).
Burnens, A.P., 2001. Aktuelle Diagnostik von Verotoxin bildenden Escherichia coli, Epidemiologie und Infektionskrankheiten, Bulletin, 2:27-30.
Burren, 2000. Persönliche Kommunikation. Firma WATECH in Thun, BE.
Buttler, R.C., Lund, V. und Carlson, D.A., 1987. Susceptibility of Campylobacter jejuni and Yersinia enterolitica to UV radiation. Appl. Environ. Microbiol, 53:375-378.
BUWAL, 1998. Praxishilfe, Kartierung der Vulnerabilität in Karstgebieten (Methode EPIK), 56.
Cali, A., 1999. Microsporidia. Waterborne Pathogens. Manual of Water Supply Practices, America Water Works Association, 1st Edition, 191-197.
Castell-Exner, C., Liessfeld, R. und Mendel, B., 2001. Die neue Trinkwasserverordnung, Teil 1. Energie Wasser Praxis, 4:31-34.
CCDR, Canada Communicable Disease Report, 2000. Waterborne outbreak of gastroenteritis associated with a contaminated municipal water supply, Walkerton, Ontario, May-June 2000. 26(20).
CDC, 1996. Surveillance of waterborne disease outbreaks – United States, 1993-1994. MMWR, 45 (SS-1):1-33.
Champ, D. R., und Schroeter, J., 1988. Bacterial transport in fractured rock – A field scale tracer test at the Chalk River Nuclear Laboratories. Water Sci. Technol., 20:81-87.
Chapelle, F.H. 1992. Ground-Water Microbiology and Geochemistry, Wiley, 424.
Chapron, C. D., Ballester, N. A., Fontaine, J. H., Frades, C. N., und Margolin, A. B, 2000. Detection of astroviruses, enteroviruses, and adenovirus types 40 and 41 in surface waters collected and

evaluated by the Information Collection Rule and an integrated cell culture-nested PCR procedure. Applied and Environmental Microbiology, 66:2520-2525.
Chrisp, C. E., und LeGendre, M., 1994. Similarities and differences between DNA of *Cryptosporidium parvum* and *C. wrairi* detected by the polymerase chain reaction. Folia Parasitologica, 41:97-100.
Chung, H., Jaykus, L. A., und Sobsey, M. D., 1996. Detection of human enteric viruses in oysters by in vivo and in vitro amplification of nucleic acids. Applied and Environmental Microbiology, 62:3772-3778.
Clancy, J. L., Hargy, T. M., Marshall, M. M. und Dyksen, J. E., 1998. UV light inactivation of cryptosporidium oocysts. Journal AWWA, 90(9):92-102.
Cleasby, J. L., 1991. Biological and physical mechanisms in slow sand filtration. Slow Sand Filtration, A report prepared by the Task Committee on Slow Sand Filtration of the Water Supply Committee of the Environmental Engineering Division of the American Society of Civil Engineers, G. S. Logsdon (ed.), American Society of Civil Engineers, New York, 69-100.
Corapcioglu, M.Y. und Haridas, A., 1984. Transport and fate of microorganisms in porous media: a theoretical investigation. J. Hydrol., 72:149-169.
Correa, C. M., Tibana, A. und Gontijo-Filho, P. P., 1991. Vegetables as a source of infection with *Pseudomonas aeruginosa* in a university and oncology hospital of Rio de Janeiro. J. Hosp. Infect., 18:301.
Courbat, R., Ramseier, S., Walther, J. L., Gaille, P., Jordan, R., Kaiser, H-P., Revelly, P., Stettler, R. und von Gunten, U., 1999. Utilisation de l'ozone pour la traitement des eaux potables en Suisse. gwa, 10:843-852.
Covert, T. C., 1999. Salmonella. Waterborne Pathogens. Manual of Water Supply Practices, America Water Works Association, 1st Edition, 107-113.
Cronholm, L. S., et al., 1976. Enteric virus survival in package plants and the upgrading of small treatment plants using ozone. Research Report No. 98, Water Resources Research Institute, University of Kentucky, Lexington, KY.
Daniel, O., 1998. Mikrobielle und chemische Risiken des Trinkwassers. Mitt. Gebiete Lebensm. Hyg., 89:684-699.
D'Antonio, R. G., Winn. R.E. und Taylor, J.P., 1985.A waterborne outbreak of cryptosporidiosis in normal hosts. Ann. Intern. Med., 312:647-648.
Deng, M. Q., Cliver, D. O., und Mariam, T. W., 1997. Immunomagnetic capture PCR to detect viable *Cryptosporidium parvum* oocysts from environmental samples. Applied and Environmental Microbiology 63:3134-3138.
Derjaguin, B.V. und Landau, L., 1941. Theory of the stability of strongly charged lyophobic soils and of the adhesion of strongly charged particles in solutions of electrolytes. Acta Physiocochim. URSS, 14:633-662.
Devriese, L. A., Pot, B., und Collins, M. D., 1993. Phenotypic identification of the genus *Enterococcus* and differentiation of phylogenetically distinct enterococcal species and species groups. Journal of Applied Bacteriology, 75:399-408.
Diaper, E. W. J., 1972. Wat. Sew. Works, 117(11):373-380.
Dictionary of Modern Medicine, 1. Auflage, 1992, Segen, J. C. (ed), Editiones Roche Verlag, 800.
di Giovanni, G. D., Hashemi, F. H., Shaw, N. J., Abrams, F. A., LeChevallier, M. W., und Abbaszadegan, M., 1999. Detection of infectious *Cryptosporidium* oocysts in surface and filter backwash water samples by immunomagnetic separation and integrated cell culture-PCR. Applied and Environmental Microbiology, 65:3427-3432.
Doerfliger, N., 1996. Advances in karst groundwater protection strategy using artificial tracer tests analysis and multiattribute vulnerability mapping (EPIK method). Diss. Uni Neuchâtel, 308.
Doerfliger, N. und Zwahlen, F., 1998. Kartierung der Vulnerabilität in Karstgebieten (Methode EPIK). BUWAL Bestellnummer VU-2504-D.

Doerfliger, N., Zwahlen, F. Meylan, B., Tripet, J.-P. und Wildberger, A., 1999. Die Vulnerabilität von Grundwasserfassungen in Karstgebieten. gwa, 7:585-592.
Dowd, S. E., Gerba, C. P., und Pepper, I. L., 1998. Confirmation of the human-pathogenic Microsporidia *Enterocytozoon bieneusi, Encephalitozoon intestinalis,* and *Vittaforma corneae* in water. Applied and Environmental Microbiology 64:3332-3335.
Drozd, C. und Schwartzbrod, J., 1997. Removal of cryptosporidium from river water by crossflow microfiltration: a pilot-scale study. Wat. Sci. Tech., 35(11-12):391-395.
Dubey, J. P., 1999. Toxoplasma gondii. Waterborne Pathogens. Manual of Water Supply Practices, America Water Works Association, 1st Edition, 213-216.
Duden Band 5, Das Fremdwörterbuch, 5. Auflage, 1990, Dudenverlag, 832.
DuPont, H. L., Chappell, C. R., Sterling, C. R. et al., 1995. The infectivity of Cryptosporidium parvum in healthy volunteers. New England J. Med., 332:855.
Dutka, B. J., Chau, A. S. Y, und Coburn, J., 1974. Relationship between bacterial indicators of water pollution and fecal sterols. Water Research, 8:1047-1055.
EC (European Community), 1998. Richtlinie 98/83/EG des Rates vom 3. Nov. 1998 über die Qualität von Wasser für den menschlichen Gebrauch. Amtsblatt der Europ. Gemeinschaften v. 5.12.98, L330/32-54 mit Berichtigung v. 19.2.99, L45/55.
Eckert, J., 1993. Träger und Ausscheider von Protozoen. Zbl. Hyg., 194:173-185.
Eckert, J., Jacquier, P. und Weber, R., 1990. Intestinale Protozoen – neue Aspekte. Internist, 31:386-398.
Eckert, J., 1998. Medizinische Mikrobiologie, Georg Thieme Verlag, Stuttgart, 504-507.
Edberg, S. C., Gallo, P. und Kontnick, C., 1996. Analysis of the virulence characteristics of bacteria isolated from botteled, water cooler, and tap water. Microb. Ecol. Health Dis., 9:67.
Elimelech, M. et al., 1995. Particle deposition and aggregation. Butterworth-Heinemann, Oxford.
Enriquez, F. J., Ditrich, O., Palting, J. D., und Smith, K., 1997. Simple diagnosis of *Encephalitozoon* sp. microsporidial infections by using a panspecific antiexospore monoclonal antibody. Journal of Clinical Microbiology 35:724-729.
Everett, D., 1992. Grundzüge der Kolloidwissenschaften. Steinkopf Verlag, Düsseldorf.
Ewald, R. 2001. pers. Kommunikation, Kantonales Laboratorium Basel-Stadt.
Exner, M., Gornik, V. und Kistemann, T., 2001. Charakterisierung, Risikoeinschätzung und Prävention wasserassoziierter Parasitosen. Bundesgesundheitsbl – Gesundheitsforsch – Gesundheitsschutz, 44:358-363.
Farnleitner, A. H., Kreuzinger, N., Kavka, G. G., Grillenberger, S., Rath, J., und Mach R. L., 2000. Simultaneous detection and differentiation of *Escherichia coli* populations from environmental freshwaters by means of sequence variations in a fragment of the β-D-Glucoronidase gene. Applied and Environmental Microbiology, 66:1340-1346.
Filkorn, R., Wiedenmann, A., und Botzenhart, K., 1994. Selective detection of viable *Cryptosporidium* oocysts by PCR. Zentralblatt für Hygiene und Umweltmedizin, 195:489-494.
Fischeder, R., Schulze-Röbbecke, R. und Weber, A., 1991. Occurence of Mycobacteria in drinking water samples. Zbl. Hyg., 192:154-158.
FIV, 1995. Fremd- und Inhaltsstoffverordnung SR 817.021.23.
Fletcher, M. und Loeb, G.I., 1979. Influence of substratum characteristics on the attachment of a marine pseudomonad to solid surfaces. Appl. Environ. Microbiol., 44:184-192.
Fontes, D. E., Mills, A. L., Hornberger, G. M. and Herman, J. S., 1991. Physical and chemical factors influencing transport of microorganisms through porous media. Applied and Environmental Microbiology, 57(9):2473-2481.
Ford, T. E., 1999. Microbiological safety of drinking water: United States and global perspectives. Environmental Health Perspectives, 107, Supplement 1:191-206.
Forster, R., 1998. Netzbetrieb ohne Netzschutz in Zürich. gwa, 78(9):706-710.

Fox, K. R. und Reasoner, D. J., 1999. Water quality in source water, treatment, and distribution systems. Waterborne Pathogens. Manual of Water Supply Practices, America Water Works Association, 1st Edition, 19-27.

Frahm, E., und Obst, U., 1995. Comparison of two rapid tests for the detection of legionellaceae in water, a microbiological-immunological method and a commercial gene probe test kit. Water Science and Technology, 31:403-406.

Frahm, E., Heiber, I., Hoffmann, S., Koob, C., Meier, H., Ludwig, W., Amann, R., Schleifer, K. H., und Obst, U., 1998. Application of 23S rDNA-targeted oligonucleotide probes specific for Enterococci to water hygiene control. Systematic and Applied Microbiology 21:450-453.

Freeze, A. und Witherspoon, W. 1967. Theoretical analysis of regional groundwater flow: 2 Effect of water-table configuration and subsurface permeability variation. Water Resources Res., (3):623-634.

Fretz, R., 2001. Epidemiologischer Report (Endbericht) vom 19. April 2001. Interner Bericht Kantonales Laboratorium BL.

Fretz, R, Svoboda, P., Ryan, U. M., Thompson, R. C. A., Tanner, M. und Baumgartner, A., 2002. Genotyping of *Cryptosporidium* spp. isolated from Human stool samples in Switzerland. Epidemiology and Infection (submitted).

Fricker, E. J., und Fricker, C. R., 1995. Detection of *Legionella* spp. using a commercially available polymerase chain reaction test. Water Science and Technology, 31:407-408.

Fricker, C. R., 1999a. Campylobacter. Waterborne Pathogens. Manual of Water Supply Practices, America Water Works Association, 1st Edition, 67-70.

Fricker, C. R., 1999b. Yersinia. Waterborne Pathogens. Manual of Water Supply Practices, America Water Works Association, 1st Edition, 131-134.

Gale, P., 1996. Developments in microbial risk assessment models for drinking water – a short review. J. Appl. Bacteriol., 81:403-410.

Gale, P. van Dijk, P. A. H. und Stanfield, G., 1997. Drinking water treatment increases micro-organism clustering; the implications for microbial risk assessment. J. Water SRT – Aqua, 46(3):117-126.

Garabedian, S.P., Gelhar, L.W. und Celia, M.A., 1988. Large-scale dispersive transport in aquifers: Field experiments and reactive transport theory, Ralph M. Parsons Lab. Rep. 315, Mass. Inst. Of Technol., Cambridge, Mass.

Geldreich, E.E., 1999. Pseudomonas. Waterborne Pathogens. Manual of Water Supply Practices, America Water Works Association, 1st Edition, 103-105.

Gelhar, L.W., 1993. Stochastic Subsurface Hydrology. Prentice-Hall, Englewood Cliffs, New Jersey.

Gerba, C. P., 1984. Microorganisms as groundwater tracers. Groundwater Pollution Microbiology, Bitton, G. and Gerba, C. P. (eds.), John Willey & Sons, New York, 225-233.

Gerba, C.P., 1999. Enteroviruses. Waterborne Pathogens. Manual of Water Supply Practices, America Water Works Association, 1st Edition, 235-239.

Gerba, C. P., Rose, J. B. und Haas, C. N., 1996. Sensitive populations: who is at the greatest risk?. Int. J. Food Microbiol., 30:113-123.

Gerba, C.P., und Rose, J.B., 1990. Viruses in source and drinking water. Drinking Water Microbiology: Progress and recent developments, G. A. McFeters (ed.) Springer-Verlag New York, 380-396.

Gerigk, K., 1993. Foodborne disease in Europe: present situation and future development of the WHO surveillance programme. Paper presented to Second UMDP Steering Committee Meeting on Food Safety, ICP/FOS 029/6, Budapest, 1990. In: Sockett, P., Social and economic aspects of food-borne disease. Food Policy, 18:110-119.

Gewässerschutzgesetz (GschG) vom 8. Oktober 1971 (Stand am 1. Juli 1990), SR 814.20.

Gewässerschutzverordnung (GschV) vom 28. Oktober 1998 (Stand am 10. August 1999), SR 814.201.

Gimbel, R., 1985. Flockung und Schnellfiltration. Flockung in der Wasseraufbereitung, DVGW-Schriftenreihe, Wasser Nr. 42:205-246.

Girdwood, R. W. A., 1995. Some clinical perspectives on waterborne parasitic protozoa. Protozoan Parasites and Water, Betts, W. B., Casemore, D., Fricker, C., Smith, H., und Watkins, J. (eds), 3-9.

Gonser, T., 1994. Die Bedeutung der physiographischen Faktoren für Fliessgewässerökosysteme. EAWAG news, 36D, März 1994.

Gorelick, S.M. et al., 1993. Groundwater Contamination: Optimal Capture and Containment. A.F. Lewis, New York.

Gornik, V., Behringer, K., Kölb, B. und Exner, M., 2001. Erster Giardiaausbruch im Zusammenhang mit kontaminiertem Trinkwasser in Deutschland. Bundesgesundheitsbl.-Gesundheitsforsch.-Gesundheitsschutz, 4:351-357.

Goyal, S. M. und Gerba, C. P., 1979. Comparative adsorption of human enteroviruses, simian rotavirus, and selected bacteriophages to soil. Applied and Environmental Microbiology, 38(2):241-247.

Grasso, D., B. F., Smets, K. A. Strevett, B. D., Machinist, C. J., VanOss, R. F., Giese und Wu, W., 1996. Impact of physiological state on surface thermodynamics and adhesion of Pseudomonas aeruginosa Environ. Sci. Technol., 30:3604-3608.

Gratacap-Cavallier, B., Genoulaz, O., Brengel-Pesce, K., Soule, H., Innocenti-Francillard, P., Bost, M., Gofti, L., Zmirou, D., und Seigneurin, J. M., 2000. Detection of human and animal rotaviruses in drinking water. Applied and Environmental Microbiology, 66:2690-2692. Gray, L. D., 1995. Escherichia, Salmonella, Shigella, and Yersinia. In: Manual of Clinical Microbiology, 6th ed. Murray, P. R., Baron, E. J., Pfaller, M. A., Tenover, F. C., and Yolken, R. H. eds. (ASM Press, Washington D. C.), 450-452.

Gribbon, L. T., und Barer, M., R., 1995. Oxidative metabolism in nonculturable *Helicobacter pylori* and *Vibrio vulnificus* cells by substrate-enhanced tetrazolium reduction and digital image processing. Applied and Environmental Microbiology, 61:3379-3384.

Haag, W. R. und Hoigné, J., 1983. Ozonatin of water containing chlorine or chloramines. Water Res., 17(10):1397-1402.

Haas, C. N., Meyer, M. A. und Paller, M. S., 1983. The ecology of acid-fast organisms in water supply, treatment and distribution systems. J. AWWA, 75:139-144.

Haas, C. N. und Rose, J., 1996. Distribution of Cryptosporidium oocystes in a water supply. Wat. Res., 30(10):2251-2254.

Haarhoff, J. und Cleasby, J. L., 1991. Biological and physical mechanisms in slow sand filtration. Slow Sand Filtration, A report prepared by the Task Committee on Slow Sand Filtration of the Water Supply Committee of the Environmental Engineering Division of the American Society of Civil Engineers, G. S. Logsdon (ed.), American Society of Civil Engineers, New York, 19-68.

Haas, C. N., Joffe, J., Anmangandla, U., Jacangelo, J. G. und Heath, M., 1996. Water quality and desinfection kinetics. Journal AWWA, 95-103.

Häfliger, D., Hübner, Ph., Lüthy, J., 2000. Outbreak of viral gastroenteritis due to sewage-contaminated drinking water. Internat. J. Food Microbiol., 54:123-126.

Hall, N. H., 1999. Legionella. Waterborne Pathogens. Manual of Water Supply Practices, America Water Works Association, 1st Edition, 171-175.

Hardalo, C. und Edberg, S. C., 1997. Pseudomonas aeruginosa: Assessment of risk from drinking water. Crit. Rev. Microbiol., 23(1):47-75.

Harvey, R.W. und George, L.H., 1987. Growth determinations for unattached bacteria in a contaminated aquifer. Appl. Environ. Microbiol., 53:2992-2996.

Harvey, R. W., Georges, Smith, LeBlanc, 1989. Transport of microspheres and indigenous bacteria through a sandy aquifer: Results of natural and forced gradient tracer experiments. Environ. Sci. Technol., 23(1):51-56.

Harvey, R. W. und Garabedian, S. P., 1991. Use of colloidal filtration theory in modelling movement of bacteria through a contaminated sandy aquifer. Environ. Sci. Technol., 25(1):178-185.

Harvey, R.W., 1991. Parameters involved in modeling movement of bacteria in groundwater: In Modeling the environmental fate of microorganisms. Edited by C.J. Hurst. American Society for Microbiology, Washington, D.C., 89-114.

Harvey, R. W., Kinner, N. E., MacDonald, D., Metge, D. W. und Bunn, A., 1993. Role of physical heterogenity in the interpretation of small-scale laboratory and field observations of bacteria, microbial-sized microsphere, and bromide transport through aquifer sediments. Water Res. Resear., 29(8):2713-2721.

Harvey, R. W., 1997. In situ and laboratory methods to study subsurface microbial transport. Manual of Environmental Microbiology, C. J. Hurst, G. R. Knudsen, M. J. Mclnerny, L. D. Stenzenback and W. W. Walter. Washington D.C., ASM, 586-599.

Harvey, R. W., 1997. Microorganisms as tracers in groundwater injection and recovery experiments: a review. FEMS Microbiology Reviews, 20:461-472.

Hawkins, E. W., und Orme, L., 1995. Rapid testing methodology for *Escherichia coli* O157:H7 using commercially available products. Proceedings of the American Society of Animal Sciences West. Sect., 46:281-283.

Hedberg, C. W. und Osterholm, M. T., 1993. Outbreaks of food-borne and waterborne viral gastroenteritis. Clin. Microbiol.Rev., 6:199-210.

Hendricks, D. W., Asce, F., Bellamy, W. D. und Asce, M., 1991. Microorganism Removals by slow sand filtration. A report prepared by the Task Committee on Slow Sand Filtration of the Water Supply Committee of the Environmental Engineering Division of the American Society of Civil Engineers, Slow Sand Filtration, G. S. Logsdon (ed.), American Society of Civil Engineers, New York, 101-121.

Hermann, J. E., 1999. Astroviruses. Waterborne Pathogens. Manual of Water Supply Practices, America Water Works Association, 1st Edition, 227-230.

Herold, Th., 1997. Räumliche Beziehungen der Karstsysteme zu den tektonisch-geologischen Strukturen im Gebiet der Weissenstein- und Farisbergantiklinale (Solothurner Jura). Dissertation ETH Zürich 12400.

Hiepe, T., 1983. Lehrbuch der Parasitologie. Bd. 2 Veterinärmedizinische Parasitologie. GF Verlag, Stuttgart, 11-15, 38-41, 139-140.

Higgo, J. J. W., Williams, G. M., Harrison, I, Warwick, P., Gardiner, M. P. und Longworth, G., 1993. Colloid transport in a glacial sand aquifer. Laboratory and field studies. Colloids and Surfaces A: Physicochemical and Engineering Aspects, 73:179-200.

Highsmith, A. K. und Abshire, R. L., 1975. Evaluation of a most-probable-number technique for the enumeration of *Pseudomonas aeruginosa*. Appl. Microbiol., 30(4):596.

Hirata, T. und Hashimoto, A., 1998. Experimental assessment of the efficacy of micro-filtration and ultrafiltration for cryptosporidium removal. Wat. Sci. Tech., 38(12):103-107.

Hobbins, M., Svoboda, P., Tanner, M. und Lüthi, T., 2001. Nachweis von Norwalk-like-Viren-Sequenzen in Umweltproben. gwa, 81(7):473-479.

Hobbins, M., 1999. An assessement of the presence and relevance of HuCV in Switzerland through a newly installed RT-PCR detection method - A first step. MSc Thesis, Swiss Tropical Institute, Basel, pp 82.

Hoff, J. C., 1987. Strength and weaknesses of using c*t values to evaluate disinfection practice. Proc. AWWA Seminar, Assurance of Adequate Disinfection, or C*t or not C*t. American Water Works Assn., Denver, CO, 49-65.

Hoigné, J. und Bader, 1994. Characterisation of water quality criteria for ozonation processes. Part II: Lifetime of added ozone. Ozone Sci. Eng., 16:113-134.

Hopkins, R.S. und Scott, A.S., 1983. Handling raw chicken as a source for sporadic Campylobacter jejuni infections (letter). J. infect. Dis., 148:770.

Hornberger, G.M. et al., 1992. Bacterial transport in porous media: evaluation of a model using laboratory observations. Water Resour.Res. 28, 915-938.

Huggenberger, P., Siegenthaler, C., Stauffer, F., 1988. Grundwasserströmung in Schottern: Einfluss von Ablagerungsformen auf die Verteilung von Grundwasserfliessgeschwindigkeit. Wasserwirtschaft 78(5):202-212.

Huggenberger, P., 1993. Radar facies: recognition of facies patterns and heterogeneities within Pleistocene Rhine gravels, NE Switzerland. In: Braided Rivers. Edited by Best, CoL., Bristow, C.S., Geological Society special Publication, 75:136-176.

Huggenberger, P., 2001. Wiese-Revitalisierung: Führen die Veränderungen der Sohlenstruktur zu einer Trinkwassergefährdung? Regio Basiliensis, 41(1):63-76.

Humphrey, T. J., 1986. Techniques for the optimum recovery of cold injured *Campylobacter jejuni* from milk or water. Journal of Applied Bacteriology 61:125-132.

Hunter, R.J., 1981. Zeta Potenzial in Colloid Science. Academic Press, London.

Hurst, C. J., 1991. Presence of enteric viruses in freshwater and their removal by the conventional drinking water treatment process. WHO Bulletin OMS, 69:113-119.

Hurst, C. J., 1997a. Detection of viruses in environmental waters, sewage, and sewage sludges. In: Manual of Environmental Microbiology. Hurst, C. J., Knudsen, G. R., McInerney, M. J., Stetzenbach, L. D., Walter, M. V., eds. (American Society for Microbiology, Washington D. C.), 168-175.

Hurst, C. J., 1997b. Sampling viruses from soil In: Manual of Environmental Microbiology. Hurst, C. J., Knudsen, G. R., McInerney, M. J., Stetzenbach, L. D., Walter, M. V., eds. (American Society for Microbiology, Washington D. C.), 400-405.

Hurst, C. J., 1999. Caliciviruses. Waterborne Pathogens. Manual of Water Supply Practices, America Water Works Association, 1st Edition, 231-234.

HyV, 1995. Verordnung des EDI über die hygienischen und mikrobiologischen Anforderungen an Lebensmittel, Gebrauchsgegenstände, Räume, Einrichtungen und Personal (Hygieneverordnung, HyV), vom 26. Juni 1995 (Stand am 7. Mai 2002), SR 817.051.

Isaac-Renton, J., Bowie, W., R., King, A., Irwin, G. S., Ong, C. S., Fung, C. P., Shokeir, M. O. und Dubey, J. P., 1998. Detection of *Toxoplasma gondii* oocystes in drinking water. Appl. Environ. Mirobiol., 64(6):2278-2280.

Jacangelo, J. G. und Olivieri, V. P., 1985. Aspects of the mode of action of monochloramine. Water Chlorination, Chemistry, Environmental Impact and Health Effects, Jolley, R. L., Bull, R. J., Davis, W. P., Katz, S., Roberts, M. H. and Jacobs, V. A. eds., Chelsea, MI, Lewis Publishers, 575-586.

Jakubowski, W., und Ericksen, T. H., 1979. Methods for the detection of *Giardia* in water supplies. In: Waterborne transmission of Giardiasis. Jakubowski, W., and Hoff, H., eds. (US EPA, Cincinnati), 193-210.

Jang, L.K. et al., 1983. Selection of bacteria with favorable transport properties through porous rock for the application of microbial enhanced oil recovery. Appl. Eviron. Microbiol., 46, 1066-1072.

Jeannin, P.-Y. und Sauter, M., 1998. Analysis of karst hydrodynamic behavior using global approaches: a review. Bull. d'Hydrogéol. 16:31-48.

Jekel, M., 1995. Beeinträchtigung des Flockungsprozesses durch organische Säuren. ed. Heinz, B., Probleme der Trinkwasseraufbereitung aus Talsperren, 599.

Johnson, C. H., Rice, E. W. und Reasoner, D. J., 1997. Inactivation of *Helicobacter pylori* by chlorination. Appl. Environ. Microbiol, 63:4969-4870.

Jury, W. A. Gardner, W. R., und Gardner, W. H., 1991. Soil Physics, 5th ed. John Wiley & Sons, Inc. New York, 328.

Jussel, P., 1992. Modellierung des Transports gelöster Stoffe in inhomogenen Grundwasserleitern. Dissertation ETH Zürich, Nr. R-29-92

Jussel, P., Stauffer, F., Dracos, T., 1994. Transport modeling in heterogeneous aquifers: 1. Statistical description and numerical generation of gravel deposits. Water Resources Research 30(6):1803-1817.

Jütte, R., 1997. Seuchen im Spiegel der Geschichte. Spektrum der Wissenschaft, Dossier Seuchen, 3:6-13.

Kaneko, M., 1998. Chlorination of pathogenic E.coli O157. Wat. Sci. Tech., 38(12):141-144.

Kapikian, A. Z., Wyatt, R. G., Dolin, R., Thornmill, T. S., Kalica, A. R. und Chanock, R. M., 1972. Visualisation by immuns electron microscopy of a 27-nm particle associated with acute infectious nonbacterial gastroenteritis. J. Virol., 10:1075-1081.

Kapperud, G., Vardund, T., Skjerve, E., Hornes, E, und Michaelsen, T. E., 1993. Detection of pathogenic Yersinia enterocolitica in foods and water by immunomagnetic separation, nested polymerase chain reactions, and colorimetric detection of amplified DNA. Applied and Environmental Microbiology, 59:2938-2944.

Karansis, P. und Seitz, H. M., 1996. Vorkommen und Verbreitung von Giardia und Cryptosporidium in Roh- und Trinkwasser von Oberflächenwasserwerken. gwf Wasser Abwasser, 137(2):94-100.

Kaucner, C, und Stinear, T., 1998. Sensitive and rapid detection of viable *Giardia* cysts and *Cryptosporidium* oocysts in large-volume water samples with wound fiberglass cartridge filters and reverse transcription-PCR. Applied and Environmental Microbiology, 64:1743-1749.

Kauffman, L.J., 1996. Dispersion of dense, viscous, miscible fluids in heterogeneous porous media: Laboratory investigation and stochastic, two-dimensional mean simulation. M.S. thesis, Drexel Univ., Philadelphia, Pa.

Käss, W., 1998. Tracing Technique in Geohydrology. A.A. Balkema, 590.

Käss, W., Schaber, J. und Schloz, W., 1994. Eine kombinierte Karstgrundwassermarkierung mit Uranin und Mikrokügelchen zwischen Geislingen-Waldhausen und Eybach. Laichinger Hohlenfreund, 61-70.

Käss, W., Ritter, R. und Sacré, C., 1983. Lebensdauer und Transport von Bakterien in typischen Grundwasserleitern - Oberrheinische Schotterebene. DVGW-Schriftenreihe Wasser Nr. 35, 127-138.

Keene, W., 1999. Entamoeba histolytica. Waterborne Pathogens. Manual of Water Supply Practices, America Water Works Association, 1st Edition, 171-175.

Kennedy, K. G., 2000. Bacteriophages as particle migration indicators in subsurface environments. Tracers and Modelling in Hydrogeology, Proceedings of the TraM'2000 Conference, Liège Belgium May 2000, IAHS Publ. Nr. 262:151-158.

Keswick B. H., Gerba, C. P., Secor, S. L., Cech, T., 1982b. Survival of enteric viruses and indicator bacteria in groundwater. J. Environmental Sci. Health, A17:903-912.

Keswick B. H., Wang, D.-S., Gerba, C. P, 1982a. The use of microorganisms as ground-water tracers: a review. Ground Water, 20(2):142-149.

Kilger, G., und Grimont, P. A. D., 1993. Differentiation of Salmonella phase 1 flagellar antigen types by restriction of the amplified fliC gene. Journal of Clinical Microbiology, 31:1108-1110.

Kimura, R., Mandrell, R. E., Galland, J. C., Doreene, H., und Riley, L. W., 2000. Restriction-site-specific PCR as a rapid test to detect enterohemorrhagic *Escherichia coli* O157:H7 strains in environmental samples. Applied and Environmental Microbiology, 66:2513-2319.

Kinoshita, T., Bales, R. C., Maguire, K. M. und Gerba, C. P., 1993. Effect of pH on bacteriophage transport through sandy soils. Journal of Contaminant Hydrology, 14:55-70.

KLBL, 1999. Jahresbericht, Kantonales Laboratorium Basel Landschaft.

KLBL, 2001. Jahresbericht, Kantonales Laboratorium Basel Landschaft.

KLBS, 1999. Jahresbericht, Kantonales Laboratorium Basel Stadt.

Knippers, R., 1995. Molekulare Genetik (Georg Thieme Verlag; Stuttgart, Germany).

Koerner, E.L. und Haws, D.A., 1979. Long-term effects of land application of domestic wastewater: Vineland, New Jersey, rapid infiltration site. Rep. EPA-600/2-79-072, Environ. Prot. Agency, Washington, D.C.

Kolch, A., 1998. Aktiver Gewässerschutz durch die Abwasserdesinfektion mit UV-Strahlen. gwa, 78(10):804-809.

Kominos, S. D., Copeland, C. E., Grosiak, B. und Postic, B., 1972. Introduction of *Pseudomonas aeruginosa* into a hospital via vegetables. Appl. Microbiol., 24(4):567.

Korich, D. G., Mead, J. R., Madore, M. S., Sinclaire, N. A. und Sterlin, C. R., 1990. Effects of ozone, chlorine Dioxid , chlorine and monochloramin on Cryptosporidium parvum oocyst viability. Appl. Env. Microbiol., 56(5):1423-1428.

Köster, W., Egli, T. und Rust, A., 2002. Krankheitserreger im (Trink)wasser? EAWAG news, 53, 26-28.

Krämer, J., 1992. Lebensmittel-Mikrobiologie. 2. Auflage, UTB für Wissenschaft, Stuttgart, 336.

Krüger, P., Wiedenmann, A., und Botzenhart, K., 1998. Detection of Cryptosporidium oocysts in water: comparison of the conventional immunofluorescence method with PCR and TaqMan PCR. Workshop of the Organisation for Economic Co-operation and Development (OECD) on „Molecular Technologies for Safe Drinking Water", 5-8 July 1998, Interlaken, Switzerland,

Kukkula, M., Maunula, L., Silvennoinen, E. und von Bonsdorff, C.-H., 1999. Outbreak of viral gastroenteritis due to drinking water contaminated by Norwalk-like viruses. J. Infectious Diseases, 180:1771-1776.

Kuhls, T. L:, Crawford, D. L. und Griffis, J., 1994. Seroprevalence of cryptpospridial antibodies during infancy, childhood and adolescence. Clin. Infect. Dis., 18:731-735.

Kurtz, J. B. und Lee, T. W., 1987. Astroviruses: human and animal. In: Ciba Foundation Symposium 128 (ed.), Novel diarrhoea viruses, Wiley, Chicester, 92-125.

Kurtz, J. B. und Lee, T. W., 1978. Astrovirus gastroenteritis: Age distribution of antibody. Med. Microbiol. Immunol., 166:227-230.

Kyle, D. E., und Noblet, G. P., 1985. Vertical distribution of potentially pathogenic free-ling amoebae in freshwater lakes. Journal of Protozoology, 32:99-105.

Langlais, B., Reckhow, D. A. und Brink, D. R. (eds), 1991. Ozone in Water Treatment, Application and Engineering, Cooperative Research Report, Lewis Publishers.

Lawrence, J.R. und Hendry, M.J., 1996. Transport of bacteria through geologic media. Can. J. Microbiol. 42, 410-422.

LeChevalier, M. W., 1987. Inactivation of attached bacteria. Proc. AWWA Seminar, Assurance of Adequate Disinfection, or C*t or not C*t. American Water Works Assn., Denver, CO, 101-113.

LeChevallier, M. W., Norton, W. D., Siegel, J. E., und Abbaszadegan, M., 1995. Evaluation of the immunofluorescence procedure for detection of *Giardia* cysts and *Cryptosporidium* oocysts in water. Applied and Environmental Microbiology, 61:690-697.

LeChevallier, M. W. und Norton, W. D., 1995. Giardia and Cryptosporidium in raw and finished water. J. Am. Water Works, Assoc., 87:54-68.

LeChevallier, M. W., 1999. Mycobacterium avium complex. Waterborne Pathogens. Manual of Water Supply Practices, America Water Works Association, 1st Edition, 99-102.

Leisinger, M., 1997. Molekularbiologischer Nachweis von Viren und Kryptosporidien im Wasser. Dissertation Universität Zürich.

Lemon, A. M., 1995. Hepatitis E Viruses. In: Mandell, Douglas and Bennett's Principles and Practices of Infectious Diseases, 4th ed. Churchill Livingstone, New York, Edingburgh, London, Melbourne, Tokyo, 1663-1666.

Lin, C. K., und Tsen, H. Y., 1999. Comparison of the partial 16S rRNA gene sequences and development of oligonucleotide probes for the detection of Escherichia coli cells in water and milk. Food Microbiology, 16:551-562.

Lipp, P und Baldauf, G., 2000. Membranfiltrationstechnik. gwa, 80(1):37-44.

Lisle, J. T. und Rose, J. B., 1995. Cryptosporidium contamination of water in the USA and UK: a minireview. J. Water SRT,44:103-107.

LMG, 1992 Bundesgesetz über Lebensmittel und Gebrauchsgegenstände vom 9. Oktober 1992. SR 817.0.

LMV, 1995. Lebensmittelverordnung (LMV), vom 1. März 1995, (Stand am 30. April 2002), SR 817.02.

Lodder, W. J., Vinjé, J., van de Heide, R., de Roda Husman, A. M., Leenen, E. J. T. M., und Koopmans, M. P. G., 1999. Molecular detection of Norwalk-like calicivirus in sewage. Applied and Environmental Microbiology, 65:5624-5627.

Lüthi, T. M., 1997. Food and waterborne viral gastroenteritis: a review of agents and their epidemiology. Mitt. Gebiete Lebensm. Hyg., 88:119-150.

Lüthi, T. und Beuret, C., 2001. Weshalb ein Virennachweis in Wasser? gwa, 81(5)283-290.

MacKenzie, W. R., Hoxie, N. J., Proctor, M. E., Gradus, M. S., Blair, K. A., Peterson, D. E., Kazmierczak, J. J., Addiss, D. G., Fox, K. R., Rose., J. B., und Davis, J. P., 1994. A massive outbreak in Milwaukee of *Cryptosporidium* infection transmitted through the public water supply. New England Journal of Medicine, 331:161-167.

Mahbubani, M. H., Bej, A. K., Perlin, M., Schaefer, F. W., Jakubowski, W., und Atlas, R. M., 1991. Detection of *Giardia* cysts by using the polymerase chain reaction and distinguishing live from dead cysts. Applied and Environmental Microbiology, 57:3456-3461.

Mahbubani, M. H., Schaefer III, F. W., Jones, D. D., und Bej, A. K., 1998. Detection of *Giardia* in Environmental Waters by immuno-PCR amplification methods. Current Microbiology, 36:107-113.

Mallevialle, J., Mandra, V., Baudin, I. und Anselme, C., 1993. Membrane filtration in drinking water treatment. European Water Filtration Congress, Proceedings, 15-17 March 1993, 17:4.1-4.16.

Manafi, M. und Kneifel, W., 1989. A combined chromogenic-fluorogenic medium for the simultaneous detection of coliform groups and *E.coli* in water. Zentralblatt für Hygiene und Umweltmedizin, 189:225-234.

Marschall, M. M., Naumovitz, D., Ortega, Y., und Sterling, C. R., 1997. Waterborne protozoan pathogens. Clinical Microbiology Reviews, 10:67-85.

Marti, F., Della Valle, G., Krech, U., Gees, R. A. und Baumgartner, E., 1979. Trinkwasserhygiene. Alimenta, 18:135-145.

Marti, F., 1975. Bakterien als Markierstoffe für Grundwasserströme. Alimenta, 14:115-121.

Martin, G.N. und Noonan, M.J., 1977. Effects of domestic wastewater disposal by land irrigation on groundwater quality of the central Canterbury Plains. Tech. Publ.7, Water and Soil Div., Minist. of Works and Dev., Christchurch, New Zealand.

Martin, R. E., Bouwer, E. J. und Hanna, L. M., 1992. Application of clean-bed filtration theory to bacterial deposition in porous media. Environ. Sci. Technol., 26(5):1053-1058.

Mathewson, J.J., 1998. Detection of bacterial pathogens. In: Microbial Pathogens within Aquifers: Principles and Protocols. Pillai, S. D., ed. (Springer Verlag), 43-57.

Mattle, N., Kinzelbach, W., Beyerle, U., und Huggenberger, P., 2000, Calibration of a 3-dim. Hydrodynamic Transport model with Tritiogenic 3He. Submitted to Journal of Hydrology.

Maurer, A. M. und Stürchler, D., 2000. A waterborne outbreak of small round structured viruses, Campylobacter and Shigella co-infections in La Neuveville, Switzerland, 1998. Epid. Infect., 125:325-332.

Maule, A., 2000. Survival of verotoxigenic *Escherichia coli* O157 in soil, water, and surfaces. J Appl. Micorbiol., 88:71S-78S.

Mayer, C. L., und Palmer, C. J., 1996. Evaluation of PCR, nested PCR, and fluorescent antibodies for detection of Giardia and Cryptosporidium species in wastewater. Applied and Environmental Microbiology, 62:2081-2085.

McDowel-Boyer, L. M., Hunt, J. R. und Sitar, N., 1986. Particle transport through porous media. Water Resources Research, 22(13):1901-1921.

McFeters, G. A., Kippin, J. S., und LeChevallier, M. W., 1986. Injured coliforms in drinking water. Applied and Environmental Microbiology, 51:1-5.

McLauchlin, J., Pedraza-Diaz, S., Amar-Hoetzeneder, C. und Nicholas, G. L., 1999. Genetic characterization of Cryptosporidium strains from 218 patients with diarrhea diagnosed as having sporadic cryptosporadiosis. J. Clin. Microbiol, 37(10):3153-3158.

Mead, P. S. Slutsker, L., Dietz, V., McCaig, L. F., Bresee, J. S., Shapiro, C., Griffin, P. M. und Tauxe, R., 1999. Food-related illness and death in the United States. Emerging Infectious Diseases., 5(5):607-625.

Meier, H., Koob, C., Ludwig, W., Amann, R., Frahm. E., Hoffmann, S., Obst, U., und Schleifer, K. H., 1997. Detection of enterococci with rRNA targeted DNA probes and their use for hygienic drinking water control. Water Science and Technology, 35:437-444.

Merkli, B., 1974. Untersuchungen über Mechanismen und Kinetik der Elimination von Bakterien und Viren im Grundwasser. Diss. ETH 5420, 89.

Metzler, A., Regli, W., Leisinger, M., Heider, H., Schweizer, K. und Tabisch, A., 1996. Viren und Parasiten im Trinkwasser: Risiken und Prävention. Mitt. Gebiete Lebensm. Hyg., 87:55-72.

Metzler, A. und Tabisch, A., 1997. Fakten und Spekulationen über die Kontamination der Umwelt mit Cryptosporidium-Oozysten. Tagungsband Workshop „Cryptosporidium“ Schweizerischer Verein des Gas- und Wasserfaches, SVGW 14. Mai 1997, 29-33.

Meyers, K.E.C., Schulman, S.L., Kaplan, B.S., 1998. Principles of the treatment of Shiga toxin-associated hemolytic uremic syndrome; pay meticulous attention to detail, and do no harm. In: Kaper, J.B. and O'Brien, A.D., (eds.) Escherichia coli 0157:H7 infections. N. Engl. J. Med. Ch., 35:364-373.

Michel, P. und Schweizer, H. U., 1998. Die Wasserressourcen. gwa, 78(6):436-440.

Miyamoto, H., Yamamoto, H., Arima, K., Fujii, J., Maruta, K., Izu, K., Shiomori, T., und Yoshida, S.-I., 1997. Development of a new seminested PCR method for detection of Legionella species and its application to surveillance of legionellae in hospital cooling tower water. Applied and Environmental Microbiology, 63:2489-2494.

MMWR, 1996. Update: Outbreaks of Cyclospora cayetanensis infection – United States and Canada. Mortality and Morbidity Weekly Report, 45(28):611-612

MMWR, 1997. Outbreak of Cyclosporiasis – Northern Virginia-Washington, D.C. – Maryland, Metropolitan area. Mortality and Morbidity Weekly Report, 46(30):689-691.

MMWR, 2000. Surveillance for waterborne-disease outbreaks – United States, 1997-1998. May 26, 49(SS04):1-35.

Molecular Probes, 2000. Produktekatalog.

Morgan, U. M., Constantine, C. C., Forbes, D. A. und Thomson, R. C. A., 1997. Differenciation between human and animal isolates of *Cryptosporidium parvum* usind rDNA sequencing and direct PCR analysis. J. Parasitol., 83(5):825-830.

Morris, R. D. und Levin, R., 1995. Estimating the incidence of waterborne infectious disease related to drinking water in the United States. In: Assessing and Managing Health Risks from Drinking Water Contamination: Approaches and Applications. eds. Reichard, E. G. und Zapponi, G. A., IAHS Publ no 233, UK: International Association of Hydrological Sciences, 75-88.

Murgia, R., Riquelme, N., Baranton, G., und Cinco, M., 1997. Oligonucleotides specific for pathogenic and saprophytic leptospira occuring in water. FEMS Microbiology Letters, 148:27-34.

Murphy, G.S. Jr, Echeverria, P., Jackson, L.R., Arness, M.K., LeBron, C., Pitarangsi, C., 1996. Ciprofloxacin- and azithromycin-resistant Campylobacter causing traveler's diarrhea in U.S. troops deployed to Thailand in 1994, Clin. Infect. Dis., 22:868-869.

Nadal, D., 2001 Virusinfektionen beim Menschen, Infektketten, Auswirkungen und Prophylaxe, gwa, 81(5):299-303.

Natsch, A., Keel, C., Troxler, J., Zala, M., Von Albertini, N. und Défago, G., 1996. Importance of preferential flow and soil management in vertical transport of a biocontrol strain of pseudomonas fluorescens in structured field soil. Applied and Environmental Microbiology, 62(1):33-40.

Nguyen, A. L., Pron, B., Quesne, G., Brusset, M. C., und Berche, P., 1998. Outbreak of nosocomial urinary tract infections due to *Pseudomonas aeruginosa* in a pediatric surgical unit associated with tap-water contaminations. Journal of Hospital Infection, 39:301-307.

Niehren, S. und Kinzelbach, W., 1998. Artificial Colloid Tracer Tests: Development of a compact On-line Microsphere Counter and Application to Soil-Column Experiments. Accepted for special issue MIGRATION'97. J. Cont. Hydrol. 1998.

Niehren, S., 1999. Neue Methode zur Grundwassermarkierung: Entwicklung eines feldtauglichen Mikrosphärenzählers. Diss. ETH Zürich Nr. 12710.

Nieminski, E. C., Ongerth, J. E., 1995. Removing giardia and cryptosporidium by conventional treatment and direct filtration. Journal AWWA, 96-106.

Nieminski, E. C., Schaefer III, F. W., und Ongerth, J. E., 1995. Comparison of two methods for detection of *Giardia* cysts and *Cryptosporidium* oocysts in water. Applied and Environmental Microbiology, 61:1714-1719.

Nime, F. A., Burek, J. D., Page, D. L., Holscher, M. A. und Yardley, J. H., 1976. Acute eterocolitis in a human being infected with the protozoan Cryptosporidium. Gastroenterology, 70:592-598.

Nolte, P., 1999. Christian Friedrich Schönbein, Entdecker des Ozons – Zum 200 Geburtstag. gwa, 10:819-823.

Noonan, M.J. und McNabb, J.F., 1979. Contamination of Canterbury groundwater by viruses In The Quality and Movement of Groundwater in Alluvial Aquifers of New Zealand. Tech. Publ.2, edited by M.M. Noonan, 195-210, Dep.of Agric. Microbiol., Lincoln Coll., Canterbury, New Zealand.

O'Brien, A. D., und Holmes, R. K., 1987. Shiga and Shiga-like toxins. Microbiological Reviews, 51:206-220.

Oehmichen, U., Schmitz, M. und Seeliger, P., 2001. Die neue Trinkwasserverordnung: Der Kommentar aus rechtlicher und technisch-wirtschaftlicher Sicht. wvgw, 337.

Ortega, Y., 1999. Cyclospora cayetanensis. Waterborne Pathogens. Manual of Water Supply Practices, America Water Works Association, 1st Edition, 163-166.

Owen, R. J., 1995. Bacteriology of *Helicobacter pylori*. Baillieres Clinical Gastroenterology, 9:415-446.

Parashar, U. D., Quiroz, E. S., Mounts, A. W., Monroe, S. S. , Fankhauser, R. L., Ando, T., Noel, J. S., BULens, S. N., Beard, R. S., Li, J-F., Bresee, J. S. und Glass, R. I., 2001. „Norwalk-like-viruses", Public health consequences and outbreak management. MMWR, 50(RR-9) pp. 20.

Parveen, S., Portier, K. M., Robinson, K., Edmiston, L. und Tamplin, M. L., 1999. Discriminant analysis of ribotype profiles of *Escherichia coli* for differentiating human and nonhuman sources of fecal pollution. Applied and Environmental Microbiology, 65:3142-3147.

Payment, P., Trudel, M. und Plante, R., 1985. Elimination of viruses and indicator bacteria at each step of treatment during preparation of drinking water at seven water treatment plants. Applied and Environmental Microbiology, 49(6):1418-1428.

Payment, P., Richardson, L., Siemiatycki, J., Dewar, R. und Franco, E., 1991. A randomized trial to evaluate the risk of gastrointestinal disease due to consumption of drinking-water meeting current microbiological standards. Am. J. Public Health, 81:703-708.

Payment, P. und Franco, E., 1993. Clostridium perfringens and somatic coliphages as indicators of the efficiency of drinking water treatment for viruses and protozoan cysts. Appl. Environ. Microbiol., 59(8):2418-2424.

Pernin, P., Pélandakis, M., Rouby, Y, Faure, A, und Siclet, F., 1998. Comparative recoveries of *Naegleria fowleri* amoebae from seeded river water by filtration and centrifugation. Applied and Environmental Microbiology, 64:955-959.

Pfuhl, E., 1897. Über die Verschleppung von Bakterien durch das Grundwasser. Z. Hygiene u. Infektionskrankheiten, 25:549-554. Leipzig: Veit & Comp.

Pochon, A. und Zwahlen, F., in Vorbereitung. Guide pratique - Délimitation des zones de protection des eaux souterraines en milieu fissuré. Schriftenreihe L'environnement pratique, BUWAL.

Pointer, S. F. B. und Slade, J. S., 1977. The removal of viruses by slow sand filtration. Prog. Wat. Tech., 9:75-88.

Pontius, F., W., 1995. Cryptosporidium: Answers to common questions. J. Amer. Water Works Assoc., 87(9):10-12.

Porter, J., Mobbs, K., Hart, C. A., Saunders, J. R., Pickup, R. W, und Edwards, C., 1997. Detection, distribution and probable fate of *Escherichia coli* O157 from asymptomatic cattle on a diary farm. Journal of Applied Microbiology, 83:297-306.

Pozio, F., Gomez Morales, M. A., Mancini, F., Barbieri, LaRosa, G., 1992. Cryptosporidium: different behaviour in calves of isolates of human origin. Transactions Roy. Soc. Trop. Med. Hyg., 86:636-638.

Puig, M., Jofre, J., Lucena, F., Allard, A., Wadell, G., und Girones, R., 1994. Detection of adenoviruses and enteroviruses in polluted waters by nested PCR amplification. Applied and Environmental Microbiology, 60:2963-2970.

Pusch, M., Fiebig, D., Brettar, I., Eisenmann, H., Ellis, B.K., Kaplan, L.A., Lock, M.a., Nägeli, M.W., and Traunsburger, W. 1998. The role of micro-organisms in the ecological connectivity of running waters. Freshwater Biology 40:453-495.

Pyle, B. H., Broadaway, S. C., und McFeters, G. A., 1999. Sensitive detection of *Escherichia coli* O157:H7 in food and water by immunomagnetic separation and solid-phase laser cytometry. Applied and Environmental Microbiology, 65:1966-1972.

Rahe, T.M., Hagedorn, C. und McCoy, E. L., 1979. Transport of antibiotic-resistant Escherichia coli through western Oregon hillslope soils under conditions of saturated flow. J. Environ. Qual., 7:487-494.

Rajagopalan,R. und Tien,C., 1976. Trajectory analysis of deep-bed filtration with the sphere-in-cell porous media model. AIChE J., 22(3):523-533.

Raso, G., 2001. Mikrobiologische Veränderungen eines Karstquellwassers: Einfluss von Regen und natürlicher Düngung. Diplomarbeit, Schweizerisches Tropeninstitut, Universität Basel und Kantonales Laboratorium BL, pp.87.

Rauber, M., Stauffer, F., Huggenberger, P., Dracos, T., 1998. A numerical three-dimensional conditioned/unconditioned stochastic facies type model applied to a remediation well system. Water Resources Research, 34(9):2225-2233.

Redman, J.A. et al., 1997. Filtration of recombinant Norwalk virus particles and bacteriophage MS2 in quartz sand: Importance of electrostatic interactions. Environ. Sci. Technol. 31, 3378-3383.

Regan, P. M., und Margolin, A. B., 1997. Development of a nucleic acid capture probe with reverse transcriptase-polymerase chain reaction to detect polioviruses in groundwater. Journal of Virological Methods, 64:65-72.

Regli, W., 1994. Verbesserte Methoden für die Isolierung und den Nachweis von Giardia – Zysten und Cryptosporidien – Oozysten in Oberflächengewässern: Flockung mit Al2(SO4)3 und fluorescence-activated cell sorting (FACS). Inaugural Dissertation Veterinär-medizinische Fakultät, Uni Zürich, pp.70.

Regli, W., Niederer, E. und Metzler, A., 1996. Giardien und Cryptosporidien in Wasserproben: Eine neue Nachweismethode. gwf / Wasser-Abwasser, 136(4):182-187.

Regli, S., Rose, J.B., Haas, C.N. und Gerba, C.P., 1991. Modeling the risk from Giardia and viruses in drinking water. J. AWWA, 83:76-84.

Rehmann, L.L.C, Welty, C., und Harvey, R.W., 1999. Stochastic analysis of virus transport in aquifers. Water Resources Res., 1987-2006.

Reichert, P., 1998. AQUASIM 2.0 - User Manual. Computer Program for the Identification and Simulation of Aquatic Systems. EAWAG Dübendorf, ISBN: 3-906484-16-5.

Regli, Ch., Huggenberger, P., Rauber, M., 2002. Interpretation of drill-core and georadar data of coarse gravel deposits. Journal of Hydrology 255, 234-252.

Renard, Ph. und de Marsily, G., 1997. Calculating equivalent permeability: a review. Advances in Water Resources 20, 253-278.

Reynolds, K. S., Gerba, C. P., und Pepper, I. L., 1997. Rapid PCR-based monitoring of infectious enteroviruses in drinking water. Water Science and Technology, 35:423-427.

Rice, E.W. und Hoff, J.C., 1981. Inactivation of Giardia lamblia cysts by ultraviolet irradiation. Appl. Env. Microbiol., 42(3):546-547.

Rice, E. W., 1999. Escherichia coli. Waterborne Pathogens. Manual of Water Supply Practices, America Water Works Association, 1st Edition, 75-78.
Riggs, J. L., Dupis, K. W., Nakamura, K., und Spath, D. P., 1983. Detection of *Giardia lamblia* by immunofluorescence. Applied and Environmental Microbiology, 45:698-700.
Rijnaarts, H. M., Norde, W., Bouwer, E. J., Lyklema, J. und Zehnder, A. J. B., 1995. Reversibility and mechanism of bacterial adhesion. Colloids and Surfaces B: Biointerfaces, 4:5-22.
RKI (Robert Koch Institut), 2000. Zwei Gastroenteritis-Ausbrüche durch *Campylobacter jejuni* nach dem Genuss von Rohmilch. Epidemiologisches Bulletin, 18:143-147.
Roche Lexikon Medizin, 2. Auflage, 1991, Hoffmann-La Roche AG und Urban & Schwarzenberg (Hrsg.), Verlag Urban & Schwarzenberg, 2026.
Rochelle, P. A., De Leon, R., Ferguson, D. M., Stewart, M. H., und Wolfe, R. L., 1997a. Optimization of an infectivity assay, combining cell culture and PCR, for waterborne *Cryptosporidium parvum*. In: 1997 International Symposium on Waterborne Cryptosporidium Proceedings. Fricker, C. R., Clancy, J. L., and Rochelle, P. A., eds., (AWWA, Denver), 31-40.
Rochelle, P. A., Ferguson, D. M, Handojo, T. J., De Leon, R., Stewart, M. H., und Wolfe, R. L., 1997b. An assay combining cell culture with reverse transcriptase PCR to detect and determine the infectivity of waterborne *Cryptosporidium parvum*. Applied and Environmental Microbiology, 63:2029-2037.
Rose, J.B. und Botzenhardt, K., 1990. Cryptosporidium und Giardia im Wasser gwf-Wasser/Abwasser, 131:563-572.
Rose, J. B. und Yates, M. V., 1995. Microbial risk assessment application for groundwater. Protozoan Parasites and Water, Betts, W. B., Casemore, D., Fricker, C., Smith, H., and Watkins, J. (eds), 113-131.
Ross, B. C., Johnson, P. D. R., Oppedisano, F., Marino, L., Sievers, A., Stinear, T., Hayman, J. A., Veitch, M. G. K., und Robins-Browne, R. M., 1997. Detection of *Mycobacterium ulcerans* in environmental samples during an outbreak of ulcerative disease. Applied and Environmental Microbiology, 63:4135-4138.
Rossi, P., 1994. Advances in biological tracer techniques for hydrology and hydrogeology using bacteriophages. Diss. Uni Neuchâtel, 197.
Rossi, P., Dörfliger, N., Kennedy, K., Müller, I. und Aragno, M., 1998. Bacteriophages as surface and groundwater tracers. Hydrology and Earth Systems Sciences, 2(1):101-110.
Rossi, P. und Käss, W., 1998. Phages. In: Tracing Technique in Geohydrology, Käss, W. (ed.). A.A. Balkema, 245-271.
Rosskamp, E., Dieter, H. H., 1999. Entstehung von Bromat bei der Aufbereitung von Schwimm- und Badewasser. Bundesgesundheitsblatt – Gesundheitsforschung – Gesundheitsschutz, 11:859-862.
Rothmeier, R., Weidenmann, A., und Botzenhart, K., 1997. Transport of *Escherichia coli* through soil to groundwater traced by Randomly Amplified Polymorphic DNA (RAPD). Water Science and Technology, 35:351-357.
Rumm, P., Schmidt, H. und Schminke, H. K., 1997. Organismenaustrag aus Langsamsandfiltern. gwf, 138(7):355-361.
Ryan, J. N., Elimelech, M., Ard, R. A., Harvey, R. W. und Johnson, P. R., 1999. Bacteriophage PRD1 and silica colloid transport and recovery in an iron oxide-coated sand aquifer. Environ. Sci. Technol., 33(1):63-73.
Sacher, F., Schmidt, W., Böhme, U. und Brauch, H.-J., 1997. Bromat – Ein Problem für die Trinkwasserversorgung in Deutschland? gwf Wasser Abwasser, 138(4):199-207.
Samonsis, G., Elting, L., Skoulika, E., Maraki, S. und Tselentis, Y., 1994. An outbreak of diarrhoeal disease attributed to Shigella sonnei. Epidemiol Infect, 112:235-245.
Sasaki, K., Tajiri, Y., Sata, M., Fujii, Y., Matsubara, F., Zhao, M., Shimizu, S., Tojonaga, A., und Tanikawa, K., 1999. *Helicobacter pylori* in the natural environment. Scandinavian Journal of Infectious Diseases, 31:275-280.

Sauter, B., 1994. Neue Wege in der Desinfektion von Trink- und Brauchwasser. Lebensmittel-Technologie, 27(9):290-295.

Schaber, J., 1993. Markierungsversuche mit fluoreszierenden Mikropartikeln. Diplomarbeit (Ms.), p. 107. Univ. Freiburg i. Br.

Schaefer, F. W., 1997. Detection of protozoan parasites in source and finished drinking water. Manual of Environmental Microbiology. Hurst, C. J., Knudsen, G. R., Mclnerney, M. J., Stetzenbach, L. D., Walter, M. V., eds. (American Society for Microbiology, Washington D. C.), 153-167.

Schäfer, A., 1997. Bacterial transport and pollutant degradation: influences of air-water interfaces and solid surfaces. PhD Thesis, Diss ETH No. 12416.

Schälchli, U., 1993. Die Kolmation von Fliessgewässersohlen: Prozesse und Berechnungsgrundlagen. VAW ETH Zürich Mitteilungen 142.

Schafter, N. und Parriaux, A., 2002. Pathogenic-bacterial water contamination in mountainous cathments. Water. Res., 36:131-139.

SchALVO, 2002. Verordnung des Ministeriums für Umwelt und Verkehr über Schutzbestimmungen und die Gewährung von Ausgleichsleistungen in Wasser- und Quellenschutzgebieten (Schutzgebiets- und Ausgleichsverordnung). Vom 20. Februar 2001, GBI. S. 145, ber. S. 414, geändert durch Verordnung vom 2. Februar 2002, GBI. S. 109.

Schaub, S.A. und Sorber, C.A., 1977. Virus and bacteria removal from wastewater by rapid infiltration through soil. Appl. Environ. Microbiol., 33:609-619.

Schiemann, D. A., 1990. Yersinia enterolitica in drinking water. In: McFeters, G.A. (ed), Drinking Water Microbiology. Series in Contemporary Bioscience. Springer Verlag, 322-339.

Schillinger, J. E. und McFeters, G. A., 1978. Survival of Escherichia coli and Yersinia enterolitica in stream and tap water. Abstracts of the Annual Meeting of the American Society of Microbiology, 175.

Schleupen, E., 1996. Cryptosporidium parvum und Giardia lamblia, gwf, 137(2):83-93.

Scholl, M.A. et al., 1990. The influence of mineralogy and solution chemistry on the attachment of bacteria to representative aquifer materials. J. Contam. Hydrol., 6, 321-336.

Scholl, M. A. und Harvey, R. W., 1992. Laboratory investigation on the role of sediment surface and groundwater chemistry in transport of bacteria through a contaminated sandy aquifer. Environ. Sci. Technol., 26(7):1410-1417.

Schuler, F. P., Ghosh, M. M. und Gopalan, P., 1991. Slow sand and diatomaceous earth filtration and other particulates. Wat. Res., 25:995-1005.

Schwab, K., 1999. Small round structured viruses (human Caliciviruses). Waterborne Pathogens. Manual of Water Supply Practices, America Water Works Association, 1st Edition, 257-261.

Schwartzbrod, L., 1991. Virologie des milieux hydriques. Lavoisier, Paris, p. 72-100.

Schweizer, K. und Metzler, A., 1997. FACS (Fluorescence Activated Cell Sorting) als Bestandteil einer Methode zur Detektion von Cryptosporidium-Oozysten in Wasser. Tagungsband Workshop „Cryptosporidium" Schweizerischer Verein des Gas- und Wasserfaches, SVGW 14. Mai 1997, 43-46.

Seiler, K.-P., 1988. Die mechanische Ausfilterung von Escherichia coli in quartären Kiesen Oberbayerns. Z. dt. geol. Ges., 139:475-484.

Severin, B. F., 1987. Kinetics of microbial inactivation. Proc. AWWA Seminar, Assurance of Adequate Disinfection, or C*t or not C*t. American Water Works Assossiation., Denver, CO, 21-48.

SGH, 2002. Praxishilfe: Einsatz künstlicher Tracer in der Hydrogeologie. Schweizerische Gesellschaft für Hydrogeologie – SGH, 92.

Shaban, A. M., El-Taweel, G. E. und Ali, G. H., 1997. UV ability to inactivate microorganisms combined with factors affecting radiation. Wat. Sci. Tech., 35(11-12):107-112.

Sharma, M.M. et al., 1985. Reversible and irreversible surface charge modifications of bacteria for facilitating transport through porous media. Colloids Surf., 16, 193-206.

Shin, G.-A., und Sobsey, M. D., 1998. Reduction of Norwalk virus, poliovirus 1 and coliphage MS2 by monochloramine disinfection of water. Wat. Sci. Tech., 38(12):151-154.

Silliman, S.E. et al., 2001. Bacterial transport in heterogeneous porous media: Observations from laboratory experiments. Water Resources Res., 37(11):2699-2707.

Simonet, M. und Catteau, M., 1997. Yersinia enterolitica et alimentation humaine. Bull. Soc. Fr. Microbiol., 12(4):359-363.

Simoni, St. F., 1999. Factors Affecting Bacterial Transport and Substrate Mass Transfer in Model Aquifers. Dissertation ETH Zürich 13232.

Sinton, L. W., Donnison, A. M., und Hastie C. M., 1993. Fecal Streptococci as fecal pollution indicators: a review. Part II: sanitary significance, survival and use. New Zealand Journal of Marine and Freshwater Research, 27:101-115.

Sluter, S. D., Tzipori, S., und Widmer, G., 1997. Parameters affecting polymerase chain reaction detection of waterborne *Cryptosporidium parvum* oocysts. Applied Microbiology and Biotechnology, 48:325-330.

Smith, M. S., Thomas, G. W., White, R. E. und Ritonga, D., 1985. Transport of Escherichia coli through intact and disturbed soil columns. J. Environmental Quality, 14(1):87-91.

Smith, H. V., und Lloyd, A., 1996. Protozoen parasites in drinking water: a UK perspective. Water Management Int., 11-17.

Snozzi, M., Haas, R., Leuker, G., Kolch, A. und Bergman, R. W., 1999. Prüfung und Zertifizierung von UV-Anlagen. gwa, 79(5):380-385.

Sobsey, M. D., 1987. Methods for recovering viruses from shellfish, seawater, and sediments. In: Methods for recovering viruses from the environment. Berg, G., ed. (CRC Press, Inc. Boca Raton, Fla.).

Sobsey, M.D., 1989. Inactivation of health-related microorganisms in water by disinfection process. Wat.Sci Tech., 21(3):179-195.

Sobsey, M., 1999. Hepatitis A. Waterborne Pathogens. Manual of Water Supply Practices, America Water Works Association, 1st Edition, 241-246.

Sockett, P.N., Cowden, J.M., Le Baigue, S., Ross, D., Adak, G. K. und Evans, H., 1993. Food-borne disease surveillance in England and Wales: 1989-1991. Communicable Disease Report Review, 3, R159-R173.

Sommer, R., Cabaj, A., Pribil, W. und Haider, T, 1997. Influence of lamp intensity and water transmittance on the UV disinfection of water. Wat. Sci. Tech., 35(11-12):113-118.

Sommer, R., Haider, T., Cabaj, A., Pribil, W. und Lhotsky, M., 1998. Time dose reciprocity in UV disinfection of water. Wat. Sci. Tech., 38(12):145-150.

Sonntag, H.-G., 1997. Infektionsgefahren nehmen zu. Spektrum der Wissenschaft, Dossier, 3:110-113.

Soule, H., Genoulaz, O., Gratacap-Cavallier, B., Chevallier, P., Liu, J.-X., und Seigneurin, J.-M., 2000. Ultrafiltration and reverse transcription-polymerase chain reaction: An efficient process for poliovirus, rotavirus, and hepatitis A virus detection in water. Water Research, 34:1063-1067.

Sparfel, J.M., Sarfati, C., Liguory, O., Caroff, B., Dumoutier, N., Gueglio, B., Billaud, E., Raffi, F., Molina, J. M., Miegeville, M., 1997. Detection of microsporidia and identification of Enterocytozoon bienuesi in surface water by filtration followed by specific PCR. J. Eukaryot. Microbiol., 44:78S.

Staehelin, J. und Hoigné, J., 1985. Decomposition of ozone in water in the presence of organic solutes acting as promoters and inhibitors of radical chain reactions. Environ. Sci. Technol., 19(12):1206-1213.

Stanford, J.A., 1998. Rivers in the landscape: introduction to the special issue on riparian and groundwater ecology. Freshwater Biology 40, 402-406.

Stauffer, F. und Rauber, M. 1998. Stochastic macrodispersion models for gravel aquifers, J. Hydraulic Res., 36(6):885-896.

Stephan, R., Ragettli, S. und Untermann, F., 2000. Vorkommen verotoxinbildender Escherichia coli (VTEC) in Kottupfern von Schlachtrindern und Schlachtschafen – eine Betrachtung aus fleischhygienischer Sicht. Schweiz. Arch. Tierheilk., 142:110-114.

Sterling, C. R., 1995. Emerging Pathogens: Cyclospora, the Microspora, and how many more? Protozoan Parasites and Water, Betts, W. B., Casemore, D., Fricker, C., Smith, H., und Watkins, J. (eds), 243-245.

Stettler, R., Courbat, R., von Gunten, U., Kaiser, H.-P., Walther, J. L., Gaille, P., Jordan, R., Ramseier, S. and Revelly, P., 1998. Utilisation de l'ozone pour le traitement des eaux potables en Suisse. gwa, 78(11):876-890.

Stoffverordnung, StoV, 1986.Verordnung über umweltgefährdende Stoffe.

Stinear, T., Matusan, A., Hines, K., und Sandery, M., 1996. Detection of a single viable *Cryptosporidium parvum* oocyst in environmental water concentrates by revers transcription-PCR. Applied and Environmental Microbiology, 62:3385-3390.

Su, C., Brandt, L.J., 1997. *Escherichia coli* 0157:H7 infections in humans, Ann. Intern. Med., 123:698-714.

Svoboda, P., Ruchti, S., Bissegger, C. und Tanner, M., 1999. Occurence of Cryptosporidium spp. Oocystes in surface, raw and drinking water samples. Mitt. Lebensm. Hyg., 90:553-563.

Sykora, J. L., Keleti, G., und Martinez, A. J., 1983. Occurrence and pathogenicity of *Naegleria fowleri* in artificially heated waters. Applied and Environmental Microbiology, 45:974-979.

Talamba, D., Joerin, C. und Musy, A., 2000. Study of subsurface flow using environmental and artificial tracers: the Haute-Mentue case, Switzerland. Tracers and Modelling in Hydrogeology (Proceedings of the TraM' 2000 Conference Liège, Belgium, IAHS Publ. no. 262:559-564.

Tamanai-Shacoori, Z., Jolivet-Gougeon, A., und Cormier, M., 1996. Comparison of direct PCR and PCR amplification after DNA extraction of viable enterotoxigenic *Escherichia coli* in laboratory microcosms. Journal of Microbiological Methods, 26:21-26

Tan, Y. et al., 1994. Transport of bacteria in an aquifer sand: experiments and model simulations. Water Resour. Res., 30:3243-3252.

Tauxe, R.V. 1997. Emerging foodborne diseases: An evolving public health challenge, Emerging Infectious Diseases, 3(4):425-434.

Theron, J., Cilliers, J., Du Preez, M., Brozel, V. S., und Venter, S. N., 2000. Detection of toxigenic Vibrio cholerae from environmental water samples by an enrichment broth cultivation-pit-stop seminested PCR procedure. Journal of Applied Microbiology, 89:539-546.

Tibbetts, J., 2000. Water World 2000. Environmental Health Perspectives, 108(2):69-72.

Timms, S., Slade, J. S. und Fricker, C. R., 1995. Removal of Cryptosporidium by slow sand filtration. Wat. Sci. Tech., 31(5-6):81-84.

Toranzos, G. A., und McFeters, G. A., 1997. Detection of indicator microorganisms in environmental freshwaters and drinking water. In: Manual of Environmental Microbiology. Hurst, C. J., Knudsen, G. R., McInerney, M. J., Stetzenbach, L. D., Walter, M. V., eds. (American Society for Microbiology, Washington D. C.), 184-194.

Tortorello, M. L., und Stewart, D. S., 1994. Antibody-direct epifluorescent filter technique for rapid, direct enumeration of *Escherichia coli* O157:H7 in beef. Applied and Environmental Microbiology, 60:3553-3559.

Tóth, J., 1963. Two-dimensional isotropic flow model showing the distribution of local, intermediate, and regional groundwater flow systems, Geophys. Res. v.68, 4795-4812.

Trebesius, K., Harmsen, D., Rakin, A, Schmelz, J., und Heeseman, J., 1998. Development of rRNA-targeted PCR and in situ hybridization with fluorescently labelled oligonucleotides for detection of *Yersinia* species. Journal of Clinical Microbiology, 36:2557-2564.

Trinkwasserverordnung (TrinkwV), 2001.

Tschäppe, H., 2000. Lebensmittelbedingte Infektionskrankheiten durch Bakterien. Bundesgesundheitsbl-Gesundheitsforsch-Gesundheitsschutz, 43(10):758-769.

Tschäppe, H., 1997. Toxinbildende *E. coli*-Stämme auf dem Vormarsch. Bundesgesundheitsbl-Gesundheitsforsch-Gesundheitsschutz, 40(6)193.

Tsen, H. Y., Lin, C. K., und Chi, W. R., 1998. Development and use of 16S rRNA gene targeted PCR primers for the identification of Escherichia coli cells in water. Journal of Applied Microbiology, 85: 554-560.

Tsen, H. Y., und Jian, L. Z., 1998. Development and use of a multiplex PCR system for the rapid screening of heat labile toxin I, heat stable toxin II and shiga-like toxin I and II genes of Escherichia coli in water. Journal of Applied Microbiology, 84:585-592.

Uhlenbrook, S., Leibundgut, C. und Maloszewski, P., 2000. Natural tracers for investigating residence times, runoff components and validation of a rainfall-runoff model. Tracers and Modelling in Hydrogeology (Proceedings of the TraM' 2000 Conference Liège, Belgium, IAHS Publ. no. 262, 465-471.

Umweltbundesamt, 1997. Human- und ökotoxikologische Bewertung von Markierungsmitteln in Gewässern. Bundesgesundheitsblatt, 10:414-416.

USEPA, 1998. Announcement of drinking water contaminant candidate list; notice. Fed. Reg., 63(40):10274.

Van Loosdrecht, M.C.M., Lyklema, L., Norde, W., Schraa, G. und Zehnder, A.J.B., 1987. The role of bacterial cell wall hydrophobicity in adhesion. Appl. Environ. Microbiol., 53:1893-1897.

Van Oss, C. J., R. J. Good, und M. K. Chaudry. 1986. The role of van der Waals forces and hydrogen bonds in 'hydrophobic interactions' between biopolymers and low energy surfaces. J. Colloid Interface Sci., 111:378-390.

Vaughn, J. und Landry, E.F., 1977. Data report: An assessement of the occurrence of human virusesin Long Island aquatic systems. Rep. BNL 50787, Brookhaven Natl. Lab., Dep.of Energy and Environ., Upton, N.Y.

Vaughn, J. M., et al., 1987. Inactivation of human and simian rotaviruses by ozone. Appl. Env. Microbiol., 53:2218-2221.

Velàzquez, M., und Feritag, J. M., 1999. *Helicobacter pylori*: characteristics, pathogenity, detection methods and mode of transmission implicating foods and water. International Journal of Food Microbiology, 53:95-104.

Verwey, E.J.W. und Overbeek, J.T.G., 1948. Theory of the stability of lyophobic colloids. Elsevier, Amsterdam.

Vesey, G., Slade, J. S., Byrne, M., Shepard, K., und Fricker, C. R., 1993. A new method for the concentration of *Cryptosporidium* oocysts from water. Journal of Applied Bacteriology, 75:82-86.

von Gunten, U. und Hoigné, J., 1994. Bromat formation during ozonation of bromide-containing waters: Interaction of ozone and hydroxyl radical reactions. Environ. Sci. Techn., 28:1234-1242.

von Gunten, U., 1998a. Ozonanwendung in der Trinkwasseraufbereitung: Möglichkeiten und Grenzen. Mitt. Gebiete Lebensm. Hyg., 89:669-683.

von Gunten, U., 1998b. Vorlesungsskript: Trinkwasser/Abwasser.

von Gunten, U. und Sahli, E., 2000. Bromat im Trinkwasser: Ein Problem in der Schweiz? gwa, 80(10):705-710.

Von Huben, H., 1991. Surface Water Treatment: the New Rules. Denver, CO, AWWA.

von Sonntag, C. und Schuchmann, H.-P., 1992. UV-Desinfection of drinking-water and by-product formation: some basic considerations. J. Water SRT - Aqua, 41:67-74.

VwV-WSG, 1994. Verwaltungsvorschrift des Ministeriums über die Festsetzung von Wasserschutzgebieten, Vom 14. November 1994, zuletzt geändert 6. Mai 1996.

Waage, A. S., Vardund, T., Lund, V., und Kapperud, G., 1999a. Detection of small numbers of Campylobacter jejuni and Campylobacter coli cells in environmental water, sewage, and food samples by a seminested PCR assay. Applied and Environmental Microbiology, 65:1636-1643.

Waage, A. S., Vardund, T., Lund, V., und Kapperud, G., 1999b. Detection of low numbers of Salmonella in environmental water, sewage and food samples by a nested polymerase chain reaction assay. Journal of Applied Microbiology, 87:418-428.

Waage, A. S., Vardund, T., Lund, V., und Kapperud, G., 1999c. Detection of low numbers of Yersinia enterocolitica in environmental water and sewage samples by nested polymerase chain reaction. Journal of Applied Microbiology, 87:814-821.

Wagner-Wiening, C., und Kimmig, P., 1995. Detection of viable *Cryptosporidium parvum* oocysts by PCR. Applied and Environmental Microbiology, 61:4514-4516.

Wallis, P. M., Erlandsen, S. L., Isaac-Renton, J. L., Olson, M. E., Robertson, W. J., and van Keulen, H., 1996. Prevalence of *Giardia* cysts and *Cryptosporidium* oocysts and characterization of *Giardia* spp. isolated from drinking water in Canada. Applied and Environmental Microbiology, 62:2789-2797.

Walter, R., 1999. Allgemeine Grundlagen der Umweltxirologie. In: Umweltvirologie, Viren in Wasser und Boden, Walter, R. (ed.), Springer, Wien, New York., 1-39.

Way, J. S., Josephson, K. L., Pillai, S. D., Abbaszadegan, M., Gerba, C. P., und Pepper, I. L., 1993. Specific detection of *Salmonella* spp. by multiplex polymerase chain reaction. Applied and Environmental Microbiology, 59:1473-1479.

Weber, R., Bryan, R. T., Schwartz, D. A., und Owen, R. L., 1994. Human microsporidial infections. Clinical Microbiology Reviews, 7:426-461.

Wegelein, M., Boller, M. und Schertenleib, R., 1987. Partical removal by horizontal-flow roughing filtration. Aqua, (2):80-90.

Weidenmann, A., Braun, M. und Botzenhart, K., 1997. Evaluation of the disinfection potential of low chlorine concentrations in tap water using immobilised *Enterococcus faecium* in a continuous flow device. Wat. Sci. Tech., 35(11-12):77-80.

Weidner, E., 1991. Closing remarks on opportunistic microsporidians in humans. J. Protozool., 38:638.

Wheeler, J. G., Sethi, D., Cowden, J. M., Wall, P. G., Rodrigues, L. C., Tompkins, D. S., Hudson, M. J. und Roderick, P. J., 1999. Study of infectious intestinal disease in England: rates in the community, presenting to general practice, and report to national surveillance. BMJ, 318:1046-1050.

White, C., 1992. Handbook of Chlorination and alternative Desinfection. 3ed. , van Nostran Reinhold, 1308.

White, G. C., 1972. The Handbook of chlorination: for potable water, wastewater, cooling water, industrial processes and swimming pools. New York, Van Nostrand Reinhold 744.

Wickramanayake, G. B., et al., 1984a. Inactivation of *Naegleria* and *Giardia lamblia* cysts in water by ozonisation. Jour. WPCF, 56(8):983-988.

Wickramanayake, G. B., et al., 1984b. Inactivation of *Giardia lamblia* cysts with ozone. Appl. Env. Microbiol., 48(3):671-672.

Widmer, 2000. Persönliche Kommunikation. Kostenberechnungen für Aufbereitungsanlagen, Firma Grüter, Zürich.

Wiedenmann, A., Krüger, P., und Botzenhart, K., 1998. PCR detection of *Cryptosporidium parvum* in environmental samples – a review of published protocols and current developments. Journal of Industrial Microbiology & Biotechnology, 21:150-166.

Wiedenmann, A., Krüger, P., Filkorn, R., und Botzenhart, K., 1997. A simple procedure for an exact evaluation of the sensitivity of the selective detection of viable *Cryptosporidium oocysts* by in vitro excystation and PCR. In: 1997 International Symposium on Waterborne Cryptosporidium Proceedings. Fricker, C. R., Clancy, J. L., and Rochelle, P. A., eds., (AWWA, Denver), 109-114.

Willshaw, G.A., Thirlwell, J., Jones, A.P., Parry, S., Salmon, R.L. und Hickey, M., 1994. Verocytotoxin-producing Escherichia coli 0157 in beefburger linked to an outbreak of diarrhoea, haemorrhagic colitis and haemolytic uraemic syndrome in Britain. Letters in Applied Microbiology, 19:304-307.

Wilson, J.B., McEwen, S.A., Clarke, R.C., Leslie, K.E., Wilson, R.A., Waltner, T.D. und Gyles, C.L., 1992. Distribution and characteristics of verocytotoxigenic Escherichia coli isolated from Ontario dairy cattle. Epidemiol. Infect., 108:423-439.

Winzenbacher, R., Schick, R. und Stabel, H.-H., 1999. Zum Mechanismus der durch Fe(III)-Zugabe unterstützten Filtration bei der Aufbereitung von Bodenseewasser. Vom Wasser, 93:215-283.
Winzenbacher, R., Schick, R. und Stabel, H.-H., 2000. Fe(III)-unterstützte Filtration ($\beta_{Fe} \leq 0,1$ mg l^{-1}) bei der Trinkwasseraufbereitung. gwa, 1:20-28.
Wood, W. W. und Ehrlich, G. G., 1978. Use of backer's yeast to trace microbial movement in ground water. Ground Water, 16:398-403.
World Health Organization (WHO), 1989. Guidelines for drinking-water quality, 2nd ed., (WHO, Geneva), p. 8-29.
World Health Organisation (WHO), 1996. Guidelines for drinking-water quality. Volume 2: Health criteria and other supporting information.
Wuhrmann, K. und Meyrath, J., 1955. Über die bakterizide Wirkung wässriger Ozonlösungen. Schweiz. J. Allgem. Pathol. Bakteriol., 18:1060.
Wüthrich, Ch. et al., 2001. Revitalisierung und Trinkwasserschutz in der dicht genutzten Wiese-Ebene. Feuchtgebiete als Reinigungsstufe. Regio Basiliensis, 42(1):97-116.
Yao, K-M., Habibian, M. T. und O'Melia, C. R., 1971. Water and waste water filtration: concepts and applications. Environmental Science & Technology, 5(11):1105-1112.
Yates, M.V. und Yates, S.R., 1988. Modelling microbial fate in subsurface environment. CRC Crit. Rev. Environ. Control, 307-344.
Yates, M. V., Gerba, C. P. und Kelly, L. M., 1985. Virus persistence in groundwater. Appl. Environ. Microbiol., 49:778-781.
Young, C. C. und Greenfield, H., 1923. Observations on the viability of the Bacterium coli group under natural and artificial conditions. Am. J. Public Health, 13:270.
Zahorsky, J., 1929. Hyperemesis hiemis or the winter vomiting disease. Arch. Pediat., 46:391-395.
Zhang, P. et al., 2001. Extended tailing of bacteria following breakthrough at the Narrow Channel focus area, Oyster, Virginia. Water Resources Res., 37(11):2687-2698.

Glossar

Absterbekinetik: Verlauf der Wirkung eines Desinfektionsmittels auf einen Mikroorganismus.

Advektion: mittlere Fliessgeschwindigkeit.

Aerosol: kleinste Wassertröpfchen in der Luft.

Agens: wirkendes Mittel, Krankheitserreger.

Alkalität: Alkalinität repräsentiert die Summe der Kapazitäten der einzelnen Wasserinhaltsstoffe starke Säuren zu neutralisieren bezüglich eines Referenzzustandes. Für natürliche Gewässer ist dieser Referenzpunkt die Kohlensäure (H_2CO_3). Alkalinität im Grundwasser entspricht in etwa [HCO_3^-].

alkalisch: basisch

Amöbiasis: in (Sub-)Tropen vorkommende Infektion durch Entamöbea histolytica nach Aufnahme von Zystenformen des Erregers mit Trinkwasser oder – fliegenvermittelt - aus Nahrungsmitteln.

animal: tierisch

AOC (Assimilable organic carbon): Durch Mikroorganismen verwertbarer organischer Kohlenstoff.

ätiologisch: die Ursache von Krankheiten betreffend.

Bakteriophage (Phagen): Viren, die nur ein spezifisches Bakterium befallen und weder auf Menschen noch Tiere eine Wirkung haben.

Benthos, benthisch: Gesamtheit der auf, in oder dicht über dem Boden der (Filterschicht) lebenden Organismen.

Bimodal (bimodale Korngrössenverteilung): Häufigkeitsverteilung mit zwei Maximas (z.B. Maximas bei den Kornklassen "Mittelsand" und "mittelkörniger Kies").

Biodosimetrie: Überprüfung der angegebenen Reduktionsraten bei einem Wasseraufbereitungsprozess mit Hilfe von Mikroorganismen.

Biotin: Vitamin; dient als nichtradioaktive Markierung von DNA-Sonden, deren Nachweis anschliessend über eine Komplexbildung zwischen Biotin und Streptavidin erfolgt.

BGMK-Zelllinie: häufigsten verwendete Zellkulturlinie für Enteroviren (Buffalo Green Monkey Nierenzellen).

BUWAL: Bundesamt für Umwelt Wald und Land in der Schweiz.

Chlorung: Zugabe von Chlor (Javelle, Chlordioxid, Chlorgas) ins Wasser zur Desinfektion.

Coccidia, coccidisch: Kokkzidien; eine Ordnung der Sporozoa (parasitäre Protozoen). Leben als intrazelluläre Parasiten; nur Jugendformen freilebend.

CPE (cytopathologische Effekte): Die pathologischen Folgen einer Virus-Infektion an einer dafür geeigneten Zellkultur werden als CPE bezeichnet.

Darmepithel: Deckschicht die den Darm auskleidet und als Schutz- und Stoffwechselorgan mit der Fähigkeit zur Resorption dient.

Dekolmation: Aufreissen der kolmatierten Flusssohle, kann eine verstärkte Infiltration von Flusswasser und darin enthaltener Mikroorganismen ins Grundwasser bewirken.

DGGE (Denaturating gradient gel electrophoresis): Trennung von doppelsträngiger DNA (z.B. PCR-Produkten) gleicher Länge, welche sich nur in einer oder wenigen Punktmutationen unterscheiden.

Dialyse: physikalisches Verfahren zur Trennung molekulardisperser und kolloidal gelöster Teilchen mittels einer semipermeablen Membran. Das Prinzip findet in der Medizin Anwendung bei ungenügender oder ausgefallener Ausscheidungsleistung der Niere.

Diarrhoe: Durchfall.

Dispersion (Dispersionstensor): Vermischung von Stoffen mit dem Grundwasser entlang einer Fliessstrecke aufgrund von Sorptionseffekten und unterschiedlichen Porenweiten und –längen in der Transportmatrix.

DLVO-Theorie: Beschreibt die Interaktionsenergie zwischen zwei Oberflächen als Folge von van der Waals Kräften, elektrostatischen Kräften und der Separationsdistanz der Oberflächen.

DOC (Dissolved organic carbon): Gelöster organischer Kohlenstoff natürlicher oder anthropogener Herkunft.

enterisch: den Darm betreffend.

Enteritis: Entzündung des Dünndarms.

***enteroaggregativ: Aggregation*:** Zusammenballen

enterohämorrhagisch: Blutungen des Darmes betreffend.

***Enteroinvasiv*:** über den Darm in den Makroorganismus eindringende Krankheitserreger

***Enterotoxin, enterotoxisch*:** Im Darmtrakt wirksames hochgiftiges von Bakterien abgesondertes Toxin (Gift).

Enzephalitis: Gehirnentzündung; die akute oder chronische Entzündung von Hirngewebe infolge einer Infektion durch Bakterien, Viren oder Parasiten.

Epidemiologie, epidemiologisch: Wissenschaft von der Entstehung, Verbreitung, Bekämpfung und den sozialen Folgen von Epidemien (zeitlich und örtlich auftretende Infektionskrankheiten).

EPIK: Multikriterien Methode zur Dimensionierung von Grundwasserschutzzonen in Karstgebieten (E: Epikarst, P: Protection, I: Infiltration, K: Karst).

Epikarst: oberflächennaher, stark verkarsteter Bereich des Kalkuntergrundes, der eine stark erhöhte Wasserdurchlässigkeit aufweist.

fäkal-oral: Krankheitserreger, die mit Fäkalien ausgeschieden werden, gelangen durch orale Aufnahme (über Wasser oder Lebensmittel) in den Menschen.

Fingerprintanalysen: Erstellung charakteristischer, für einzelne Organismen spezifischer DNA-Fragmentmuster (genetischer Fingerabdruck), basierend auf variablen Regionen im Erbgut.

FIV: Fremd- und Inhaltsstoffverordnung in der Schweiz.

Gastritis: durch äussere oder innere Faktoren verursachte, akute oder chronische Magenschleimhautentzündung.

Gastroenteritis: Magen-Darm-Entzündung.

gastrointestinale Erkrankungen: *Erkrankungen des Magen-Darm-Traktes.*

Georadar: Instrument zur geophysikalischen Erkundung des Untergrundes.

Grenzwert: Höchstkonzentration, bei deren Überschreitung ein Lebensmittel für die menschliche Ernährung als ungeeignet gilt (Art. 2 Abs. 4 FIV).

Grundwasser: Wasser, das die Hohlräume im Untergrund zusammenhängend ausfüllt; das Grundwasser wird durch versickernde Niederschläge und Infiltration von Oberflächengewässern gebildet und bewegt sich ausschliesslich unter dem Einfluss der Schwerkraft.

HACCP: Hazard Analysis of Critical Controll Points. Konzept zum Aufbau eines Qualitätssicherungssystems.

hämolytisch-urämisches Syndrom: v.a. im Säuglings- und Kleinkindalter im Rahmen von Infekten auftretende akute hämolytische Anämie (Blutarmut durch die Auflösung der roten Blutkörperchen) mit gleichzeitig auftretender Niereninsuffizienz.

Hepatitis: infektiöse Leberentzündung.

humanpathogen: krankheitserregend für den Menschen.

Hybridisierung: Anlagerung eines einzelsträngigen markierten DNA-Moleküls (Sonde) an die komplementäre, ebenfalls einzelsträngige Zielsequenz, die dadurch zum stabilen DNA-Doppelstrang ergänzt und ebenfalls markiert wird.

Hydrogramm: Aufzeichnung der Schüttungsmessung einer Karstquelle.

Immunomagnetische Trennung: Verfahren zur Abtrennung antigener Strukturen z.B. auf Zelloberflächen. Beruht auf Antikörpern, die an paramagnetische Partikel gekoppelt werden und somit die Abtrennung ermöglichen.

Immunsupprimierte: Menschen mit geschwächtem Abwehrsystem des Körpers.

in vitro: im Reagenzglas ablaufend oder durchgeführt.

in vivo: im oder am lebenden Organismus ablaufend oder durchgeführt.

Inaktivierung: Unwirksammachen oder Ruhigstellung (z.B. Verlust der Virulenz bei Mikroorganismen).

Inaktivierungsleistung: Gibt an, wieviele Mikroorganismen durch ein bestimmtes Desinfektionsmittel innerhalb einer bestimmten Zeit nicht mehr nachweisbar sind, da sie ihre Virulenz verlieren.

Indikatorkeime: Stellvertretende Keime zur Beurteilung der mikrobiologischen Qualität des Grund- und Trinkwassers.

Infektiosität: Ansteckungsfähigkeit eines Krankheitserregers.

Interaktion (Fluss-Grundwasser): Austausch von Fluss- und Grundwasser. Der Prozess findet vor allem bei Hochwasser oder nicht kolmatierter (durch organisches Material abgedichteter) Flusssohle statt.

intermediäre Fliessysteme: Fliessysteme im Massstab mehrere 10 bis mehrere 100 Meter.

Interstitial (hyporheisches Interstitial): biologisch aktive Zone im oberen Teil der Sedimente von Flüssen oder bei Langsam-Sandfiltern.

Interzeption: Anlagerung von suspendierten Partikeln an Oberflächen aufgrund von laminarer Strömung.

Invertebraten: wirbellose Tiere.

Inzidenz: Anzahl neuer Krankheitsfälle pro Zeiteinheit.

Ionenstärke (einer Lösung): Die Art und Anzahl Ionen im Wasser haben einen Einfluss auf die Adhäsion von Partikeln an Oberflächen. Angabe der Ionenstärke in molar [M = Molar].

ISH: Die In-Situ-Hybridisierung (ISH) mit Gensonden beruht in den meisten Fällen auf den Nachweis der 16S- oder der 23S-ribosomalen Ribonukleinsäure (rRNA). Die Gensonden sind entweder mit fluoreszierenden Farbstoffen markiert, oder mit Enzymen, die ein nachträglich zugegebenes gefärbtes Substrat in der Zelle abbauen und präzipitieren. Die so markierten Zellen können je nach verwendeter Markierung entweder im Fluoreszenz- oder Lichtmikroskop nachgewiesen werden.

Kolloid: Kleines suspendiertes Partikel im Wasser.

Kolmation (der Flusssohle): Abdichten der Flusssohle durch kleinste Partikel, organisches Material und Mikroorganismen.

Koloniehybridisierung: Methode zur Identifizierung von Bakterienkolonien, die nach Wachstum auf Agarplatten auf eine geeignete Trägerfolie, z.B. Nitrocellulose- oder Nylon-Membran transferiert werden.

konservative Tracer: Markierstoffe, die auf dem Transport von Inputstelle zu Outputstelle weder abgebaut werden noch sich verändern und auch nicht mit dem Transportmedium oder der Transportmatrix reagieren.

Korrelationslänge: Distanz über welche zwei oder mehrere Messpunkte miteinander in Beziehung stehen (bsp. K-Wert).

kritischer Kontrollpunkt (CCP): Ort im Prozess der Trinkwassergewinnung, -aufbereitung und -verteilung, wo eine Messgrösse bestimmt und in den Ablauf eingegriffen werden kann (z.B. Überwachung der Leitfähigkeit im Rohwasser um bei plötzlichen Veränderungen Förderung einzustellen, Messung der Druckdifferenz in Sandfiltern um eine Rückspülung auszulösen).

K-Wert: hydraulische Leitfähigkeit [m/s].

Lithofazies: Definiert Sedimentabfolgen aufgrund von Farbe, Zusammensetzung, Gefüge und Sedimentstrukturen.

Lithologie: optische, physikalische und chemische Eigenschaften eines Gesteins.

LMG: Lebensmittelgesetz in der Schweiz.

LMV: Lebensmittelverordnung in der Schweiz.

ln K: Logarithmus naturalis von K, K bezeichnet die hydraulische Leitfähigkeit.

Management (Wasserfassungen): Überwachung und Steuerung der Wasserentnahmen aus dem Grundwasser zur Trinkwassergewinnung mittels Systemparametern wie Niederschlag, Schüttung, Pegelstand, Trübungen, etc..

Matrix: Festsubstanz des Bodens oder des geologischen Untergrundes ohne den Porenraum.

Meningitis: Entzündung der Hirn- und/oder Rückenmarkshäute.

Messenger-RNA (mRNA): RNA-Molekül, das einem DNA-Abschnitt mit Genen, die exprimiert werden, komplementär ist.

Mikrosphären: künstliche Partikel aus Polystyrol, denen bei der Herstellung ein Fluoreszenzfarbstoff eingegeben werden kann.

Morbidität: Häufigkeit der Erkrankungen innerhalb einer Bevölkerungsgruppe.

Mortalität: Sterblichkeit; Verhältnis der Zahl der Todesfälle zur Gesamtzahl der berücksichtigten Personen.

Most probable number (MPN)-Verfahren: Statistisches Verfahren, bei dem aus der Anzahl bewachsener Zell- bzw. Flüssigkulturen in mehreren parallelen Verdünnungsreihen die Anzahl Viren oder Bakterien bestimmt wird.

Multibarrierensystem: Trinkwasseraufbereitung mittels mehrere Stufen (z.B. Flockung, Filtration, Ozonung, Aktivkohlefilter, Chlorung).

Multiplex PCR: Diese PCR-Anwendung erlaubt die gleichzeitige Analyse unterschiedlicher Zielsequenzen mit mehreren Primern in einer einzigen PCR-Reaktion.

neonatal: Neugeborene (neonatale) Diese PCR-Anwendung erlaubt die gleichzeitige Analyse unterschiedlicher Zielsequenzen mit mehreren Primern in einer einzigen PCR-Reaktion.Mäuse werden zur Abklärung der Infektiosität bestimmter Mikroorganismen verwendet. Da die Neugeborenen noch keine Immunabwehr haben, erkranken sie besonders rasch.

Nested PCR: Zur Erhöhung der Spezifizität und der Sensitivität wird oft ein zweiter PCR-Schritt angeschlossen. Dabei werden die Produkte der ersten PCR mit Primern, die komplementär sind zur Zielsequenz innerhalb des amplifizierten PCR-Produktes,

nochmals vervielfältigt. Bei der Verwendung nur eines internen Primers spricht man von einer *seminested PCR*.
Niereninsuffizienz: unzureichende Nierenfunktion bis zu Nierenversagen.
nummerische Modellierung: Computersimulation z.B. vom Transport eines Stoffes im Grundwasser.
Oozysten: umweltresistente Dauerstadien (z.B. bei Cryptosporidien enthalten die O. die Sporozoiten, die im Wirt freigesetzt werden und die Infektion verursachen).
Openframework Zone: Rollkieslagen.
Ozonung: Zugabe von Ozon ins Wasser zur Desinfektion.
Parasit: Schmarotzer; tierisches Lebewesen, das sich auf oder in einem anderen Lebewesen ernährt, mit oder ohne Verursachung von Krankheitserscheinungen.
pathogen: krankheitserregend.
PCR = polymerase chain reaction: Einzelne DNA Abschnitte können mit der Polymerasekettenreaktion (PCR) vervielfältigt und dadurch sichtbar gemacht werden.
Percoll-Dichtegradienten-Zentrifugation: Percoll besteht aus Körnchen verschiedener Dichte, die sich während der Zentrifugation ihrer Dichte entsprechend anordnen. Eine beigefügte Substanz begibt sich in die ihrer Dichte entsprechende Zone des Gradienten.
Persistenz, persistieren: Bestehenbleiben eines Zustandes über längere Zeiträume.
Poliomyelitis: Kinderlähmung, Schädigung des Zentralen Nervensystems.
Prävalenz: epidemische Häufigkeit aller Fälle einer bestimmten Krankheit in einer Population zum Zeitpunkt der Untersuchung.
primäre Porosität: ursprüngliche Porosität z.B. eines Gesteins oder einer geologischen Einheit.
Primer: Ein kurzes, synthetisch hergestelltes einzelsträngiges DNA-Molekül, das als Startpunkt für die Neusynthese eines komplementären DNA-Stranges während der PCR dient. Wird von allen natürlichen DNA-Polymerasen einschließlich der reversen Transkriptase benötigt. Zwei Primer flankieren jeweils einen definierten DNA-Abschnitt, so dass nur dieser Abschnitt in der PCR vervielfältigt wird.
Quelle: Austrittsstelle von Grundwasser, das frei zu Tage tritt.
RAPD (random amplified polymorphic DNA): Anwendung der PCR mit verschiedenen Primer, deren Bindungsort innerhalb der DNA unbekannt ist. Somit ist jedoch auch die Länge der PCR-Fragmente unbekannt, aber für bestimmte Mikroorganismen reproduzierbar. Nach der anschliessenden Gelelektrophorese können Mikroorganismen auf Grund unterschiedlicher Bandenmuster voneinander abgegrenzt werden.
Reinwasser: aufbereitetes Trinkwasser.
Reservoir: Ort oder Wirt wo der Mikroorganismus überleben kann (Kap. Mikroorganismen).
Reverse Transkriptase PCR (RT-PCR): Mit Hilfe einer RNA-abhängigen Polymerase wird einzelsträngige RNA (ribosomale oder messenger RNA) in ihre cDNA (complementary DNA) umgeschrieben. Diese cDNA dient anschliessend als Ausgangsmaterial für die PCR.
Ribotyping: Genomische DNA wird mit spezifischen Enzymen, den sogenannten Restriktionsendonukleasen in kleine Fragmente geschnitten, mittels Gelelektrophorese in spezifische „Bandenmuster“ aufgetrennt und anschliessend auf eine geeignete Trägermatrix übertragen. Das Operon für die ribosomalen Gene wird in diesem Fall als Probe verwendet.

Riparian Zone: Übergangszone vom Oberflächen- ins Grundwasser in welcher komplexe biogeochemische Prozesse stattfinden und wo sich Oberflächenwasser- und Grundwasserfauna mischen.
Rohwasser: Grund- oder Oberflächenwasser vor der Aufbereitung zu Trinkwasser.
Sanitärhygiene: Zustand und Einrichtungen von Abwasserentsorgungs- bzw. Trinkwasserversorgungssystemen im Bezug auf den Gesundheitsschutz.
SchALVO: Schutzgebiets- und Ausgleichsverordnung in Deutschland.
Schmutzdecke: biologische aktive Filterschicht mit partikulärem Sediment auf Langsam-Sandfiltern.
sekundäre Porosität: durch Verwitterungs- und Lösungsprozesse, Kluftbildung, etc. entstandene Porosität.
Serogruppe: über das Blutserum durch ein spezifisches Antiserum identifizierbare Gruppe von Mikroorganismen derselben Spezies oder Gattung, z.B. Salmonellen, mit gemeinsamem gruppenspezifischem Antigenmuster.
Serotyp: Mikroorganismus innerhalb einer Serogruppe, die anhand eines typspezifischen Antigenmusters identifizierbar sind.
Serotyping: Das Serotyping beruht auf der An- oder Abwesenheit von Antigenen auf der Zelloberfläche, von Geisseln und von Kapseln und deren Reaktion mit spezifischen Antiseren.
Sohlendurchlässigkeit: Wasserdurchlässigkeit der Sohle eines Fliessgewässers.
Spezies: Bezeichnung für eine Tier- oder Pflanzenart.
Sporen: Bei einigen Bakterienarten (z.B. Bacillus, Clostridium) vorkommende, gegen Umwelteinflüsse sehr widerstandsfähige Dauerformen.
Sporozoiten: der infektiöse "Sichelkeim" der Kokzidien als Produkt der Sporogonie (Vermehrungsweise der 1. Generation der Sporozoen, bsp. Oozysten bei Cryptosporidium). Die Infektion erfolgt entweder durch die Aufnahme von Oo- bzw. Sporozysten durch den Mund in den Verdauungstrakt.
Sporulation: Sporenbildung.
SRSV (Small round structured Virus): Kleine runde Viren, die als humanpathogen erkannt wurden und über das Wasser übertragbar sind.
Taxierung: Systematische Einordnung von Mikroorganismen, Tieren oder Pflanzen aufgrund ihrer Eigenschaften.
Tektonik: Gebirgsbau, grossräumige geologische Strukturen wie Faltenjura und Brüche.
Toleranzwert: Höchstkonzentration, bei deren Überschreitung ein Lebensmittel als verunreinigt oder sonst im Wert vermindert gilt (Art. 2 Abs. 3 FIV). Massnahmen zur Verbesserung der Situation müssen ergriffen werden.
Tracer: künstlicher oder natürlicher Markier- oder Farbstoff oder Partikel zur Erkundung von ober- und unterirdischen Wasserwegen.
Transmission: Durchgang von Licht (UV-Strahlung) durch ein Medium (Wasser) ohne Frequenzänderung.
Transmission: Übertragung.
ubiquitär: allgegenwärtig.
ulzerativ: geschwürartig.
ungesättigte Zone: Deckschicht über dem Grundwasserspiegel.
Untergrund, nicht wassergesättigter: Deckschicht über dem Grundwasserspiegel.

Upscaling: ausgehend von Detailkentnissen im kleinmassstäblichem Bereich wird versucht Zusammenhänge im Feldmassstabsbereich zu verstehen (Bildung von regionalen Mittelwerten).
UV-Absorption: Absorption von Licht mit einer Wellenlänge von 254 nm durch organische Substanz im Wasser (insbesondere Doppelbindungen von Molekülen).
UV-Anlage: Ultraviolet-Bestrahlung zur Aufbereitung des Wassers (meist bei einer Wellenlänge von 254 nm).
Vakzin: Impfstoff.
VBNC-Zellen: Abkürzung für "viable, but not cultivable cells" oder deutsch: lebende, aber nicht kultivierbare Zellen.
Vektor: "Träger", Überträger eines Krankheitserregers.
viral: durch Viren verursacht.
virulent: infektionsfähig, ansteckend.
Vulnerabilität: Empfindlichkeit des Grundwasservorkommens bezüglich natürlicher oder anthropogener Verschmutzungen aufgrund geringer Filterwirkung des Bodens und des geologischen Untergrundes oder der Heterogenität der hydraulischen Leitfähigkeit im Grundwasserleiter.
VwV- WSG: Verwaltungsvorschrift des Umweltministeriums über die Festsetzung von Wasserschutzgebieten in Deutschland.
Wahrscheinlichkeitsdichtefunktion: Häufigkeitsfunktion, z.B. Häufigkeit des Auftretens charakteristischer Eigenschaften wie Porosität oder hydraulische Leitfähigkeit.
wassergesättigter Untergrund: Wassergesättigter Teil eines Grundwasserleiters.
Wiederfindungsrate: Anzahl oder Menge des gemessenen Tracers z.B. in einer Quelle im Verhältnis zur eingegebenen Anzahl oder Menge des Tracers.
Wiederverkeimung: Wachstum von Mikroorganismen im aufbereiteten Trinkwasser
Zellparasit: Schmarotzer, der die Zelle befällt.
Zoonose, zoonotisch: von Tieren auf Menschen übertragbare Infektionskrankheit.
Zuströmbereich: Derjenige Teil des Einzugsgebietes einer Grundwasserfassung, aus dem rund 90% des geförderten Trinkwassers stammen.
Zysten: umweltresistente Dauerform bei Giardia.
Zytoplasma: "Zellflüssigkeit", in der verschiedene Stoffwechselvorgänge ablaufen.

Index

GPSR Compliance

The European Union's (EU) General Product Safety Regulation (GPSR) is a set of rules that requires consumer products to be safe and our obligations to ensure this.

If you have any concerns about our products, you can contact us on ProductSafety@springernature.com

In case Publisher is established outside the EU, the EU authorized representative is:

Springer Nature Customer Service Center GmbH
Europaplatz 3
69115 Heidelberg, Germany

Batch number: 10584556

Printed by Printforce, the Netherlands